Saas-Fee Advanced Course 43

More information about this series at http://www.springer.com/series/4284

Nickolay Y. Gnedin · Simon C.O. Glover
Ralf S. Klessen · Volker Springel

Star Formation in Galaxy Evolution: Connecting Numerical Models to Reality

Saas-Fee Advanced Course 43

Swiss Society for Astrophysics and Astronomy
Edited by Yves Revaz, Pascale Jablonka, Romain Teyssier and Lucio Mayer

Springer

Nickolay Y. Gnedin
Department of Astronomy and Astrophysics
The University of Chicago
Chicago, IL
USA

Simon C.O. Glover
Institute for Theoretical Astrophysics
University of Heidelberg
Heidelberg
Germany

Ralf S. Klessen
Institute for Theoretical Astrophysics
University of Heidelberg
Heidelberg
Germany

Volker Springel
ZAH, ARI
Heidelberg University
Heidelberg
Germany

Volume Editors
Yves Revaz
Laboratoire d'astrophysique
École Polytechnique Fédérale de Lausanne (EPFL)
Observatoire de Sauverny
Versoix
Switzerland

Pascale Jablonka
Laboratoire d'astrophysique
École Polytechnique Fédérale de Lausanne (EPFL)
Observatoire de Sauverny
Versoix
Switzerland

Romain Teyssier
Center for Theoretical Astrophysics and Cosmology
Institute for Computational Science
University of Zurich
Zurich
Switzerland

Lucio Mayer
Center for Theoretical Astrophysics and Cosmology
Institute for Computational Science
University of Zurich
Zurich
Switzerland

This Series is edited on behalf of the Swiss Society for Astrophysics and Astronomy: Société Suisse d'Astrophysique et d'Astronomie, Observatoire de Genève, ch. des Maillettes 51, CH-1290 Sauverny, Switzerland

Cover figure: Cygnus-X is an extremely active region of massive-star birth some 4500 light-years from Earth in the constellation of Cygnus, the Swan. This picture, taken by Herschel's far-infrared camera (Credit: ESA), illustrate the complexity of the star formation out of turbulent clouds. It is superimposed on the lower right by the Mönch supercomputer hosted at the Swiss National Supercomputing Center (CSCS) in Lugano, Switzerland (Credit: CSCS), and on the upper left by the spiral galaxy NGC1232 (Credit: ESO) obtained by the FORS (FOcal Reducer and low dispersion Spectrograph) instrument on the 8 meters Very Large Telescope at Paranal, Chile (Montage, Credit: EPFL/Yves Revaz).

ISSN 1861-7980 ISSN 1861-8227 (electronic)
Saas-Fee Advanced Course
ISBN 978-3-662-47889-9 ISBN 978-3-662-47890-5 (eBook)
DOI 10.1007/978-3-662-47890-5

Library of Congress Control Number: 2015944146

Springer Heidelberg New York Dordrecht London

Printed on acid-free paper

Springer-Verlag GmbH Berlin Heidelberg is part of Springer Science+Business Media (www.springer.com)

Preface

In the last few years it has become clear that the modelling of star formation and feedback processes is central to the quest of finding solutions to the key problems of galaxy formation. A complete story of star formation and its connection with the cosmic evolution of our Universe requires to study physical processes taking place at very different scales.

At cosmological scales the gas is sparse and highly ionized. Its temperature depends on the subtle balance between photo-heating and adiabatic as well as radiative cooling. In the densest regions, following the growth of the cosmological fluctuations, it may be accreted onto forming and evolving galaxies. At the galactic scale, the physics becomes extremely complex, non-linear and far from equilibrium. The interstellar medium is composed of a mixture of charged particles, atoms, molecules and dust grains. There, turbulent cascades drive the formation of cloud complexes of various sizes and masses, from which stars may eventually form. The collapse of these complexes is ultimately halted by star formation, resulting in a system intricately linked together through a variety of feedback loops.

Due to the numerous complex and interleaved process involved, modelling the star formation is challenging. The theory and numerical models of star formation have traditionally evolved independently from those of galaxy evolution, because they act at different spatial scales. We are now at a point however, where substantial steps forward can only arise from the combined knowledge of these two research fields.

The goal of the 43rd *Saas-Fee Advanced Course* was to bring together, in a single place, these two fields. It aimed to take an inventory of the physical processes related to the star formation involved at different scales and also to provide an overview of the major computational techniques used to solve the equations governing self-gravitating fluids, essential to galactic modelling. Together this provides a unique framework essential to developing and improving the simulation techniques used to understand the formation and evolution of galaxies.

The lack of a textbook joining these different fields motivated the members of the Swiss Society for Astrophysics and Astronomy to vote in favour of the organisation of a winter Saas-Fee course on the star formation in galaxies and its modelling techniques.

The three selected lecturers—Nickolay Gnedin, Ralf Klessen and Volker Springel—succeeded in bringing to the 95 participants a very rich and interesting review of the fields related to the star formation in galaxies. An invaluable additional contribution came later from Simon Glover who participated in the writing of the chapter dedicated to the physical processes in the interstellar medium. The reader can revel now those lectures in the following pages. The present book is supplemented with the complete video recordings of the lectures, which are accessible online, via the 43rd *Saas-Fee Advanced Course* website: http://lastro.epfl.ch/conferences/sf2013/.

We are very grateful to the lecturers for their invaluable live lectures as well as for their written version presented here. We are particularly thankful to Prof. Georges Meylan, Head of the Laboratory of Astrophysics at EPFL, who supported the organisation at all stages, making this course a success. We are extremely grateful to Matthew Nichols, who video-recorded the lectures and post-processed the movies. Olivier Genevay has been at the heart of all practical arrangements without counting his time; we want to thank him very warmly. We also thank the course secretaries, Carol Maury and Claire Schatzmann, as well as our colleagues, Malte Tewes, Vivien Bonvin, Alexis Arnaudon and Daniel Pfenniger for all their help in the practical organisation of the course.

The course took place during winter in the village of Villars-sur-Ollon in the Alps of Switzerland. While benefiting from superb weather after some snowy days, a conference picture was kindly taken by Ievgen Vovk and is displayed below.

Yves Revaz
Pascale Jablonka
Romain Teyssier
Lucio Mayer

Contents

Modeling Physical Processes at Galactic Scales and Above

Nickolay Y. Gnedin

1 In Lieu of Introduction

What should these lectures be? The subject assigned to us is so broad that many books can be written about it. So, in planning these lectures I had several options.

One would be to focus on a narrow subset of topics and to cover them in great detail. Such a subset necessarily would be highly personal and useful to a few readers at best. Another option would be to give a very shallow overview of the whole field, but then it won't be very much different from a highly compressed version of a university course (which anyone can take if they wish so).

So, I decided to be selfish and to prepare these lectures as if I was teaching my own graduate student. Given my research interests, I selected what the student would need to know to be able to discuss science with me and to work on joint research projects. So, the story presented below is both personal and incomplete, but it does cover several subjects that are poorly represented in the existing textbooks (if at all).

Some of topics I focus on below are closely connected, others are disjoint, some are just side detours on specific technical questions. There is an overlapping theme, however. Our goal is to follow the cosmic gas from large scales, low densities, (relatively) simple physics to progressively smaller scales, higher densities, closer relation to galaxies, and more complex and uncertain physics. So, we (you—the reader, and me—the author) are going to follow a "yellow brick road" from the gas well beyond any galaxy confines to the actual sites of star formation and stellar feedback. On the way we will stop at some places for a tour and run without looking back through some others. So, the road will be uneven, but I hope that some readers find it useful.

N.Y. Gnedin (✉)
Department of Astronomy and Astrophysics, The University of Chicago,
Chicago, IL, USA
e-mail: gnedin@fnal.gov

Y. Revaz et al. (eds.), *Star Formation in Galaxy Evolution: Connecting Numerical Models to Reality*, Saas-Fee Advanced Course 43,
DOI 10.1007/978-3-662-47890-5_1

2 Physics of the IGM

Most of the volume of the universe is occupied by gas outside galaxies, the so-called intergalactic medium (IGM). It may seem this gas is located far from galaxies, and should not be relevant to formation of galaxies and stars. Wrong!—IGM is the gas that eventually gets accreted by galaxies and turns into stars. After all, before the first galaxy formed, the whole universe was just IGM.

Hence, as we follow the "yellow brick road" to our goal of modeling star formation in galaxies, we pass through the IGM land first …

2.1 Linear Hydrodynamics in the Expanding Universe

Linear dynamics of the non-relativistic cold dark matter is almost trivial, density fluctuation $\delta_X(t,k)$ with a spatial wavenumber k satisfies a simple ordinary differential equation (ODE),

$$\frac{d^2}{dt^2}\delta_X(t,k) + 2H\frac{d}{dt}\delta_X(t,k) = 4\pi G\bar{\rho}\delta_{\rm tot}(t,k), \quad (1)$$

where $a(t)$ is the cosmic scale factor, $H(t) \equiv \dot{a}/a$ is the Hubble parameter and $\bar{\rho}$ is the mean density of the universe. If the universe only contained cold dark matter, then $\delta_{\rm tot} = \delta_X$. A second order ODE has two solutions, one of them is always growing with time,

$$\delta_X(t,k) = D_+(t)\delta_0(k), \quad (2)$$

where D_+ is called "the linear growing mode".

In reality, the universe contains gas, which is also subject to pressure forces. Hence, in the linear regime the evolution of the dark matter and gas fluctuations (δ_X, δ_B) is described by a system of two coupled equations,

$$\frac{d^2\delta_X}{dt^2} + 2H\frac{d\delta_X}{dt} = 4\pi G\bar{\rho}\left(f_X\delta_X + f_B\delta_B\right), \quad (3)$$

$$\frac{d^2\delta_B}{dt^2} + 2H\frac{d\delta_B}{dt} = 4\pi G\bar{\rho}\left(f_X\delta_X + f_B\delta_B\right) - \frac{c_S^2}{a^2}k^2\delta_B, \quad (4)$$

where $f_X \approx 0.84$ and $f_B \approx 0.16$ are the mass fractions of dark matter and baryons respectively, and c_S is the speed of sound in the gas.

This system of equations is coupled, but if high precision is not required, one can assume $f_B \ll f_X$ and ignore the baryonic contribution in the gravity terms in both equations. In that case the solution for the dark matter fluctuation is still given by Eq. (2), while the equation for the baryonic fluctuation reduces to

$$\frac{d^2\delta_B}{dt^2} + 2H\frac{d\delta_B}{dt} = 4\pi G\bar{\rho}\delta_X - \frac{c_S^2}{a^2}k^2\delta_B. \tag{5}$$

Notice the difference between this equation and an equation for baryonic fluctuations in a static reference frame (a = const, no expansion of the universe) in the absence of dark matter:

$$\frac{d^2\delta_B}{dt^2} = 4\pi G\bar{\rho}\delta_B - \frac{c_S^2}{a^2}k^2\delta_B.$$

We know that in the latter case the characteristic scale over which baryonic fluctuations are suppressed by the pressure force is the Jeans scale,

$$k_J \equiv \frac{a}{c_S}\sqrt{4\pi G\bar{\rho}}.$$

Equation (5) cannot be solved analytically in a general case, but the important physics we are after is how baryonic fluctuations deviate from the dark matter ones. Hence, a quantity of interest is the ratio of two fluctuations, which can be expanded in the Taylor series of powers of k^2,

$$\frac{\delta_B(t,k)}{\delta_X(t,k)} = r - \frac{k^2}{k_F^2} + O(k^4), \tag{6}$$

where r = const and we will call $k_F(t)$ a *filtering scale*. Because dark matter is expected to be more clustered that baryons (it is not a subject of the pressure force in the linear regime), we can expect that, in a general case $k_F > k_J$ (in the presence of dark matter baryonic fluctuations are less suppressed than in a purely baryonic case).

In the following we will only consider the case of $r = 1$ (baryons trace the dark matter on large scales), since this is an excellent approximation for $z < 10$. However, at higher redshifts this is not the case any more (Naoz and Barkana 2007), as the different evolution of baryons and dark matter during the recombination epoch is not completely forgotten at these high redshifts (for example, $r \ll 1$ at $z > 1000$).

Substituting Eq. (6) into (5), it is possible to obtain an expression for k_F in a closed form (Gnedin and Hui 1998),

$$\frac{1}{k_F^2(t)} = \frac{1}{D_+(t)}\int_0^t dt' a^2(t')\frac{\ddot{D}_+(t') + 2H(t')\dot{D}_+(t')}{k_J^2(t')}\int_{t'}^t \frac{dt''}{a^2(t'')}.$$

While this expression is long and ugly, for reasonable thermal histories of the universe a good rule of thumb at $z \sim 2$–4 is $k_F \approx 2 \times k_J$ (the filtering scale is about half the Jeans scale).

Figure 1 gives an example of scale-dependence of $\delta_B(t,k)/\delta_X(t,k)$ for a representative thermal history of the universe at several redshifts (see Gnedin et al. 2003

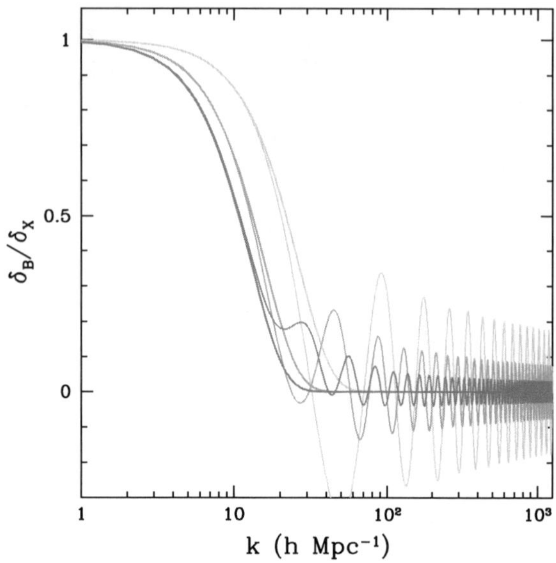

Fig. 1 Solutions to Eq. (4) for a representative thermal history of the universe at $z = 4$ (*light gray*), $z = 1.5$ (*medium gray*), and $z = 0$ (*dark gray*); *thin lines* show the exact solutions, *thick lines* give the approximation $\delta_B/\delta_X = \exp(-k^2/k_F^2)$ (adopted from Gnedin et al. (2003))

for details). Fluctuations on small scales, where the pressure force dominates, are simple sound waves, and the transition to the baryons-trace-the-dark-matter regime is well described by the filtering scale.

Brain teaser #1: Pressure generates sound waves, and sounds waves in the ideal gas do not dissipate. Why, then, are fluctuations "suppressed" by the pressure force?

2.2 Lyman-α Forest

A well known empirical fact is that the IGM is highly ionized at low and intermediate redshifts, $z < 6$ (we will come back to that fact). To keep the cosmic gas ionized, the universe must be filled with ionizing radiation, the so-called "Cosmic Ionizing Background" (CIB).

Since most of the IGM is hydrogen, let us consider hydrogen first. The ionization balance equation for hydrogen in the expanding universe is simple,

$$\dot{n}_{\rm HI} = -3Hn_{\rm HI} - n_{\rm HI}\Gamma + R(T)n_e n_{\rm HII},$$

where $n_{\rm HI}$, $n_{\rm HII}$, and n_e are number densities of neutral hydrogen, ionized hydrogen, and free electrons respectively, Γ is the *photoionization rate* and $R(T)$ is a (temperature-dependent) recombination coefficient.

Often it is more convenient to consider not the actual number density of neutral or ionized hydrogen, but the *neutral fraction* $x \equiv n_{\rm HI}/n_{\rm H}$, because then the Hubble expansion term cancels out,

$$\dot{x} = -x\Gamma + R(T)n_e(1-x). \tag{7}$$

In the ionization equilibrium $\dot{x} = 0$, hence

$$x_{\rm eq} = \frac{R(T)}{\Gamma} n_e (1 - x_{\rm eq}),$$

and since the IGM is highly ionized ($x \ll 1$),

$$x_{\rm eq} = \frac{R(T)}{\Gamma}(\bar{n}_{\rm H} + 2\bar{n}_{\rm He})(1+\delta),$$

where we assumed that Helium is fully ionized, $\bar{n}_e = \bar{n}_{\rm H} + 2\bar{n}_{\rm He}$ (denser gas is more neutral).

Let us now consider a light source somewhere in the universe (a quasar, a galaxy, a gamma-ray burst, etc.); the light source is at redshift z_e in our reference frame. Let us also imagine that a photon with wavelength λ_e is emitted by the source. As it propagates through the universe, the photon is going to be redshifted. At a redshift $z_a < z_e$ (from our reference frame) the photon has a wavelength

$$\lambda_e \frac{1+z_a}{1+z_e}.$$

Hence, for any $1216\,\text{Å}(1+z_e) < \lambda_e < 1216\,\text{Å}$ there is such z_a that

$$\lambda_e \frac{1+z_a}{1+z_e} = 1216\,\text{Å}.$$

When a photon with wavelength of 1216 Å (= Lyman-α) hits a neutral hydrogen atom, it can get absorbed and excite the atom to $n = 2$ level.

Indeed, this is exactly what happens in the real universe. Figure 2 shows a spectrum of a typical $z \sim 3$ quasar. The broad emission line in the middle is the Lyman-α of the quasar itself, and blue envelope for the observed spectrum is the continuum—i.e. the light that the quasar itself emitted. Black absorption lines come from the gas between us and the quasar, and the numerous forest of them at shorter wavelength is the hydrogen Lyman-α absorption from the neutral gas in the IGM, the so-called *Lyman-α Forest*.

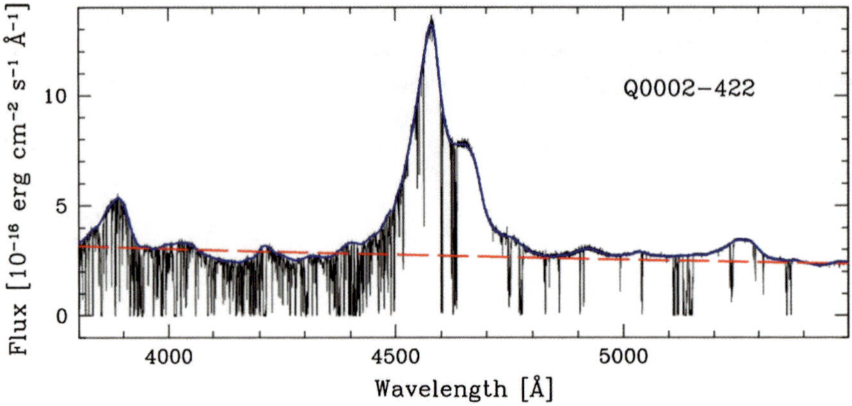

Fig. 2 Typical $z \sim 3$ quasar spectrum together with the power law continuum fit (*dashed red line*) and the local continuum fit (*blue line*; adopted from Dall'Aglio et al. (2008))

Figure 3 illustrates how fluctuations in the neutral hydrogen density and in the gas temperature combine to produce the Lyman-α forest of absorption features in the spectrum. In order to understand how one goes from the lower two panels to the top one in that figure, we need to refresh the basics of resonant line absorption in the expanding universe.

Brain teaser #2: Hydrogen atoms do not sit forever in $n = 2$ state, they decay back into $n = 1$ and a Lyman-α photon is re-created back. How can there be any Lyman-α absorption?

2.2.1 Introduction to Resonant Line Absorption

The cross-section for an atom at rest to absorb a photon in the frequency range from ν to $\nu + \Delta\nu$ to the energy level with the energy $h\nu_0$ is

$$\sigma(\nu) = \frac{\pi e^2}{m_e c \nu_0} f \phi(\nu) \equiv \sigma_0 \phi(\nu),$$

where f is the oscillator strength for the transition and

$$\phi(\nu) = \frac{1}{\pi} \frac{w \nu_0}{(\nu - \nu_0)^2 + w^2} \approx \nu_0 \delta(\nu - \nu_0),$$

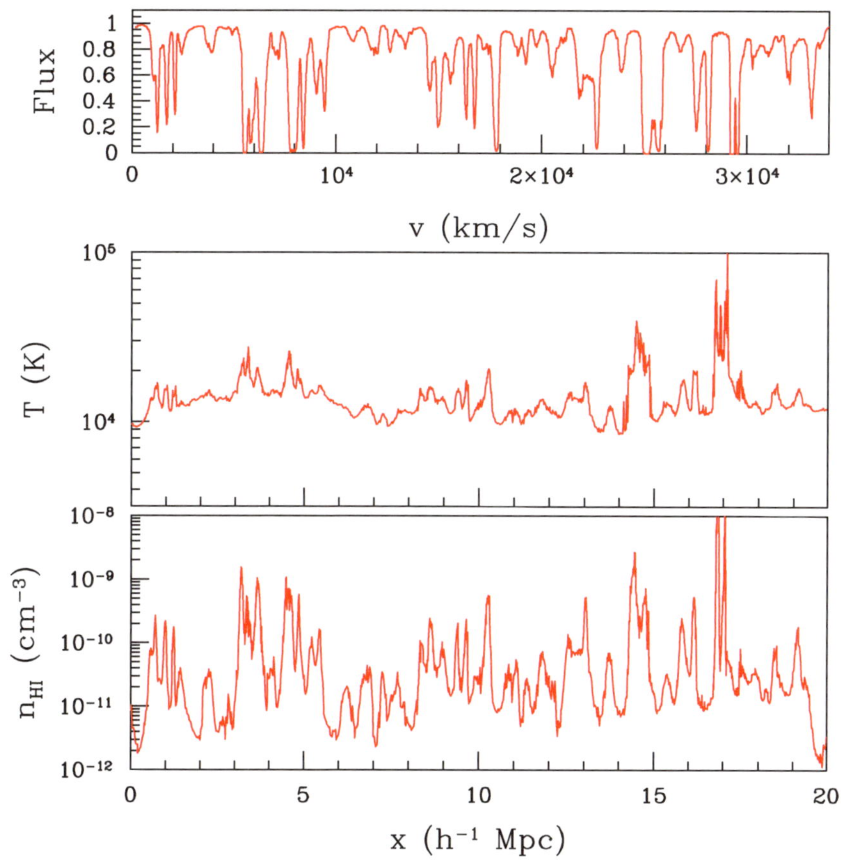

Fig. 3 Runs of neutral hydrogen density (*bottom*) and gas temperature (*middle*) along one line of sight in a numerical simulation of Lyman-α forest at $z \sim 3$. The resultant absorption spectrum is shown in the *top panel*

where w is the *natural line width* in frequency units. For hydrogen Lyman-α the combination of fundamental constants

$$\sigma_0 = \frac{\pi e^2}{m_e c \nu_0} f = 4.5 \times 10^{-18}\,\mathrm{cm}^2.$$

Atoms, though, are social creatures and rarely live alone. For a cloud of gas of density n, size L, and temperature T we need to integrate over all atoms to compute the optical depth of the transition at any frequency ν,

$$\tau(\nu) = nL \int \sigma_0 \phi(\nu') \frac{1}{\sqrt{\pi} b} e^{-\frac{(u_\nu - u')^2}{b^2}} du',$$

where $\nu' = \nu_0(1 + u'/c)$ and u_ν is defined via the expression $\nu = \nu_0(1 + u_\nu/c)$. The quantity

$$b = \left(2\frac{k_B T}{m_H}\right)^{1/2}$$

is called the *Doppler parameter* and the product nL is the *column density*.

In an expanding universe it is not enough just to multiply by the cloud size L, since different locations along the line-of-sight are redshifted relative to the observer and project to different locations in the velocity (or frequency) space. Hence, we must integrate along the line-of-sight,

$$\tau(\lambda) = \sigma_0 \int n(x) \frac{c}{\sqrt{\pi} b_x} e^{-\frac{(u_\lambda - u_x)^2}{b_x^2}} \frac{dx}{1 + z_x}, \tag{8}$$

where we switch to from the frequency to the wavelength (as almost all observers tend to live in the wavelength space), and we integrate over the comoving distance x (as almost all theorists tend to live in the comoving space); both b_x and u_x are, in general, functions of position, since the temperature and velocity vary in space. The wavelength is related to the velocity along the line-of-sight through the usual Doppler effect,

$$\lambda = \lambda_0 \left(1 + z_x + \frac{u_\lambda}{c}\right)$$

and z_x is the redshift of location x along the line-of-sight.

The spectrum shown on the top panel of Fig. 3 is just $\exp(-\tau(\lambda))$ with $\tau(\lambda)$ computed with Eq. (8) from the two bottom panels of the same figure.

Now we are ready to figure out why the IGM must be highly ionized at $z < 6$. From Fig. 2 we notice that the forest absorbs about 50 % of the quasar flux, so the average optical depth is $\tau \sim 0.5$–1. Considering one absorption system stretching for about $\Delta x \sim 100$ kpc and having temperature of 10^4 K (or $b \sim 10$ km/s), a crude estimate for τ is

$$\begin{aligned}
\tau &\sim \sigma_0 \frac{c}{\sqrt{\pi} b} x_{\rm HI} n_{\rm H} a \Delta x \\
&= 4.5 \times 10^{-18}\,\text{cm}^2 \frac{3 \times 10^5\,\text{km/s}}{\sqrt{\pi} 10\,\text{km/s}} x_{\rm HI} 1.3 \times 10^{-5}\,\text{cm}^{-3} (4a)^{-3} 0.75 \times 10^{23}\,\text{cm} (4a) \\
&= 7 \times 10^4 \frac{x_{\rm HI}}{(4a)^2}
\end{aligned}$$

at the cosmic scale factor a. To get $\tau \sim 1$ the neutral fraction $x_{\rm HI}$ must be $x_{\rm HI} \sim 10^{-5}$.

2.2.2 Temperature

The final component in modeling the IGM is to know what the temperature of the gas is.

Since the IGM is highly ionized, a process of photo-heating (heating by ionizing radiation) is important. When a high energy photon hits a hydrogen atom, 13.6 eV of its energy goes into ionizing the atom, the rest goes into the energy of the ejected electron. If the electron is not super-energetic (less that ~40 eV), it thermalizes and adds its energy to the thermal energy of the gas. For more energetic electrons the situation may be more complex, as it can ionize another atom by colliding with it (a so-called *secondary ionization*). That, in turn, produces an energetic electron which may ionize another atom etc. Usually, these secondary ionizations are only important if the gas is substantially (more than a few percent) neutral; for the low redshift IGM with $x_{\rm HI} \sim 10^{-5}$ secondary ionizations are completely unimportant.

If all the excess energy of an ionizing photon goes into heat, the rate of internal energy increase in the gas due to photo-heating is

$$\frac{3}{2}\frac{d}{dt}(nk_BT)\bigg|_{\rm PH} = cn_{\rm HI}\int_{E_0}^{\infty}(E-E_0)\sigma_{\rm HI}(E)n_E dE,$$

where $E_0 = 13.6\,$eV is the hydrogen ionization threshold, $\sigma_{\rm HI}(E)$ is the hydrogen ionization cross-section, and n_E is the radiation spectrum measured in photons per unit volume per unit energy.

The photoionization rate of hydrogen is

$$\Gamma = c\int_{E_0}^{\infty}\sigma_{\rm HI}(E)n_E dE,$$

hence

$$\frac{3}{2}\frac{d}{dt}(nk_BT)\bigg|_{\rm PH} = n_{\rm HI}\Gamma\langle\Delta E\rangle,$$

where $\langle\Delta E\rangle$ is the average excess energy (over 13.6 eV) of an ionizing photon,

$$\langle\Delta E\rangle \equiv \frac{\displaystyle\int_{E_0}^{\infty}(E-E_0)\sigma_{\rm HI}(E)n_E dE}{\displaystyle\int_{E_0}^{\infty}\sigma_{\rm HI}(E)n_E dE}. \tag{9}$$

Let us ignore helium for a moment: $n_e = (1-x)n_{\rm H}$, $n = n_{\rm H} + n_e = (2-x)n_{\rm H}$. Then the thermal balance equation together with the ionization balance equation become

$$\frac{3}{2}\frac{d}{dt}((2-x)k_BT) = x\Gamma\langle\Delta E\rangle, \tag{10}$$

$$\frac{d}{dt}x = -x\Gamma + R(T)n_{\rm H}(1-x)^2. \tag{11}$$

Let us start with cold neutral IGM ($x = 1, T = 0$, like at very high redshift, before cosmic reionization), and assume that the ionizing radiation pops out of nowhere instantaneously at a cosmic time t_R (a favorite approximation of your CMB friends),

$$\Gamma \propto \theta(t-t_R)$$

($\theta(x)$ is a Heaviside function). In the ionization equilibrium

$$x_{\rm eq} = \frac{Rn_{\rm H}}{\Gamma}(1-x_{\rm eq})^2.$$

Hence, until the ionization equilibrium is established (i.e. while $x \gg x_{\rm eq}$) $x\Gamma \gg Rn_{\rm H}(1-x)^2$. In that limit Eq. (11) becomes simply

$$\frac{d}{dt}x = -x\Gamma$$

and its solution for $t > t_R$ is

$$x(t) = e^{-\Gamma(t-t_R)}.$$

That solution is valid until x becomes small enough ($\sim Rn_{\rm H}/\Gamma$) for the ionization equilibrium to get established.

Equation (10) can also be solved easily in the same limit,

$$(2-x)k_BT = \frac{2}{3}\langle\Delta E\rangle\left(1-e^{-\Gamma(t-t_R)}\right),$$

and in the limit of small x the gas temperature becomes constant (i.e. gas becomes *isothermal*),

$$T_\infty = \frac{\langle\Delta E\rangle}{3k_B} \tag{12}$$

and independent of density or the photoionization rate. This is an important lesson: **if a region of space is ionized rapidly, its temperature does not depend on the strength of the radiation field**. I.e., you cannot heat up the IGM by cranking up the ionizing source, only by making the source spectrum *harder*.

It is also instructive to plug some numbers into Eq. (12). For example, for a power-law energy spectrum for ionizing photons, $n_E \propto E^{-\alpha}$, and using the fact that just beyond the ionization edge $\sigma_{\rm HI}(E) \propto E^{-3}$, we find

$$\langle \Delta E \rangle = \frac{\int_{E_0}^{\infty} (E - E_0)\sigma_{\rm HI}(E) n_E dE}{\int_{E_0}^{\infty} \sigma_{\rm HI}(E) n_E dE} = \frac{E_0}{1+\alpha}$$

and

$$T_\infty = \frac{52{,}000\,{\rm K}}{1+\alpha} = \left[\begin{array}{l} 26{,}000\,{\rm K}\ (\alpha = 1) \\ 5{,}000\,{\rm K}\ (\alpha = 9) \end{array} \right.$$

In other words, the temperature of the photo-ionized gas is about 10,000 K, give-or-take a factor of 2.

Let us now consider what happens next. A region of space was ionized to $x = x_{\rm eq}$ at $t = t_R$ $(a = a_R)$, and the temperature of the gas is at this moment constant at $T = T_\infty$. Another important effect is plain adiabatic cooling due to the expansion of the universe, so that the full equation that governs the temperature evolution after ionization equilibrium is established is

$$\frac{dT}{dt} = T\frac{2\dot{n}_{\rm H}}{3n_{\rm H}} + T_\infty x_{\rm eq}\Gamma = T\frac{2\dot{n}_{\rm H}}{3n_{\rm H}} + T_\infty R n_{\rm H}. \tag{13}$$

The recombination coefficient can be well approximated as a power-law function of gas temperature, $R(T) \approx 4.3 \times 10^{-13} T_4^{-0.7}\,{\rm cm}^3/{\rm s}$ $(T_4 \equiv T/10^4\,{\rm K})$. It is easy to solve Eq. (13) for the temperature T_0 at the cosmic mean density, $\bar{n}_{\rm H} \propto a^{-3}$,

$$T_0(a) = T_\infty \left(\frac{a_R}{a}\right)^2 \left[1 + 1.34 R(T_\infty)\bar{n}_{{\rm H},R} t_R \left(\left(\frac{a}{a_R}\right)^{1.9} - 1\right)\right]^{1/1.7}. \tag{14}$$

At late times $(a \gg a_R)$ the asymptotic behavior of the temperature at the mean density is

$$T_0(a) \propto \left(\frac{a_R}{a}\right)^{1.5/1.7}.$$

It is less rapid than pure adiabatic expansion $T \propto a^{-2}$ because photo-heating off the residual neutral hydrogen fraction remains non-negligible at all times.

It turns out that for densities other than the mean a power-law ansatz provides a decent approximation for moderate overdensities, $\delta \lesssim 10$,

$$T(\rho) \approx T_0(1+\delta)^{\gamma-1}, \tag{15}$$

where both T_0 (as is given above) and γ are functions of time (Hui and Gnedin 1997). Expression for $\gamma(a)$ is rather ugly, but its main important properties are that $\gamma = 1$ right after instantaneous reionization and $\gamma \to 1.62$ at late times (notice, it is 1.62 and *not* $2/3$).

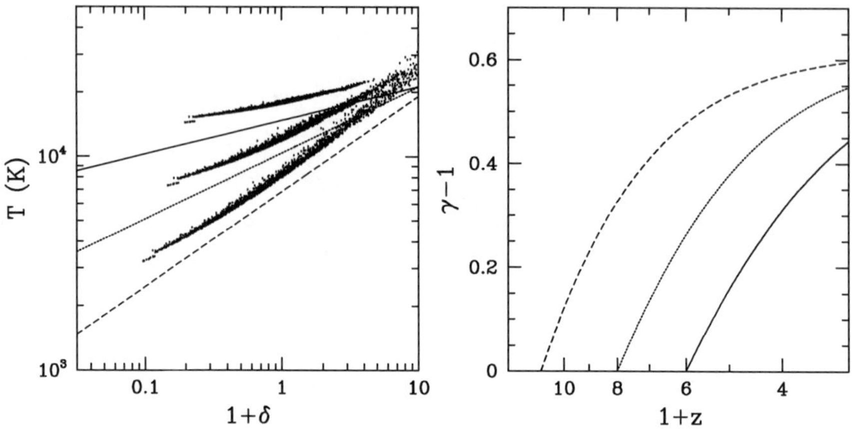

Fig. 4 *Left* Temperature-density relation for a sudden reionization model at $z = 6$: *dots* show the results of a full calculation at $z = 4, 3, 2$ (from top down) while *lines* are approximation (15) with T_0 given by Eq. (14); the approximate solution slightly underestimates the temperature because it ignores the heat input from helium ionizations. *Right* Time evolution of γ for $z_R = 10, 7, 5$ respectively

Figure 4 compares the temperature-density relation in the IGM from a real calculation (by following heating and cooling of individual fluid elements, Hui and Gnedin 1997) with the approximate solution above. One effect that we ignored is photoheating of helium—heat input from the ionizations of the residual neutral helium will heat the gas a bit more than is given by Eq. (14), but, overall, our analytical calculation does rather well.

"Hey", a meticulous reader will exclaim, "what about radiative cooling?" After all, gas does cool by emitting radiation. A story of gas cooling, with all its gory details, awaits us in the future, but here let us estimate how important radiative cooling actually is in the Lyman-α forest.

In a highly ionized gas the dominant radiative cooling mechanism is recombination cooling,

$$\left.\frac{dU}{dt}\right|_{\rm RC} = -\frac{3}{4}k_B T R(T) n_e n_{\rm HII}.$$

If we compare this term to photoionization heating in ionization equilibrium,

$$\left.\frac{dU}{dt}\right|_{\rm PH} = n_{\rm HI}\Gamma\langle\Delta E\rangle = \langle\Delta E\rangle R(T) n_e n_{\rm H},$$

we see that the radiative cooling is lower by a factor of

$$\frac{3}{4}\frac{k_B T}{\langle\Delta E\rangle} = \frac{T}{4T_\infty}.$$

Hence, radiative cooling makes at most a 25 % correction, and well after reionization ($T < T_\infty$) the correction is even smaller.

Is this the complete story? Alas, no, the reality is always more complicated than we are ready to accept and you need to be aware of several caveats when using the temperature-density relation.

- The temperature-density relation is an *approximation*, with 5–10 % scatter at low densities and progressively larger scatter as one moves up the density axis, because it misses a major hydrodynamic effects—shocks. Gas motions in the IGM will cause shock waves that will lead to additional gas heating.
- There may exist other heating and cooling mechanisms. For example, Puchwein et al. (2012) argued that heating of the Lyman-α forest by ultra-high energy gamma rays from a population of blazars is important at very low densities. The jury is still out on whether such an effect is important or not, but we should always be aware that we do not known everything.
- Our analysis assumed that the gas is optically thin to ionizing radiation. While this is the case for hydrogen at $z \lesssim 6$, helium is believed to be reionized the second time (from HeII to HeIII) at $z \sim 3$, in which case gas is *not* optically thin to helium ionizing radiation at $z \gtrsim 3$. Non-trivial opacity to ionizing radiation normally leads to increasing photo-heating rates in the gas.

Brain teaser #3: The temperature-density relation is sometimes called an "equation of state" (occasionally even without quotes). Do not fall into that trap—it relates the gas temperature and density, but it is *not an equation of state*. Can you explain why?

2.3 Modeling the IGM

The most straightforward model of Lyman-α forest is a hydrodynamic simulation with ionization balance. In the 1990s several approximate methods have been used, such as a log-normal approximation, Zel'dovich approximation, a pure N-body simulation, Hydro-Particle-Mesh (HPM) approximation. None of these methods is competitive any more and their use can be hardly justified.

The assumption of the ionization equilibrium is very good in the IGM, but it does break down in a few special cases (quasar proximity zones, helium reionization, etc.). Hence, the most accurate simulation of the IGM includes (a model of) Cosmic Ionizing Background (CIB), radiative transfer, non-equilibrium ionization, separate fields for each of ionizing species (HI, HII, HeI, HeII, ...). Such a simulation, however, is usually an overkill, except when it is used for a special purpose like modeling non-equilibrium effects in quasar proximity zones.

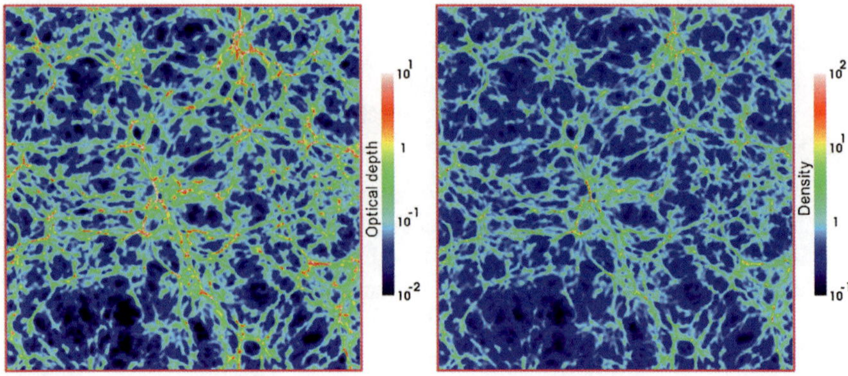

Fig. 5 Slices of optical depth (*left*) and gas density (in units of the cosmic mean, *right*) in a simulation of Lyman-α forest at $z \sim 3$. The box size is $20\,h^{-1}$ comoving Mpc

A standard approach is to include CIB and ionization equilibrium and follow radiative heating (and, optionally, cooling) "on-the-fly" (a-la Eq. 13). For simulations of the Lyman-α forest alone the temperature-density relation may be assumed, but the computational savings in that case will be modest and it is rarely worth it.

Example of a numerical simulation of the forest is shown in Fig. 5. The right panel shows the gas density, and looks like a usual image of large-scale structure. The left panel shows the Lyman-α optical depth that would be observed in the corresponding position along the absorption spectrum towards a high redshift quasar. The main thing to take from that figure is that the actual absorption lines we see clearly in the spectra (those with $\tau \gtrsim 0.5$) come from filaments: weaker ones tend to cluster around stronger ones, although a few of the weakest ones do occur in the voids. The higher optical depth systems, those that lead to saturated lines with $\tau \gtrsim 2$ tend to occur at the intersections of filaments, nearer to galaxies.

2.3.1 Density—Column Density Correlation

What is clear from Fig. 5 is that the gas density and the optical depth of the corresponding absorption feature are well correlated. Crudely, the relation is

$$\tau \sim (1 + \delta)^{1.5},$$

although the slope and normalization of this correlation are redshift dependent.

This correlation is so good, especially on large scales, that it is often used to match directly the gas density into the opacity along the line of sight—such an *ansatz* is called *Fluctuating Gunn-Peterson Approximation*, or FGPA. FGPA is useful for modeling the forest on large scales, but one has always keep in mind that the absorption spectrum is in the velocity space, while the density is sampled in real, physical space, hence FGPA breaks down on sufficiently small scales (roughly less than 1 Mpc).

2.4 What Observations Tell Us

For a long time since the Lyman-α forest was discovered in the 60s, it was treated by observers as just another absorption spectrum—as a collection of individual absorption lines, each having a fixed column density N and the Doppler parameter b, as if the absorption was coming from discrete clouds. Now we know that this is not a good description—the density, temperature, and neutral fraction fields are continuous, and it is impossible to decompose the realistic spectrum into a set of (N, b) pairs uniquely.

The modern view of the forest is that $\tau(\lambda)$ is a continuous field and should be treated as such. However, there is one application where the (N, b) decomposition is still useful—measuring the temperature-density relation. If we think about a segment of the spectrum that has an "absorption line", the width of the feature is determined by the temperature of the gas plus any velocity gradient across the region that may exist. In some cases that velocity gradient will be very small, so the narrowest features at each column density should be those that are broadened by temperature alone. Hence, looking at the distribution of fitted b parameters at given column density, one can measure $T(N)$ in the forest and, by virtue of the strong correlation between ρ and τ (and, hence, N), translate that measurements into the measurement of $T - \rho$ correlation.

Figure 6 shows an example of such distribution from a single high resolution quasar spectrum (Rudie et al. 2012). The cutoff in the distribution of Doppler parameters for a given N is clearly visible, and the value of the cutoff is well fit by the power-law in N, demonstrating the fact that the power-law temperature-density relation is indeed a good approximation.

A compilation of the majority of existing measurements is shown in Fig. 7 (Lidz et al. 2010; McDonald et al. 2001; Ricotti et al. 2000; Rudie et al. 2012; Schaye et al.

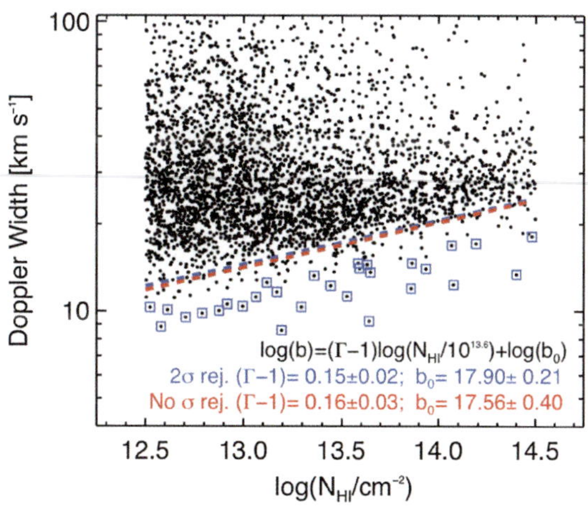

Fig. 6 Example of the (N, b) distribution for a quasar spectrum at $z = 2.4$. The measurement points labeled by *blue squares* are contamination from heavy elements. A relatively sharp edge of the distribution of Doppler parameters at a given N is apparent in the figure (adopted from Rudie et al. (2012))

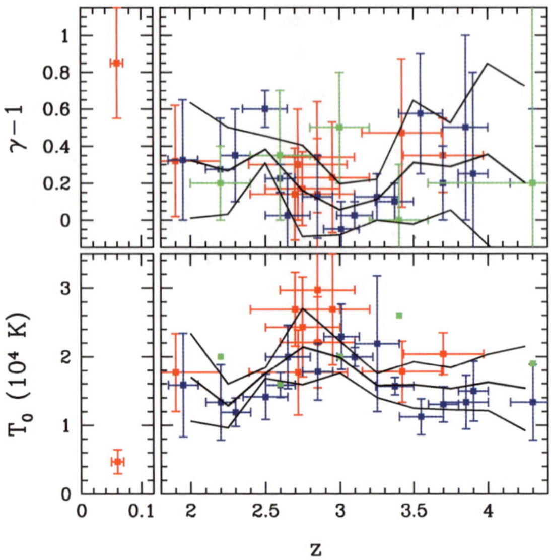

Fig. 7 Evolution of the temperature-density relation $T \approx T_0(1+\delta)^{\gamma-1}$. *Red*, *blue*, and *green* points show individual measurements from Ricotti et al. (2000), Schaye et al. (2000), Lidz et al. (2010) respectively; *thick* and *thin black lines* shows the average values and 1σ dispersion for all measurements. Possible increase in temperature and dip in γ at $z \sim 3$ is attributed to HeII → HeIII reionization

2000). The data seem to indicate (albeit rather vaguely) an increase in the temperature and a decrease in γ at $z \approx 3$—a behavior reminiscent of cosmic reionization (Eq. 12). Indeed, this may correspond to the second reionization of helium (HeII going into HeIII), thought to occur at redshifts around 3.

An even simpler quantity is the *column density distribution*—a distribution of all N values irrespectively of what their b values are. Altay et al. (2011) show how modern cosmological simulations can match the observed distribution over 10 orders of magnitude in column density (see Fig. 8).

The column density distribution is a useful observational measurement for other types of hydrogen absorbing systems, such as *Lyman-limit system* ($10^{17}\,\mathrm{cm}^{-2} < N_{\mathrm{HI}} < 10^{20}\,\mathrm{cm}^{-2}$) and *Damped Lyman-$\alpha$ systems* (DLA) ($10^{20}\,\mathrm{cm}^{-2} < N_{\mathrm{HI}}$), but has not been particularly constraining for the forest.

Brain teaser #4: The photo-ionization cross-section for neutral hydrogen at the ionization edge (13.6 eV) is $\sigma_{\mathrm{ion}} = 6.3 \times 10^{-18}\,\mathrm{cm}^2$. Hence, a column density of $N_{\mathrm{HI}} = 1.7 \times 10^{17}\,\mathrm{cm}^{-2}$ has an optical depth of $\tau_{\mathrm{ion}} = \sigma_{\mathrm{ion}} N_{\mathrm{HI}} = 1$. Never-the-less, Lyman-$\alpha$ absorbers remain ionized almost all the way to Damped Lyman-α systems, $N_{\mathrm{HI}} \gtrsim 10^{19}\,\mathrm{cm}^{-2}$ ($\tau_{\mathrm{ion}} \sim 500$). Can you explain why?

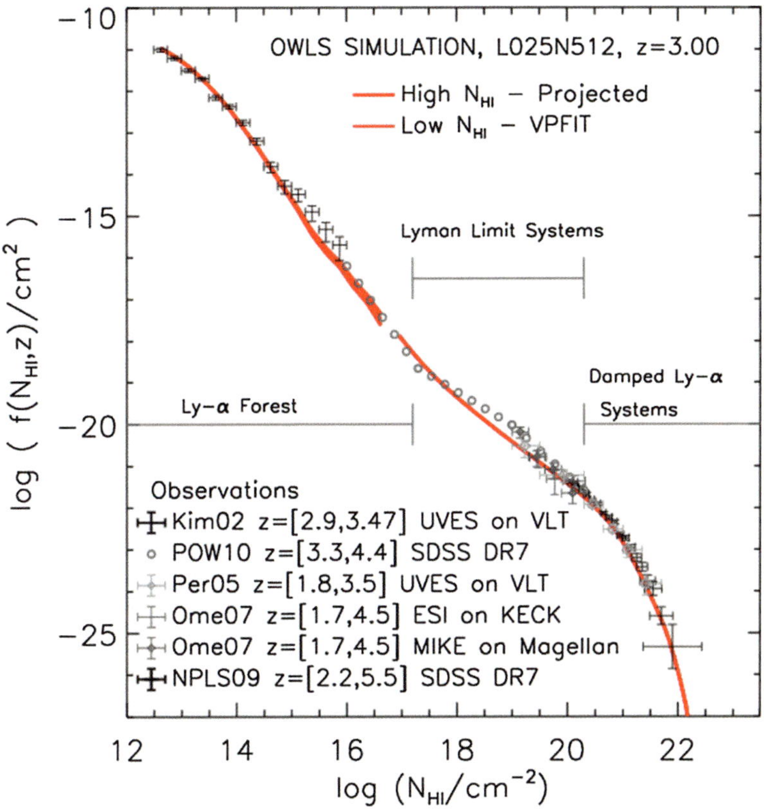

Fig. 8 Distribution of column densities of Lyman-α absorbing systems (adopted from Altay et al. (2011))

2.4.1 Lyman-α Power Spectrum

Perhaps the most important use of Lyman-α forest in cosmology is in measuring the evolution of the matter power spectrum. Observations of the forest cover a wide redshift range, from $z \gtrsim 2$ to $z \lesssim 5$; since the observed optical depth is well correlated with the gas density, which, in turn, traces the matter density on large scales (above the filtering scale), the observed spectra of the forest contain hidden information about the clustering of matter and its evolution over the redshift range $2 \lesssim z \lesssim 5$.

Measuring the matter power spectrum is exactly the application for which the Fluctuating Gunn-Peterson Approximation (FGPA) is most suitable. In the theory of large scale structure formation there is a theorem that states that if a locally non-linear field is a function of matter density only ($f = f(\rho)$), then on sufficiently large scales the field f is *linearly biased* with respect to the density field, i.e. for sufficiently small k

$$P_f(k) = b_f^2 P(k),$$

where the bias factor b_f is independent of k. Hence, one can measure the matter power spectrum $P(k)$ in a few simple steps:

1. measure the 1D power spectrum of the transmitted Lyman-α flux, $P_{1D}(k)$, directly from the observed spectra;
2. convert from a 1D to a 3D flux power spectrum,

$$P_F(k) = -\frac{2\pi}{k}\frac{dP_{1D}}{dk};$$

3. determine the flux bias factor, b_F, from numerical simulations,
4. and, finally, compute the matter power spectrum

$$P(k) = \frac{P_F(k)}{b_F^2}. \tag{16}$$

Such a program was first completed by Croft et al. (1998) and later repeated many times with better data. For example, the largest set of observed Lyman-α spectra was obtained as part of the Sloan Digital Sky Survey (SDSS), and is shown in Fig. 9. A little bit of nuisance is that the flux power spectrum is measured in the velocity space, so the units of k in Eq. (16) are (km/s)$^{-1}$. That makes it hard to compare with other measurements of matter power spectrum without knowing cosmological parameters. But a good piece of news is that the power spectrum grows (or the plotted quantity, $\Delta(k)^2 = k^3P(k)/2\pi^2$, decreases) with redshift with the rate prescribed by the standard cosmology, so you have not studied your Introduction To Cosmology in vain …

On a more serious note, this measurement provides extremely powerful constraints on the matter power spectrum at the smallest scales—in fact, the forest probes the smallest scales currently accessible to any observational measurement. Many important cosmological and physical studies use these measurements, from determining cosmological parameters to constraining neutrino masses (but that is a field I am not going to review in these lectures).

2.4.2 Where the Forest Ends

The Lyman-α forest is a small-scale scale counterpart of the large-scale structure—but how small is "small"? In other words, what are the smallest spatial scales on which there is structure in the IGM?

This question is not moot—indeed, the filtering scale tells us where the baryonic fluctuations lag behind the dark matter, but it only applies to linear evolution. The forest is nonlinear, and nonlinear evolution may drive new fluctuations on a variety of scales.

One way to measure structure in any distribution is the, familiar to us already, power spectrum. Using high resolution spectra from 8m-class telescopes one can extend the SDSS measurement to much smaller scales, as is illustrated in Fig. 10.

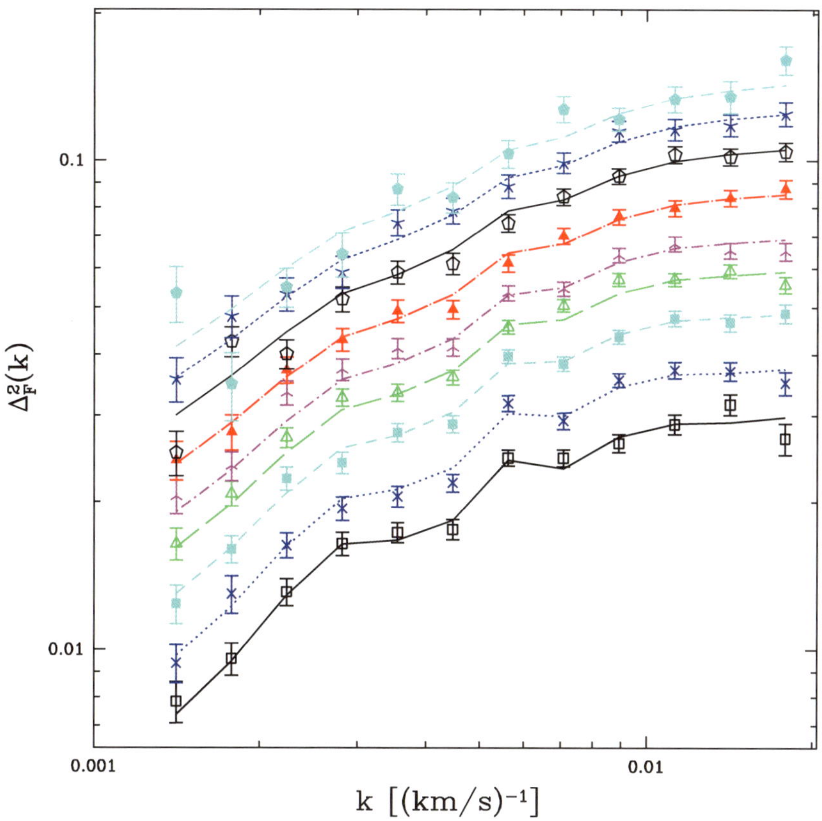

Fig. 9 Matter power spectra measured from SDSS Lyman-α measurements at a range of redshifts from $z = 2.2$ (*bottom*) to $z = 4.2$ (*top*) (adopted from McDonald et al. (2006))

The decrease in the clustering amplitude is clearly visible at $k > 0.03$ s/km, but is it really the end of the forest? The answer is "unfortunately, no"—unfortunately, because the roll over in the flux power spectra has nothing to do with the actual matter clustering—it is merely an artifact of the thermal broadening of the spectra (the exponential factor in Eq. 8). Alternatively, one can think of it as the break up of the linear biasing approximation (Eq. 16).

So, how would one approach the question of studying the smallest scale structure in the forest? One option is offered by spectra of double or gravitationally lensed quasars—if the two quasar images are not too far on the sky, their sightlines probe small spatial scales. Unfortunately, this approach has not been particularly popular among observers—in the only study I am aware of Rauch et al. (2001) demonstrated that, in fact, there is not that much structure in the forest on scales below a kpc. For example, Fig. 11 shows Lyman-α spectra along two lines of sight to two images of a gravitationally lensed quasar separated by about 0.5 comoving kpc at $z \sim 3$.

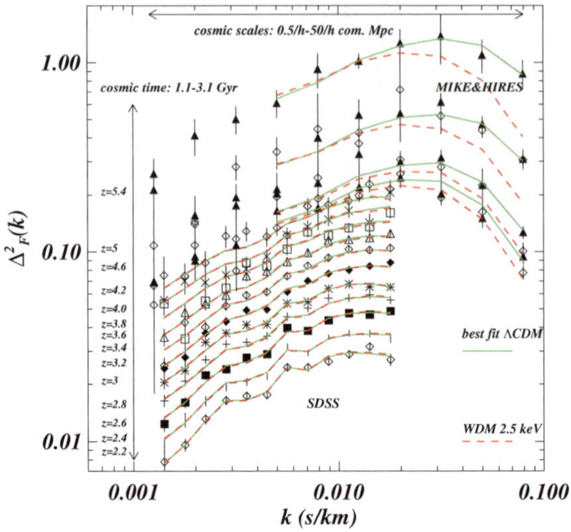

Fig. 10 Matter power spectra measured from SDSS Lyman-α measurements (as in Fig. 2) combined with data from high resolution spectra of several quasars (adopted from Viel et al. (2013))

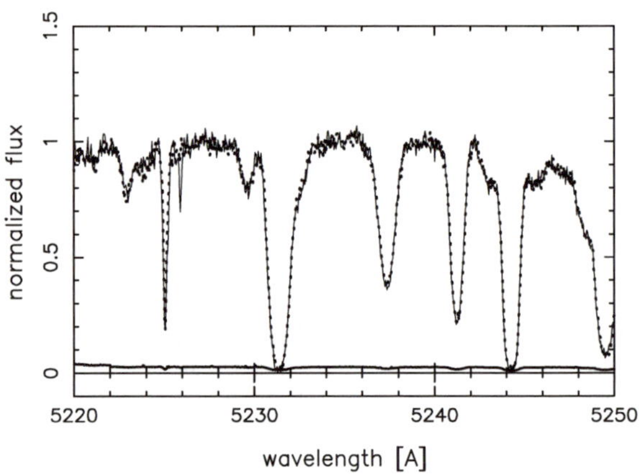

Fig. 11 Lyman-α spectra along lines of sight to the gravitationally lenses quasar Q1422+231 (images A and C). One spectrum is shown with the *solid line*, another one with the *dotted line* beaded with *dots*. The two spectra are identical to within the observational errors (adopted from Rauch et al. (2001))

Using this measurement, Rauch et al. (2001) placed a strict constraint on the density variation in the forest on small scales,

$$\sqrt{\langle(\Delta \ln \rho)^2\rangle} < 3 \times 10^{-2} \text{ for } \langle \Delta x\rangle = 0.6\,\text{kpc},$$

or, alternatively,

$$\sqrt{\left\langle\left(\frac{\Delta \ln \rho}{\Delta x}\right)^2\right\rangle} < 0.05\,\text{kpc}^{-1}.$$

A scientifically interesting question is whether the IGM is turbulent on small scales. The Rauch et al. (2001) constraint implies that either the forest is *not* turbulent on these small scales, or that any turbulence that is present is highly sub-sonic (i.e. incompressible). The latter option is possible but is not too likely—density fluctuations in the sub-sonic turbulence scale as Mach number squared, with the flow in the forest becoming transonic at scales 100–200 kpc. If we take a Kolmogorov-like scaling law,

$$\sqrt{\langle(\Delta \ln \rho)^2\rangle} \approx 1 \left(\frac{\Delta x}{200\,\mathrm{kpc}}\right)^{1/3},$$

then on scale of 0.6 kpc we find the rms density fluctuation of $\sqrt{\langle(\Delta \ln \rho)^2\rangle} \approx 0.15$, 5 times higher than the actual observed upper limit. Of course this is not a formal derivation, and factors of several may be lurking here and there, but the estimate serves to demonstrate that the forest is remarkably quiet on scales below a kpc.

Brain teaser #5: It is well known in classical hydrodynamics that any flow with Reynolds number in excess of about 1000 becomes turbulent. The viscosity in the IGM is very small, and Reynolds number in the forest is of the order of 10^6. Hence, the naive expectation is that the IGM must be very turbulent on small scales, but the Rauch et al. (2001) observations suggest it is not. Can you think of an explanation?

3 From IGM to CGM

Circumgalactic medium, or CGM, is often understood as the gas within the galactic dark matter halo. I am taking a broader view here, since some of the structures in the universe, like filaments, fall in the border zone between the IGM and CGM, they are not always considered to be part of the Lyman-α forest, but they also are not related to galaxies. They do produce absorption lines in the quasar spectra, but they also stream gas into galactic halos.

3.1 Large Scale Structure

Probably everyone has seen a picture of the large-scale structure of the universe by now (if you have not, check out excellent visualizations of the Millennium simulation at www.mpa-garching.mpg.de/galform/virgo/millennium). Since the large-scale structure forms as a result of gravitational clustering from the linear Gaussian fluctuations, it is fully characterized by the linear matter power spectrum. Hence, various scales that we see in the pictures are all related to features in the power spectrum. For example, typical size of voids corresponds to about $1/2$ of the scale at which the power spectrum peaks (which is about $100h^{-1}$ Mpc in comoving units). Hence, in comoving reference frame void sizes do not change—they are as large at $z = 10$ as they are at $z = 0$ (although, these largest voids are, of course, not nearly as empty at $z = 10$ as they are at $z = 0$). Filaments that surround voids are highly non-linear structures and their width is controlled by the nonlinear scale at each epoch, i.e. the scale at which the amplitude of linear fluctuations reaches unity. Finally, material that makes the largest objects at any time (clusters of galaxies today, galaxies at $z \gg 2$) is assembled from regions roughly the nonlinear scale in size, so masses of these objects are about $4 \times$ (mean density) $\times$ (nonlinear scale)^3.

Since our main interest is how gas flows from low to high density regions, the actual motion of matter is of particular importance to us. With time voids become deeper as matter (both dark and gaseous) flows from them onto filaments, and then along the filaments into the galaxies. This pattern of flows is illustrated in Fig. 12 from a numerical simulation of a local region around the Local Group by Klypin et al. (2003).

As gas flows into a filament from opposite directions, it gets shocked, and the gas temperature is expected to rise above that maintained by photo-heating and adiabatic expansion/contraction—a complication that eventually destroys nice and tight density-temperature relation that exists in the lower density IGM.

The actual structure of the filaments received surprisingly little attention in the literature. In a classical review Shandarin and Zeldovich (1989) showed the profiles of one-dimensional collapse onto a 2D pancake (Fig. 13). Collapse onto a 1D filament

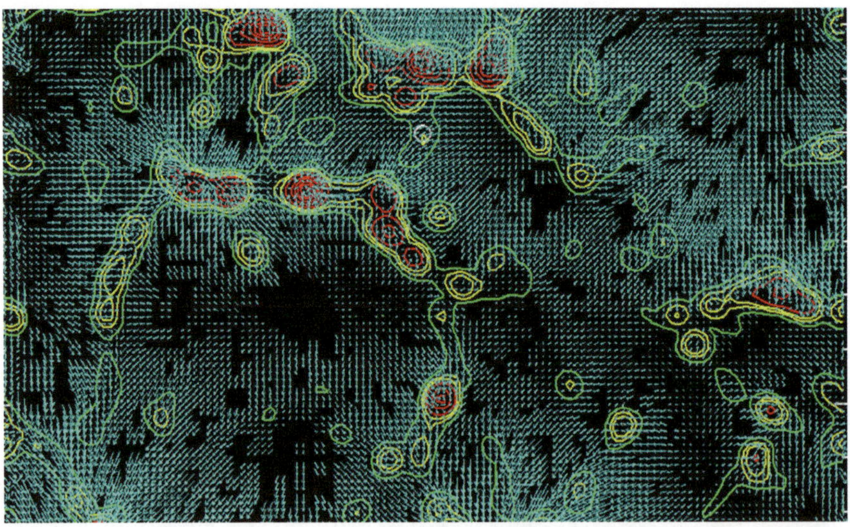

Fig. 12 Large-scale flows (*cyan arrows*) on top of the density contours (*green*, *yellow*, and *red*). Flows of matter onto (almost all of the) filaments are clearly visible in this visualization of a numerical simulation (adopted from Klypin et al. (2003))

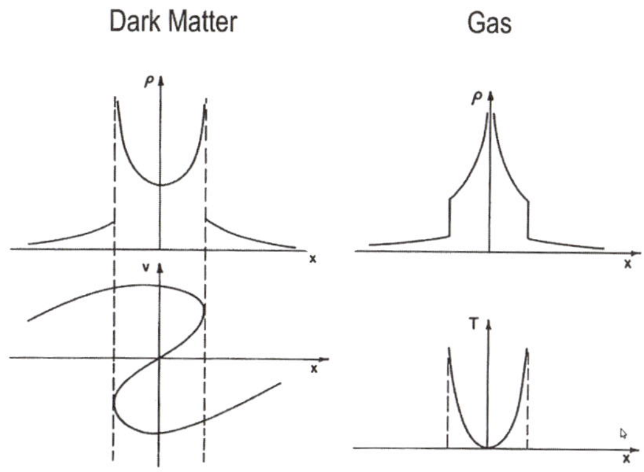

Fig. 13 Profiles of one-dimensional collapse of dark matter and gas onto a 2D pancake (adopted from Shandarin and Zeldovich (1989))

is qualitatively similar, because the physics is the same—gas gets piled up at the center, where the entropy is the lowest, while the dark matter from each side flows through the upcoming stream, creating density caustics on the outside. What happens next is determined by whether the filaments are self-gravitating—but since they have widths comparable to the nonlinear scale, we know that they, on average, *are* self-gravitating. In a self-gravitating filament the dark matter will stop streaming, turn around, and fall on itself once again and again, increasing the number of intersecting streams as time goes on.

In order to illustrate the large- (and not-so-large-) structure further, I will use a cosmological simulation from Gnedin and Kravtsov (2010). This simulation is not very big, and focuses on the environs of a single, Milky-Way like galaxy, but it will suffer for our purpose. Figure 14 shows the gas density and the gas temperature around the main galaxy at $z = 2$.

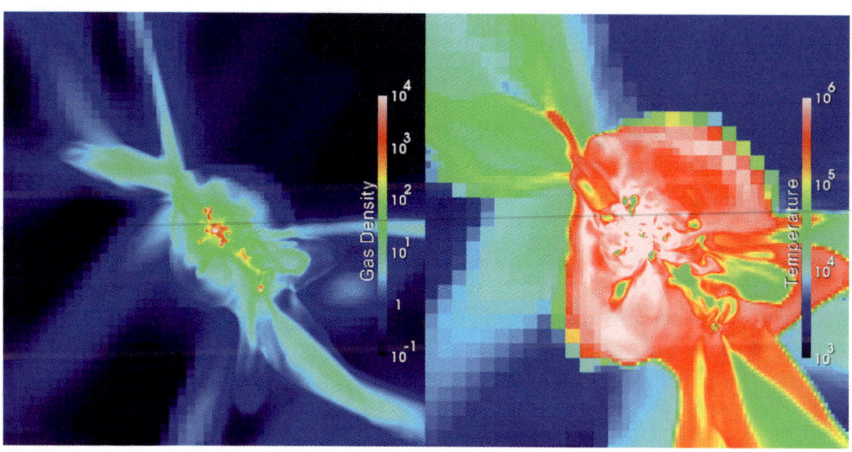

Fig. 14 Thin slices through cosmological simulation that show the gas density (*left*) and the gas temperature (*right*) around a typical galaxy at $z \approx 2$

There are a few features to note. First of all, the gas filaments do appear to be denser and cooler in the middle, similarly to the 1D collapse. Second, in the temperature plot we see really hot (million degrees) gas. Most of that hot gas is concentrated around the galaxy, in the dark matter halo and beyond, but some of it extends way into the filaments—those are the temperature spikes that we see in Fig. 13.

3.2 How Gas Gets onto Galaxies

Everyone knows that dense enough regions of the large-scale structure will collapse and *virialize* (i.e. reach, or, at least, approach, the virial equilibrium). The simplest model of such collapse is a *top-hat*,

$$\rho(\mathbf{x}) = \begin{cases} \bar{\rho}(1+\delta_i), & r < r_i \\ \bar{\rho}, & r > r_i \end{cases}$$

where r_i and δ_i are the initial radius and amplitude of the perturbation. The overdense perturbation collapses, and the evolution of the radius of the perturbation can be solved analytically in the matter-dominated regime ($a \propto t^{2/3}$), albeit parametrically with a parametric variable θ:

$$r = \frac{GM}{\delta_i \dot{r}_i^2}(1-\cos\theta),$$
$$t = \frac{GM}{\delta_i^{3/2}\dot{r}_i^3}(\theta - \sin\theta).$$

The moment of collapse is defined as $r = 0$, which occurs at the time when $\theta = 2\pi$. A remarkable property of the top-hat solution is that at the moment of collapse t_f the linear density fluctuation

$$\delta_L(t) = \frac{D_+(t)}{D_+(t_i)}\delta_i$$

is just a number, independent of the initial overdensity, size, or the mass of the overdense region,

$$\delta_L(t_f) = \frac{3}{5}\left(\frac{9\pi^2}{4}\right)^{1/3} = 1.69.$$

A perturbation cannot collapse to a point—that would be even less likely than making a pencil stand on a sharp end. A standard assumption is that the collapsing perturbation virializes—i.e. reaches the virial equilibrium—at around the time t_f. In that case the average overdensity δ_v of the final virialized object is $1+\delta_v = 18\pi^2 \approx 178 \approx 180 \approx 200$.

The virial radius of the dark matter halo in Fig. 14 is roughly the green roundish region in the density panel (overdensity $\gtrsim 100$), while the million-degree gas extends well beyond it. The virial radius serves as a good approximation of a boundary beyond which any, even imaginable, resemblance of spherical symmetry totally vanishes! As gas falls into potential wells of dark matter halos, it gets shocked and heated to around the virial temperature (also deviations can easily be a factor of 2–3 in each direction). Shocks never stand still (in the reference frame of the gas behind them), so the accretion shock propagates outward. For typical cosmological objects, be it star-forming galaxies at $z \sim 2$ or galaxy clusters at $z = 0$ (or anything in between), it is not uncommon to find the accretion shock extending to 3 virial radii. Since it goes so much beyond the quasi-spherical region, it is highly asymmetric and non-spherical, with some of its protrusions reaching well into voids, up to ~ 10 virial radii, while along filaments the accretion shock may not even exist (or do not reach to even a modest fraction of the virial radius).

3.3 Cool Streams

A story of the "cold streams" is a real-life story of an elephant-in-the-room. In a gesture of non-conformity, I am going to call them "cool streams", because in the ISM-speak (which we are going to use for the most of this course) the term "cold" refers to truly cold gas, below 100 K. Strictly speaking, they should be called "warm streams", since 10^4 K gas is "warm" in the ISM-speak, but that would confuse too many people …

Every practicing simulator knew about cool streams, but no one paid any attention to them until in 2005 in an influential paper Kereš et al. (2005) showed that at intermediate redshifts—the epoch where galaxies make most of their stars—cool streams deliver significant, or even dominant, fraction of gas onto the galactic disks, where stars actually form. Hence, from the point of view of a galaxy as a gas consumer, cool streams are the primary consumption channel.

Examples of cool steams in cosmological simulations from Overwhelmingly Large Simulation project (OWLS, van de Voort et al. 2011) are shown in Fig. 15. As in a weird monster movie, the blue "tentacles" of cool gas try to reach the central galaxy; they break up into individual blobs for a massive one ($M = 10^{12.5}\,M_\odot$), remain as thin streams for a $M = 10^{12}\,M_\odot$ one, and completely swamp gas accretion for a Milky-Way type galaxy ($M = 10^{11.5}\,M_\odot$ at $z = 2$). Images like that can be made from almost any cosmological simulation, and from any modern simulation code, be it an SPH code, an AMR, or a moving mesh code like AREPO[1] (Springel 2010). All simulations agree that the cool flows dominate the gas accretion for halos above about $M = 10^{11.5}\,M_\odot$, with this mass being only weakly (if at all) redshift dependent.

[1]For these and other curious abbreviations check out Volker Springel's lectures in this volume.

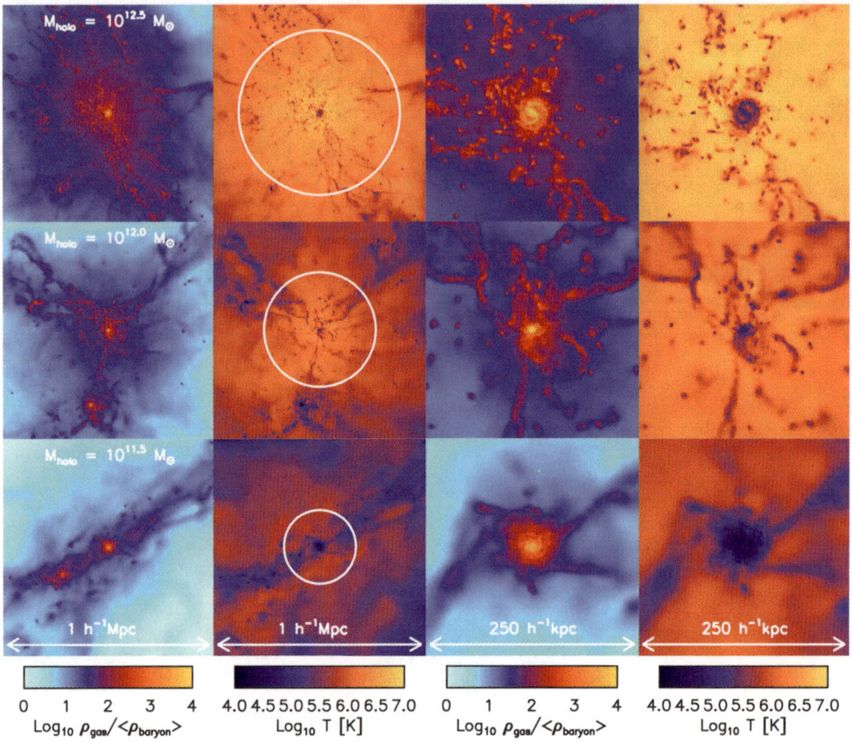

Fig. 15 Density and temperature images of galaxies of different masses at $z \approx 2$. Cool streams are clearly visible in temperature images as *blue blobs* and filaments (adopted from van de Voort et al. (2011))

Since most of the gas accretion occurs in low mass galaxies at all times, most of gas that ends up in galactic disks enters the halo as "cool"—significantly below the virial temperature, but it may still be well above the "ISM warm" of 10^4 K—at all cosmic times up to the present epoch. The contribution of cool streams is, however, diminishing with time, so by $z = 0$ they, on average, only deliver about half of the accreting gas onto galactic disks.

A happy concordance is broken, however, when the fate of cool streams inside the halo is explored further. In a recent study, a carefully designed comparison between GADGET (Springel 2005) and AREPO (Springel 2010) codes found some disturbing differences (Nelson et al. 2013, shown in Fig. 16). The two codes have the same gravity and dark matter solvers, but differ in the way gas dynamics is treated (for details, check Volker Springel's lectures in this volume). While in the SPH GADGET simulation the cool streams remain cool inside the halo and reach all the way to the galactic disk, in the mesh-based simulation with AREPO the cools streams heat up as they approach the disk. This discrepancy reflects the well-known dichotomy between SPH and mesh codes—the former do not have enough diffusion (without

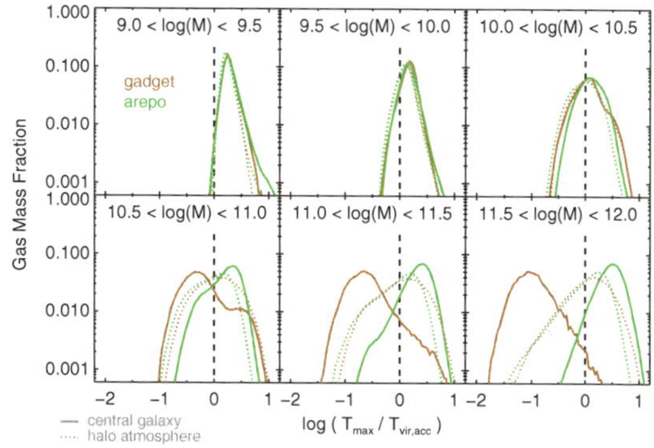

Fig. 16 Temperature distribution function for galaxies in different mass bins simulated with GADGET and AREPO. The two codes predict significantly different distributions for high mass galaxies (adopted from Nelson et al. (2013))

special fixes), while the latter may have too much numerical diffusion, especially in the poorly resolved regions. Which of the two codes is closer to reality is not yet clear; the progress in this field, though, happens at a relativistic speed, so as you are reading these lectures, the ambiguity may have been already resolved.

3.4 Galactic Halos

Few sane people doubt the existence of dark matter halos. Whether galaxies have gaseous halos is an entirely different matter.

Cosmological simulations generically predict that galaxies like the Milky Way (MW) should be surrounded by hot gaseous halos in quasi-virial equilibrium. For two decades the actual existence of these hot halos was an even more hotly debated topic. The point of contention was the simple fact that hot gas emits X-rays, hence hot halos must be detectable in X-rays. The cruel reality is that the halo gas is rather tenuous, and for galaxies like the Milky Way it is expected to have temperatures that are very hard to detect observationally.

How much gas one expects to reside in the Milky Way halo actually depends on the halo mass, which has been notoriously difficult to estimate. Proposed values range from $\sim 7 \times 10^{11}\,M_\odot$ to over $2 \times 10^{12}\,M_\odot$ (values outside this range are considered extremist and will provoke a French military intervention or an American bombing campaign). For the fiducial value of $10^{12}\,M_\odot$ the cosmic share of baryons in the MW is $1.6 \times 10^{11}\,M_\odot$. The stellar mass of the MW is about $6 \times 10^{10}\,M_\odot$ (although, values up to $8 \times 10^{10}\,M_\odot$ are sometimes used) and the disk gas mass is $\lesssim 1 \times 10^{10}\,M_\odot$. Hence, the gaseous halo may contain up to $10^{11}\,M_\odot$ (it may, of course, be much less if some of the gas is expelled from the Galaxy by stellar feedback and other energetic processes).

The contention about the existence of the hot halo finally has been resolved by *Chandra*—not the brilliant man who resolved so many other contentions, but the remarkably successful space mission named after him. In a ground-breaking observation the Chandra team finally detected the X-ray emission from the hot gas around

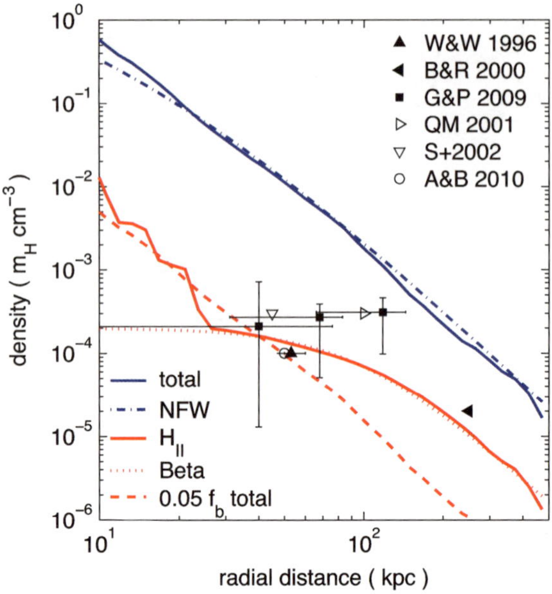

Fig. 17 Radial density profile at $z = 0$ of the same galaxy shown in Fig. 14. *Blue* and *red solid lines* shows the actual simulated profiles of dark matter and gas, while *dashed lines* give the best-fit NFW (for dark matter) and rescaled by $0.05 f_b$ NFW (for gas) profiles respectively. The *red dotted line* is the best-fit beta profile for the gaseous halo. Filled and open symbols are pre-Chandra observational constraints (Anderson and Bregman 2010; Blitz and Robishaw 2000; Grcevich and Putman 2009; Quilis and Moore 2001; Stanimirović et al. 2002; Weiner and Williams 1996)

the Milky Way (Gupta et al. 2012). While measuring the total mass of the halo from Chandra observations is very challenging (try measuring the mass of a giant monster that swallowed you), the limits that the Chandra team has been able to place on the gas mass in the halo are consistent with our estimate of 10^{11} $M_\odot$.

X-ray detection of the halo is important, because it is a *direct* evidence for the existence of a massive (from the point of view of the disk) gaseous halo. Historically, however, a large number of indirect constraints existed that all pointed out towards the same conclusion. In Fig. 17, I show the $z = 0$ dark matter and gas profiles for the same galaxy we met in Fig. 14. The hot halo (solid red line) in that simulation is consistent with the existing pre-Chandra observational constraints as well as with the actual Chandra measurement. What is remarkable is that in the simulation all stellar feedback processes were switched off (see Gnedin (2012), for details about the actual simulation). The galactic disk in the simulation is overly massive and has incorrect density profile, but the halo seems to be ok (at least within the precision of observational constraints). There is, actually a simple reason for it—the main physical process that matters for the gas in the halo is radiative cooling, it is cooling that determines which gas can rain on the disk and which remains in the halo in the hot phase.

Hence, the physics of radiative cooling is our next stop.

3.5 Diversion: Cooling of Rarefied Gases

Before we proceed further along our yellow brick road, let's step aside for a short while and consider how cosmic gas cools—the process we have already met in the IGM segment of our journey, and which we will be meeting over and over again in the future.

Radiative cooling is an "umbrella" name for diverse physical processes through which gas transforms its thermal energy into the radiation that leaves the system. At low enough density, three processes dominate, and all three of them involve a collision of a free electron or an atom/ion with a neutral atom or a partially neutral ion. These three processes are

line excitation: a collision excites the neutral atom into a higher energy state, the state decays and the resultant photon leaves the system;
collisional ionization: a collision ionizes the neutral atom and the binding energy of the freed electron is charged against the thermal energy account;
recombination: an ion captures a free electron and the sum of the kinetic energy of the electron and the binding energy of the neutral atom is emitted as a photon.

All these collisional processes depend on the square of the density, so it is convenient (and customary) to factor out that density dependence explicitly in the cooling rate of the gas,

$$\left.\frac{dU}{dt}\right|_{\text{cool}} = -n_b^2 \Lambda(T, \ldots),$$

where n_b is the number density of baryons (I prefer it to another commonly used parametrization that factors out the hydrogen nucleus number density n_{H}, because n_b is directly proportional to the gas mass density for any value of helium abundance or gas metallicity) and Λ is commonly called a *Cooling Function.*

In the simplest case of gas in pure collisional equilibrium (no external or internal radiation of any kind—the so-called *collisional ionization equilibrium*, or CIE) the cooling function is called "standard". If the relative abundance of various chemical elements is fixed and small variations in the helium abundance are neglected, the cooling function only depends on the gas temperature T and the total metallicity Z,

$$\Lambda_{\text{CIE}} = \Lambda_{\text{CIE}}(T, Z).$$

Examples of this function for $Z = 0$ and $Z = Z_{\odot}$[2] are plotted in Fig. 18.

[2]Throughout these lectures I define "solar metallicity" as the metallicity of our galactic neighborhood, $Z_{\odot} = 0.199$ in absolute units, rather than metallicity of an average-looking single star somewhere in the outskirts of the Galaxy.

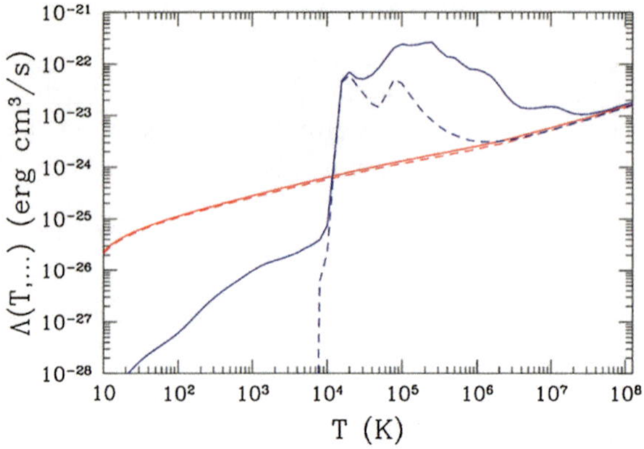

Fig. 18 Cooling functions for the primordial gas ($Z = 0$, *dashed lines*) and the gas at solar metallicity (*solid lines*). *Blue lines* show the "standard" CIE cooling functions, while *red lines* show the cooling functions for the fully ionized gas (the only cooling process is Bremsstrahlung)

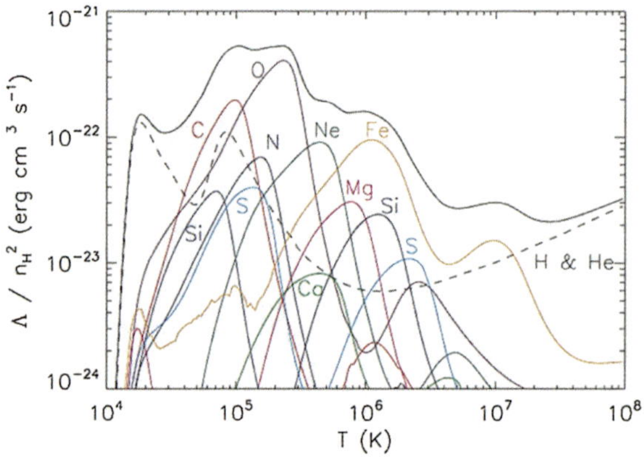

Fig. 19 Contributions of individual chemical elements to the "standard" CIE cooling function (adopted from Wiersma et al. (2009))

The specific shape of the CIE cooling function, with its "bumps and wiggles", is determined by the interplay between contributions of over a dozen various chemical elements. A good recent review is given by Wiersma et al. (2009), an illustration from which is reproduced here in Fig. 19. In particular, one has to be aware that many of the atomic cooling rates used to construct the cooling function are know rather poorly, not better than a factor of 2, and that uncertainty propagates into the actual value of the cooling function. In realistic galactic and cosmological simulations this uncertainty is often, however, unimportant: the cooling time-scale is so much shorter than any other physical time-scale in the problem that it does not need to be known very precisely (all gas that can cool will indeed cool rapidly).

Wiersma et al. (2009) paper offers another, much more important lesson, though. As they show, the actual cooling function in the IGM, CGM, and even ISM of galaxies at low and high redshifts may deviate from the "standard" one quite substantially. In other words, the "standard" CIE cooling function is actually highly non-standard and is almost never realized in nature. The reason for that is that low density cosmic gas is always affected by external radiation field.

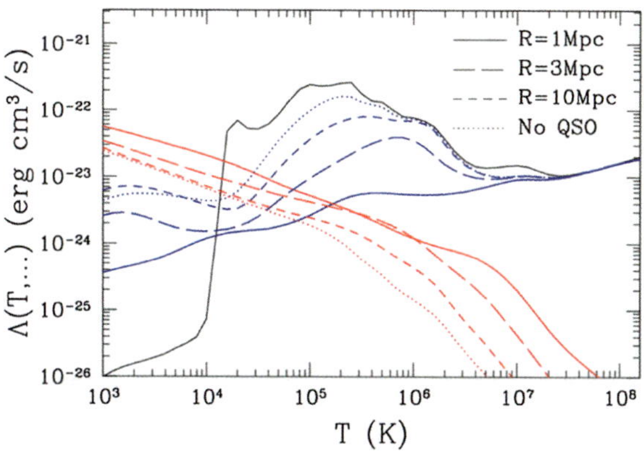

Fig. 20 Illustration for the role of radiation field in suppressing cooling (*blue lines*) and enhancing heating (*red lines*). A gas in the galactic halo (at density 340 over the cosmic mean) is shined upon by the $10^{12}L_{\odot}$ quasar. Sufficiently close, the quasar radiation modifies the cooling and heating functions in a major way

Figure 20 is a simple illustration of this process. Cooling (and heating) processes in a gaseous halo can be modified in a major way if it straddles too close to a strong source of ionizing radiation, such a bright quasar. Within 1 Mpc from the quasar, the equilibrium temperature in the halo goes all the way up to 200,000 K, twenty times above our "canonical" 10,000 K.

So, let us review the cooling function from the very beginning, this time being careful. In a most general case in addition to cooling there is also radiative heating by the radiation field. Hence, the change of the gas internal energy due to radiative processes has two terms with opposite signs,

$$\left.\frac{dU}{dt}\right|_{\text{rad}} = n_b^2 \left(\Gamma - \Lambda\right),$$

where Λ is our old acquaintance the cooling function and Γ is the *heating function*. Both of them depend on a multitude of parameters,

$$[\Gamma, \Lambda] = \mathcal{F}\left(T, n_b, X_{ijl}, J_\nu, \tau_{ijl}\right), \tag{17}$$

where the density dependence reappears because not all processes are two-body, X_{ijl} is the abundance of the chemical element $i = $ H, He, ... in the ionization state state $j = $ neutral, single ionized, ... in the quantum state l, J_ν is the spectrum of the incident radiation field that shines on a given (formally infinitesimally small) parcel of gas, and τ_{ijl} are opacities in each radiative transition (gas may be optically thick to some of its own cooling radiation if our parcel is embedded deep inside a huge cloud). For the sake of brevity in notation, we will use $\mathcal{F}$ to label either Γ or Λ, since both functions always depend on the same set of arguments.

In order to compute the cooling and heating functions in such a detail one needs a highly sophisticated computer code that, in its complexity, rivals modern cosmological simulation codes. Fortunately, such codes exist, and the most famous and widely used of them is *Cloudy*.[3] Conceived by Gary Ferland from the University of Kentucky and contributed to by many people, *Cloudy* is freely available from its website, `nublado.org`, and is well-documented for a fast start-up curve.

There is one problem only with *Cloudy*—it is way too complex to be used in modern simulation codes for computing cooling functions "on the fly". Perhaps in the future, in the era of exa-scale computing, it will be possible to run *Cloudy* as a "sub-grid" model in real simulations, but for now we need to seek approximate short-cuts.

So, what one can do? Unless densities are very high (hence our focus on *low density gas*), the gas will be optically thin to its own cooling radiation, so the dependence of cooling and heating functions on τ_{ijl} disappears—for this to be exactly true, we also should exclude all cooling and heating processes due to molecules, since those always require radiative transfer to be followed properly. Thus, if you need to follow molecular cooling/heating as well, you will have to add them "manually" to the cooling and heating functions that we discuss below.

Second, in almost all galactic and cosmological simulations the assumption of the *ionization and excitation equilibrium* is not a bad one. In the ionization equilibrium the distribution of a given chemical element over various ionization states is uniquely determined by density, temperature, and the radiation field. The same is true about various quantum levels in the local thermodynamic equilibrium. If, in addition, we assume that relative abundances of chemical elements are fixed (say, to the solar abundance pattern), then the dependence on X_{ijl} reduces to the simple dependence on the overall gas metallicity Z,

$$\mathcal{F} = \mathcal{F}(T, n_b, Z, J_\nu). \tag{18}$$

Very often this latter expression is what actually called a "cooling/heating function". But even the latter form is unusable in modern simulations codes, because it includes an explicit dependence on the radiation spectrum, which is an arbitrary function of frequency (in a strict mathematical sense $\mathcal{F}$ in Eq. (18) is actually an *operator*, not a function). Hence, we still need to account for that dependence in an approximate manner.

One particular short-cut has been used in many cosmological codes for over a decade. Wiersma et al. (2009) paper again serves as a good reference, although the first known (to me) example of such approach is used by Kravtsov (2003). In the most of the volume of the universe the dominant source of external radiation is the cosmic background that we already met in the previous chapter. The cosmic background is uniform in space and is a function of the cosmic redshift only, hence in the limit when J_ν can be approximated by the cosmic background, cooling and heating functions become functions of 4 arguments, temperature, density, gas metallicity, and cosmic redshift, and hence can be easily tabulated and used in simulation codes efficiently via a simple table look-up.

[3]Notice the convention, *Cloudy* is a name, not an abbreviation.

Unfortunately, most of the volume in the universe contains only a modest fraction of the mass, and even smaller fraction of action. The radiation field in the ISM (and, in at least part of the CGM) of galaxies is dominated by local radiation sources (for example, the UV radiation field in the solar neighborhood is 500 times higher than the cosmic background; at the center of the galaxy that ratio jumps to 5,000). Since stars form in the ISM, any galactic or cosmological simulation that attempts to model star formation cannot use cooling and heating functions which only account for the cosmic background.

How one can attempt to construct a more accurate short-cut? After all, the effect of external radiation is in ionizing some of the chemical elements and/or exciting particular levels, and ionization and excitation rates are all integrals over the radiation spectrum with some cross-sections, which are broad and relatively slowly varying functions. Let's imagine the following thought experiment: we take a given spectrum and increase the radiation intensity in a narrow frequency bin between some ν_0 and $\nu_0 + \Delta\nu$. If the increase is large, the cooling and heating functions will be affected. Now shift the frequency bin to $\nu_0 - \Delta\nu$ to ν_0. Most of ionization and excitation rates will be barely affected (unless we choose ν_0 very carefully to correspond exactly to the ionization/excitation threshold of an important cooling channel), since cross-sections of most physical processes will not change significantly between the two narrow bins. Hence, in order to compute the cooling and heating functions accurately, we do not need to know the radiation spectrum in excessive detail (say, in hundreds of frequency bins), but it may be sufficient to describe it by several "broadband filters".

There can be infinitely many choices for these filters. In a specific implementation of this idea, Nick Hollon and I decided to use photoionization rates of several chemical elements as "broadband filters". After all, the ionization balance is controlled by photoionization rates, so it makes sense from the atomic physics perspective. We have explored over 20 various chemical elements and their ionization states, and the best approximation that we have been able to come up depends on just 4 ionization rates (Gnedin and Hollon 2012).

Specifically, we adopt the approximation in which the metallicity dependence of the cooling and heating functions is expanded into the Taylor series in gas metallicity,

$$\mathcal{F}(T, n_b, Z, J_\nu) = \sum_{i=0}^{n} \left(\frac{Z}{Z_\odot}\right)^i \mathcal{F}_i(T, n_b, J_\nu), \tag{19}$$

with $n = 2$ providing a highly accurate approximation for $Z < 5Z_\odot$. Each of the expansion coefficients is approximated as

$$\mathcal{F}_i(T, n_b, J_\nu) \approx \mathcal{F}_i(T, \{r_j\}, n_b), \tag{20}$$

with several parameters r_j encapsulating the full dependence of the cooling and heating functions on the external radiation field.

A parameter set that we found to work well is defined as follows:

$$
\begin{aligned}
r_1 &= \frac{P_{\rm LW}}{n_b}, \\
r_2 &= \left(\frac{P_{\rm HI}}{P_{\rm LW}}\right)^{0.353} \left(\frac{P_{\rm HeI}}{P_{\rm LW}}\right)^{0.923} \left(\frac{P_{\rm CVI}}{P_{\rm LW}}\right)^{0.263}, \\
r_3 &= \left(\frac{P_{\rm HI}}{P_{\rm LW}}\right)^{-0.103} \left(\frac{P_{\rm HeI}}{P_{\rm LW}}\right)^{-0.375} \left(\frac{P_{\rm CVI}}{P_{\rm LW}}\right)^{0.976},
\end{aligned}
\tag{21}
$$

where $P_{\rm LW}$ is the rate of photo-destruction of molecular hydrogen (molecules are excluded from the cooling and heating functions, since they cannot be treated without radiative transfer, so we use $P_{\rm LW}$ just as a convenient "broadband filter" for the radiation below the hydrogen ionization threshold) and $P_{\rm HI}$, $P_{\rm HeI}$ and $P_{\rm CVI}$ are photoionization rates of HI (ionization edge of 1 Ry), HeI (ionization edge of 1.8 Ry), and CVI (ionization edge of 36 Ry). These rates sample a large range of photon energies, and serve as a good set of more-or-less independent "broadband filters".[4]

The main problem with approximation (19)–(21) is that it occasionally results in "catastrophic errors"—for example, if you choose the radiation field, gas temperature, density, and metallicity at random, in about 1 case out of the million the approximate cooling or heating function will deviate from the actual Cloudy calculation by a factor of several (that is a consequence of not being able to fully represent all possible variations in the radiation field by just 3 coefficients r_j, $j = 1, 2, 3$). Figure 21 demonstrates the worst-case catastrophic error of the approximation.

The good news is that these catastrophic errors occur for either highly implausible or completely irrelevant values of parameters—for example, the large error in the heating function at $T \sim 10\,$K in the bottom panel of Fig. 21 is not very important because the heating function there is much larger than the cooling function, and the equilibrium temperature (blue and red lines cross) is $T_{\rm eq} \approx 2 \times 10^6\,$K. If the gas at 10 K finds itself suddenly in such conditions, it will be heated to above million Kelvins rapidly, quickly leaving the parameter space where the approximation is inaccurate.

Similarly, the large error in the cooling function at $T \sim 10\,$K in the top panel of Fig. 21 is irrelevant, because the heating function in those conditions is more than 3 orders of magnitude larger than the cooling function, hence it is not important to know the cooling function at all.

Undoubtedly, a better approximation for the cooling and heating functions is possible, but in the absence of such, Eqs. (19)–(21) provide a practical way to fully account for the effect of the radiation field in modern cosmological and galactic simulations.

---

[4]They are not fully independent, of course—a photon ionizing CVI can also ionize neutral hydrogen, but it is convenient to use photoionization rates rather that some other, arbitrary filter shapes, since the same rates can be useful in the simulation code for other purposes—for example, for computing the ionization balance of hydrogen, helium, or other chemical elements.

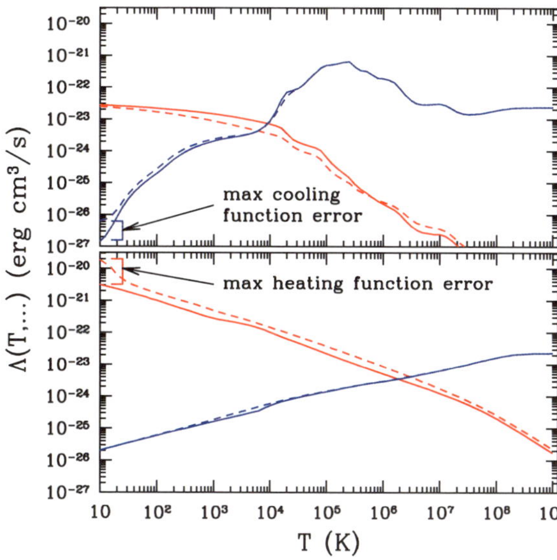

Fig. 21 Cooling (*blue lines*) and heating (*red lines*) functions for our test models that maximize the error in the cooling function (*top panel*) and the heating function (*bottom panel*). Approximate functions from Eqs. (19)–(21) are shown as *dashed lines*, while exact calculations from Cloudy are shown with *solid lines*

3.6 Back to Galactic Halos

Armed with the understanding of the cooling and heating functions, we can now return to the fate of gas in galactic halos. As gaseous halos are expected to become denser at the center, the cooling time will decrease towards the center. Hence, there must exist a *cooling radius* R_C at which the cooling time is equal to the age of the halo. Gas inside R_C is able to cool efficiently and condense towards the halo center, while the gas outside R_C cools too slowly and will remain in the (quasi-) hydrostatic equilibrium.

A detailed analysis of the cooling process is well presented in Maller and Bullock (2004), although they were not the first group who considered that process. In Fig. 22, adopted from that paper, the final profile of the hot gas is shown with the dashed red line. The density profile is cored—all the gas above some threshold density is able to cool, and the core density is set by the requirement that the cooling time in the remnant of the core gas is longer than the age of the halo.

The gas that is able to cool will stream towards the center and will settle into a galactic disk. It can do that, however, in two distinct ways: it can either develop a cooling flow and smoothly flow in a quasi-spherical way all the way to the center, or it can experience thermal instability, split into individual dense clouds, which then fall onto the disk along parabolic orbits like rain drops fall on the ground. Which of these two ways dominates is still a completely open question, with the observational evidence being sparse and inconclusive.

Clouds of neutral hydrogen (hence dense and cool) are indeed detected in the halo of the Milky Way, they are commonly known as "high velocity clouds" (HVC), since they are detected in radio observations as neutral hydrogen at velocities significantly

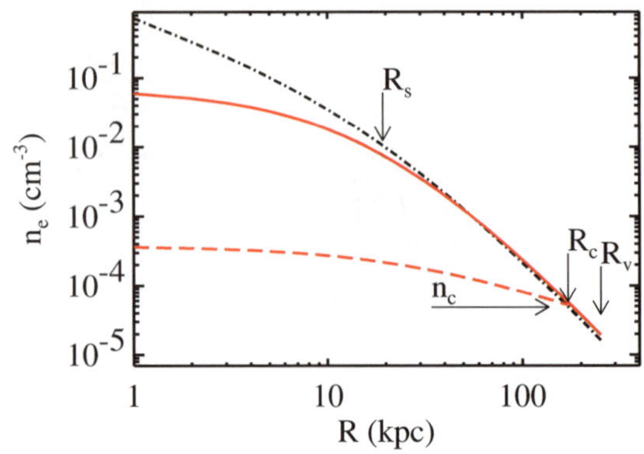

Fig. 22 Density profiles of the hot phase of halo gas in Maller and Bullock (2004) model in the absence of cooling (*solid red line*) and with cooling properly accounted for (*dashed red line*) for a Milky Way like galaxy at $z = 0$. The *dot-dashed black line* shows the NFW profile (adopted from Maller and Bullock (2004))

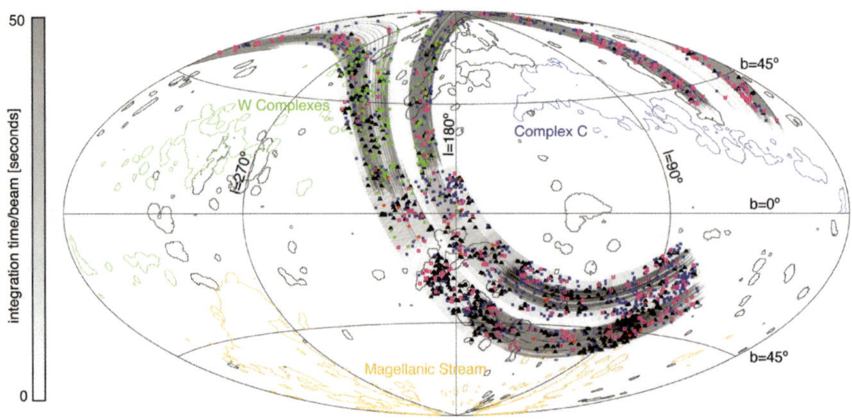

Fig. 23 Neutral hydrogen clouds in the halo of the Milky Way discovered by the GALPHA-HI survey. *Various colors* mark the cloud type, with HVC plotted in *black* (adopted from Saul et al. (2012))

offset from the gas in the galactic disk (clouds that are not offset in the velocity would not be distinguishable from the disk itself).

For example, the recent GALPHA-HI survey by Arecibo telescope uncovered a large number of new clouds (Saul et al. (2012) as shown in Fig. 23). Unfortunately, from the radio observations alone it is very hard to determine the distances to those clouds. Perhaps, they are not located in the halo but form the so-called "galactic fountain", with the gas being thrown up by stellar feedback.

One way to resolve the ambiguity is to search for high velocity clouds in external galaxies. Alas, even in our neighbor Andromeda galaxies none have been found.

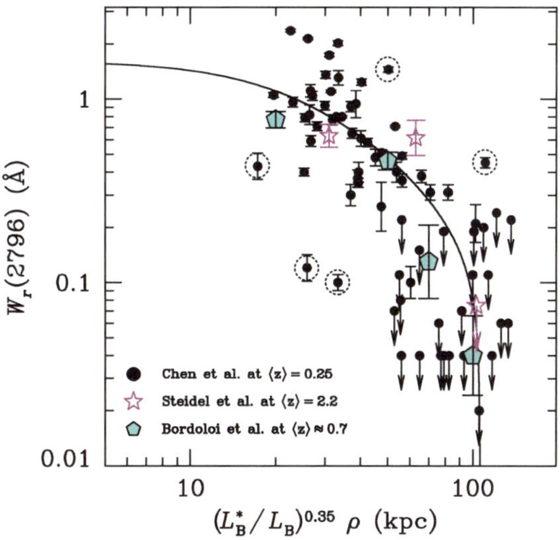

Fig. 24 Strength of MgII absorption as a function of distance from the host galaxy. There is a sharp drop in absorption for distances in excess of about 100 kpc, probably indicating the cooling radius for halos (adopted from Chen (2012))

Andromeda is sufficiently far away for sufficiently small clouds to remain undetected, so the jury is still out on whether HVCs are indeed the halo gas raining onto the galactic disks or the disk gas pushed (temporarily) into the halo.

One part of the problem is that the 21 cm line that is used to detect neural hydrogen in radio observations is one of the weakest lines in this universe. Neutral hydrogen also has one of the strongest lines—Lyman-α. However, it is not easy to excite $n = 2$ level in the hydrogen atom, hence Lyman-α is usually seen in absorption.

That is where other chemical elements come to rescue. Even while we are primarily after hydrogen, a trace amount of heavy elements may produce enough absorption in some of their, more easily excitable and observable lines. One such element is Magnesium, the ionization threshold of its singly ionized state is just 15 eV, very close to the hydrogen ionization threshold of 13.6 eV. Because of that, MgII has been used as a proxy for neutral hydrogen in absorption studies of galaxies for several decades. Figure 24 shows a plot from a recent compilation of observational constraints in several ions by Chen (2012). A general feature of all observations is that MgII drops precipitously further away that about 100 kpc from a galaxy (with a mild dependence on the galaxy luminosity). It is highly tempting to associate this drop with the cooling radius for the halo, and MgII with the cool clouds formed by thermal instability, but in the absence of additional evidence such a proposition will remain no more than a plausible conjecture.

One way or the other the gas from the halo (and beyond) ends up in the galactic disk, making up the Interstellar Medium (ISM) of galaxies. This is where our yellow brick road leads us next.

4 ISM: Gas in Galaxies

The field of Interstellar Medium takes easily a quarter of all of Astronomy. Any attempt to review it at any reasonable level will result in me still writing these lectures on my deathbed. Hence, our journey through the ISM realm will be brief and highly focused—we will be mainly concerned with "gas in galaxies", i.e. gas as a medium (forget about chemistry, except for one very specific topic), and gas as a galaxy component (i.e. not small-scales behavior of gas, but rather the role of gas as a citizen of a galaxy). Even with these restrictions, the journey that lays ahead is extremely biased towards my own research interests and topics I find fascinating.

4.1 Galaxy Formation Lite

Galaxies are rather complex creatures; understanding galaxy formation and evolution is the current frontier of extragalactic astronomy and cosmology. Never-the-less, the basic sketch of how galaxies form and evolve has been developed—it is captured by the Mo et al. (1998) model (hereafter MMW98).

The cornerstone assumption of MMW98 model is that the cool ($\sim 10^4$ K) gas is delivered to the bottom of the potential well of a dark matter halo—either by radiative cooling in the halo or by inflow along cool flows. The specific way by which gas is delivered is unimportant; what matters is that the angular momentum is conserved, and hence the cool gas settles into a rotationally-supported disk.

It is convenient to parametrize the mass of the disk M_d as a fraction m_d of the halo mass M_h,

$$M_d = m_d M_h,$$

and the disk angular momentum J_d as a fraction j_d of the halo angular momentum J_h,

$$J_d = j_d J_h.$$

For an exponential disk with constant circular velocity V_c and the surface density profile

$$\Sigma(R) = \Sigma_0 \exp(-R/R_d),$$

$M_d = 2\pi\Sigma_0 R_d^2$ and $J_d = 4\pi\Sigma_0 R_d^3 V_c$. From these two equations the disk density profile (parameters Σ_0 and R_d) can be expressed as functions of m_d, j_d, and V_c.

The distribution of angular momenta for dark matter halos is usually quantified by the *spin parameter*

$$\lambda = \frac{J_h |E_h|^{1/2}}{G M_h^{5/2}},$$

where E_h is the binding energy of the halo (which depends on the actual adopted density profile). In hierarchically clustered universe spins of dark matter halos are induced by tidal torques from the surrounding material (Heavens and Peacock 1988). The distribution of spin parameters of halos of various masses turns out to be surprisingly independent of anything else (halo mass, shape of the matter power spectrum, cosmological parameters, redshift, etc.) and is approximately lognormal,

$$p(\lambda)d\lambda = \frac{1}{\sqrt{2\pi}\sigma_\lambda} \exp\left(-\frac{\ln^2(\lambda/\bar{\lambda})}{2\sigma_\lambda^2}\right)\frac{d\lambda}{\lambda},$$

with $\bar{\lambda} \approx 0.05$ and $\sigma_\lambda \approx 0.5$—that result remains unchanged from the first N-body simulations (Barnes and Efstathiou 1987) to the present day (Trowland et al. 2013).

The final step in the MMW98 model is the connection between the disk circular velocity V_c and the virial velocity of the halo,

$$V_{\rm vir} = \left(\frac{GM_h}{r_{\rm vir}}\right)^{1/2}.$$

In the original MMW98 model the coefficient of proportionality between V_c and $V_{\rm vir}$ was assumed to be 1, but it does not have to be. For example, for the NFW profile

$$\left(\frac{V_c(r)}{V_{\rm vir}}\right)^2 = \frac{1}{x}\frac{\ln(1+cx) - cx/(1+cx)}{\ln(1+x) - c/(1+c)}$$

where $x \equiv r/r_{\rm vir}$ and c is the concentration of the halo. In this case, however, V_c is a function of radius and is not constant, so which one should we use? One solution is to consider the "maximal" disk, i.e. take the largest value of V_c for any radius, commonly referred to as $V_{\rm max}$, as the disk circular velocity. That value is mildly dependent on the halo concentration c,

$$\begin{aligned} V_{\rm max} &= 1.0\, V_{\rm vir} \text{ for } c = 3, \\ V_{\rm max} &= 1.2\, V_{\rm vir} \text{ for } c = 10, \\ V_{\rm max} &= 1.6\, V_{\rm vir} \text{ for } c = 30. \end{aligned}$$

The MMW98 model is controlled by two main parameters, m_d and j_d. In principle, they can be arbitrary. However, recently an interesting property of real galaxies has been noticed by Kravtsov (2013): disk sizes (both for stellar disks and gaseous disks) are linearly proportional to the virial radii, with the scatter in the relation entirely consistent with the distribution of λ parameters for halos of a given mass. In other words, parameters m_d and r_d must be such that for stellar disks $R_d \approx 0.01 R_{\rm vir}$ and for gaseous disks it is about a factor of 2.5 larger.

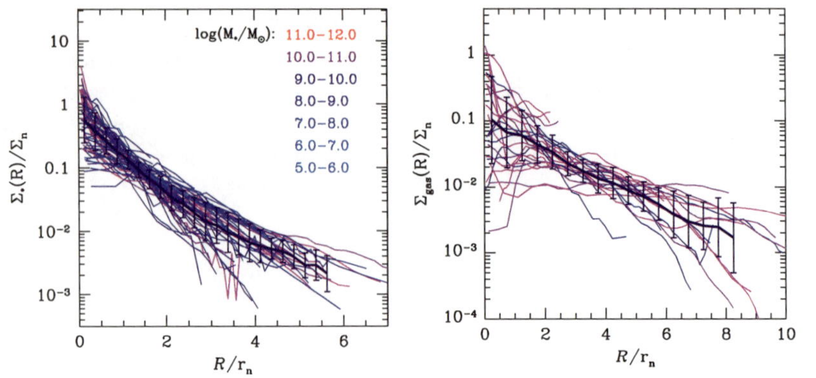

Fig. 25 Normalized surface density profiles of stars and neutral gas for late-type galaxies (adopted from Kravtsov (2013))

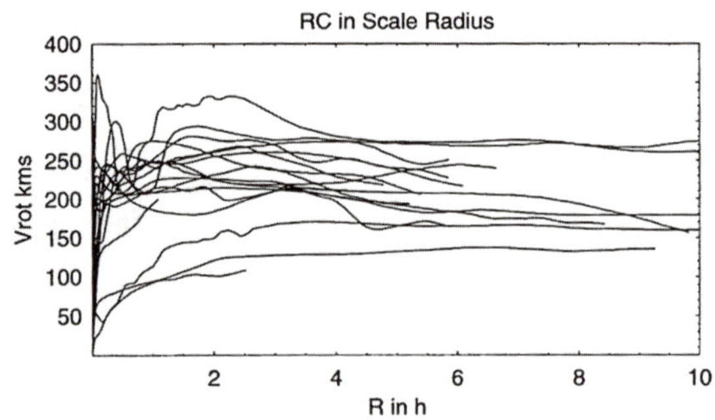

Fig. 26 Rotation curves of several *spiral* galaxies from Sofue et al. (1999)

4.2 Galactic Disks

We now descend into the actual galactic disks. The common lore is that disks are exponential, rotationally supported, and have flat rotation curves. While all these statements are kind of true, they are very far from being exact.

Disks come with a variety of density profiles and a variety of rotation curves. For example, Fig. 25 shows surface density profiles for stars and gas for several samples of disk galaxies (Kravtsov 2013). On average profiles are indeed exponential, but deviations of individual galaxies from the mean can easily reach a factor of several.

Similarly, rotation curves of individual galaxies (Fig. 26) show large deviations from the canonical flat shape—some rotation curves are rising, some are falling, some remain truly flat all the way to the outer edge of the disk.

Disk dynamics in general is a very complex affair. A large number of various disturbances and waves can propagate over the disks—in addition to spiral arms, there exit bending modes, bars, warps, etc. All these perturbations cause orbits of stars and gas to deviate from spherical symmetry. For example, spiral arms are shock waves, gas changes its velocity abruptly by a large factor (up to several times its sound speed) as it crosses the shock, and hence the gas in front of and behind the spiral arm shock cannot remain on the same circular orbit—one of the sides has to deviate substantially. For example, in the classical example of the grand design spiral, M51, the deviations of the gas rotational velocity from the circular velocity reach 20 km/s *almost everywhere in the disk* (Hitschfeld et al. 2009).

Such deviations, in fact, may be responsible, at least partially, for the notorious cusp-core controversy. Some of the "observed" cusps may, in fact, be just an erroneous consequence of the incorrect assumption that the rotational velocity is equal to the circular velocity for gas (Valenzuela et al. 2007).

4.2.1 Disk Stability

How one would investigate such waves and features? Nonlinear treatment would require numerical simulations, but some widely known (and not so widely known) results can be obtained analytically for the linear stability of disk systems. A standard approach to studying linear stability of any system is to impose small fluctuations on the system and derive their dispersion relation. For an infinitely thin disk one can represent the radially perturbed (i.e. a perturbation remains azimuthally symmetric) surface density $\Sigma(t, R)$ as

$$\Sigma(t, R) = \bar{\Sigma}(R) + \Delta\Sigma(t, R),$$

where the perturbation $\Delta\Sigma(t, R)$ is assumed to be a collection of linear waves, each wave characterized by the frequency ω and the wavevector $\mathbf{k} = (k_R, k_\phi)$. Let's first focus on purely radial perturbations, $k_\phi = 0$. In that case the dispersion relation for the gaseous disk becomes (Binney and Tremaine 1987)

$$\omega^2 = \kappa^2 - 2\pi G\bar{\Sigma}|k_R| + c_s^2 k_R^2, \tag{22}$$

where $\kappa^2 \equiv R(d\Omega^2/dR) + 4\Omega^2$ is the so-called *epicyclic frequency* and $\Omega(R)$ is the disk angular velocity, $V_c(R) = R\Omega$.

The disk is stable when the right hand side is always positive, which is achieved if and only if

$$Q \equiv \frac{c_s\kappa}{\pi G\bar{\Sigma}} > 1. \tag{23}$$

This condition is universally known as *Toomre stability criterion*, although for gaseous disks it has been obtained earlier by Safronov (1960), while Alan Toomre derived a similar relation for stellar disks (Toomre 1964), a much more difficult exercise.

When $Q < 1$, some of the radial modes in the disk become unstable,

$$\frac{\kappa}{Qc_s}\left(1 - \sqrt{1 - Q^2}\right) < k_{\text{unstable}} < \frac{\kappa}{Qc_s}\left(1 + \sqrt{1 - Q^2}\right).$$

An interesting property of this relation is that only a limited range of wavenumbers become unstable, the disk remains stable at very large ($k \rightarrow 0$) and very small ($k \rightarrow \infty$) scales.

4.2.2 Beyond Toomre

Toomre stability criterion is often used in galactic and extragalactic studies. However, it is, unfortunately, often forgotten that it is incomplete. No disk is infinitely thin, and no perturbation is perfectly radial.

A case of arbitrary, not necessarily radial, perturbations was considered by Polyachenko and Polyachenko (1997), who found that the critical value for the Q parameter is actually larger than 1. This is not surprising—at $Q = 1$ radial perturbations go unstable; however, for the disk to become unstable it is only enough for *some* waves to become unstable, and these first unstable waves do not have to be radial. Thus, some of the non-radial (i.e. non-axially-symmetric) perturbations may become unstable when all radial perturbations remain stable with $Q > 1$.

The critical value of the Q parameter turns out to depend on the disk density profile,

$$Q_{\rm crit}^2 = \frac{3\alpha^2 - 3}{2\alpha^2 - 3} > 1,$$

where

$$\alpha^2 = \frac{2\Omega(R)}{R|d\Omega/dR|}.$$

For example, for a flat rotation curve ($\Omega \propto R^{-1}$) $\alpha^2 = 2$ and

$$Q_{\rm crit} = \sqrt{3}.$$

This is the reason why most actively star-forming (and, thus, instability-developing) disk galaxies have Q parameters above unity but not significantly greater than 2 (Leroy et al. 2008).

Another generalization of the Toomre stability criterion is obtained when the finite thickness of a disk is taken into account. In that case the dispersion relation has been introduced by Begelman and Shlosman (2009), although in a highly convoluted form it has been derived earlier by Safronov (1960),

$$\omega^2 = \kappa^2 - 2\pi \frac{G\bar{\Sigma}|k_R|}{1 + |k_R|h} + c_s^2 k_R^2, \qquad (24)$$

where h is the disk *scale height*, $\bar{\Sigma}(z) \propto \exp(-z/h)$. For a non-exponential vertical profile the dispersion relation becomes more complex and is not presentable analytically in a closed form.

Relation (24) is remarkable in that in the limit of very small scales, well below the disk scale height, $kh \gg 1$ (in which case the disk cannot be considered as a flattened system any more), it reduces to

$$\omega^2 = -4\pi G\bar{\rho} + c_s^2 k_R^2,$$

(with $\bar{\Sigma} = 2\bar{\rho}h$), which is nothing else as a usual Jeans stability dispersion relation, familiar to any astrophysicist since kindergarten.

4.2.3 Modeling Disks

Modeling disks numerically is a subject of itself, and cannot be covered in these lectures. However, a word of caution is in order here. Let's imagine one is trying to model a galactic disk (or, for that matter, a disk around a supermassive black hole, or any other self-gravitating disk). A natural setup is to start with an axially-symmetric disk and let the instabilities develop.

So, you prepared your symmetric disk as the initial condition for your powerful numerical code that includes all kind of important physical processes (cooling, star formation, feedback, etc.). To be specific, let's say you set the gas temperature to 10^4 K in the disk with the circular velocity of 200 km/s.

You press the magic button, simulation starts, and in an instant your disk cools off to the lowest temperatures your cooling module allows (indeed, cooling times in astrophysical environments are often very short), the Q parameters plunges to very small values, and your disk fragments into tiny clumps of size comparable to the wavelength of fastest growing instability mode $\lambda_{\rm fast}$,

$$R \sim \lambda_{\rm fast} = 2\pi Q \frac{c_s}{\kappa}.$$

Such a state, however—cold homogeneous disk—is *unphysical*, there is no plausible physical process that can create such a system: after all, you started with an artificial initial condition; try running it backward in time, the disk is still cooling, so shortly before your initial moment it should have been blazingly hot, at 10^7–10^8 K, and how would you propose to keep 10^8 K plasma in a disk with 200 km/s circular velocity?

Ok, that does not work. Let's now start with an initially stable disk ($Q \gg 1$) and let it become unstable gradually (either by artificially introducing cooling gradually, or disabling cooling below 10^4 K, or, even better, gradually adding mass to the disk). As Q decreases gradually, at some moment it will reach a critical value $Q_{\rm crit} > 1$. At that moment some non-radial perturbations become unstable and start growing, turning into non-linear waves; any non-linear wave in the gas steepens to a shock; any shock in a differentially-rotating disk becomes an oblique spiral wave; oblique shocks are known to generate an energy cascade a-la turbulence (although it may not be turbulence in the exact meaning of that word). Turbulence will provide extra support to the gas, replacing the sound speed c_s in Eq. (22) with $\sqrt{c_s^2 + \sigma_t^2}$ and will limit the fragmentation scales to $R \sim 2\pi\sqrt{c_s^2 + \sigma_t^2}/\kappa$.

In other words, in the latter scenario the Q parameter never had a chance to become much lower than the critical value, but must linger at around it, maintaining the disk in the just-unstable-enough state to generate enough turbulence. Hence the conclusion that the author arrived at himself after much suffering and erring: if your disk simulation has $Q \ll 1$, you are doing something wrong …

4.3 Ionized, Atomic, and Molecular Gas in Galaxies

Everyone knows that ISM consists of several gas phases. The ionized gas comes in two flavors, as hot ($\sim 10^6$ K) coronal gas and warm/cool ($\sim 10^4$ K) ionized gas (known under many names: warm ionized medium (WIM), diffuse ionized gas (DIG), Reynolds Layer); atomic gas exists as warm/cool ($\sim 10^4$ K) and cold ($\sim 10^2$ K) neutral media (WNM and CNM respectively); finally, molecular gas is almost always cold ($< 10^2$ K).

4.3.1 Ionized Gas

A story of coronal gas is misty and messy—it is not even clear how much of it there is in the Milky Way ISM, or what fraction of it comes from stellar feedback processes and what fraction is merely halo gas intermixed into the ISM due to various disk instabilities. Warm ionized medium is understood better because it is primarily located at the outer edges of the disk.

What causes WIM? We can get a hint on its origin from its temperature—gas at 10^4 K is likely to be photo-ionized. If we recall that only gods have the power to switch off the Cosmic Ionizing Background, the ionizing source is there too—plus whatever ionizing radiation escapes from star-forming regions inside the Milky Way disk.

An example of how the relative distribution of neutral and ionized gas may look like in the Milky Way galaxy (or other similar galaxies) is shown in Fig. 27. The WIM contribution stays more-or-less constant at about $0.5\,\mathrm{M}_\odot/\,\mathrm{pc}^2$ (column density $N_\mathrm{H} = 6 \times 10^{19}\,\mathrm{cm}^{-2}$) in the outer disk, but increases to several $\mathrm{M}_\odot/\,\mathrm{pc}^2$ inside the solar radius because of the increased radiation field and a contribution of coronal gas. Broadly, such behavior is consistent with actual observations of the ionized gas in the Milky Way and other galaxies. For example, in the Milky Way the contribution of ionized gas at the solar radius is about $1\,\mathrm{M}_\odot/\,\mathrm{pc}^2$.

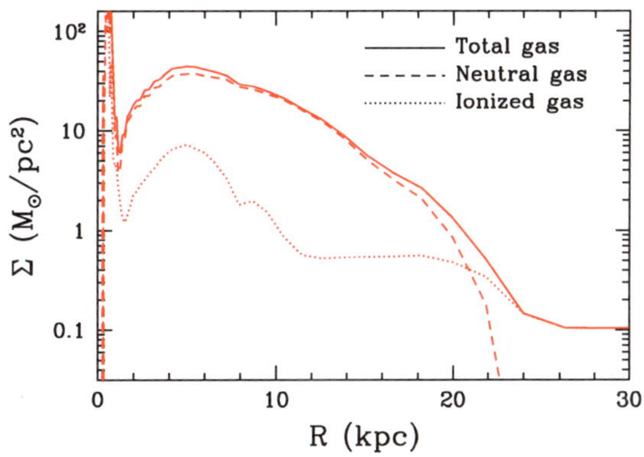

Fig. 27 Surface density profiles for the total, ionized, and neutral (atomic and molecular) gas for a model Milky-Way-like galaxy (from the simulation described in Gnedin (2012))

The outer parts of the disk are consistent with being ionized by Cosmic Ionizing Background, and the transition from neutral to ionized gas is often very sharp. However, consistency does not imply causality. There could be other ionizing sources, such as stellar radiation escaping from star-forming regions or cosmic rays. Since stars do not form in the ionized gas (as far as we can tell), we leave the WIM-land on our way to denser and colder domains; interested readers should check an excellent recent review by Haffner et al. (2009).

4.3.2 From Atomic to Molecular Gas

Stars (at least most of them) form from molecular gas. Few astronomers would question this conjecture. While a minority of all stars may form in the atomic gas (at least Pop III stars certainly form in gas that is 99 % atomic), on this journey we are chasing the bulk of star formation. Hence, the transition from atomic to molecular gas is a necessary condition for (the bulk of) star formation.

Chemistry of molecular hydrogen is not particularly complex; H_2 forms through two physically distinct channels: in numerous reactions in the gaseous phase, from rare ions H− and H_2^+ (the best reference for these processes is Glover and Abel (2008)), and on the surface of cosmic dust, which serves as a catalyst. The gas processes are slow exactly because H− and H_2^+ are rare; fraction of molecular hydrogen forming in the gas phase saturates at 10^{-3}–10^{-2} and only jumps to close to 1 when 3-body reactions become sufficiently efficient (which only happens at densities above about $10^{12}\,\mathrm{cm}^{-3}$). This channel of H_2 formation does not require any metals and can proceed in the primordial gas (indeed, this is how Pop III stars form).

Formation of H_2 on dust grains is not fully understood. It is usually assumed that atomic hydrogen accumulates on grains where two atoms can find each other much more easily (young couples tend to live in cities). The formation rate R_D, defined as

$$\left.\frac{dn_{\mathrm{H}_2}}{dt}\right|_{\mathrm{dust}} = R_D n_{\mathrm{H}} n_{\mathrm{HI}},$$

has been modeled (some what inconclusively) theoretically and measured observationally by Wolfire et al. (2008):

$$R_D = D_{\mathrm{MW}} R_0,$$

with $R_0 \approx 3.5 \times 10^{-17}\,\mathrm{cm}^3/\,\mathrm{s}$, where from now on I will use a convenient parameter D_{MW} that measures the abundance of dust relative to the solar neighborhood; i.e. $D_{\mathrm{MW}} = 1$ implies the same abundance of dust per unit mass of gas as in the Milky Way ISM around us.

It is not, however, enough to know the formation rate to determine the abundance of molecular hydrogen—like predator and prey, ultraviolet radiation plays with H_2 the game of life and death. Particularly deadly for molecular hydrogen is radiation in the so-called Lyman and Werner bands, at energies between 11.3 and 13.6 eV (actually, the bands extends further, but hydrogen ionizing radiation is often well shielded by neutral atomic ISM). In addition, molecular hydrogen is destroyed by collisions with atoms and other molecules when gas temperatures raise above about 5000 K. Hence, in order to predict the abundance of molecular hydrogen in specific conditions, we need to know the Interstellar Radiation Field (ISRF).

ISRF is not measured directly, but rather modeled based on the observations of various line ratios in the ISM. Two canonical references to such models are Draine (1978) and Mathis et al. (1983), which are perfectly consistent with each other. In the solar neighborhood $J_0 \approx 10^6\,\mathrm{phot/cm^2/s/eV/rad}$, but in the Galaxy the radiation field changes with the distance from the center. At the center it is up to 10 times higher than around the Sun.

Just like masses and luminosities are convenient to measure in solar units, in galactic studies it is convenient to measure the radiation field and other quantities (like dust abundance) in the Milky Way units. Hence, hereafter we will also use $U_{\mathrm{MW}} \equiv J_{\mathrm{LW}}/J_0$ (where J_{LW} is the average radiation field in the Lyman and Werner bands). By definition, $U_{\mathrm{MW}} = 1$ in the solar neighborhood, but in high redshift galaxies it can be large, $U_{\mathrm{MW}} = 30$–300 at $z \sim 2$ (Chen et al. 2009).

Even the Milky Way radiation field is extremely strong from the molecular hydrogen point of view—if it could shine on typical molecular clouds unimpeded, the molecular fraction would only be 10^{-6}–10^{-5}. The only reason molecular clouds exist in the universe is because all that radiation is *shielded*.

There are two distinct shielding processes: dust shielding and molecular self-shielding. Dust absorbs radiation over a very large range of wavelengths, from infrared to X-rays. Dust opacity is a smooth function of wavelengths, and in the first approximation it can be considered constant over narrow Lyman and Werner bands (for detailed plots of dust opacity see Weingartner and Draine (2001)). In different galaxies the dust opacity is different, but in the three galaxies it was studied best—Milky Way and two Magellanic clouds—it is roughly proportional to the dust-to-gas ratio,

$$\sigma_{LW} = D_{MW}\sigma_0$$

with $\sigma_0 = 1.7 \times 10^{-21}$ cm^2 for the Milky Way ($D_{MW} = 1$), $\sigma_0 = 1.6 \times 10^{-21}$ cm^2 for the LMC ($D_{MW} \approx 0.5$), and $\sigma_0 = 2.2 \times 10^{-21}$ cm^2 for the SMC ($D_{MW} \approx 0.2$). Thus, it is possible to simply take σ_0 as a universal constant,

$$\sigma_0 \approx 2 \times 10^{-21}\ \text{cm}^2.$$

Accounting for continuum shielding over a narrow band is easy; the molecular hydrogen photo-destruction rate Γ is then simply

$$\Gamma = c\sum_j \int_{\nu_1}^{\nu_2} \sigma_j(\nu)\,\underbrace{e^{-\sigma_d(\nu)N_H} n_\nu}_{\text{radiation field}}\,d\nu \approx e^{-\bar{\tau}_d}\Gamma_{LW},$$

where N_H is the total hydrogen column density, $\bar{\tau}_d \equiv \bar{\Sigma}_d N_H$ is the average dust opacity in the Lyman and Werner bands, Γ_{LW} is the so-called "free space" photo-destruction rate (i.e. photo-destruction rate in the absence of any shielding), and the sum is taken over all H_2 lines in the Lyman and Werner bands. It is convenient to define a *shielding factor* S_D that parametrizes the suppression of the free space field by dust shielding, $\Gamma = S_D\Gamma_{LW}$, with

$$S_D(D_{MW}, N_H) = e^{-D_{MW}\sigma_0 N_H}.$$

Self-shielding of molecular hydrogen is much more complicated. Lyman and Werner bands consist of numerous lines of various strengths (Fig. 28). Absorbing a

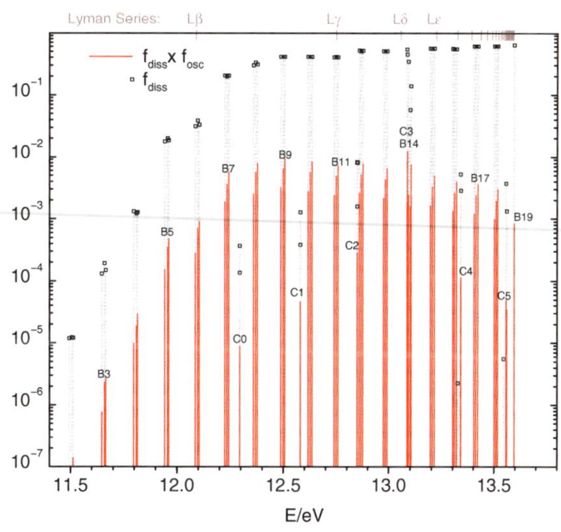

Fig. 28 Molecular lines in the Lyman and Werner bands. A hydrogen molecule has a non-zero probability to be photo-dissociated f_{diss} when it is excited into any of these states (adopted from Haiman et al. (2000))

photon in one of those lines may or may not lead to the destruction of the hydrogen molecule, and the probability of dissociation varies significantly for different lines.

Hence, the shielded photo-destruction rate can be represented as a sum over individual lines, each with its own cross section $\sigma_j(\nu)$,

$$\Gamma = c\sum_j \int_{\nu_1}^{\nu_2} \sigma_j(\nu)\,\underbrace{e^{-\sigma_j(\nu)N_{\mathrm{H}_2}} n_\nu}_{\text{radiation field}}\, d\nu \approx \sum_j e^{-\bar{\tau}_j}\,\Gamma_{\mathrm{LW},j} = S_{\mathrm{H}_2}(N_{\mathrm{H}_2})\Gamma_{\mathrm{LW}}. \quad (25)$$

The self-shielding factor $S_{\mathrm{H}_2}(N_{\mathrm{H}_2})$ is much harder to compute, but its general behavior may be guessed. As individual Lyman and Werner bands lines become optically thick, some of the terms in the sum in Eq. (25) become small, but weaker lines will remain optically thin and un-shielded for much higher column densities than the stronger lines, thus allowing the destructing radiation to sneak deeper into a molecular cloud. Hence, as the column density of molecular hydrogen increases, the self-shielding factor will fall at a rate, which is much slower than the exponential decline of an individual line.

The self-shielding of molecular hydrogen has been modeled extensively; a specific approximation for the self-shielding factor that is most commonly used is due to Draine and Bertoldi (1996),

$$S_{\mathrm{H}_2} = \frac{0.965}{(1+x/b_5)^\alpha} + \frac{0.035}{\sqrt{1+x}}\exp\left(-\frac{\sqrt{1+x}}{1180}\right), \quad (26)$$

where $x \equiv N_{\mathrm{H}_2}/5\times 10^{14}\,\mathrm{cm}^{-2}$, $b_5 \equiv b/\,\mathrm{km/s}$, and in the original approximation $\alpha = 2$. Wolcott-Green et al. (2011) suggested that at higher temperatures a better fit is $\alpha = 1.1$, but the first term in Eq. (26) is not important anyway.

Figure 29 shows the Draine and Bertoldi (1996) approximation as a function of the molecular column density. A gradual decline of the self-shielding factor (S_{H_2} going approximately as $N_{\mathrm{H}_2}^{-0.75}$) is apparent for almost 8 orders of magnitude. However, at very high column densities, $N_{\mathrm{H}_2} > 10^{22}\,\mathrm{cm}^{-2}$, the fall-off becomes steeper, with the last factor in Eq. (26) dominating. What could cause such a steep decline?

Our deduction above that the weaker lines remain optically thin and serve as avenues for the radiation to sneak into a molecular cloud remain correct for as long as each absorption line can be treated as independent. However, just like in human society neighbors sooner or later will put a stop on a weak person misbehaving, so in the society of Lyman and Werner bands stronger lines begin to interfere in the affairs of weaker one at sufficiently high column densities. Since each excited state in an atom or molecule lives for a finite time, lines have non-trivial *natural width* (see Sect. 2.2.1). In the high column density limit the natural width dominates, and the equivalent width of a line (the area of the spectrum the line takes out) grows as $N_{\mathrm{H}_2}^{1/2}$. As the strongest lines begin to overlap, the nature of self-shielding changes—instead of individual lines absorbing UV radiation each by itself, the absorption cross-section now becomes a continuous function of frequency, with cross-sections

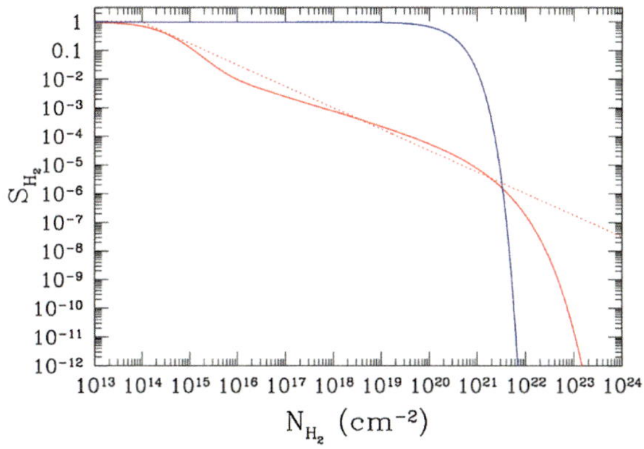

Fig. 29 Draine and Bertoldi (1996) molecular self-shielding factor as a function of H_2 column density (*solid red line*). For comparison, exponentially falling off shielding factor (dust shielding with Milky Way dust and fully molecular gas, $N_{H_2} = 2N_{H_2}$) is shown as a *blue line*. *Red dotted line* is a power-law approximation for the self-shielding factor, $S_{H_2} \propto N_{H_2}^{-0.75}$ that has been also used in the past

of individual lines all blending together into a single, continuum-like absorption. Hence, self-shielding becomes much stronger, and that is manifested in the drop-off in the Draine and Bertoldi (1996) formula at $N_{H_2} > 10^{22}\,\mathrm{cm}^{-2}$.

Finally, we need to figure out what N_{H_2} actually is. Let's imagine that we have a line-of-sight through a molecular cloud with the total hydrogen column density N_H. The first inclination is to simply use $N_{H_2} = 0.5N_H$ (let's assume the cloud is fully molecular), but that is actually *wrong*!

Equation (26) is suitable for the idealized case of a slab of gas with no internal motions. Real molecular clouds are, however, supersonically turbulent on scales above the *sonic length*, $l_s \lesssim 1$ pc. In other words, if you take two parcels of gas inside a molecular cloud separated by a distance l, the rms velocity dispersion between them satisfies what is known as Larson's law,

$$\delta v(l) \approx c_s \left(\frac{l}{l_s}\right)^{0.5}$$

with c_s being the gas sound speed (in fact, the definition of the sonic length is that $\delta v(l_s) = c_s$). For $l \gg l_s$, the velocity difference between them would be much larger than the width of each Lyman and Werner bands line $b \sim c_s$. Hence, these two fluid elements would shield each other only if they happen accidentally to fall at the same line-of-sight velocity, which would occur with the probability $b/\delta v$.

This is illustrated in a cartoon fashion in Fig. 30. Hence, a fluid element inside a molecular cloud sees a column density of about $N_{H_2} \sim \langle n_{H_2}\rangle_s L_{MC} b/\delta v \approx N_{H_2} \sim \langle n_{H_2}\rangle_s (l_s L_{MC})^{1/2}$, where $\langle n_{H_2}\rangle_s$ is the average molecular hydrogen density on a sonic scale at the location of interest and L_{MC} the width of the whole molecular cloud. This

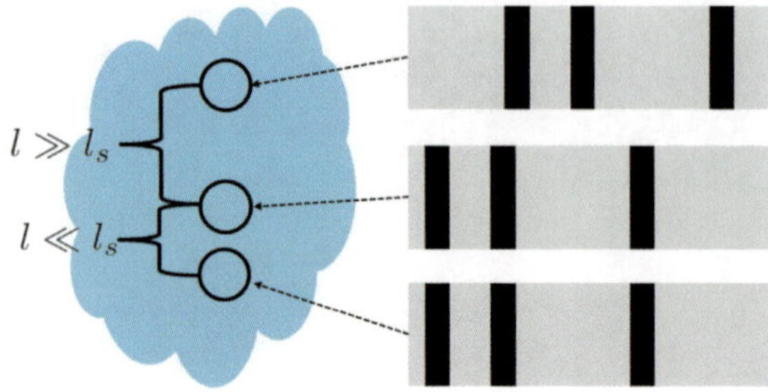

Fig. 30 A cartoon illustrating the role of ISM turbulence in suppressing self-shielding of molecular hydrogen on scales above the sonic length

is valid, however, only until individual lines do not overlap. With line overlap relative velocity shifts between different fluid elements become unimportant (lines overlap anyway). In other words, at sufficiently large column densities line radiative transfer in the Lyman and Werner bands effectively behaves as continuum radiative transfer, and the effective length over which the column density is accumulated approaches L_{MC}.

In Eq. (26) the line overlap is described by the last exponential factor. To account for the supersonic turbulence inside the molecular cloud, Eq. (26) can be modified as

$$S_{H_2} = \frac{0.965}{(1+x_1/b_5)^2} + \frac{0.035}{\sqrt{1+x_1}} \exp\left(-\frac{\sqrt{1+x_2}}{1180}\right), \tag{27}$$

where $x_1 \equiv \langle n_{H_2} \rangle_s (l_s L_{MC})^{1/2}/5 \times 10^{14}\,\text{cm}^{-2}$ is proportional to the H_2 column density over the sonic length, while $x_2 \equiv \langle n_{H_2} \rangle_{MC} L_{MC}/5 \times 10^{14}\,\text{cm}^{-2}$ accounts for the column density of the whole molecular cloud. Obviously, $x_2 \gg x_1$.

Armed with understanding of dust and self-shielding, we can consider some interesting limiting cases. In the kinetic equilibrium the rates of photo-destruction and molecular hydrogen formation balance, hence

$$\Gamma_{LW} S_{H_2} e^{-\sigma_{LW} N_H} n_{H_2} = R_D n_H n_{HI}.$$

The free-space radiation field is parametrized by the introduced above U_{MW} parameter, $U_{MW} \equiv \Gamma_{LW}/\Gamma_0$. Hence,

$$\frac{f_{H_2}}{(1-f_{H_2})} = \frac{D_{MW}}{U_{MW}} \frac{R_0}{S_{H_2}\Gamma_0} e^{D_{MW}\sigma_0 N_H} n_H. \tag{28}$$

As we already know, in low metallicity environments self-shielding is expected to dominate over dust shielding,

$$\frac{f_{H_2}}{(1-f_{H_2})} = \frac{D_{MW}}{U_{MW}} \frac{R_0}{S_{H_2}\Gamma_0} n_H.$$

Let's say we are interested in densities at which the gas becomes 50 % molecular ($f_{\rm H_2} = 0.5$). In that case

$$S_{\rm H_2} \propto \frac{D_{\rm MW}}{U_{\rm MW}},$$

and for high enough column density, when

$$S_{\rm H_2} \sim e^{-{\rm const}\times N_{\rm H_2}^{1/2}},$$

we find

$$N_{1/2} \equiv N_{\rm H}(f_{\rm H_2} = 1/2) \propto \ln^2\left(\frac{U_{\rm MW}}{D_{\rm MW}} \times {\rm const}\right),$$

i.e. the column density of the atomic-to-molecular transition depends only weakly on the dust abundance or the interstellar radiation field.

In the opposite extreme, in high radiation fields the dust shielding dominates,

$$\frac{f_{\rm H_2}}{(1 - f_{\rm H_2})} = \frac{D_{\rm MW}}{U_{\rm MW}}\frac{R_0}{\Gamma_0}e^{D_{\rm MW}\sigma_0 N_{\rm H}} n_{\rm H},$$

hence

$$N_{1/2} \propto \frac{1}{D_{\rm MW}} \ln\left(\frac{U_{\rm MW}}{D_{\rm MW}} \times {\rm const}\right).$$

As could have been easily guessed, higher dust abundance pushes the atomic-to-molecular transition towards lower (column) densities.

How should we go now from shielding factors for individual parcels of gas to the factors that should be used in actual numerical simulations? Modern cosmological or galactic scale simulation may not resolve molecular clouds at all or may resolve them down to parsec scales. Hence, in the most general case we can imagine whole space being tessellated into regions (say, simulation cells) of size L some of which include pieces of molecular clouds. Each such piece has a distribution of column density inside it, $\phi_j(N_{\rm H_2})$, where j refers to a given piece. Hence, the average shielding factor is

$$\langle S_{\rm H_2}\rangle_j = \int S_{\rm H_2}(N_{\rm H_2})\phi_j(N_{\rm H_2})dN_{\rm H_2} = S_{\rm H_2}(N_{{\rm eff},j}) \int \phi_j(N_{\rm H_2})dN_{\rm H_2} = S_{\rm H_2}(N_{{\rm eff},j})$$

since $\int \phi_j(N_{\rm H_2})dN_{\rm H_2} = 1$ by definition. If the distribution $\phi_j(N_{\rm H_2})$ was known, one can also compute $N_{{\rm eff},j}$, but at present there are no models that attempt to determine ϕ_j. Hence, we need to come up with an ansatz for $N_{{\rm eff},j}$. For example, in the absence

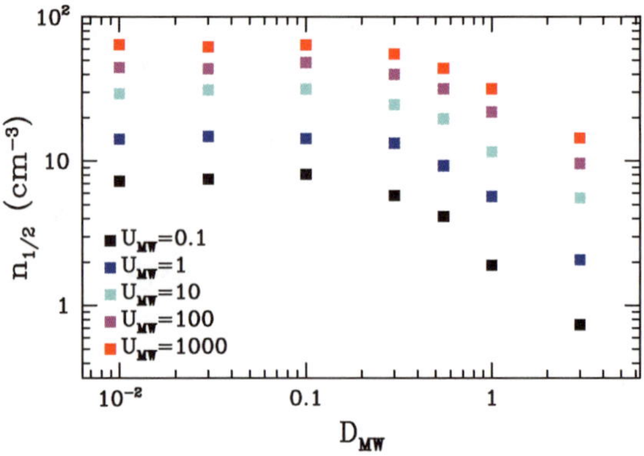

Fig. 31 Average total hydrogen number density of atomic-to-molecular gas transition (defined as $f_{H_2} = 1/2$) as a function of the dust-to-gas ratio D_{MW} and the interstellar radiation field U_{MW} on scales $L = 65\,\text{pc}$

of a better alternative, we can simply take Eq. (27) with the sonic length l_s being fixed to some small value (0.1–1 pc) and a model for the size of molecular cloud L_{MC}.

Perhaps the simplest such model is a "Sobolev-like" approximation that Andrey Kravtsov and I introduced a few years ago (Gnedin et al. 2011),

$$L_{MC} \equiv \frac{\rho}{2|\nabla\rho|}.$$

With such an approximation the complete set of equations is obtained. The dependence of the characteristic density of the transition on the environmental parameters on the particular spatial scale $L = 65\,\text{pc}$ (read "resolution of your simulation") is shown in Fig. 31—the two limiting regimes are easily noticeable in the figure.

In Fig. 32, I show how such a model fares in matching the observed surface densities of atomic and molecular gas on larger scales, where they are actually measured. The main achievement of models like this one is that they capture the observed saturation of the atomic surface density at about $10\,M_\odot/\,\text{pc}^2$ (for $D_{MW} = U_{MW} = 1$ case; the saturation level does depend on the environment, just like $n_{1/2}$). A detailed description of the latest edition of Gnedin et al. (2011) model is presented in Gnedin, Kravtsov, and Draine (2013, in preparation).

An alternative model for the atomic-to-molecular transition is due to Krumholz et al. (2009)—that model is simpler to implement, but does not account for line overlap, and, hence, breaks down for metallicities (or, rather, dust-to-gas ratios) below about 20 % of the Milky Way value.

4.4 Molecular ISM

Ok, we arrived into the molecular ISM. Now what? Why do we even care about the molecular gas? After all, many experts in star formation will tell you that molecules

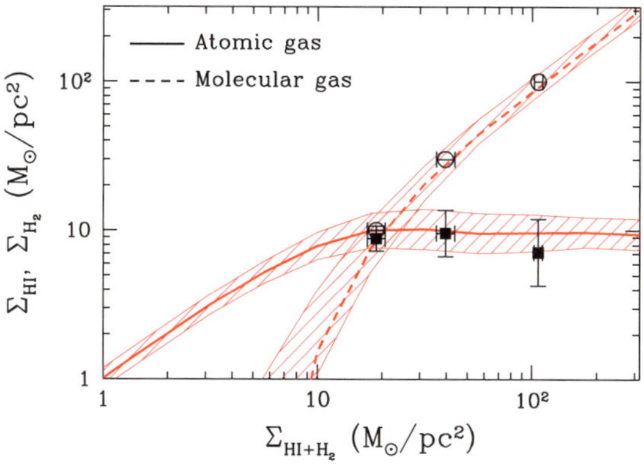

Fig. 32 Average atomic and molecular gas surface densities as functions of the total (neutral) hydrogen gas surface density averaged over 500 pc scale for the ($D_{MW} = 1$, $U_{MW} = 1$) simulation case (*red lines*/bands for mean/rms). *Filled squares* and *open circles* with error bars mark the observed average and rms atomic and molecular hydrogen surface densities from Wong and Blitz (2002)

are *not* required for star formation. Now we know this is not quite true—line overlap makes H_2 self-shielding important at low dust abundances, and, hence, in that regime molecules *are required* for star formation.

A second answer to that question is offered by Krumholz et al. (2011) (and by the nature herself, but that story is still well ahead). Shielding in molecular gas actually performs two functions at once—it protects hydrogen molecules from photo-destruction by Lyman and Werner bands photons, but it also allows gas to cool to the state that is properly called *cold* (100 K and below)—without shielding, UV and optical photons can eject energetic electrons from dust grains by photoelectric effect (the one Einstein got the Nobel prize for); these energetic electrons thermalize in the gas, effectively transferring the energy of radiation into the gas thermal energy. With shielding, this process becomes much less efficient and the gas can cool to low temperatures—and, hence, fragment into small clumps from which stars can form.

Thus, even in the regime when molecular self-shielding is not important, molecular gas plays a role of a "paint" someone poured into the ISM—the "painted" (i.e. molecular) gas is cold and can form stars, while gas without "paint" is too hot for star formation to take place there. Think of this as a lucky "coincidence", if you like.

4.4.1 Thermodynamics of H_2

Before we move further down the yellow brick road towards star forming regions, let us pause for a short while and refresh what we know about the hydrogen molecule. After all, it is the simplest molecule one can imagine, containing just two atoms (hence *diatomic*), and its thermodynamics can be solved (almost) exactly.

If you recall some college thermodynamics, you may remember that the diatomic gas has a polytropic index of $7/5$ (or, equivalently, specific heat $c_V = 5/2$). If you have forgotten that, it should not be hard to re-derive that result! After all, the partition function for the diatomic molecule is simply

$$Z = e^{-E_n/(k_B T)} Z_{\text{rot}} Z_{\text{vib}},$$

where the vibrational part is

$$Z_{\text{vib}} = \sum_{v=0}^{\infty} e^{-\hbar\omega(v+1/2)/(k_B T)}.$$

The rotational part is a bit tricky, but really just a bit—since H_2 is a symmetric molecule and two protons are indistinguishable, two nuclear states (with the spins aligned, total nuclear spin is 1 and the spins anti-aligned, total nuclear spin is 0) behave almost like two different molecules (transitions between the two states are possible, but highly suppressed and only occur at high enough densities). The state with the nuclear spin of 1 is called an *ortho*-hydrogen molecule, and only allows odd values for the total angular momentum J, while the state with the 0 nuclear spin is a *para*-hydrogen molecule and only has even values of the angular momentum. Ortho-H_2 has a higher statistical weight than the para-state, hence

$$Z_{\text{rot}} = \frac{3}{4} Z_{\text{ortho}} + \frac{1}{4} Z_{\text{para}},$$

where

$$\begin{aligned} Z_{\text{ortho}} &= \sum_{J=1,3,\ldots} (2J+1) e^{-\hbar^2 J(J+1)/(2 I k_B T)}, \\ Z_{\text{para}} &= \sum_{J=0,2,\ldots} (2J+1) e^{-\hbar^2 J(J+1)/(2 I k_B T)}. \end{aligned} \tag{29}$$

The partition function Z is a magic wand of thermodynamics, all other quantities are derived from it: free energy

$$F = -k_B T \ln \left[\frac{V}{N!} \left(\frac{m k_B T}{2\pi\hbar^2} \right)^{3/2} Z \right],$$

internal energy

$$E = F - T \left. \frac{\partial F}{\partial T} \right|_V,$$

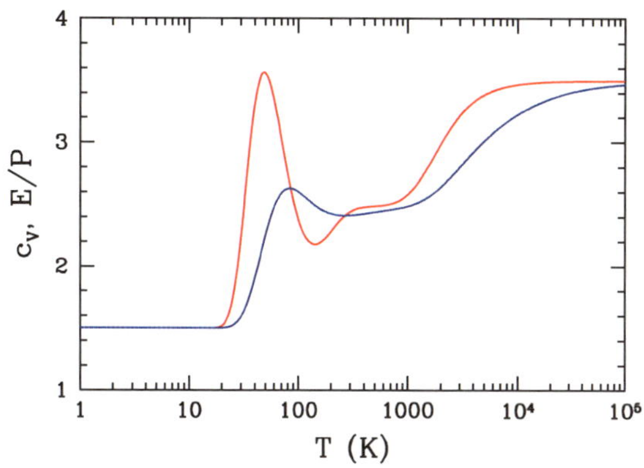

Fig. 33 Specific heat c_V (*red*) and the internal energy over pressure (*blue*) for molecular hydrogen gas as a function of temperature. Notice that H_2 never behaves as classic diatomic gas ($c_V = E/P = 5/2$)

specific heat

$$c_V = \frac{1}{k_B N} \left. \frac{\partial E}{\partial T} \right|_V,$$

etc.

For example, Fig. 33 shows c_v and $E/(k_b T)$ for H_2 gas with 3 : 1 ratio of ortho-to-para molecules. If that plot does not surprise you, then you are a true expert in quantum thermodynamics—**molecular hydrogen actually never behaves as classic diatomic gas with** $c_V = 5/2$ **(or, equivalently,** $\gamma = 7/5$ **).** More than that, it does not even behave as *polytropic* gas with $E = P/(\gamma - 1)$ except for very low temperatures ($T < 20\,\mathrm{K}$) where it behaves as *monoatomic* gas with $c_V = 3/2$! If you did not know that, you can be excused—some of highly distinguished astrophysicists made that error too …

4.4.2 Cosmic Pandora Box: The X-Factor

We are now approaching the most confused, abused, and misused subject in the studies of molecular ISM—CO emission and the X_{CO} factor.

Molecular hydrogen is a great example of a classic catch-22 - H_2 has to be shielded from the outside to exist, hence the outside (i.e. us observing it) cannot actually see its emission in the Lyman and Werner bands. And to add insult to injury, the same dust obscures background sources, making absorption spectroscopy extremely difficult. Historically, by far the most common method to observe molecular gas was via its CO emission.

Rotational transitions of the CO molecule are equally spaced in frequency, $\nu_J = \hbar J/(2\pi I)$ (the molecule is asymmetric, so we do not need to worry about ortho/para mess). For the most common $^{12}C^{16}O$ isotope the first ($1 \rightarrow 0$) transition is located at $\nu_{1\rightarrow 0} = 115\,\mathrm{GHz}$ (or $\lambda_{1\rightarrow 0} = 0.26\,\mathrm{cm}$). This is a major convenience, since CO emission lines are easy to identify (just look for a uniform fence in the millimeter

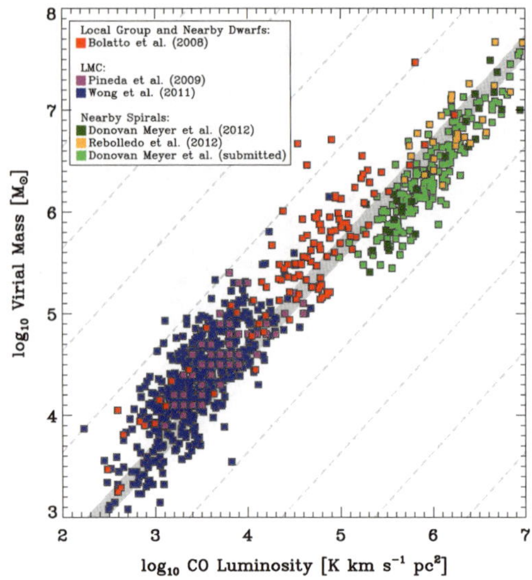

Fig. 34 CO luminosity versus the virial mass for extragalactic molecular clouds. The *gray band* shows the average values in the Milky Way (adopted from Bolatto et al. (2013))

wavelengths). On the other hand, there is no a priori reason why CO should be a good tracer of H_2: CO needs higher dust shielding to form and it gets saturated at too high column densities. Hence, CO emission comes from a narrow range of column densities, both cloud outskirts and cloud centers emit little.

Never-the-less, whenever a mass of molecular gas can be estimated by other means (usually the virial theorem), observations show a good correlation between the CO luminosity and the gas mass, albeit with substantial scatter from one cloud to another (Fig. 34).

In galactic studies the relevant conversion factor between the molecular gas and CO luminosity is the infamous *X-factor*,

$$X_{CO} \equiv \frac{N_{H_2}}{W_{CO}},$$

where W_{CO} is the equivalent width of a CO emission line (which will be different for different transitions),

$$W_{CO} = \int T_A(v)dv$$

with T_A being the antenna temperature of the radio emission. The canonical Milky Way value for the X-factor is $X_{CO} = 2 \times 10^{20}\,\mathrm{cm}^{-2}\,\mathrm{K}^{-1}\,(\mathrm{km/s})^{-1}$ (enjoy the elegance of units!). The reason for this particular combination is that a measurement of the equivalent width in your telescope beam can be directly converted into the column density of molecular hydrogen along the line-of-sight.

In extra-galactic studies most of the time a galaxy is not spatially resolved (at least until the full ALMA comes online), so a single observation measures the total CO luminosity L_{CO} of a galaxy, and a convenient quantity is

$$\alpha_{CO} \equiv \frac{1.36 M_{H_2}}{L_{CO}}$$

(the factor of 1.36 is a contribution of Helium, and it really should be $1/(1-Y)$, since Y does depend slightly on the metallicity). The Milky Way value is $\alpha_{CO} = 4.3\,\mathrm{M}_\odot/\,\mathrm{pc}^2/\,\mathrm{K}/\,(\mathrm{km/s})$. Notice, that the connection between X_{CO} and α_{CO} is somewhat non-trivial; α_{CO} can be re-written as

$$\alpha_{CO} \propto \frac{\int N_{H_2} dA}{\int W_{CO} dA} = \frac{\langle N_{H_2} \rangle}{\langle W_{CO} \rangle},$$

where A is the area on the sky. Hence, α_{CO} is *not* directly proportional to the average X_{CO} for a galaxy. Rather, it is proportional to the ratio of average N_{H_2} to the average W_{CO}. Alternatively, we can re-interpret the averaging procedure for X_{CO} in a highly non-trivial way,

$$\bar{X}_{CO}^{-1} \equiv \frac{\langle W_{CO} \rangle}{\langle N_{H_2} \rangle} = \frac{\langle (W_{CO}/N_{H_2}) N_{H_2} \rangle}{\langle N_{H_2} \rangle} = \left\langle \frac{1}{X_{CO}} \right\rangle_{N_{H_2}}.$$

i.e., X_{CO} should be averaged harmonically and with the H_2 column density weighing.

So, how should we approach modeling CO emission in modern cosmological or galactic-scale simulations? Scales on which CO emission originates are not yet resolvable in modern simulations, hence, it needs to be followed with a sub-grid model. However, since CO emission is not important dynamically, it can be modeled in post-processing, after the simulation had been completed. There exist many approaches to constructing a sub-grid model, and the best (at least in principle) sub-grid model is a someone's else simulation!

The field of modeling internal structure of molecular clouds with sufficient physics is rather new, with only a few attempts made so far, but it certainly developing rapidly. One example of how CO emission can be modeled is the work that was led by Robert Feldmann in two series of paper in 2012 (Feldmann et al. 2012a, b). This is just an illustration, one can follow a similar path with newer, better small-scale simulations for an undoubtedly better result.

One of the very first attempt to model CO emission directly in GMC-scale simulations was done by Simon Glover and collaborators (Glover et al. 2011; Shetty et al. 2011a, b). Images of H_2 and CO column densities, CO equivalent width W_{CO}, the X_{CO} factor from these simulations are shown in Fig. 35. As can be expected in a turbulent ISM, there are large variations in the X_{CO} factor on very small, sub-pc scales. Never-the-less, when averaged over the whole simulated region, X_{CO} dependence on the properties of the molecular cloud exhibits remarkable regularity—Glover et al.

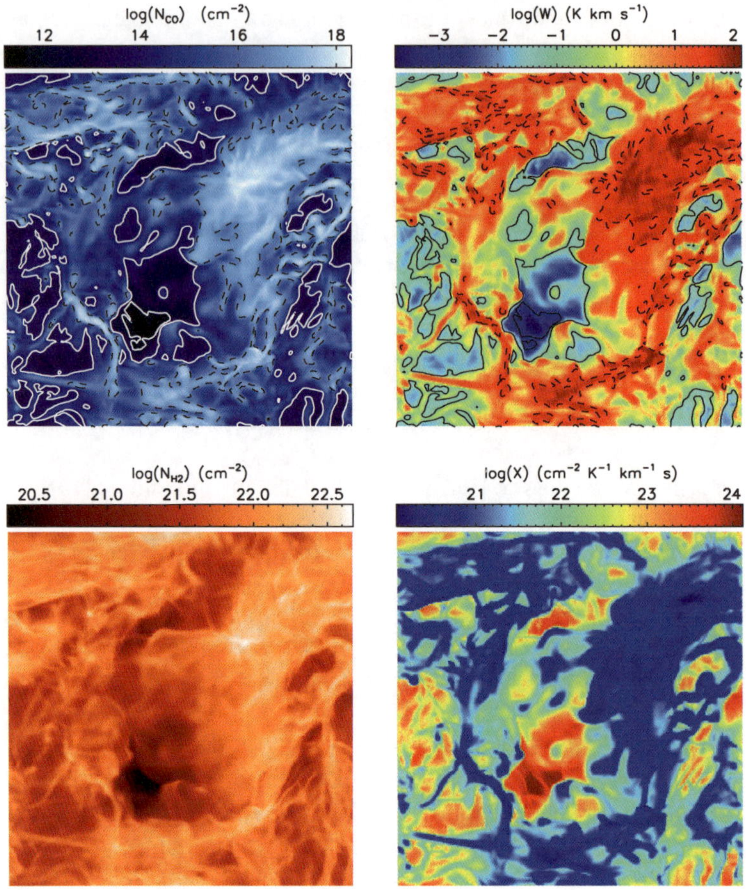

Fig. 35 Images of H_2 and CO column densities, CO equivalent width W_{CO}, the X_{CO} factor from Shetty et al. (2011a) simulations

(2011) found that the main parameter that controls the X_{CO} factor is (surprise!) the dust opacity (sometimes parametrized as the visual extinction $A_V \sim \tau_D$).

Figure 36 shows the mass-weighted molecular and CO fractions from Glover et al. (2011) simulations. Using these tabulated values, Feldmann et al. (2012a, b) developed a sub-grid model that can be used in cosmological and galactic-scale simulations for computing the X_{CO} factor in each simulation cell. Realistic simulated galaxies have complex ISM, with gas densities, metallicities, dust abundances, and interstellar radiation field varying from place to place. Hence, one can and *should* expect the X_{CO} factor to vary significantly inside a given galaxy and from galaxy to galaxy.

As the result, the Feldmann et al. (2012a, b) model predicts a range of values for X_{CO} even for a given metallicity and U_{MW}, not a single number, as is shown in

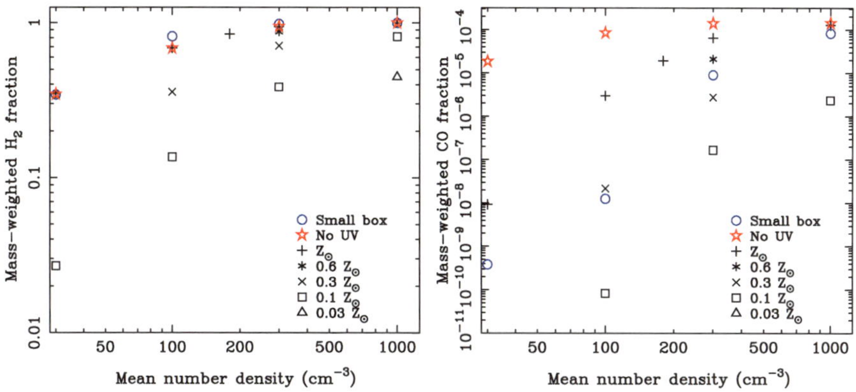

Fig. 36 Mass-weighted fractions of H_2 and CO as a function of average gas density in Glover et al. (2011) simulations

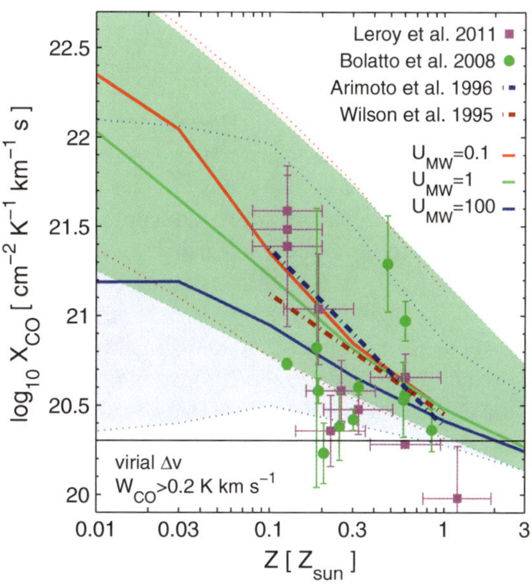

Fig. 37 Dependence of the X_{CO} factor on the environmental parameters: gas metallicity Z and the interstellar radiation field U_{MW}, on $\sim$50 pc scales. *Colored bands* show the variation over different locations in a single simulated galaxy (adopted from Feldmann et al. (2012a))

Fig. 37. Overall, the predictions of the model are within the existing observational measurements, although observations are still too imprecise to provide a serious constraint on the theoretical models.

Using the Feldmann et al. (2012a, b) model, we can explore why observers are often extremely stubborn in using a constant value for X_{CO} (or, alternatively, for α_{CO}). In Fig. 38, I show the dependence of the X_{CO} factor on the molecular hydrogen column density on small (GMC) scales and on large (galactic) scales. Averaging over large scales performs a miracle—almost all the complicated variations in the X_{CO} factor with various environmental parameters disappear (except for the mild

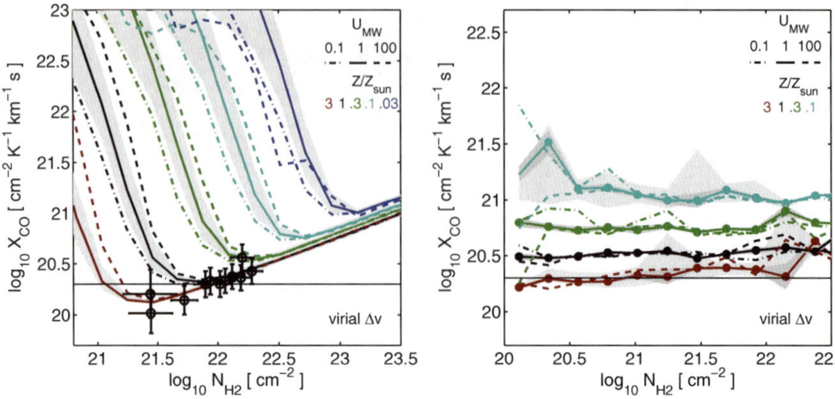

Fig. 38 X_{CO} factor as a function of H_2 column density for a variety of values for the gas metallicity and the interstellar radiation field U_{MW} on small, 50 pc scale (*left*) and large, ~1 kpc scale (*right*). The data points on the *left panel* are from Heiderman et al. (2010). The role of large-scale averaging in making X_{CO} approximately constant is apparent (adopted from Feldmann et al. (2012a))

residual dependence on the metallicity) and $\bar{X}_{CO} \propto \alpha_{CO}$ becomes a surprisingly robust conversion factor from the observed CO luminosity to the total mass of the molecular gas in a distant galaxy (this is indeed nothing short of a miracle).

Before we depart from the domain of sub-grid modeling of the X_{CO} factor, a word of caution is in order. Such modeling is, obviously, not unique. In addition, the existing observational constraints that can be used to calibrate such modeling are still in their infant stage. Hence, any sub-grid model for the X_{CO} factor will remain highly imprecise for some time. For example, an alternative model was proposed by Narayanan et al. (2011) in which X_{CO} is a *decreasing* function of H_2 column density—the dependence that has the *opposite* sign to the left panel of Fig. 38. That does seem somewhat inconsistent with the data from Heiderman et al. (2010), but the measurements are not yet fully constraining. In any event it is clear that if two different models predict opposite signs, there is a large amount of work laying ahead …

4.4.3 Cosmic Pandora Box, Level 2: The X-Factor in ULIRGS

Cosmic Pandora boxes are like Russian Matrioshka dolls, inside one there is always another one …

The remarkable property of the X_{CO} factor to average out on large scales has been used extensively in many extragalactic studies. From an observer's point of view, it is very convenient to be able to determine the molecular gas mass of a distant galaxy by a simple multiplication. There is, however, a complication. For an optically thick emission, like CO, the equivalent width of the line $W_{CO} = T_B \int \beta(v) dv$, where T_B is the brightness temperature of the emitting gas and $\beta(v)$ is the escape probability from a parcel of gas with velocity v. The second factor can be thought of as an effective line width $\Delta v_{\rm eff}$, so that

$$X_{\rm CO} = \frac{N_{\rm H_2}}{T_B \Delta v_{\rm eff}}.$$

Variations in $N_{\rm H_2}$ and $\Delta v_{\rm eff}$ do average out, so it is T_B that we are concerned with now. In LTE brightness temperature is equal to the gas temperature. In normal galaxies molecular gas is very cold, $T_B \sim 10\,{\rm K}$, but dust is usually warmer than the gas, $T_{\rm dust} = 40\text{–}60\,{\rm K}$. Hence, if dust and gas couple thermally, T_B can increase systematically in at least some molecular clouds, causing a systematic decrease in the $X_{\rm CO}$ factor that will not average out.

For dust and gas to couple, densities must be really high, significantly higher than is achieved in normal molecular clouds, so in normal galaxies coupling occurs only in a tiny fraction of the most dense molecular gas. The situation is different in Ultra-Luminous IR Galaxies (ULIRG), which are major merger of large galaxies. In mergers substantial fraction of the total gas in both galaxies gets channeled towards the center, where it gets extremely dense, piling up to many thousands of solar masses per square parsec (versus a few tens for galaxies like the Milky Way). At such high densities (and column densities) dusts starts coupling to (and, hence, heating) the gas.

In a classical study Solomon et al. (1997) explored that effect in several nearest ULIRG, and concluded that the $X_{\rm CO}$ factor (or, rather, $\alpha_{\rm CO}$, since we are talking about external galaxies) could be as low as $\alpha_{\rm CO,min} = 0.8\,{\rm M}_\odot/\,{\rm pc}^2/\,{\rm K}/\,({\rm km/s})$. That value, however, was only a strict lower limit, as their results depended on several assumptions that all added a factor of 2 factors on top of $\alpha_{\rm CO,min}$. Alas, in an ironic mis-interpretation of the Solomon et al. (1997) paper many observers took that lower limit as the actual value, and for almost 2 decades it was quite common to hear a fairy tale of "two modes of star formation", each with its own value of $\alpha_{\rm CO}$ (0.8 and 4.3).

Obviously, such a "bimodality" makes no physical sense—a miracle of nature may make $\alpha_{\rm CO}$ a universal constant, but if it is not, then there must be either a distribution of $\alpha_{\rm CO}$ for different galaxies or a systematic trend of the average $\alpha_{\rm CO}$ value with some of galaxy properties, like the mean surface density or IR luminosity.

Fortunately, the dust settled (or, more precisely, was observed directly), thanks to Herschel (again, not a somewhat eccentric, clever, and compassionate man but a space telescope). Measurements of dust emission over several bands between 100 and 1000 microns, when taken together with optical and sub-mm observations from the ground, allow to fit detailed models of dust spectral energy distribution and, hence, derive dust temperature and mass, in a substantial sample of ULIRG over a wide redshift range, all the way to $z \sim 3$ (c.f. Magdis et al. 2012).

These observations can then be combined with measurements of gas metallicities and CO luminosities in the same galaxies in two different ways.

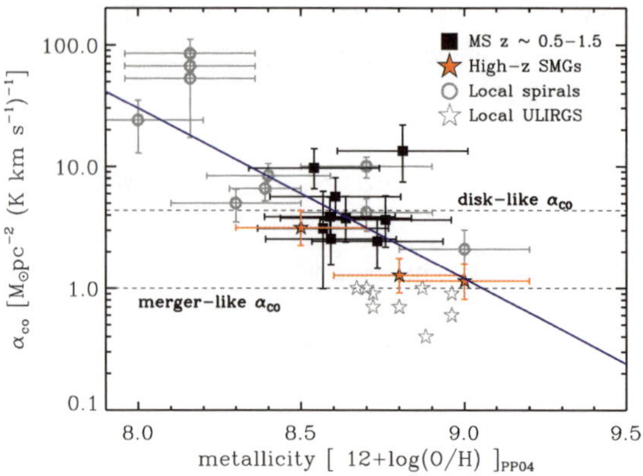

Fig. 39 α_{CO} as a function of gas metallicity for several samples of local and high redshift galaxies (adopted from Magdis et al. (2012))

1. If one assumes dust-to-gas ratio as a function of metallicity $M_{gas}/M_{dust}(Z)$ (for example, by calibrating from the measurements of nearby galaxies), then from M_{dust} one gets M_{gas}, and under the assumption that all gas is molecular, M_{gas} and L_{CO} give α_{CO}.
2. Alternatively, if one adopts a value for α_{CO}, the dust-to-gas ratio can be derived in the reverse order of steps.

The measurements of α_{CO} vs. Z for Magdis et al. (2012) sample and other available samples are shown in Fig. 39. The data are inconclusive—a trend with metallicity, a wide distribution, even bimodality cannot yet be excluded, but the main conclusion is clear—the X_{CO} *is not universal.*

4.4.4 Cosmic Pandora Box, Level 3: Which Transition Dominates?

If, by now, you are totally disenchanted with the X_{CO} factor, here is an insult to your injury—in Fig. 40, I show a distribution of CO emission over the rotational transitions $J \rightarrow (J-1)$ for several galaxies. Even in our own Milky Way CO emits most of its energy in the $2 \rightarrow 1$ transition, in more active/merging galaxies the peak of the emission is shifted to even high transitions (i.e. higher gas temperatures). Hence, the X_{CO} factor is different for different $J \rightarrow (J-1)$ transitions, so to compare apples to apples, we need to convert different observed transitions to one baseline one (say, $1 \rightarrow 0$). These new conversions factors will also depend on the galactic environment, dust temperature, perhaps redshift, etc. A hierarchy of nested Pandora boxes never ends …

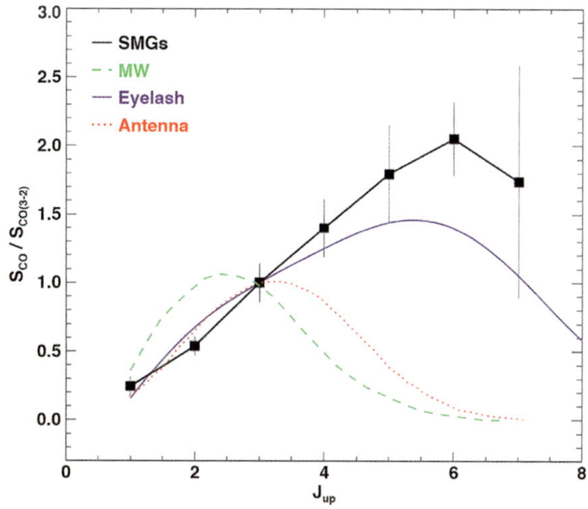

Fig. 40 Distribution of CO emission over the rotational transition J for several galaxies (*colors*) and the average distribution for high-redshift sub-millimeter galaxies (SMG). For most galaxies the $1 \rightarrow 0$ transition is not the dominant one (adopted from Bothwell et al. (2013))

5 Star Formation

If the field of ISM is large, what one can say about star formation—it is at least another quarter of all Astronomy research. So, we must thread very carefully, or we will be lost forever in the jungle of clouds, disks, and outflows. We will attempt to stay on largest scales, and will look only on the most generic relations between gas and stars. We are not even going to paint the broad picture, we will just look at the frame …

5.1 Kennicutt-Schmidt and All, All, All

For us, looking down on star formation from galactic scales and above, the story of star formation begins in March 1959, with the classical paper by Schmidt (1959), who noticed that the *surface* density (and let us be precise here, we still have very little observational clues on what the *volumetric* density of star formation is doing) of star formation correlates with the surface density of gas approximately as a power-law,

$$\Sigma_{\rm SFR} \propto \Sigma_{\rm gas}^n,$$

with $n = 1$–2.

This relationship was firmed up later by Kennicutt (1989, 1998), resulting in what is nowadays commonly referred to as the Kennicutt-Schmidt (KS) relation,[5]

[5]God save you from calling it a "law" in the presence of a devout physicist!

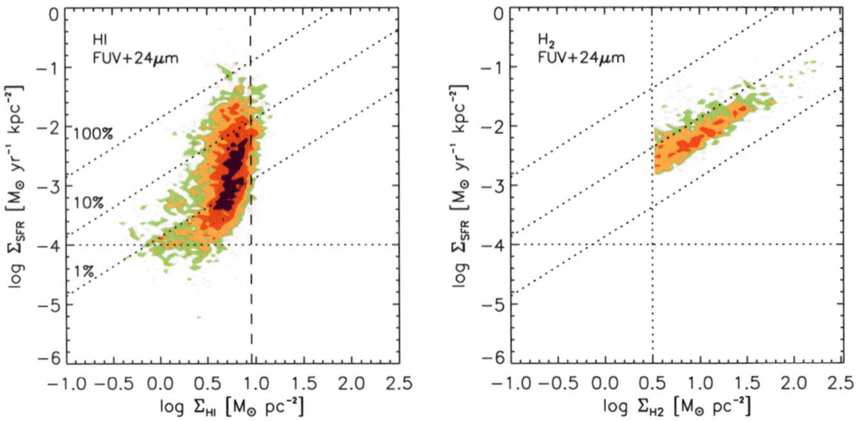

Fig. 41 Star formation surface density as a function of HI (*left*) and H_2 surface densities from the THINGS survey (adopted from Bigiel et al. (2008))

$$\Sigma_{\rm SFR} = (2.5 \pm 0.7) \times 10^{-4} \frac{\rm M_\odot}{\rm kpc^2\, yr} \left(\frac{\Sigma_{\rm gas}}{1\, \rm M_\odot/\, pc^2}\right)^{1.4\pm0.15} . \tag{30}$$

In this form the KS relation survived for 10 years. But THINGS does matter (orthography is correct), The Nearby HI Galaxies Survey was an important step in shaping our modern understanding and interpretation of the KS relation, in large part because in addition to HI, the THINGS team assembled a large amount of other data on their target galaxies, from CO emission to UV and H-α measurements of star formation rates.

The THINGS survey unambiguously proved what everyone knew in their hearts: stars form from molecular gas.[6] In Fig. 41, there is a clear strong correlation between $\Sigma_{\rm SFR}$ and the surface density of the molecular gas, but there is almost no correlation with the atomic gas. Hence, we have not wasted our time discussing the atomic-to-molecular transition, it is one of the bottlenecks that controls star formation in galaxies.

Historically, it was common to represent the KS relation as the relation between the star formation surface density and the surface density of the "total" gas, which actually meant the sum of atomic and molecular (i.e. *neutral*) gas. The left panel of Fig. 42 shows this "classical" form of KS relation from the THINGS data, together with the original measurements from Kennicutt (1998) (although the latter are tricky to interpret, since they use a different value of $\alpha_{\rm CO}$ to convert CO emission to the molecular gas surface density).

In order to illustrate how that particular shape appears, the right panel shows the KS relation from a numerical simulation of the Milky-Way-like galaxy that we already met several times in two previous chapters. In the simulation the star formation rate

[6] At least, the vast majority of them—by itself, the THINGS result does not exclude a possibility of a small fraction of stars forming in the atomic gas.

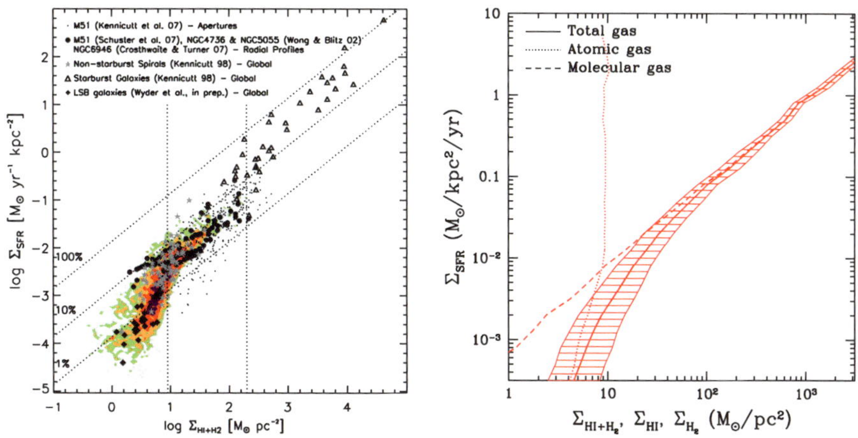

Fig. 42 *Left* "Classical" Kennicutt-Schmidt relation from THINGS (adopted from Bigiel et al. (2008)). *Right* Separate average KS relations for the atomic (*dotted*), molecular (*dashed*), and neutral (*solid*) gas from the already familiar to us cosmological simulation of the Milky Way like galaxy. The *solid line* is the sum of the *dotted* and *dashed* along the horizontal direction

is postulated to be linearly proportional to the molecular gas surface density,

$$\Sigma_{\rm SFR} = \frac{1.36\Sigma_{\rm H_2}}{\tau_{\rm SF}}, \tag{31}$$

where the factor 1.36 is, again, to account for Helium, and $\tau_{\rm SF}$ is the gas *depletion time*, assumed to be constant $\tau_{\rm SF} = 1.5\,$Gyr in the simulation (we will come back to that number shortly).

The "classical" KS relation then forms from the separate atomic $\Sigma_{\rm HI}$ and molecular $\Sigma_{\rm H_2}$ surface densities as

$$\begin{aligned} \Sigma_{\rm SFR} &= \frac{1.36}{\tau_{\rm SF}}\Sigma_{\rm H_2} + 0 \times \Sigma_{\rm HI}, \\ \Sigma_{\rm HI+H_2} &= \Sigma_{\rm H_2} + \Sigma_{\rm HI}. \end{aligned}$$

The steepening of the KS relation at low surface densities is simply due to gas becoming predominantly atomic, and is fully explained by the physics of the atomic-to-molecular transition. Indeed, observations support this interpretation (but we won't dive into that question here, it is too wide and deep for us to linger in it, as we are rushing along our yellow brick road).

5.1.1 How We Should Think About Star Formation

If star formation correlates well with molecular gas, it is useful to think about Eq. (31) as our primary ansatz, and consider how τ_{SF} may depend on other properties (for example, density). That thinking, however, is *totally wrong*!

A simple fact that we often forget is that density is *not even defined* without a particular scale. After all, $\rho = M/V$, and if there is no V, there is no ρ. Hence, both theoretically and observationally, we need to explicitly consider the range of spatial scales that is relevant for our problem.

Let us take some spatial scale L. One can imagine the whole universe divided into boxes of size L, like in a super-huge numerical simulation, or the universe observed with a given telescope resolution. If we average gas densities on scale L, they become meaningfully defined. Thus, Eq. (31) should really be replaced with

$$\langle \dot{\rho}_* \rangle_L = \frac{\langle \rho_{\rm mol} \rangle_L}{\tau_{\rm SF}}, \tag{32}$$

where $\rho_{\rm mol} = 1.36\rho_{\rm H_2}$ is the density of the molecular gas, and averaging is done over the spatial scale L. In that case depletion time becomes the function of other gas properties on scale L,

$$\tau_{\rm SF} = \tau_{\rm SF}(L, \langle \rho_{\rm mol} \rangle_L, \ldots).$$

In other words, we need to explicitly think of star formation relation as (at least) a two-dimensional relation on the plane $(L, \langle \rho_{\rm mol} \rangle_L)$, or, perhaps, even a higher-dimensional relation if chemistry, magnetic fields, properties of ISM turbulence, etc. are also important.

Armed with this understanding, we can now reinterpret the existing observational constraints on various scales on a uniform basis.

In Fig. 43, I show the molecular KS relation for normal star forming galaxies (i.e. not ULIRGs, with their complicated CO $\rightarrow$ H_2 conversion) in the local universe and at high redshift. These observations sample star formation on large scales (from many hundreds of parsecs to several kilo-parsecs). They all are consistent with roughly linear KS relation,

$$\tau_{\rm SF} \approx {\rm const}(L \gg 100\,{\rm pc}),$$

although with substantial *intrinsic* (i.e. exceeding the formal observational error) scatter, and the actual value for $\tau_{\rm SF}$ noticeably different at $z = 0$ ($\tau_{\rm SF} \approx 2$ Gyr) and high redshift ($\tau_{\rm SF} \approx 0.7$ Gyr).

At the present moment (Sep 2013) it seems difficult to make any further inference from these measurements. For example, is $2 = 0.7$? In fact, they might, since local and high redshift measurements probe very different, non-overlapping ranges of gas surface density: few local observations reach $100\,{\rm M_\odot/\,pc^2}$, while all high redshift measurement are way above that limit. The Feldmann et al. (2012a) model for the $X_{\rm CO}$ factor predicts that $X_{\rm CO}/\alpha_{\rm CO}$ factor increases gradually with the gas surface

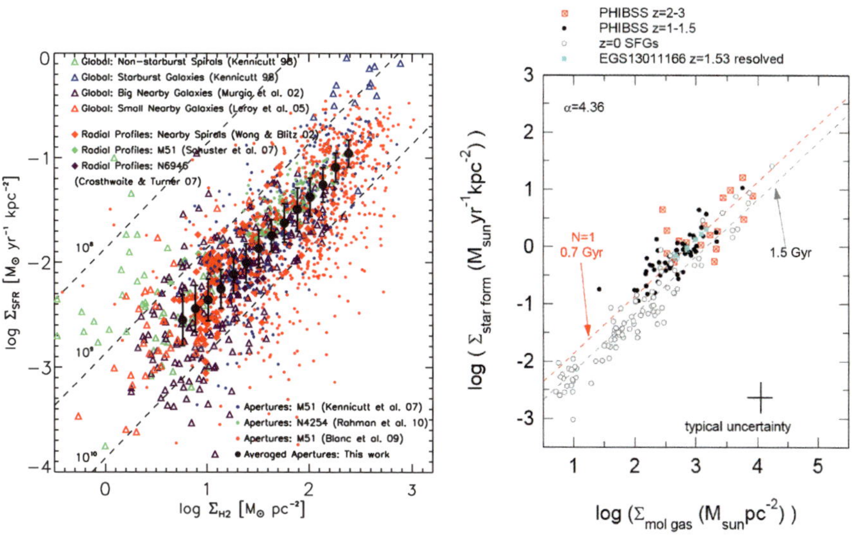

Fig. 43 Star formation rate surface density versus the surface density of the molecular gas for local galaxies (*left*, Bigiel et al. (2011)) and high redshift normal star forming galaxies (*right*, Tacconi et al. (2013))

density; Feldmann et al. (2012b) present, as an example, a cosmological simulation with constant in time and space $\tau_{SF} = 1.5$ Gyr, which is consistent with both the low and high redshift measurements. We do not yet know how accurate the Feldmann et al. (2012a) model is, but, at the very least, there exists a plausible counterexample to any potential claim that low- and high-redshift KS relation are inconsistent with each other.

The same uncertainty applies to the scatter around the mean KS relation. We do know that the X_{CO} factor varies across single galaxies and between different galaxies, so some fraction of the scatter should be due to scatter in X_{CO}. In addition, there is scatter from the time dimension that we have so far ignored: CO emission from the molecular gas is essentially instantaneous, but observational estimates of star formation are not. For example, Schruba et al. (2010) show that the depletion time is systematically higher around peaks of CO emission (molecular clouds where star formation is just starting) than around peaks of H-α emission (star forming regions where star formation is well under way).

This difference is purely due to the fact that we do not measure the instantaneous rate of star formation, but use observational indicators that return a time-averaged star formation rate over some characteristic time-scale ($\sim$20 Myr for UV light, $\sim$5 Myr for H-α). Hence, if we point our telescope on a freshly formed molecular cloud, we will see a lower star formation rate than the actual instantaneous one—if the cloud has been forming stars for only 1 Myr and we use H-α, we will measure a 5 Myr/1 Myr $=$ 5 times lower star formation rate than the true one. Now, if we point it at a mature star forming region, we will measure a higher time-averaged star formation rate than the instantaneous one, because 5 Myr ago the region contained more molecular gas (and, hence, higher instantaneous star formation rate) than it has right now.

The combined scatter due to variations in the X_{CO} factor and the finite time-averaging is easily quantifiable, though, and appears to be less than the actual observed scatter in the KS relation (Feldmann et al. 2012b). This should not be particularly surprising, though—it is hard to imagine that the nature is so kind to us as to make each region of space with the same surface density of molecular gas to have *exactly* the same star formation rate, surely there must be random or systematic variation from place to place that affect star formation rate, and that will appear as the true intrinsic scatter in Eq. (32).

There exist several other constraints we can place on $\tau_{SF}(L, \langle\rho_{mol}\rangle_L, \ldots)$. Lada et al. (2010) found that on the scale of individual star-forming cores ($\sim$1 pc) the depletion time is also constant (i.e. independent of density) and is about 20 Myr, but only if the density is above $\rho_{min} = 700\,M_\odot/\,pc^3$. A threshold must exist in that case, since any small-scale relation must be consistent with the large-scale one. Namely, if on 1 pc scale we have

$$\langle\dot{\rho}_*\rangle_1 = \begin{cases} \dfrac{\langle\rho_{mol}\rangle_1}{20\,\text{Myr}}, & \rho_{mol} > \rho_{min}, \\ 0, & \rho_{mol} < \rho_{min}, \end{cases}$$

and on 500 pc scale we have a usual molecular KS relation,

$$\langle\dot{\rho}_*\rangle_{500} = \frac{\langle\rho_{mol}\rangle_{500}}{2\,\text{Gyr}},$$

then these two relations can be mutually consistent if and only if exactly 1 % of the molecular gas sits above the small-scale density threshold ρ_{min}—after all,

$$\langle\dot{\rho}_*\rangle_{500} = \left\langle\langle\dot{\rho}_*\rangle_1\right\rangle_{500}.$$

Another commonly used ansatz for the star formation rate is constant efficiency per free-fall time,

$$\tau_{SF}(L, \langle\rho_{mol}\rangle_L, \ldots) = \frac{\tau_{ff}(\langle\rho_{mol}\rangle_L)}{\epsilon_{SF}} = \epsilon_{SF}^{-1}\sqrt{\frac{3\pi}{32G\langle\rho_{mol}\rangle_L}},$$

or, in a more familiar form,

$$\langle\dot{\rho}_*\rangle_L = \epsilon_{SF}\frac{\langle\rho_{mol}\rangle_L}{\tau_{ff}} = \epsilon_{SF}\frac{\langle\rho_{mol}\rangle_L^{3/2}}{\sqrt{3\pi/(32G)}}. \tag{33}$$

The origin of that formula disappears in the depths of time; it is often used without any attention to the scale under consideration. In an influential paper, Krumholz and Tan (2007) argued that many observational constraints are consistent with that ansatz[7] with $\epsilon_{SF} \approx 1$–$2\,\%$ for a wide array of molecular densities, from average molecular clouds to molecular cores.

However, observational constraints used by Krumholz and Tan (2007) sample not only various densities, but also *various spatial scales*; namely, they all fall along a particular track $L^2 \times \langle\rho_{mol}\rangle_L \approx 10^4\,\text{cm}^{-3}\,\text{pc}^2$ in the two-dimensional plane $(L, \langle\rho_{mol}\rangle_L)$. In other words, observational constraints that support the "constant efficiency per free-fall time" are equally well support the "constant efficiency per unit scale",

$$\langle\dot{\rho}_*\rangle_L = \epsilon_{SF}\frac{\langle\rho_{mol}\rangle_L}{\tau_{ff}} \approx \epsilon_{SF}\frac{\langle\rho_{mol}\rangle_L}{\tau_{ff}(10^4\,\text{cm}^{-3})(L/1\,\text{pc})}.$$

The two alternatives cannot be distinguished at present without additional observational constraints.

In fact, I am going to make a bold claim (and challenge anyone to refute it) that *all of the existing observational constraints are consistent with the linear (in density) star formation ansatz* in which the depletion time is function of scale only,

$$\langle\dot{\rho}_*\rangle_L = \begin{cases} \dfrac{\langle\rho_{mol}\rangle_L}{\tau_{SF}(L)}, & \rho_{mol} > \rho_{min}(L), \\ 0, & \rho_{mol} < \rho_{min}(L), \end{cases} \tag{34}$$

with

$$\tau_{SF}(L) \sim 2\,\text{Gyr} \times \min\left(1, \frac{L}{L_0}\right),$$

$$\rho_{min}(L) \sim \rho_0 \times \min\left(1, \frac{L_0^2}{L^2}\right),$$

and L_0 is in the range of a few hundred parsecs (for example, the scale height of the gaseous disk). In the Milky Way galaxy ρ_0 is such that the Lada et al. (2010) result is matched ($\rho_{min}(L) \approx 700\,\text{M}_\odot/\,\text{pc}^3$ at $L \sim 1\,\text{pc}$), but in other galaxies it may be different (for example, being proportional to the density of the atomic-to-molecular transition).

Figure 44 shows two alternatives (linear star formation relation (34) and constant-efficiency-per-free-fall-time star formation relation (33)) in a cartoon fashion. At present, either one is a sensible model, as well as any other, intermediate or more complex model, that still matches the observational constraints.

[7] One should never forget that the "constant efficiency per free-fall time" model is no more than an ansatz; molecular clouds are turbulent and the free-fall time has no physical relevance on scales above the sonic length.

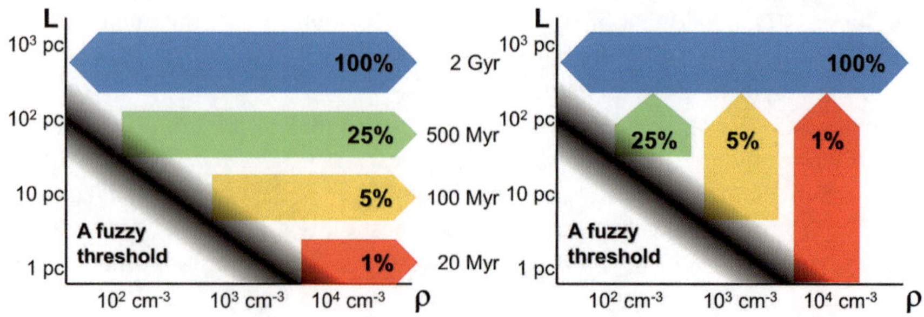

Fig. 44 Cartoon version of contours of constant depletion time (shown as *different colored bands*) on the $(L, \langle \rho_{\rm mol} \rangle_L)$ plane. The *left panel* shows the linear star formation relation (34), while the *right panel* shows the constant-efficiency-per-free-fall-time star formation relation (33). In the latter case the depletion function must transition to a constant on the largest scale somehow to be consistent with large-scale KS relation

5.2 Excursion Set Formalism in Star Formation

The idea of using Excursion Set formalism in star formation is based on a well established fact: in isothermal supersonic turbulence the density PDF is lognormal, in a direct analogy with the Gaussian distribution of the linear overdensity δ. Such an approach was first attempted by Padoan and Nordlund (2002), picked up later by Hennebelle and Chabrier (2008) and developed much further by Phil Hopkins in a recent series of papers (Hopkins 2012a, b, 2013).

5.2.1 Refresher: Excursion Set Formalism

Excursion Set formalism (sometimes also called "Press-Schechter formalism") deals with a *Gaussian random field* $\delta(\mathbf{x})$ (and δ can be anything, for supersonic turbulence it will be $\ln(\rho/\rho_0)$). For a Gaussian random field different wavenumbers of the Fourier transform

$$\delta_{\mathbf{k}} \equiv \int d^3x \delta(\mathbf{x}) e^{i\mathbf{k}\mathbf{x}}$$

of the field are uncorrelated,

$$\langle \delta_{\mathbf{k}_1} \delta^*_{\mathbf{k}_2} \rangle = P(k_1) \delta^3_D(\mathbf{k}_1 - \mathbf{k}_2).$$

One can reverse the Fourier transform,

$$\delta(\mathbf{x}) = \int d^3k \sqrt{P(k)} \lambda_{\mathbf{k}} e^{-i\mathbf{k}\mathbf{x}}, \quad (35)$$

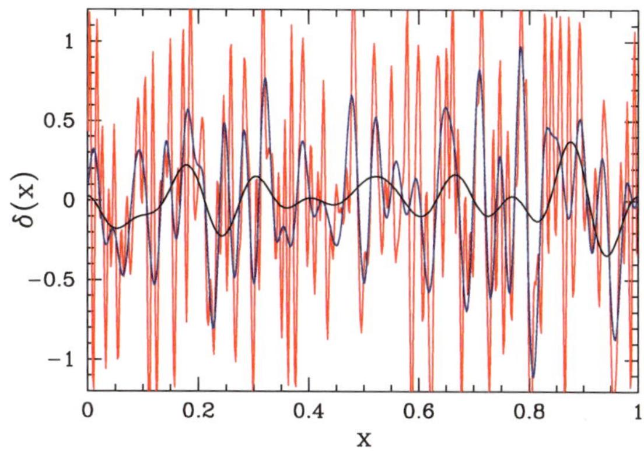

Fig. 45 Example of the Gaussian random field at three different values for the smoothing scale R

with uncorrelated, normally distributed random numbers $\lambda_{\mathbf{k}}$ satisfying the relation $\langle\lambda_{\mathbf{k}_1}\lambda^*_{\mathbf{k}_2}\rangle = \delta^3_D(\mathbf{k}_1 - \mathbf{k}_2)$. Equation (35) should be considered in a generalized sense (similar to Dirac delta-function), because for some $P(k)$ the integral may actually diverge. In that case $\delta(\mathbf{x})$ should be considered as a limit of the smoothed density field,

$$\delta(\mathbf{x}) \equiv \lim_{R\to 0} \delta(\mathbf{x}; R) = \int d^3k\sqrt{P(k)}\lambda_{\mathbf{k}} W(kR)e^{-i\mathbf{kx}},$$

where $W(kR)$ is a low-pass filter ($W(0) = 1$, $W(\infty) = 0$). An example of a Gaussian random field at 3 different resolutions is shown in Fig. 45.

Excursion Set formalism considers $\delta(\mathbf{x}; R)$ as a function of the filter scale R and compares it with some barrier function $b(R)$. Obviously, $\delta(\mathbf{x}, R = \infty) = 0$. As R decreases, $\delta(\mathbf{x}; R)$ starts deviating from zero. For some value of R it may cross the barrier for the first time. The fraction of all $\delta(\mathbf{x}; R)$ that cross the barrier at R is called the first crossing distribution. For example, in the canonical Press-Schechter formalism the barrier is constant, $b = \delta_L(t_f) = 1.69$. Then the first crossing distribution becomes (half) the mass function of dark matter halos with $M_h = 4\pi\bar{\rho}_m R^3/3$.

In modeling star formation Excursion Set formalism can be used for several goals:

- First crossing distribution gives the mass function of largest bound objects—molecular clouds.
- Last crossing distribution gives the mass function of smallest bound objects—molecular cores/stars.
- It is useful for other purposes too: distribution of holes in the ISM, clustering of stars, etc.

One only needs to define a barrier—but we have already considered it! After all, gas collapses when it becomes gravitationally unstable, hence the barrier is simply the stability criterion for the disk with finite thickness, our Eq. (24)—with a minor

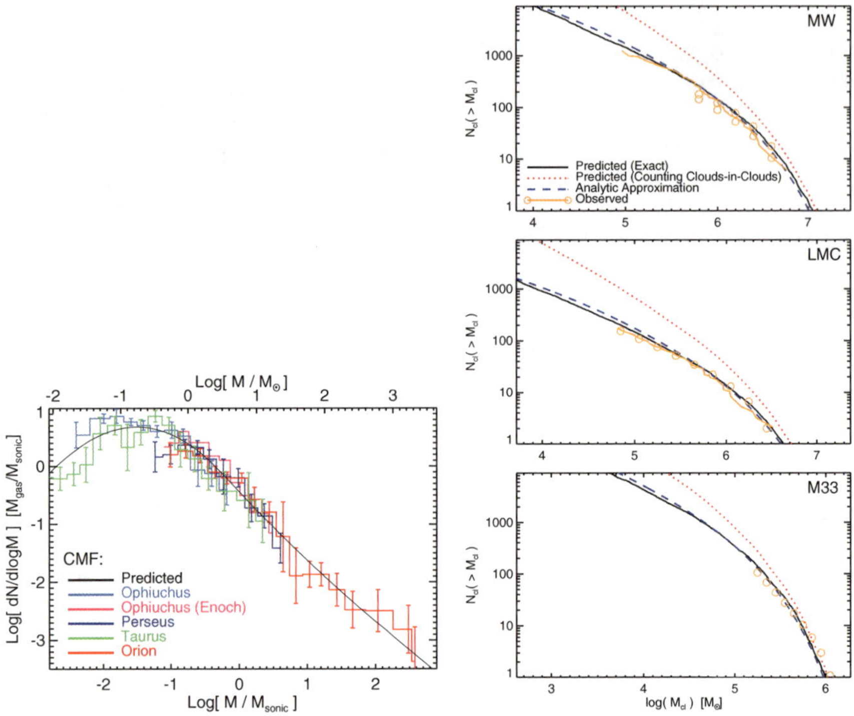

Fig. 46 Several predictions of the Excursion Set formalism as a theory of star formation: GMC mass functions for the Milky Way, LMC, and M31 (*right*, adopted from Hopkins (2012a)) and clump mass function (*bottom*, adopted from Hopkins (2012b))

correction of adding turbulent velocity dispersion σ_t^2 to the gas sound speed c_S^2, since turbulence also provides support against gravitational collapse.

Excursion Set formalism makes predictions that are computable analytically and match a large variety of observations unexpectedly well (see Fig. 46). The final verdict on this novel approach is still pending, with opinions ranging from "it should never work" to "it solves all the problems". So, if you are bold enough, make your bet …

6 Stellar Feedback

Stars affect their environments by their *feedback*—anyone reading these lines knows that well, without the stellar feedback we would not even exist (as there would not exist planets made out of heavy elements).

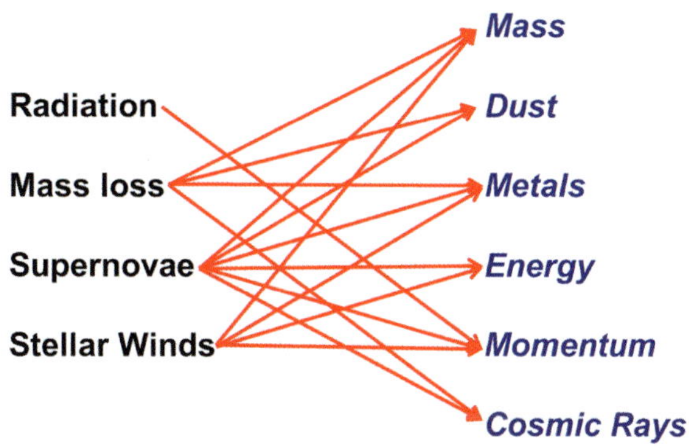

Fig. 47 Various pathways for stellar feedback

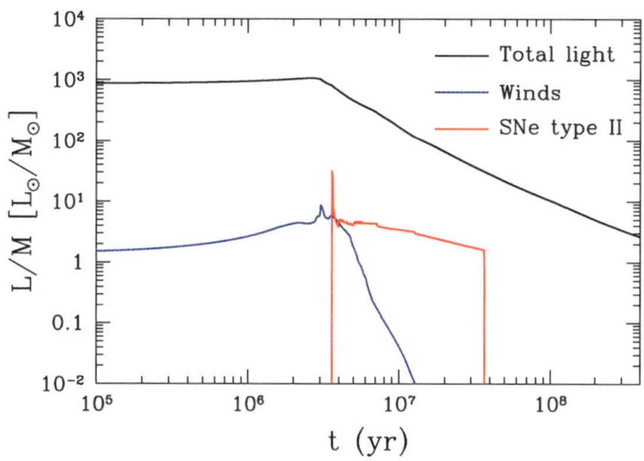

Fig. 48 Energy injection rate (=luminosity) per unit mass for the total radiation, winds, and supernovae as a function of time for a normal stellar population at solar metallicity

6.1 What Escapes from Stars

However, stellar feedback is not just supernovae (sometimes that's the impression one gets by reading simulation papers). Stars affect their environments in several ways: *supernovae* (both type II and type Ia), *stellar winds* from massive stars, *mass loss* from AGB and planetary nebulae, and, of course, *radiation*. Each of these modes inject into surrounding gas *energy*, *momentum*, *mass*, *metals*, *dust*, and *cosmic rays*. The various pathways the inputs and outputs are connected are shown in Fig. 47. If you think the feedback is complicated, then you are right!

It is easy to get lost in this maze of feedback pathways. But one important fact should light our way (literally)—by far the largest energy output of stars is light! Just as an illustration, I show in Fig. 48, the energy production rate as a function of time for a normal stellar population at solar metallicity. The bolometric luminosity of stars dwarfs all other feedback channels at all times. And we know that at least half of that energy is re-radiated in the infrared by dust, hence a substantial fraction of stellar light is indeed absorbed by the surrounding gas. We should, therefore, consider that feedback channel very seriously.

6.1.1 Radiation Pressure

Massive (hence, young) stars spend a substantial fraction of their lives heavily embedded into the surrounding gas and dust; for heavily obscured stars most of their light is absorbed. Since photons have momentum, absorbing all light from a star/star cluster of luminosity L injects momentum into the surrounding gas,

$$\dot{p}_1 = \frac{L}{c}.$$

The energy, however, is conserved—the absorbed bolometric luminosity of the star must be re-emitted by dust in the infrared. If there is enough dust around a young massive star (and, at least in the Milky Way, there is), the dust will be optically thick to its own IR radiation. That radiation will do work on the surrounding gas, i.e. will inject extra momentum, so that the total momentum injection rate is

$$\dot{p}_{\rm tot} = (1 - f_{\rm esc} + \tau_{\rm IR})\,\frac{L}{c}, \qquad (36)$$

where, in order to be completely general, I included the fraction $f_{\rm esc}$ of stellar radiation (of all frequencies) that escapes the star forming region. The new factor $\tau_{\rm IR}$ is easy to derive for a homogeneous medium (Gayley et al. 1995). Since energy is conserved, the radiation flux at each radius R from the star is still

$$F_R = \frac{L}{4\pi R^2}.$$

Hence, the momentum (in the radial direction) imparted on the gas between R and $R + dR$ is simply

$$d\dot{p}_{\rm IR} = 4\pi R^2 \frac{F_R}{c}\kappa dR = \frac{L}{c}d\tau,$$

and, hence,

$$\dot{p}_{\rm IR} = \tau_{\rm IR}\frac{L}{c}.$$

In the infrared dust opacity is

$$\kappa_{\rm IR} \approx 3\frac{{\rm cm}^2}{\rm g}\left(\frac{T_d}{100\,{\rm K}}\right)^2$$

(Semenov et al. 2003). Observational estimates of $\tau_{\rm IR}$ at $T_d = 100\,$K are shown in Fig. 49; radiation pressure is particularly important for large stellar clusters.

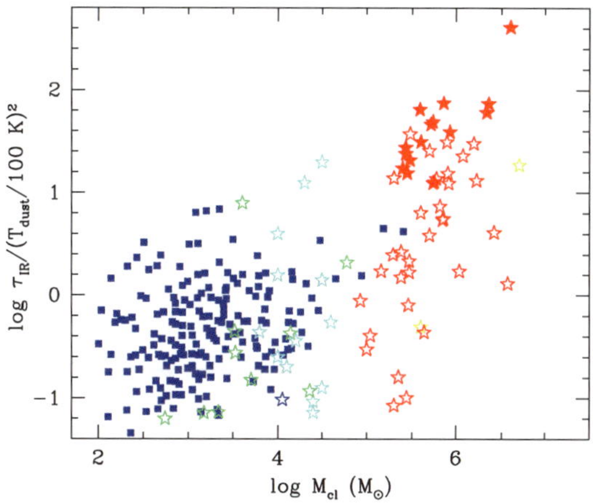

Fig. 49 Observational estimates of τ_{IR} for GMC clumps (*blue squares*) and young stellar clusters (*stars*). Adopted from Agertz et al. (2013)

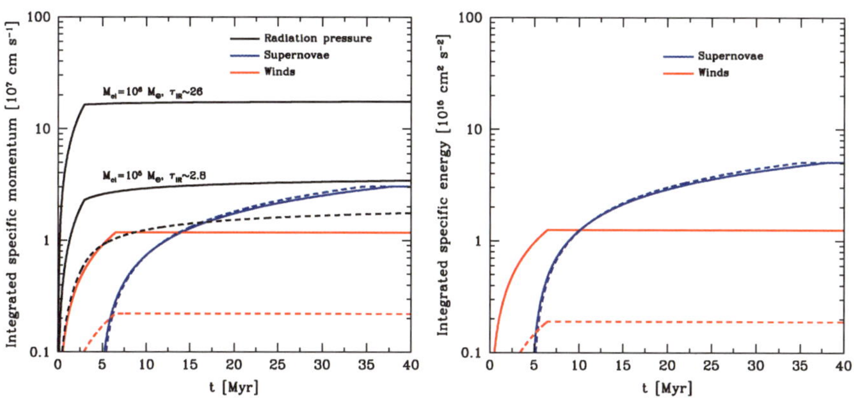

Fig. 50 Time evolution of the cumulative injected specific momentum (*left*) and specific energy (*right*) for various feedback channels at solar metallicity (*solid lines*) and at $Z = 0.01 Z_\odot$ (*dashed lines*). Notice that supernovae kick in after all other feedback channels already fired. Adopted from Agertz et al. (2013)

6.1.2 All Feedback, All the Time

Stars, of course, do not have a freedom to selectively fire only some of their feedback channels, they all work all the time. In Fig. 50, the time evolution of the momentum and energy injection is shown for several dominant feedback modes. Supernovae form the last episode in the feedback fireworks—by the time they start in earnest (after about 10 Myr after the onset of star formation) all other feedback channels have already finished.

In fact, with all likelihood, supernovae are not important to actually destroying molecular clouds (and, hence, controlling the efficiency of star formation, Fall et al. 2010). They may be important for heating the overall ISM, for stabilizing the disk, for driving galactic winds, but star formation they control not.

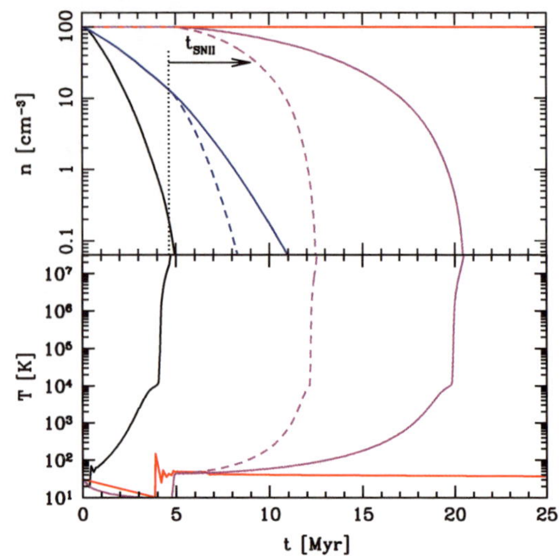

Fig. 51 Density and temperature in a computational cell evolving in response to various feedback channels: only energy from supernovae (*red*), energy and momentum from supernovae (*magenta*), only momentum feedback from all channels (*blue*), and all feedback channels acting together (*black*). Adopted from Agertz et al. (2013)

The relative role of different feedback channels can be understood even better with a simple numerical exercise—a single computational cell "simulation". Figure 51 shows the fate of a such a cell when various feedback channels are switched on and off.

The first important lesson is that purely thermal feedback—injecting all of the supernova energy as thermal energy into the parent cell (or particle in case of SPH) of even a large stellar cluster does not do anything, the cooling times are always so short that the thermal energy is quickly radiated away. This is not a new result, it has been known since the dawn of numerical galaxy formation, and re-discovered independently by many research groups; but it does pose a dilemma for cosmological and even galactic-scale simulations—the only direct way of implementing stellar feedback does not work, and one has to use a *sub-grid model*, i.e. a specific recipe about how to implement the feedback in a numerical code.

Figure 51 may offer a clue how such a sub-grid model may be implemented: other feedback channels produce a large effect on the dynamics of a single cell, and, hence, may have a significant effect on the dynamics of larger scales as well. There is just one problem with that approach—actual stellar feedback like Eq. (36) is operating on scales of molecular cores and their very vicinities, on sub-parsec scales. Whenever we use, say, the radiation pressure formula in cosmological (or even galactic-scale) simulations, we are injecting the momentum on scales of many tens, even hundreds of parsecs, well beyond the range of scales where it is actually operating. Hence, using Eq. (36) in a cosmological code is also a *sub-grid model*, an ansatz that is a priori as good or bad as any other sub-grid model. Not surprising, then, that the radiation pressure is gradually falling out of fashion.

Before we trash all sub-grid models or pick one of them and place it on the throne, it is worth taking a step back and re-thinking what we are actually trying to achieve.

6.2 Unconventional Marriage: Feedback and Star Formation

If you did not skip the previous chapter, my dear reader, then you know that star formation is *inefficient*—the depletion time τ_{SF} is of the order of a Gyr (give or take a factor of 2–3), while molecular clouds are short-lived (10–20 Myr). During their lifetimes molecular clouds convert only a small fraction, mere percents, of their gas into stars. A natural conclusion from that fact is, since star formation is inefficient, then so must be the feedback. And that conclusion is utterly wrong!

Do you recall a simulation of a Milky Way like galaxy that I used to illustrate the properties of the gaseous halo (Fig. 17)? Have you wondered why I never showed you the circular velocity profile for that galaxy? There is a good reason I have not—I am ashamed to! While the gaseous halo for that galaxy may look ok, the disk is totally wrong, it has an extremely dense spike at the center, with the circular velocity peaking at 450 km/s, more than twice the rotation velocity of the Milky Way. The reason for such a huge discrepancy is the absence of any feedback process in the simulation.

If we learned anything after 20+ years of modeling galaxy formation, then it is that the central spikes in circular velocity (caused by unrealistic central mass concentrations) can only be destroyed by strong feedback. No other physics can do the trick—in fact, as simulations grew more sophisticated, included more physics, and reached higher resolution, the central mass concentration problem became worse. It is a real, physical problem, not a numerical one— the high-redshift progenitors of normal galaxies are too dense, and these early dense gaseous concentrations survive all the subsequent adventures of galaxy evolution; if not blown out, they will become large stellar bulges.

Indeed, that was commonly occurring in simulations until only a few years ago—for example, check out beautiful pictures of center-heavy galaxies in Stinson et al. (2010). At the same time, as observers figured out the difference between the real bulges and pseudo-bulges (central features formed by secular evolution from barred disks), they realized that a significant fraction, perhaps as much as 50 %, of galaxies are actually *bulgeless*, pure disks.

All these examples illustrate one crucially important conclusion about star formation and feedback—while start formation is *inefficient*, the feedback is actually *strong*. These two facts may be deeply connected, but we are not going to dive into the connection between star formation and feedback, for our purpose what is important is this apparent dichotomy in behavior.

A good illustration of that dichotomy comes from the so-called "abundance matching" exercise—a match between the *observationally derived*[8] stellar mass and the theoretically known mass function of dark matter halos. Such a match results in a one-to-one correspondence between the stellar mass and the halo mass for individual halos (or, in a more complex implementation of the abundance matching idea, a distribution of stellar masses for a given halo mass). Figure 52 shows a comparison

[8]Never forget that stellar masses are *not* observed, they are always derived from observations of luminosity functions, with all the inherent in spectral synthesis uncertainties and biases.

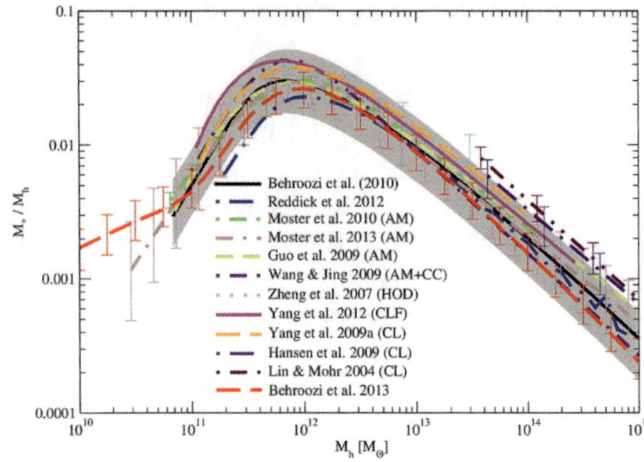

Fig. 52 Abundance matching in action: stellar vs. total mass for dark matter halos from several independent groups; they all agree that star formation is inefficient. Adopted from Behroozi et al. (2013)

(and uncertainty) of the stellar mass—halo mass relation for several independent applications of that approach from Behroozi et al. (2013).

The most important result of the abundance matching exercise is that stellar masses of low mass halos are very small, roughly $M_* \propto M_h^{2.5}$ for $M_h \ll 10^{12}\,\mathrm{M}_\odot$. To get such a behavior, it is not enough to make star formation inefficient—that would still result in stellar mass being proportional to halo mass, the inefficiency would only make the coefficient of proportionality small—but it also requires the feedback to be progressively more efficient in lower mass galaxies to sculpt the inferred $M_* \propto M_h^{2.5}$ relation.

Attempts to model the feedback as a sub-grid model are as old as the galaxy formation simulations themselves. It is not too instructive to review all of them, as until 2010 none of the sub-grid models were particularly successful. The important thing to remember about any sub-grid model is that it would only work over a finite range of spatial scales. If the model is good, that range would be sufficiently large (say, a decade in spatial scale); if the model is bad, the range may be zero. Even if the model is good, but its range of validity does not match the resolution of simulations, then it would not work well.

Indeed, that is what have happened with one simple sub-grid feedback model. In 1997 in his Ph.D. thesis, Jeroen Gerritsen proposed a simple way to make feedback strong—simply to disable cooling in star forming regions for several tens of Myr (we now call this method "delayed cooling"). The model did not work too well with the spatial resolution simulations were able to reach in 1997. However, miracles do happen—as the resolution improved, the delayed cooling model appeared to work better and better, until, finally, in 2010 it was declared to be a panacea for galaxy formation (Governato et al. 2010)!

Figure 53 gives two examples of how well modern simulations with delayed cooling feedback reproduce observations, but similarly impressive examples for various observational constraints are abound.

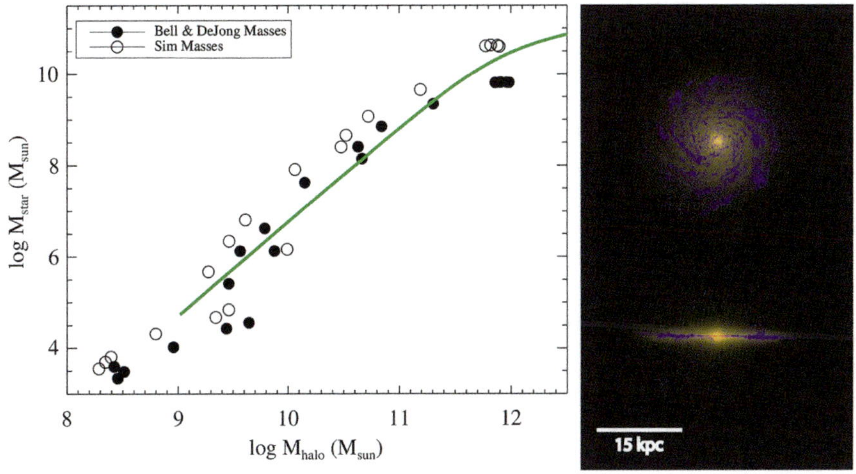

Fig. 53 *Left* Stellar versus total mass from abundance matching (*green line*) and modern galaxy formation simulations (*open* and *filled circles*—adopted from Munshi et al. (2013)). *Right* Face- and edge-on projections of Eris simulation of the Milky Way galaxy (adopted from Guedes et al. (2011))

6.2.1 Why Delayed Cooling Works

While it is easy to declare the delayed cooling a success, it is much harder to understand what it actually means. As such, it is just a numerical trick, without any serious physical justification. The fact that it works may be a pure coincidence; alternatively, it can be a manifestation of a real physical process that operates on sub-parsec scales, but its consequences on ∼100 pc scales appear as if cooling was switched off. In fact, it is easy to come up with several real physical processes that will all manifest themselves as delayed cooling on large scales:

- radiation pressure from massive stars (we now know it is important) provides support for gas that "does not cool", i.e. if treated as an effective additional pressure, that pressure would not be affected by the cooling processes in the gas, but will diminish after about 10 Myr;
- coronal gas—the hot, million-degree gas produced in supernova explosions may accumulate in regions of low density in a supersonically turbulent IGM; cooling times in such gas will depend on its density, but generally will be of the order of several to several tens of Myr;
- as stellar feedback continue to stir supersonic turbulence in molecular clouds on small scales, the energy of the kinetic motions will accumulate to the point at which the dissipation rate will approximately equal the production rate; while the dissipation time-scale is likely to be short, the supersonic turbulence (i.e. highly super-thermal additional pressure in the gas) will be maintained for the duration of stellar feedback, several tens of Myr;

- cosmic rays produced in supernova explosions are observationally known to provide significant additional support in the magnetized molecular clouds; cosmic rays diffuse out of GMC on time-scales of tens of Myr.

I am sure that list can be easily extended, but it already serves our purposed well—numerous real small-scale physical processes may hide themselves under the large-scale mask of "delayed cooling", and one, several, all of them, or different combinations of them in different environments may be the actual feedback process(es) that is/are responsible for making the real galaxies as they are …

6.3 Toward the Future

So, where do we go from there? If only we could figure out which of the actual feedback channels hides behind the mask of delayed cooling, the galaxy formation (of normal disk galaxies—the AGN feedback is entirely different story) will be essentially solved (well, hopefully you do not take me as being too optimistic).

These feedback processes are more-or-less understood, as, hopefully, I persuaded you in the beginning of this chapter. Actually modeling them in the cosmological and galactic-scale simulations is not trivial, but big strides in that directions have been already made. We may still argue occasionally how to do it better, or what the most appropriate value for, say, τ_{IR} should be, but the importance of the modeling feedback correctly is not a subject of debate any more.

The good piece of news is that even if various feedback channels are tuned to match the basic observational constraints like the stellar mass vs. total halo mass, Kennicutt-Schmidt relation, rotational velocity curves, etc., simulated galaxies in runs with different feedback channels still look amazingly different (Fig. 54), and there lies the key to the eventual success.

Hence, the plan for the future is to identify the best observational probes that will help us understand which of all of the potential feedback channels are important in which environments and on what spatial scales. This is left as an exercise for the reader...

7 Answers to Brain Teasers

1. Sound waves indeed do not dissipate. However, they also do not grow with time, since they are stable perturbations, while large-scale, unstable perturbations in both dark matter and gas grow. Hence, relative to the large-scale perturbations, the small-scale sound waves become smaller and smaller, i.e. they appear to be "suppressed".
2. The proper term for "Lyman-α absorption" is *resonant scattering*. A Lyman-α photon is re-emitted by the atom, but in the meantime that atom experienced

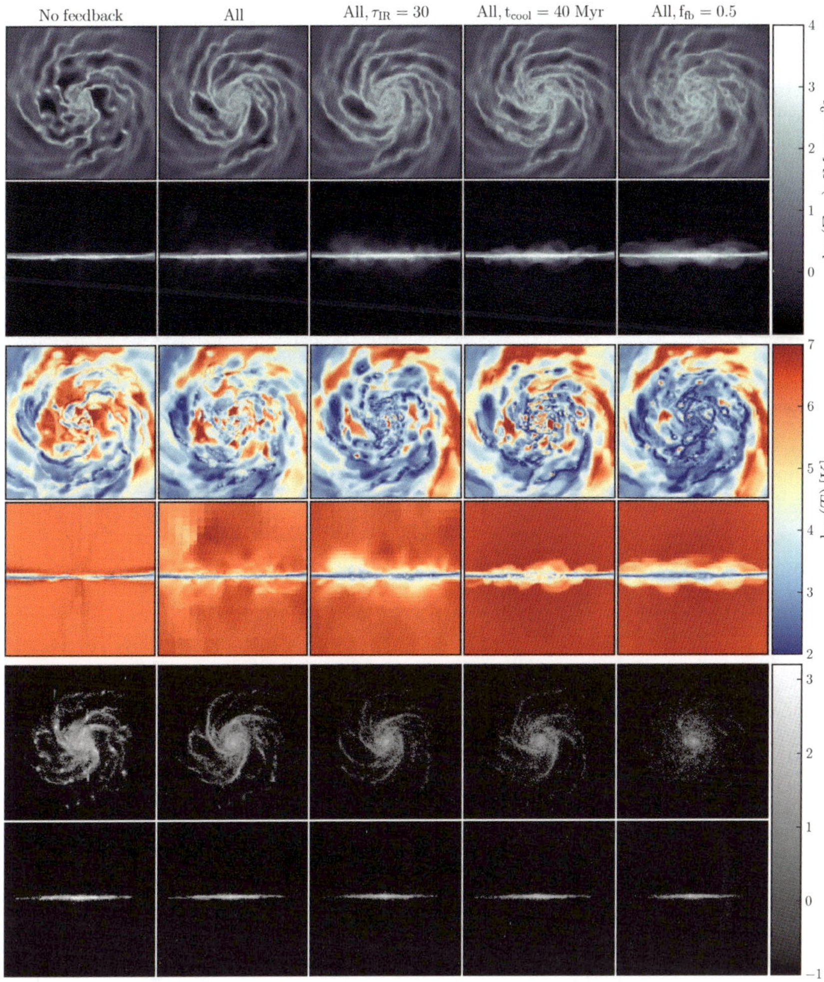

Fig. 54 Face-on and edge-on maps of simulated galactic disks with different small-scale feedback models. Separate rows show gas surface density (*top*), mass weighted average gas temperature (*middle*), and stellar surface density (*bottom*). Columns from *left* to *right* are no feedback, all feedback channels from Fig. 47, all feedback with extra radiation pressure, delayed cooling feedback, feedback model with extra energy variable (adopted from Agertz et al. (2013))

a large number of collisions with other atoms and ions, so its momentum is now unrelated to the momentum it had at the moment of absorption. Hence, the re-emitted Lyman-α photon will be send out into a random direction in the frame of the atom, and will not reach our telescope. For us, that photon is lost, hence we, sometimes, call it absorption.

3. The term "equation of state" relates the perturbations in the gas pressure (or temperature) to those of the density. If we impose (adiabatically) a perturbation $\delta\rho$ to the gas density, the instantaneous response to the pressure will be identical to the ideal gas, $\delta P = c_S^2 \delta\rho$. Only with time adiabatic expansion and photoheating will bring that perturbation back to the temperature-density relation.
4. A typical ionizing photon is not sitting at the Lyman edge, it has the energy of $E_0 + \langle \Delta E \rangle$ (see Eq. 9), which is about 40–50 eV for the cosmic background. The ionizing cross-section falls off with energy as E^{-3}, hence the typical cross-section is $\sim$(1–2) $\times 10^{-19}$ cm^2 instead of 6.3×10^{-18} cm^2. In addition, the typical ionization level in the forest is 10^{-5}, which requires $\tau = \ln(10^5) \approx 10$ to neutralize. Hence, hydrogen absorbers only become fully neutral at column densities of $N_H \sim (0.5\text{–}1) \times 10^{20}$ cm^{-2}.
5. This one is really tricky. In fact, I do not know the full answer to it. One possible reason why Lyman-α forest is not turbulent was suggested to me by Andrea Ferrara: for turbulence to develop, the gas needs to have vorticity, but in the linear regime vorticity in cosmic gas decays, so there should be no vorticity at $\delta \approx 0$ in the forest. Non-linear evolution will generate some vorticity, but since most of the forest is not extremely non-linear, it is plausible that the vorticity generated in the forest may not be enough to create a full turbulent cascade.

References

Agertz, O., Kravtsov, A. V., Leitner, S. N., & Gnedin, N. Y. 2013, ApJ, 770, 25
Altay, G., Theuns, T., Schaye, J., Crighton, N. H. M., & Dalla Vecchia, C. 2011, ApJL, 737, L37.
Anderson, M. E. & Bregman, J. N. 2010, ApJ, 714, 320
Barnes, J. & Efstathiou, G. 1987, ApJ, 319, 575
Begelman, M. C. & Shlosman, I. 2009, ApJL, 702, L5
Behroozi, P. S., Wechsler, R. H., & Conroy, C. 2013, ApJ, 770, 57
Bigiel, F., Leroy, A., Walter, F., Brinks, E., de Blok, W. J. G., Madore, B., & Thornley, M. D. 2008, AJ, 136, 2846
Bigiel, F., Leroy, A. K., Walter, F., Brinks, E., de Blok, W. J. G., Kramer, C., Rix, H. W., Schruba, A., Schuster, K., Usero, A., & Wiesemeyer, H. W. 2011, ApJL, 730, L13+.
Binney, J. & Tremaine, S. 1987, Galactic dynamics.
Blitz, L. & Robishaw, T. 2000, ApJ, 541, 675
Bolatto, A. D., Wolfire, M., & Leroy, A. K. 2013, ArXiv e-prints.
Bothwell, M. S., Smail, I., Chapman, S. C., Genzel, R., Ivison, R. J., Tacconi, L. J., Alaghband-Zadeh, S., Bertoldi, F., Blain, A. W., Casey, C. M., Cox, P., Greve, T. R., Lutz, D., Neri, R., Omont, A., & Swinbank, A. M. 2013, MNRAS, 429, 3047
Chen, H.-W. 2012, MNRAS, 427, 1238
Chen, H.-W., Perley, D. A., Pollack, L. K., Prochaska, J. X., Bloom, J. S., Dessauges-Zavadsky, M., Pettini, M., Lopez, S., Dall'aglio, A., & Becker, G. D. 2009, ApJ, 691, 152
Croft, R. A. C., Weinberg, D. H., Katz, N., & Hernquist, L. 1998, ApJ, 495, 44
Dall'Aglio, A., Wisotzki, L., & Worseck, G. 2008, A&Ap, 491, 465
Draine, B. T. 1978, ApJS, 36, 595
Draine, B. T. & Bertoldi, F. 1996, ApJ, 468, 269
Fall, S. M., Krumholz, M. R., & Matzner, C. D. 2010, ApJL, 710, L142
Feldmann, R., Gnedin, N. Y., & Kravtsov, A. V. 2012a, ApJ, 747, 124

Feldmann, R., Gnedin, N. Y., & Kravtsov, A. V. 2012b, ApJ, 758, 127
Gayley, K. G., Owocki, S. P., & Cranmer, S. R. 1995, ApJ, 442, 296
Glover, S. C. O. & Abel, T. 2008, MNRAS, 388, 1627
Glover, S. C. O. & Mac Low, M.-M. 2011, MNRAS, 412, 337.
Gnedin, N. Y. 2012, ApJ, 754, 113
Gnedin, N. Y., Baker, E. J., Bethell, T. J., Drosback, M. M., Harford, A. G., Hicks, A. K., Jensen, A. G., Keeney, B. A., Kelso, C. M., Neyrinck, M. C., Pollack, S. E., & van Vliet, T. P. 2003, ApJ, 583, 525
Gnedin, N. Y. & Hollon, N. 2012, ApJS, 202, 13
Gnedin, N. Y. & Hui, L. 1998, MNRAS, 296, 44
Gnedin, N. Y. & Kravtsov, A. V. 2010, ApJ, 714, 287
Gnedin, N. Y. & Kravtsov, A. V. 2011, ApJ, 728, 88
Gnedin, N.Y., & Draine, B.T. (2014), ApJ, 795, 37
Governato, F., Brook, C., Mayer, L., Brooks, A., Rhee, G., Wadsley, J., Jonsson, P., Willman, B., Stinson, G., Quinn, T., & Madau, P. 2010, Nature, 463, 203
Grcevich, J. & Putman, M. E. 2009, ApJ, 696, 385
Guedes, J., Callegari, S., Madau, P., & Mayer, L. 2011, ApJ, 742, 76
Gupta, A., Mathur, S., Krongold, Y., Nicastro, F., & Galeazzi, M. 2012, ApJL, 756, L8
Haffner, L. M., Dettmar, R.-J., Beckman, J. E., Wood, K., Slavin, J. D., Giammanco, C., Madsen, G. J., Zurita, A., & Reynolds, R. J. 2009, Reviews of Modern Physics, 81, 969
Haiman, Z., Abel, T., & Rees, M. J. 2000, ApJ, 534, 11
Heavens, A. & Peacock, J. 1988, MNRAS, 232, 339
Heiderman, A., Evans, II, N. J., Allen, L. E., Huard, T., & Heyer, M. 2010, ApJ, 723, 1019
Hennebelle, P. & Chabrier, G. 2008, ApJ, 684, 395
Hitschfeld, M., Kramer, C., Schuster, K. F., Garcia-Burillo, S., & Stutzki, J. 2009, A&Ap, 495, 795
Hopkins, P. F. 2012a, MNRAS, 423, 2016
Hopkins, P. F. 2012b, MNRAS, 423, 2037
Hopkins, P. F. 2013, MNRAS, 430, 1653
Hui, L. & Gnedin, N. Y. 1997, MNRAS, 292, 27
Kennicutt, Jr., R. C. 1989, ApJ, 344, 685
Kennicutt, Jr., R. C. 1998, ApJ, 498, 541
Kereš, D., Katz, N., Weinberg, D. H., & Davé, R. 2005, MNRAS, 363, 2
Klypin, A., Hoffman, Y., Kravtsov, A. V., & Gottlöber, S. 2003, ApJ, 596, 19
Kravtsov, A. V. 2003, ApJL, 590, L1
Kravtsov, A. V. 2013, ApJL, 764, L31
Krumholz, M. R., Leroy, A. K., & McKee, C. F. 2011, ApJ, 731, 25
Krumholz, M. R., McKee, C. F., & Tumlinson, J. 2009, ApJ, 693, 216
Krumholz, M. R. & Tan, J. C. 2007, ApJ, 654, 304
Lada, C. J., Lombardi, M., & Alves, J. F. 2010, ApJ, 724, 687
Leroy, A. K., Walter, F., Brinks, E., Bigiel, F., de Blok, W. J. G., Madore, B., & Thornley, M. D. 2008, AJ, 136, 2782
Lidz, A., Faucher-Giguère, C.-A., Dall'Aglio, A., McQuinn, M., Fechner, C., Zaldarriaga, M., Hernquist, L., & Dutta, S. 2010, ApJ, 718, 199
Magdis, G. E., Daddi, E., Béthermin, M., Sargent, M., Elbaz, D., Pannella, M., Dickinson, M., Dannerbauer, H., da Cunha, E., Walter, F., Rigopoulou, D., Charmandaris, V., Hwang, H. S., & Kartaltepe, J. 2012, ApJ, 760, 6
Maller, A. H. & Bullock, J. S. 2004, MNRAS, 355, 694
Mathis, J. S., Mezger, P. G., & Panagia, N. 1983, A&Ap, 128, 212
McDonald, P., Miralda-Escudé, J., Rauch, M., Sargent, W. L. W., Barlow, T. A., & Cen, R. 2001, ApJ, 562, 52
McDonald, P., Seljak, U., Burles, S., Schlegel, D. J., Weinberg, D. H., Cen, R., Shih, D., Schaye, J., Schneider, D. P., Bahcall, N. A., Briggs, J. W., Brinkmann, J., Brunner, R. J., Fukugita, M., Gunn, J. E., Ivezić, Ž., Kent, S., Lupton, R. H., & Vanden Berk, D. E. 2006, ApJS, 163, 80.

Mo, H. J., Mao, S., & White, S. D. M. 1998, MNRAS, 295, 319
Munshi, F., Governato, F., Brooks, A. M., Christensen, C., Shen, S., Loebman, S., Moster, B., Quinn, T., & Wadsley, J. 2013, ApJ, 766, 56
Naoz, S. & Barkana, R. 2007, MNRAS, 377, 667
Narayanan, D., Krumholz, M., Ostriker, E. C., & Hernquist, L. 2011, MNRAS, 418, 664
Nelson, D., Vogelsberger, M., Genel, S., Sijacki, D., Kereš, D., Springel, V., & Hernquist, L. 2013, MNRAS, 429, 3353
Padoan, P. & Nordlund, Å. 2002, ApJ, 576, 870
Polyachenko, V. L. & Polyachenko, E. V. 1997, Soviet Journal of Experimental and Theoretical Physics, 85, 417
Puchwein, E., Pfrommer, C., Springel, V., Broderick, A. E., & Chang, P. 2012, MNRAS, 423, 149
Quilis, V. & Moore, B. 2001, ApJL, 555, L95
Rauch, M., Sargent, W. L. W., Barlow, T. A., & Carswell, R. F. 2001, ApJ, 562, 76
Ricotti, M., Gnedin, N. Y., & Shull, J. M. 2000, ApJ, 534, 41
Rudie, G. C., Steidel, C. C., & Pettini, M. 2012, ApJL, 757, L30
Safronov, V. S. 1960, Annales d'Astrophysique, 23, 979
Saul, D. R., Peek, J. E. G., Grcevich, J., Putman, M. E., Douglas, K. A., Korpela, E. J., Stanimirović, S., Heiles, C., Gibson, S. J., Lee, M., Begum, A., Brown, A. R. H., Burkhart, B., Hamden, E. T., Pingel, N. M., & Tonnesen, S. 2012, ApJ, 758, 44
Schaye, J., Theuns, T., Rauch, M., Efstathiou, G., & Sargent, W. L. W. 2000, MNRAS, 318, 817
Schmidt, M. 1959, ApJ, 129, 243
Schruba, A., Leroy, A. K., Walter, F., Sandstrom, K., & Rosolowsky, E. 2010, ApJ, 722, 1699
Semenov, D., Henning, T., Helling, C., Ilgner, M., & Sedlmayr, E. 2003, A&Ap, 410, 611
Shandarin, S. F. & Zeldovich, Y. B. 1989, Reviews of Modern Physics, 61, 185
Shetty, R., Glover, S. C., Dullemond, C. P., & Klessen, R. S. 2011a, MNRAS, 412, 1686
Shetty, R., Glover, S. C., Dullemond, C. P., Ostriker, E. C., Harris, A. I., & Klessen, R. S. 2011b, MNRAS, 415, 3253
Sofue, Y., Tutui, Y., Honma, M., Tomita, A., Takamiya, T., Koda, J., & Takeda, Y. 1999, ApJ, 523, 136
Solomon, P. M., Downes, D., Radford, S. J. E., & Barrett, J. W. 1997, ApJ, 478, 144
Stanimirović, S., Dickey, J. M., Krčo, M., & Brooks, A. M. 2002, ApJ, 576, 773
Stinson, G. S., Bailin, J., Couchman, H., Wadsley, J., Shen, S., Nickerson, S., Brook, C., & Quinn, T. 2010, MNRAS, 408, 812
Tacconi, L. J., Neri, R., Genzel, R., Combes, F., Bolatto, A., Cooper, M. C., Wuyts, S., Bournaud, F., Burkert, A., Comerford, J., Cox, P., Davis, M., Förster Schreiber, N. M., García-Burillo, S., Gracia-Carpio, J., Lutz, D., Naab, T., Newman, S., Omont, A., Saintonge, A., Shapiro Griffin, K., Shapley, A., Sternberg, A., & Weiner, B. 2013, ApJ, 768, 74.
Toomre, A. 1964, ApJ, 139, 1217
Trowland, H. E., Lewis, G. F., & Bland-Hawthorn, J. 2013, ApJ, 762, 72
Valenzuela, O., Rhee, G., Klypin, A., Governato, F., Stinson, G., Quinn, T., & Wadsley, J. 2007, ApJ, 657, 773
van de Voort, F., Schaye, J., Booth, C. M., Haas, M. R., & Dalla Vecchia, C. 2011, MNRAS, 414, 2458.
Viel, M., Becker, G. D., Bolton, J. S., & Haehnelt, M. G. 2013, ArXiv e-prints.
Weiner, B. J. & Williams, T. B. 1996, AJ, 111, 1156
Weingartner, J. C. & Draine, B. T. 2001, ApJ, 548, 296
Wiersma, R. P. C., Schaye, J., & Smith, B. D. 2009, MNRAS, 393, 99
Wolcott-Green, J., Haiman, Z., & Bryan, G. L. 2011, MNRAS, 18, 838
Wolfire, M.G., Tielens, A.G.G.M., Hollenbach, D., & Kaufman, M.J. (2008), ApJ, 680, 384–397
Wong, T. & Blitz, L. 2002, ApJ, 569, 157
Springel, V., 2010, MNRAS, 401, 791
Springel, V., 2005, MNRAS, 364, 1105

Physical Processes in the Interstellar Medium

Ralf S. Klessen and Simon C.O. Glover

1 Introduction

Understanding the physical processes that govern the dynamical behavior of the interstellar medium (ISM) is central to much of modern astronomy and astrophysics. The ISM is the primary galactic repository out of which stars are born and into which they deposit energy, momentum and enriched material as they die. It constitutes the anchor point of the galactic matter cycle, and as such is the key to a consistent picture of galaxy formation and evolution. The dynamics of the ISM determines where and when stars form. Similarly, the properties of the planets and planetary systems around these stars are intimately connected to the properties of their host stars and the details of their formation process.

When we look at the sky on a clear night, we can notice dark patches of obscuration along the band of the Milky Way. These are clouds of dust and gas that block the light from distant stars. With the current set of telescopes and satellites we can observe dark clouds at essentially all frequencies possible, ranging from low-energy radio waves all the way up to highly energetic γ-rays. We have learned that all star formation occurring in the Milky Way and other galaxies is associated with these dark clouds that mostly consist of cold molecular hydrogen and dust. In general, these dense clouds are embedded in and dynamically connected to the larger-scale and less dense atomic component. Once stellar birth sets in, feedback becomes important. Massive stars emit copious amounts of ultraviolet photons and create bubbles of hot ionized plasma, thus converting ISM material into a hot and very tenuous state.

We shall see in this lecture that we cannot understand the large-scale dynamics of the ISM without profound knowledge of the underlying microphysics. And vice

R.S. Klessen (✉) · S.C.O. Glover
Zentrum Für Astronomie der Universität Heidelberg, Heidelberg, Germany
e-mail: klessen@uni-heidelberg.de

S.C.O. Glover
e-mail: glover@uni-heidelberg.de

Y. Revaz et al. (eds.), *Star Formation in Galaxy Evolution: Connecting Numerical Models to Reality*, Saas-Fee Advanced Course 43,
DOI 10.1007/978-3-662-47890-5_2

versa, we will argue that dynamical processes on large galactic scales determine the local properties of the different phases of the ISM, such as their ability to cool and collapse, and to give birth to new stars. ISM dynamics spans a wide range of spatial scales, from the extent of the galaxy as a whole down to the local blobs of gas that collapse to form individual stars or binary systems. Similarly, it covers many decades in temporal scales, from the hundreds of millions of years it takes to complete one galactic rotation down to the hundreds of years it takes an ionization front to travel through a star-forming cloud. This wide range of scales is intricately linked by a number of competing feedback loops. Altogether, characterizing the ISM is truly a multi-scale and multi-physics problem. It requires insights from quantum physics and chemistry, as well as knowledge of magnetohydrodynamics, plasma physics, and gravitational dynamics. It also demands a deep understanding of the coupling between matter and radiation, together with input from high-resolution multi-frequency and multi-messenger astronomical observations.

By mass, the ISM consists of around 70 % hydrogen (H), 28 % helium (He), and 2 % heavier elements. The latter are generally termed metals in the sometimes very crude astronomical nomenclature. We give a detailed account of the composition of the ISM in Sect. 2. Because helium is chemically inert, it is customary, and indeed highly practical, to distinguish the different phases of the ISM by the chemical state of hydrogen. Ionized bubbles are called HII regions, while atomic gas is often termed HI gas, in both cases referring to the spectroscopic notation. HII regions are best observed by looking at hydrogen recombination lines or the fine structure lines of ionized heavy atoms. The properties of HI gas are best studied via the 21 cm hyperfine structure line of hydrogen. Dark clouds are sufficiently dense and well-shielded against the dissociating effects of interstellar ultraviolet radiation to allow H atoms to bind together to form molecular hydrogen (H_2). They are therefore called molecular clouds.

H_2 is a homonuclear molecule. Its dipole moment vanishes and it radiates extremely weakly under normal Galactic ISM conditions. Direct detection of H_2 is therefore generally possible only through ultraviolet absorption studies. Due to atmospheric opacity these studies can only be done from space, and are limited to pencil-beam measurements of the absorption of light from bright stars or active galactic nuclei (AGN). We note that rotational and ro-vibrational emission lines from H_2 have indeed been detected in the infrared, both in the Milky Way and in other galaxies. However, this emission comes from gas that has been strongly heated by shocks or radiation, and it traces only a small fraction of the overall amount of molecular hydrogen. Due to these limitations, the most common tool for studying the molecular ISM is radio and sub-millimeter emission either from dust grains or from other molecules that tend to be found in the same locations as H_2. By far the most commonly used molecular tracer is carbon monoxide with its various isotopologues. The most abundant, and hence easiest to observe is $^{12}C^{16}O$, usually referred to simply as ^{12}CO or just CO. However, this isotopologue is often so abundant that its emission is optically thick, meaning that it only traces conditions reliably in the surface layers of the dense substructure found within most molecular clouds. The next most abundant isotopologues are $^{13}C^{16}O$ (usually written simply as ^{13}CO) and $^{12}C^{18}O$ (usually just

$C^{18}O$). Their emission is often optically thin and can freely escape the system. This allows us to trace the full volume of the cloud. As CO has a relatively low critical density and also freezes out on dust grains at very high densities, other tracers such as HCN or N_2H^+ need to be used to study conditions within high density regions such as prestellar cores. We discuss the microphysics of the interaction between radiation and matter and the various heating and cooling processes that determine the thermodynamic response of the various phases of the ISM in Sect. 3.

A key physical agent controlling the dynamical evolution of the ISM is turbulence. For a long time it was thought that supersonic turbulence in the interstellar gas could not produce significant compressions, since this would result in a rapid dissipation of the turbulent kinetic energy. In order to avoid this rapid dissipation of energy, appeal was made to the presence of strong magnetic fields in the clouds, which were thought to greatly reduce the dissipation rate. However, it was later shown in high-resolution numerical simulations and theoretical stability analyses that magnetized turbulence dissipates energy at roughly the same rate as hydrodynamic turbulence. In both cases, the resulting density structure is highly inhomogeneous and intermittent in time. Today, we think that ISM turbulence plays a dual role. It is energetic enough to counterbalance gravity on global scales, but at the same time it may provoke local collapse on small scales. This apparent paradox can be resolved when considering that supersonic turbulence establishes a complex network of interacting shocks, where converging flows generate regions of high density. These localized enhancements can be sufficiently large for gravitational instability to set in. The subsequent evolution now depends on the competition between collapse and dispersal. The same random flows that create high-density regions in the first place may also destroy them again. For local collapse to result in the formation of stars, it must happen rapidly enough for the region to decouple from the flow. Typical collapse timescales are found to be comparable to dispersal times of shock-generated density fluctuations in the turbulent gas. This makes the outcome highly unpredictable and theoretical models are based on stochastic theory. In addition, supersonic turbulence dissipates quickly and so needs to be continuously driven for the galaxy to reach an approximate steady state. Finding and investigating suitable astrophysical processes that can drive interstellar turbulence remains a major challenge. We review the current state of affairs in this field in Sect. 4.

We think that molecular clouds form by a combination of turbulent compression and global instabilities. This process connects large-scale dynamics in the galaxy with the localized transition from warm, tenuous, mostly atomic gas to a dense, cold, fully molecular phase. The thermodynamics of the gas, and thus its ability to respond to external compression and consequently to go into collapse, depends on the balance between heating and cooling processes. Magnetic fields and radiative processes also play an important role. The chemical reactions associated with the transition from H to H_2, the importance of dust shielding, and the relation between molecular cloud formation and the larger galactic context are discussed in Sect. 5.

These clouds constitute the environment where new stars are born. The location and the mass growth of young stars are therefore intimately coupled to the dynamical properties of their parental clouds. Stars form by gravitational collapse of shock-compressed density fluctuations generated from the supersonic turbulence ubiquitously observed in molecular clouds. Once a gas clump becomes gravitationally unstable, it begins to collapse and its central density increases considerably until a new stars is born. Altogether, star formation in molecular clouds can be seen as a two-phase process. First, supersonic turbulence creates a highly transient and inhomogeneous molecular cloud structure that is characterized by large density contrasts. Some of the high-density fluctuations are transient, but others exceed the critical mass for gravitational contraction, and hence begin to collapse. Second, the collapse of these unstable cores leads to the formation of individual stars and star clusters. In this phase, a nascent protostar grows in mass via accretion from the infalling envelope until the available gas reservoir is exhausted or stellar feedback effects become important and remove the parental cocoon. In Sect. 6, we discuss the properties of molecular cloud cores, the statistical characteristics of newly born stars and star clusters, and our current theoretical models of dynamical star formation including the distribution of stellar masses at birth.

Finally, we conclude these lecture notes with a short summary in Sect. 7.

2 Composition of the ISM

2.1 Gas

The gas in the ISM is composed almost entirely of hydrogen and helium, with hydrogen accounting for around 70 % of the total mass, helium for 28 %, and all other elements for the remaining 2 %. The total gas mass in the Milky Way is difficult to estimate, but is probably close to $10^{10}\,M_{\odot}$ (Kalberla and Kerp 2009). The majority of the volume of the ISM is occupied by ionized gas, but the total mass associated with this component is not more than around 25 % of the total gas mass. The majority of the mass is located in regions dominated by neutral atomic gas (H, He) or molecular gas (H_2). Much of the atomic gas and all of the molecular gas is found in the form of dense clouds that occupy only 1–2 % of the total volume of the ISM.

The thermal and chemical state of the ISM is conventionally described in terms of a number of distinct phases. An early and highly influential model of the phase structure of the ISM was put forward by Field et al. (1969), who showed that if one assumes that the atomic gas in the ISM is in thermal equilibrium, then there exists a wide range of pressures for which there are two thermally stable solutions: one corresponding to cold, dense gas with $T \sim 100$ K that we can identify with the phase now known as the Cold Neutral Medium (CNM), and a second corresponding to warm, diffuse gas with $T \sim 10^4$ K that we can identify with the phase now known as the Warm Neutral Medium (WNM). In the Field et al. (1969) model, gas at intermediate temperatures is thermally unstable and depending on its density will either cool down and increase its density until it joins the CNM, or heat up and reduce its density until it joins the WNM.

This two-phase model of the ISM was extended by McKee and Ostriker (1977), who pointed out that supernovae exploding in the ISM would create large, ionized bubbles filled with very hot gas ($T \sim 10^6$ K). Although this gas would eventually cool, the temperature dependence of the atomic cooling curve at high temperatures is such that the cooling time around $T \sim 10^6$ K is considerably longer than the cooling time in the temperature range $10^4 < T < 10^6$ K (see Sect. 3.4 below). Therefore, rather than this hot gas having a wide range of temperatures, one would instead expect to find most of it close to 10^6 K. This hot, ionized phase of the ISM has subsequently become known as the Hot Ionized medium (HIM).

Evidence for an additional phase, the so-called Warm Ionized Medium (WIM), comes from a variety of observations, including free-free absorption of the Galactic synchrotron background (Hoyle and Ellis 1963), the dispersion of radio signals from pulsars (Reynolds 1989; Gaensler et al. 2008), and faint optical emission lines produced by ionized species such as O^+ and N^+ (Reynolds et al. 1973; Mierkiewicz et al. 2006). This ionized phase has a density comparable to that of the WNM, and has a scale-height of the order of 1 kpc (see e.g. Reynolds 1989). Its volume filling factor is relatively small in regions close to the Galactic midplane, but increases significantly as one moves away from the midplane (see e.g. Gaensler et al. 2008). Overall, 90 % or more of the total ionized gas within the ISM is located in the WIM (Haffner et al. 2009). It should be noted that the gas in classical HII regions surrounding O stars is generally not considered to be part of the WIM.

Finally, a distinction is often drawn between the dense, molecular phase of the ISM, observed to be distributed in the form of discrete molecular clouds of various masses and sizes (see e.g. Blitz et al. 2007) and the lower density, cold atomic gas surrounding these clouds, which is part of the CNM. The distribution of this molecular gas in our Galaxy is of particular interest, as star formation is observed to correlate closely with the presence of molecular gas. The distribution of molecular gas with Galactocentric radius can be measured by combining data from CO observations, which trace clouds with high concentrations of both H_2 and CO, and C^+ observations, which trace so-called "dark molecular gas", i.e. clouds with high H_2 fractions but little CO (see e.g. Pineda et al. 2013). The molecular gas surface density shows a pronounced peak within the central 500 pc of the Galaxy, a region known as the Central Molecular Zone (CMZ). It then falls off sharply between 0.5 and 3 kpc, possibly owing to the influence of the Milky Way's central stellar bar (Morris and Serabyn 1996), before peaking again at a Galactocentric radius of around 4–6 kpc in a structure known as the Molecular Ring. Outside of the Molecular Ring, the surface density of molecular gas declines exponentially, but it can still be traced out to distances of at least 12–13 kpc (Heyer et al. 1998).

An overview of the main physical properties of these different phases is given in Table 1. The information on the typical density and temperature ranges was taken from the review by Ferriére (2001), while the information on the typical fractional ionization of the various phases is based on Caselli et al. (1998); Wolfire et al. (2003), and Jenkins (2013).

Although gas in the ISM is often classified purely in terms of these five different phases, the question of how distinct these phases truly are remains open. For example,

Table 1 Phases of the ISM

Component	Temperature (K)	Density (cm^{-3})	Fractional ionization
Molecular gas	10–20	$>10^2$	$<10^{-6}$
Cold neutral medium (CNM)	50–100	20–50	$\sim 10^{-4}$
Warm neutral medium (WNM)	6000–10000	0.2–0.5	$\sim$0.1
Warm ionized medium (WIM)	$\sim$8000	0.2–0.5	1.0
Hot ionized medium (HIM)	$\sim 10^6$	$\sim 10^{-2}$	1.0

Adapted from Ferriére (2001); Caselli et al. (1998); Wolfire et al. (2003), and Jenkins (2013)

in the classical Field et al. (1969) model and the many subsequent models inspired by it, the CNM and WNM are two completely distinct phases in pressure equilibrium with each other, and all neutral atomic hydrogen in the ISM belongs to one phase or the other. However, observations of HI in the ISM suggest that the true picture is more complicated, as there is good evidence that a significant fraction of the atomic gas has a temperature intermediate between the CNM and WNM solutions, in the thermally unstable regime (Heiles and Troland 2003; Roy et al. 2013). This gas cannot be in equilibrium, and cannot easily be assigned to either the CNM or the WNM.

One important reason why this picture of the ISM appears to be an oversimplification is that the ISM is a highly turbulent medium. Turbulence in the ISM is driven by a number of different physical processes, including thermal instability (Kritsuk and Norman 2002a), supernova feedback (see e.g. Mac Low and Klessen 2004), and the inflow of gas onto the disk (Klessen and Hennebelle 2010; Elmegreen and Burkert 2010), and acts to mix together what would otherwise be distinct phases of the ISM (see e.g. Joung et al. 2009; Seifried et al. 2011). We discuss the role that turbulence plays in structuring the ISM together with the various driving mechanisms proposed at much greater length in Sect. 4.

Finally, it is useful to briefly summarize what we know about the metallicity of the ISM, i.e. of the fractional abundance of elements heavier than helium, since this plays an important role in regulating the thermal behavior of the ISM. In the Milky Way, the metallicity can be measured using a variety of methods (Maciel and Costa 2010). Measurements of the optical emission lines OII and OIII together with Hα and Hβ can be used to constrain the oxygen abundance in Galactic HII regions (see e.g. Deharveng et al. 2000), from which the total metallicity Z follows if we assume that the oxygen abundance scales linearly with Z. Alternatively, the abundances of carbon, nitrogen, oxygen and many other elements can be measured using ultraviolet (UV) absorption lines in the spectra of bright background stars (see e.g. Cowie and Songaila 1986; Savage and Sembach 1996; Sofia 2004). The metallicity can also be measured using stars, specifically by studying the spectra of young, massive B-type stars (see e.g. Rolleston et al. 2000). Technically, stellar measurements constrain the metallicity at the time that the star formed, rather than at the present day, but since B stars have short lifetimes, this distinction does not turn out to be particularly important in practice.

None of these techniques gives us a completely unbiased picture of the metallicity of the ISM. Emission line measurements are sensitive to the temperature distribution within the HII regions, which is difficult to constrain accurately. UV absorption line measurements are much less sensitive to excitation effects, but can only be carried out from space, and also require the presence of a UV-bright background source. Therefore, although they can give us information on the composition of the more diffuse phases of the ISM, including the WNM and CNM, they cannot be used to probe denser regions, such as molecular clouds, as the extinction in these regions is typically far too high for us to be able to detect the required background sources in the UV. In addition, these measurements tell us only about the gas-phase metals and not about the metals that are locked up in dust grains (see Sect. 2.2). Finally, stellar measurements provide us with good tracers of the total metallicity, but do not tell us how much of this was formerly in the gas phase, and how much was in dust.

Nevertheless, by combining the information provided by these different methods, we can put together a pretty good picture of the metallicity distribution of the gas in the ISM. Measurements of the metallicities of B stars and of HII regions both show that there is a large-scale radial metallicity gradient in the ISM, with a value of around $-0.04\ \mathrm{dex\ kpc^{-1}}$ (Maciel and Costa 2010). The metallicity of the gas in the CMZ is therefore around twice the solar value (Ferrière et al. 2007), while in the outer Galaxy, metallicities are typically somewhat sub-solar (Rudolph et al. 2006).

Comparison of the abundances of individual elements derived using B stars and those derived using UV absorption lines shows that most elements are depleted from the gas phase to some extent, a finding that we can explain if we suppose that these elements are locked up in interstellar dust grains. Support for this interpretation comes from the fact that the degree to which elements are depleted generally correlates well with their condensation temperature, i.e. the critical gas temperature below which a solid form is the favored equilibrium state for the elements (Lodders 2003). Elements with high condensation temperatures are more easily incorporated into dust grains than those with low condensation temperatures, and so if the observed depletions are due to dust formation, one expects the degree of depletion to increase with increasing condensation temperature, as observed (see e.g. Fig. 15 in Jenkins 2009). In addition, the values of high condensation temperature elements such as iron, nickel or silicon, also seem to correlate with the mean gas density (Jenkins 2009), and so are higher in the CNM than in the WNM (Welty et al. 1999). A plausible explanation of this fact is that dust growth in the ISM is an ongoing process that occurs more rapidly in cold, dense gas than in warm, diffuse gas (see e.g. Zhukovska et al. 2008).

2.2 Dust

The reddening of starlight in the ISM, and the fact that this effect correlates closely with the hydrogen column density rather than with distance, points towards there being an additional component of the ISM, responsible for absorbing light over a wide range of frequencies. Measurements of the strength of the absorption at different frequencies show that when there are distinct features in the extinction curve—e.g. the 217.5 nm bump—they tend to be extremely broad, quite unlike what we expect from atoms or small molecules. In addition, measurement of elemental abundances in the local ISM show that a number of elements, notably silicon and iron, are considerably less abundant in the gas-phase than in the Sun. Finally, mid-infrared and far-infrared observations show that there is widespread continuum emission, with a spectrum close to that of a black-body, and an intensity that once again correlates well with the hydrogen column density. Putting all of these separate pieces of evidence together, we are lead to the conclusion that in addition to the ionized, atomic and molecular constituents of the ISM, there must also be a particulate component, commonly referred to simply as dust.

Our best evidence for the nature of this dust comes from detailed measurements of the spectral shape of the extinction curve that it produces. To a first approximation, individual dust grains absorb only those photons with wavelengths smaller than the physical size of the grain. Therefore, the fact that we see a large amount of absorption in the ultraviolet, somewhat less in the optical and even less at infrared wavelengths tells us immediately that there are many more small dust grains than there are large ones. In addition, we can often associate particular spectral features in the extinction curve, such as the 217.5 nm bump or the infrared bands at 9.7 and 18 μm, with particular types of dust grain: graphite in the case of the 217.5 nm bump (Mathis et al. 1977) and amorphous silicates in the case of the infrared bands (e.g. Draine and Lee 1984; Draine and Li 2007); see also Fig. 1.

This argument can be made more quantitative, and has been used to derive detailed constraints on the size distribution of interstellar dust grains. One of the earlier and still highly influential attempts to do this was made by Mathis et al. (1977). They were able to reproduce the then-extant measurements of the ISM extinction curve between 0.1–1 μm with a mixture of spherical graphite and silicate grains with a size distribution

$$N(a)da \propto a^{-3.5} da \,, \qquad (1)$$

where a is the grain radius, and where the distribution extends over a range of radii from $a_{\rm min} = 50$ nm to $a_{\rm max} = 0.25$ μm. Subsequent studies have improved on this simple description (see e.g. Draine and Lee 1984; Weingartner and Draine 2001a), but it remains a useful guide to the properties of interstellar dust. In particular, it is easy to see that for grains with the size distribution given by Eq. (1)—commonly known as the MRN distribution—the total mass of dust is dominated by the contribution made by large grains, while the total surface area is dominated by the contribution made by small grains. This general behavior remains true in more recent models (see the detailed discussion by Draine 2011).

The total mass in dust is difficult to constrain purely with absorption measurements, but if we combine these with measurements of elemental depletion patterns in the cold ISM, then we can put fairly good constraints on how much dust there is. In the local ISM, we find that the total mass of metals locked up in grains is roughly

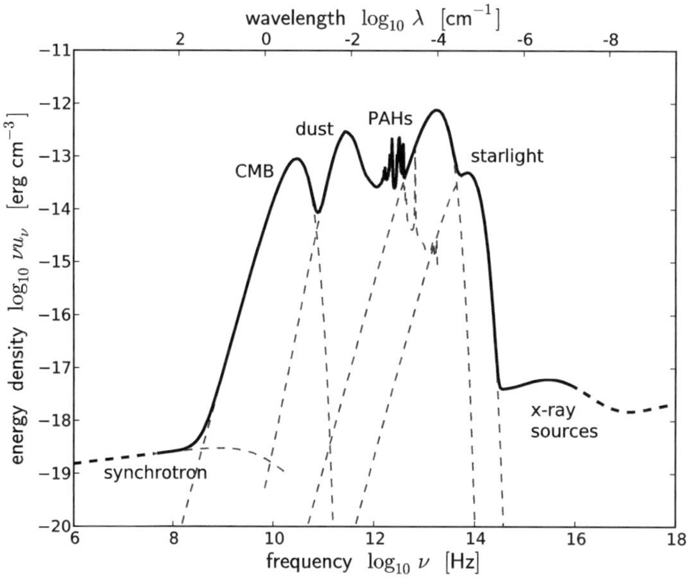

Fig. 1 Schematic sketch of the energy density of the interstellar radiation field at different frequencies. The contributions of the cosmic microwave background (CMB) as well as of old, low-mass and young, high-mass stars are taken to be perfect blackbodies with temperatures 2.73, 3500, and 18,000 K, respectively (see Chakraborty and Fields 2013). The contributions from dust and PAHs are obtained from Draine and Li (2007). The estimate for the Galactic synchrotron emission is taken from Draine (2011) and the one for the X-ray flux from Snowden et al. (1997). Note that in the vicinity of massive star clusters, the contributions from massive stars can be orders of magnitude larger than the numbers provided here. For further discussions, see for example Draine (2011)

the same as the total mass in the gas phase. The dust therefore accounts for around 1 % of the total mass of the ISM. Therefore, when we attempt to model the behavior of the ISM—particularly its thermal and chemical behavior—the dust can play a role that is as important or more important than the gas-phase metals (see e.g. Sect. 3.5).

2.3 Interstellar Radiation Field

The chemical and thermal state of the gas in the ISM is determined in large part by the interaction of the gas and the dust with the interstellar radiation field (ISRF). Several processes are important. First, the chemical state of the gas (the ionization fraction, the balance between atomic and molecular gas, etc.) depends on the rate at which molecules are photodissociated and atoms are photoionized by the radiation field. Second, the thermal state of the gas depends on the photoionization rate, and also on the rate of a process known as photoelectric heating: the ejection of an energetic electron from a dust grain due to the absorption of a UV photon by the grain. And

Table 2 Energy densities in different components of the ISRF

Component of ISRF	Energy density (erg cm^{-3})
Synchrotron	2.7×10^{-18}
CMB	4.19×10^{-13}
Dust emission	5.0×10^{-13}
Nebular emission (bf, ff)	4.5×10^{-15}
Nebular emission (Hα)	8×10^{-16}
Nebular emission (all other bb)	10^{-15}
Starlight, $T_1 = 3000$ K	4.29×10^{-13}
Starlight, $T_2 = 4000$ K	3.19×10^{-13}
Starlight, $T_3 = 7000$ K	2.29×10^{-13}
Starlight, power-law	7.11×10^{-14}
Starlight, total	1.05×10^{-12}
Soft X-rays	10^{-17}

Adapted from Draine (2011)

finally, the thermal state of the dust is almost entirely determined by the balance between the absorption by the grains of radiation from the ISRF and the re-emission of this energy in the form of thermal radiation.

In the solar neighborhood, the ISRF is dominated by six components, (1) galactic synchrotron emission from relativistic electrons, (2) the cosmic microwave background (CMB), (3) infrared and far-infrared emission from dust grains heated by starlight, (4) bound-bound (bb), bound-free (bf) and free-free (ff) emission from 10^4 K ionized plasma (sometimes referred to as nebular emission), (5) starlight, and finally (6) X-rays from hot (10^5–10^8 K) plasma. The energy densities of each of these components are summarized in Table 2 (adapted from Draine 2011); see also Fig. 1.

We see that most of the energy density of the ISRF is in the infrared, where thermal dust emission and the CMB dominate, and in the optical and UV, where starlight dominates. It is these components that play the main role in regulating the properties of the ISM, and so we focus on them below.

2.3.1 Cosmic Microwave Background

At wavelengths between $\lambda = 600$ μm and $\lambda = 30$ cm, the energy budget of the ISRF is dominated by the CMB. This has an almost perfect black-body spectrum with a temperature $T_{\rm CMB} = 2.725$ K (Fixsen and Mather 2002). This temperature is significantly lower than the typical temperatures of the gas and the dust in the local ISM, and so despite the high energy density of the CMB, energy exchange between it and these components does not substantially affect their temperature (Black 1994). The CMB therefore does not play a major role in the overall energy balance of the ISM in the Milky Way or in other local galaxies. In high-redshift galaxies, however, the CMB temperature and energy density are both much larger, with a redshift dependence of $T_{\rm CMB} \propto (1+z)$ and $u_{\rm CMB} \propto (1+z)^4$, respectively. The CMB can therefore play a much more significant role in regulating the thermal evolution of the gas and the dust. The extent to which this affects the outcome of the star formation process in high-redshift galaxies, and in particular the stellar initial mass function (IMF) remains very unclear. Some authors have suggested that as the CMB essentially imposes a temperature floor at $T_{\rm CMB}$, it can potentially affect the form of the IMF by suppressing low-mass star formation when the CMB temperature is large (see e.g. Clarke and Bromm 2003; Schneider and Omukai 2010). However, the observational evidence for a systematic change in the IMF as one moves to higher redshift remains weak (Bastian et al. 2010; Offner et al. 2014), and although some simulations find evidence that a high CMB temperature can suppress low-mass star formation (see e.g. Smith et al. 2009), other work suggests that low-mass stars can form even at very high redshifts (see e.g. Clark et al. 2011; Greif et al. 2011, 2012; Dopcke et al. 2013) by the fragmentation of the accretion disk surrounding the central star (see also Sect. 6.4).

2.3.2 Infrared and Far-Infrared Emission from Dust

Infrared emission from dust grains dominates the spectrum of the ISRF between $\lambda = 5$ and $\lambda = 600\,\mu$m. About two-thirds of the total power is radiated in the mid and far-infrared, at $\lambda > 50\,\mu$m. This emission is largely in the form of thermal emission from dust grains: the spectrum is that of a modified black-body

$$J_{\nu} \propto B_{\nu}(T_0) \left(\frac{\nu}{\nu_0} \right)^{\beta}, \tag{2}$$

where J_{ν} is the mean specific intensity of the radiation field, $B_{\nu}(T_0)$ is the Planck function, T_0 is the mean temperature of the dust grains, and β is the spectral index. In the Milky Way, the mean dust temperature $\langle T_{\rm d} \rangle \approx 20$ K, and the spectral index is typically around $\beta \approx 1.7$ (Planck Collaboration 2014). The question of whether β depends on temperature is somewhat controversial. Many studies find an apparent anti-correlation between β and $T_{\rm d}$ (see e.g. Dupac et al. 2003; Désert et al. 2008). However, these studies generally fit the spectral energy distribution (SED) with the χ^2 linear regression method, which is known to produce an artificial anti-correlation from uncorrelated data in some cases simply due to the presence of noise in the observations (Shetty et al. 2009a, b). When using hierarchical Bayesian methods for determining β and $T_{\rm d}$, Shetty et al. (2009b) and the Planck Collaboration (2014) instead find a slight positive correlation between the two parameters.

The remaining one-third of the dust emission is largely concentrated in a series of distinct peaks at wavelengths $\lambda = 3.3, 6.2, 7.7, 8.6, 11.3$ and $12.7\,\mu$m. These peaks correspond to vibrational emission bands produced by so-called polycyclic aromatic hydrocarbons, or PAHs for short. These are large organic molecules, containing one or more benzene rings (hence 'aromatic').

Although dust grains are large enough that we can usually treat them as macroscopic objects without distinct radiative transitions, the same is not true for the much smaller PAH molecules. The rate at which individual PAH molecules absorb photons is small, but each photon causes a significant change in the internal energy of the molecule. Their "temperature" therefore varies greatly with time—they are very hot (i.e. have a large internal energy) immediately after they absorb a photon, but spend much of their time being very cold. Physically, what happens is actually a form of fluoresence—the PAHs absorb UV photons, putting them into a highly excited state, and then cascade back to the ground state via a large number of infrared transitions. An important implication of this is that PAH emission dominates at short wavelengths (i.e. in the near and mid-infrared) unless the other grains are also very hot. Since the strength of the PAH emission depends on the strength of the UV radiation field, it is therefore a useful tracer of the formation of massive stars.

2.3.3 Starlight

Stars produce energy primarily at near infrared, visible and soft ultraviolet wavelengths. However, in neutral regions of the ISM, stellar photons with energies greater than the ionization energy of hydrogen, 13.6 eV, are largely absent—they are absorbed by hydrogen atoms, ionizing them, and hence cannot penetrate deeply into neutral regions.

Mathis et al. (1983) showed that in the solar neighborhood, the starlight component of the ISRF could be represented at long wavelengths as the sum of three diluted black-body spectra. At wavelengths $\lambda > 245$ nm, the radiation energy density is

$$\nu u_{\nu} = \sum_{i=1}^{3} \frac{8\pi h\nu^4}{c^3} \frac{W_i}{e^{h\nu/k_{\rm B}T_i} - 1} \ {\rm erg\ cm^{-3}} \,. \tag{3}$$

As usual $h = 6.626 \times 10^{-27}$ erg s and $k_{\rm B} = 1.381 \times 10^{-16}$ erg K^{-1} are Planck's and Boltzmann's constants. The quantities W_i and T_i are the dilution factor and temperature of each component, with

$$T_1 = 3000\ {\rm K}, \quad W_1 = 7.0 \times 10^{-13} \,, \tag{4}$$

$$T_2 = 4000\ {\rm K}, \quad W_2 = 1.65 \times 10^{-13} \,, \tag{5}$$

$$T_3 = 7500\ {\rm K}, \quad W_3 = 1.0 \times 10^{-14} \,. \tag{6}$$

At wavelengths $\lambda < 245$ nm, the starlight contribution to the ISRF has been estimated by a number of authors. The earliest widely-cited estimate was made by Habing (1968). He estimated that $\nu u_{\nu} \approx 4 \times 10^{-14}$ erg cm^{-3} at $\lambda = 100$ nm, corresponding to a photon energy of 12.4 eV. It is often convenient to reference other estimates to this value, which we do via the dimensionless parameter

$$\chi \equiv \frac{(\nu u_{\nu})_{100\mathrm{nm}}}{4 \times 10^{-14}\ \mathrm{erg\ cm^{-3}}}. \tag{7}$$

Alternatively, we can reference other estimates to the Habing (1968) field by comparing the total energy density in the range 6–13.6 eV. In this case, we define a different dimensionless parameter

$$G_0 \equiv \frac{u(6 - 13.6\ \mathrm{eV})}{5.29 \times 10^{-14}\ \mathrm{erg\ cm^{-3}}}. \tag{8}$$

If we are interested in e.g. the photodissociation of H_2 or CO, which requires photons with energies above 10 eV, then χ is the appropriate parameter to use. On the other hand, if we are interested in e.g. the photoelectric heating rate, which is sensitive to a wider range of photon energies, then G_0 is more appropriate.

Two other estimates of the UV portion of the ISRF are in widespread use: one due to Draine (1978) and the other due to Mathis et al. (1983). Draine (1978) fit the field with a polynomial function:

$$\lambda u_{\lambda} = 6.84 \times 10^{-14}\ \lambda_2^{-5} \left(31.016\ \lambda_2^2 - 49.913\ \lambda_2 + 19.897\right)\ \mathrm{erg\ cm^{-3}}, \tag{9}$$

where $\lambda_2 \equiv \lambda/100$ nm. This field has a normalization, relative to the Habing field, of $\chi = 1.71$ and $G_0 = 1.69$.

Mathis et al. (1983) instead used a broken power-law fit:

$$\lambda u_{\lambda} = \begin{cases} 2.373 \times 10^{-14} \lambda^{-0.6678} & \text{for } 0.134\ \mu\mathrm{m} < \lambda < 0.245\ \mu\mathrm{m} \\ 6.825 \times 10^{-13} \lambda & \text{for } 0.110\ \mu\mathrm{m} < \lambda \leq 0.134\ \mu\mathrm{m} \\ 1.287 \times 10^{-9} \lambda^{4.4172} & \text{for } 0.091\ \mu\mathrm{m} < \lambda \leq 0.110\ \mu\mathrm{m} \end{cases} \tag{10}$$

Here, all wavelengths are in units of μm, and the energy densities are in units of erg $\mathrm{cm^{-3}}$. This estimate has $\chi = 1.23$ and $G_0 = 1.14$. The available observational evidence (see e.g. Henry et al. 1980; Gondhalekar et al. 1980) is better fit by the Mathis et al. (1983) field than by the Draine (1978) field, but the latter estimate is probably in wider use in models of the ISM.

2.4 Cosmic Rays

The final part of our inventory of the ISM are the cosmic rays. These are high energy, relativistic particles, mostly being nuclei (~99 %) with a small fraction of electrons (~1 %). The nuclei are primarily protons, but with about 10 % being alpha particles and ~1 % being metal nuclei. Their energy spans a wide range, from 100 MeV up to more than 1 TeV (Fig. 2). The total energy density in cosmic rays is approximately $2\,\mathrm{eV\,cm^{-3}}$, within a factor of a few of the mean thermal energy density of the ISM. Cosmic rays therefore play an important role in the overall energy balance of the gas.

All but the most highly energetic cosmic rays are tied to the magnetic field of the galaxy and therefore scatter repeatedly within the disk. The expectation is therefore that the local energy density in cosmic rays should be relatively uniform. Observations of cosmic rays with TeV energies, which are not significantly affected by interactions with the solar wind, find that their intensity in the solar rest-frame is almost isotropic, consistent with this picture of a uniform energy density (Amenomori et al. 2006).

The spectrum of the cosmic rays (i.e. the flux per unit energy) decreases sharply with increasing energy, and so the majority of the heating and ionization that they provide comes from the least energetic cosmic rays, with energies of $\sim$100 MeV or below. Unfortunately, it is precisely this part of the cosmic ray energy spectrum that we know the least about. At this energy, cosmic rays are unable to penetrate within the heliosphere, owing to interactions with the solar wind. Our determination of the cosmic ray ionization rate is therefore indirect, based on chemical constraints.

An important example of this kind of constraint is provided by the abundance of the $\mathrm{H_3^+}$ ion. It is formed in the diffuse ISM via the reaction chain

$$\mathrm{H_2 + CR \rightarrow H_2^+ + e^-}, \quad (11)$$

$$\mathrm{H_2^+ + H_2 \rightarrow H_3^+ + H}, \quad (12)$$

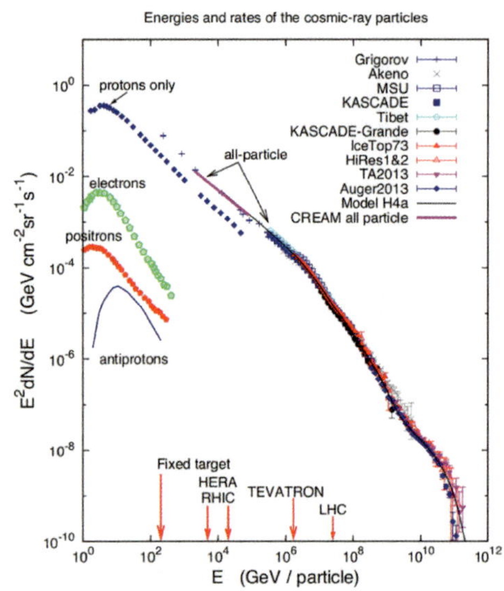

Fig. 2 Energy spectrum of cosmic rays as observed with different instruments and telescopes. Plot from Blasi (2014); see also Gaisser (2006) for further information

where the first reaction is the rate-limiting step. It is destroyed by dissociative recombination,

$$H_3^+ + e^- \rightarrow \text{various products.} \tag{13}$$

In a diffuse cloud, with $n_{H_2} > n_H$, the equilibrium number density of H_3^+ that results from these reactions is given approximately by

$$n_{H_3^+} = \frac{\zeta_{H_2}}{k_{dr}} \frac{n_{H_2}}{n_e}, \tag{14}$$

where k_{dr} is the dissociative recombination rate coefficient and ζ_{H_2} is the cosmic ray ionization rate of H_2, and where n_H, n_{H_2}, $n_{H_3^+}$, and n_e are the number densities of H, H_2, H_3^+, and free electrons, respectively.

If we assume that the temperature and the values of the H_2-to-electron ratio do not vary greatly along the line of sight, then we can convert Eq. (14) into an expression relating the column densities of H_2, H_3^+ and electrons,

$$N_{H_3^+} = \frac{\zeta_{H_2}}{k_{dr}} \frac{N_{H_2}}{N_e}. \tag{15}$$

Next, we note that within diffuse molecular clouds, the main source of electrons is ionized carbon, C^+. We therefore assume that $N_{C^+} = N_e$ and write

$$N_{H_3^+} = \frac{\zeta_{H_2}}{k_{dr}} \frac{N_{H_2}}{N_{C^+}}. \tag{16}$$

Finally, this can be rearranged to give

$$\zeta_{H_2} = \frac{N_{H_3^+} N_{C^+} k_{dr}}{N_{H_2}}. \tag{17}$$

Since all of the column densities on the right-hand side of this expression can be measured observationally (see e.g. Savage et al. 1977; Cardelli et al. 1996; McCall et al. 2002), and k_{dr} can be measured experimentally (McCall et al. 2004), we can use this expression to constrain ζ_{H_2}.

In practice, a slightly more sophisticated version of this technique is used that accounts for the fact that not all of the H_2^+ ions produced by cosmic ray ionization of H_2 survive for long enough to form H_3^+, and that also includes several additional destruction processes for H_3^+ (Indriolo and McCall 2012). Applying this technique to many lines-of-sight in the diffuse ISM, one finds that the resulting mean value for the cosmic ray ionization rate of H_2 is $\zeta_{H_2} = 3.5 \times 10^{-16}\ \text{s}^{-1}$, but also that there is a very substantial scatter around this mean, with some lines-of-sight having $\zeta_{H_2} \sim 10^{-15}\ \text{s}^{-1}$ or more, and others having $\zeta_{H_2} \sim 10^{-17}\ \text{s}^{-1}$ (Indriolo and McCall 2012).

In dense clouds of gas, the chemical abundances of other observable species such as OH or HCO^+ are also sensitive to the cosmic ray ionization rate, and measurements of the abundances of these species relative to CO can therefore be used to provide additional constraints on ζ_{H_2}. These other techniques typically find that in dense gas, $\zeta_{H_2} \sim 10^{-17}\ s^{-1}$ (see e.g. Williams et al. 1998; van der Tak and van Dishoeck 2000). This is consistent with the low end of the range of values found using H_3^+, but not with the higher values found in many diffuse clouds. This difference between diffuse and dense clouds may indicate that the cosmic rays that dominate the heating and ionization of the local ISM have energies of only a few MeV, allowing them to penetrate low column density, diffuse atomic or molecular gas, but not high column density clouds (Padovani et al. 2009). Alternatively, purely magnetic effects, such as the interaction between low energy cosmic rays and their self-generated MHD waves (Padoan and Scalo 2005) may explain the apparent inability of low energy cosmic rays to travel into dense molecular cloud regions. In either case, these mechanisms do not explain the large scatter in ζ_{H_2} seen in purely diffuse clouds, which may indicate that the energy density in very low energy cosmic rays is significantly less uniform than has been previously supposed (Indriolo and McCall 2012).

3 Heating and Cooling of Interstellar Gas

3.1 Optically-Thin Two-Level Atom

A convenient starting point for understanding how radiative cooling operates in the ISM is the two-level atom. This toy model allows us to illustrate many of the most important concepts involved without unnecessarily complicating the mathematical details.

Picture an atomic system with two bound states, a lower level l and an upper level u, with statistical weights g_l and g_u, separated by an energy E_{ul}. We will write the number density of the atoms in the lower level as n_l and the number density in the upper level as n_u. The total number density of the atoms then follows as $n_{\rm atom} = n_l + n_u$. For simplicity, we consider for the time being monoatomic gases at rest, so that $dn_{\rm atom}/dt = 0$. Even if the number density $n_{\rm atom}$ remains constant, the values of n_l and n_u will change over time, as individual atoms transition from the lower to the upper level due to collisional excitation or the absorption of a photon, and transition from the upper to the lower level due to collisional de-excitation or the emission of a photon. For the time being, we consider optically thin conditions, which means that the emitted photon can leave the region of interest unimpeded. In the opposite, optically thick case, it would very likely be absorbed by a neighboring atom.

We note that in general, the atomic species under consideration is just one amongst many. Typically, the medium is a mixture of different atomic or molecular species i, with the total number density being $n = \sum n_i$. In this case our atoms will not only collide with each other, but also with particles of the other species. Chemical reactions lead to further complications as the total number (and consequently the number density) of atoms is no longer conserved, but instead may change with time. If we also consider gas motions, even for monoatomic gases the local density will vary with time, so that again $dn_{\rm atom}/dt \neq 0$.

For a static system without chemical reactions, we can write the rates of change of n_l and n_u at fixed $n_{\rm atom}$ as

$$\frac{dn_u}{dt} = C_{lu}n_l - C_{ul}n_u - A_{ul}n_u - B_{ul}I_{ul}n_u + B_{lu}I_{ul}n_l\ , \tag{18}$$

$$\frac{dn_l}{dt} = -C_{lu}n_l + C_{ul}n_u + A_{ul}n_u + B_{ul}I_{ul}n_u - B_{lu}I_{ul}n_l\ . \tag{19}$$

Here, C_{lu} and C_{ul} are the collisional excitation and de-excitation rates, which we discuss in more detail below, A_{ul}, B_{ul} and B_{lu} are the three Einstein coefficients for the transition (describing spontaneous emission, stimulated emission and absorption, respectively), and I_{ul} is the specific intensity of the local radiation field at the frequency $\nu_{ul} = E_{ul}/h$. For a detailed derivation of this equation and the parameters involved, see e.g. the textbook by Rybicki and Lightman (1986).

Typically, the radiative and/or collisional transitions occur rapidly compared to any of the other timescales of interest in the ISM, and so it is usually reasonable to assume that the level populations have reached a state of statistical equilibrium in which

$$\frac{dn_u}{dt} = \frac{dn_l}{dt} = 0. \tag{20}$$

In this case, n_l and n_u are linked by a single algebraic equation,

$$(C_{lu} + B_{lu}I_{ul})n_l = (C_{ul} + A_{ul} + B_{ul}I_{ul})\,n_u. \tag{21}$$

A further simplification that we can often make is to ignore the effects of the incident radiation field. This is justified if the gas is optically thin and the strength of the interstellar radiation field at the frequency ν_{ul} is small. In this regime, we have

$$C_{lu}n_l = (C_{ul} + A_{ul})\,n_u\ , \tag{22}$$

which we can rearrange to give

$$\frac{n_u}{n_l} = \frac{C_{lu}}{C_{ul} + A_{ul}}\ . \tag{23}$$

The collisional excitation rate C_{lu} describes the rate per atom at which collisions with other gas particles cause the atom to change its quantum state from level l to level u. In principle, collisions with any of the many different chemical species present in the ISM will contribute towards C_{lu}, but in practice, the main contributions come from only a few key species: H, H_2, H^+, He, and free electrons. We can write C_{lu} as a sum of the collisional excitation rates due to these species,

$$C_{lu} = \sum_i q^i_{lu} n_i \ , \tag{24}$$

where $i = \mathrm{H}, \mathrm{H_2}, \mathrm{H^+}, \mathrm{He}, \mathrm{e^-}$ and q^i_{lu} is the collisional excitation rate coefficient for collisions between our atom of interest and species i. The collisional excitation rate coefficients themselves can be computed using the tools of quantum chemistry or measured in laboratory experiments. Values for many atoms and molecules of astrophysical interest can be found in the LAMDA database[1] (Schöier et al. 2005).

Given the excitation rate C_{lu}, it is straightforward to obtain the de-excitation rate C_{ul} by making use of the principle of detailed balance. This states that in local thermal equilibrium (LTE), the rate at which collisions cause transitions from level l to level u must be the same as the rate at which they cause transitions from level u to level l. This is a consequence of microscopic reversibility, i.e. the fact that the microscopic dynamics of particles and fields are time-reversible, because the governing equations are symmetric in time.

In thermal equilibrium, we know that the ratio of atoms in level u to those in level l is simply given by the Boltzmann distribution,

$$\frac{n_u}{n_l} = \frac{g_u}{g_l} e^{-E_{ul}/k_{\mathrm{B}}T} \ . \tag{25}$$

The principle of detailed balance tells us that for any collisional transition the equilibrium condition reads as

$$q^i_{lu} n_l n_i = q^i_{ul} n_u n_i \ . \tag{26}$$

It therefore follows that

$$C_{lu} n_l = C_{ul} n_u \ , \tag{27}$$

and consequently, we can write

$$\frac{C_{lu}}{C_{ul}} = \frac{n_u}{n_l} = \frac{g_u}{g_l} e^{-E_{ul}/k_{\mathrm{B}}T} \tag{28}$$

for a system in thermal equilibrium. The true power of the principle of detailed balance becomes clear once we realize that the values of C_{lu} and C_{ul} depend only on the quantum mechanical properties of our atoms, and not on whether our collection of atoms actually is in thermal equilibrium or not. Therefore, although we have assumed thermal equilibrium in deriving Eq. (28), we find that the final result holds even when the system is not in equilibrium.

---

[1] http://home.strw.leidenuniv.nl/~moldata/.

We can use this relation between C_{lu} and C_{ul} to write our expression for n_u/n_l in the form

$$\frac{n_u}{n_l} = \frac{(g_u/g_l)e^{-E_{ul}/k_B T}}{1 + A_{ul}/C_{ul}} . \tag{29}$$

In the limit that collisions dominate the behavior, i.e. for $C_{ul} \gg A_{ul}$, we recover the Boltzmann distribution. On the other hand, if radiative de-excitation is more important than the collisional one, that is in the limit $C_{ul} \ll A_{ul}$, we find that

$$\frac{n_u}{n_l} \approx \frac{C_{ul}}{A_{ul}} \frac{g_u}{g_l} e^{-E_{ul}/k_B T} . \tag{30}$$

Together with Eq. (28) we arrive at

$$\frac{n_u}{n_l} \approx \frac{C_{lu}}{A_{ul}} . \tag{31}$$

We see therefore that when collisions dominate over radiative decays, the level populations approach their LTE values, while in the other limit, collisional excitations are balanced by radiative de-excitations, and collisional de-excitations are unimportant.

In the simple case in which collisions with a single species dominate C_{ul}, we can write the collisional de-excitation rate as $C_{ul} = q^i_{ul} n_i$, where n_i is the number density of the dominant collision partner. Since the key parameter that determines whether collisions or radiative decays dominate is the ratio A_{ul}/C_{ul}, we can define a critical density for the collision partner, such that this ratio is one,

$$n_{\mathrm{cr},i} \equiv \frac{A_{ul}}{q^i_{ul}} . \tag{32}$$

When $n_i \gg n_{\mathrm{cr},i}$, collisions dominate and the level populations tend to their LTE values. On the other hand, when $n_i \ll n_{\mathrm{cr},i}$, radiative decay dominates and most atoms are in their ground states.

In the more general case in which collisions with several different species make comparably large contributions to C_{ul}, we can define the critical density in a more general fashion. If we take n to be some reference number density (e.g. the number density of H nuclei, which has the benefit that it is invariant to changes in the ratio of atomic to molecular hydrogen), then we can define a critical density with the following expression,

$$\frac{A_{ul}}{C_{ul}} \equiv \frac{n_{\mathrm{cr}}}{n} . \tag{33}$$

Here, n_{cr} is the critical value of our reference density, rather than that of a specific collision partner. In terms of the individual fractional abundances and collisional de-excitation rates, we have

$$n_{\mathrm{cr}} = \frac{A_{ul}}{\sum_c q^i_{ul} x_i} \tag{34}$$

where $x_i \equiv n_i/n$ is the relative abundance of the species i. Alternatively, if we divide through by A_{ul}, we can easily show that

$$n_{\rm cr} = \left[\sum_i \frac{x_i}{n_{{\rm cr},i}} \right]^{-1}, \tag{35}$$

where the critical densities for the individual colliders are given by Eq. (32) above.

Using our general definition of the critical density, we can write the ratio of the level populations of our two level atom as

$$\frac{n_u}{n_l} = \frac{(g_u/g_l)e^{-E_{ul}/k_{\rm B}T}}{1 + n_{\rm cr}/n}. \tag{36}$$

We can now use the fact that for our species of interest in the two-level approximation the density $n_{\rm atom} = n_l + n_u$, and rewrite this equation as

$$\frac{n_u}{n_{\rm atom} - n_u} = \frac{(g_u/g_l)e^{-E_{ul}/k_{\rm B}T}}{1 + n_{\rm cr}/n}. \tag{37}$$

Further rearrangement gives

$$\frac{n_u}{n_{\rm atom}} = \frac{(g_u/g_l)e^{-E_{ul}/k_{\rm B}T}}{1 + n_{\rm cr}/n + (g_u/g_l)e^{-E_{ul}/k_{\rm B}T}}. \tag{38}$$

The radiative cooling rate Λ_{ul} of our collection of atoms is simply the rate at which they emit photons multiplied by the energy of the photons, i.e.

$$\Lambda_{ul} = A_{ul} E_{ul} n_u. \tag{39}$$

If we make use of the expression derived above for n_u, this becomes

$$\Lambda_{ul} = A_{ul} E_{ul} n_{\rm atom} \frac{(g_u/g_l)e^{-E_{ul}/k_{\rm B}T}}{1 + n_{\rm cr}/n + (g_u/g_l)e^{-E_{ul}/k_{\rm B}T}}. \tag{40}$$

It is informative to examine the behavior of this expression in the limits of very low and very high density. At low densities, $n \ll n_{\rm cr}$, Eq. (40) reduces to

$$\Lambda_{ul,\, n\to 0} = A_{ul} E_{ul} n_{\rm atom} \frac{(g_u/g_l)e^{-E_{ul}/k_{\rm B}T}}{n_{\rm cr}/n}. \tag{41}$$

We can use the equation of detailed balance in the form (28) together with the definition of the critical density as given by Eq. (34) to derive the more useful expression

$$\Lambda_{ul,\,n\to 0} = E_{ul}\left(\sum_i q^i_{lu} n_i\right) n_{\rm atom} = E_{ul} C_{lu} n_{\rm atom} \ . \tag{42}$$

Physically, this expression has a simple interpretation. At low densities, every collisional excitation is followed by radiative de-excitation and hence by the loss of a photon's worth of energy from the gas. The cooling rate in this limit therefore depends only on the excitation rate of the atom, and is independent of the radiative de-excitation rate. Moreover, this rate is proportional to the total number density of the gas, $C_{lu} \propto n$, and in addition $n_{\rm atom} \propto n$, if the fractional abundance of our atomic coolant is independent of density. As a consequence, the cooling rate scales with the density squared in the low-density regime,

$$\Lambda_{ul,\,n\to 0} \propto n^2 \ . \tag{43}$$

The behavior is different at high densities, $n \gg n_{\rm cr}$. The expression for the cooling rate now becomes

$$\Lambda_{ul,\,\rm LTE} = A_{ul} E_{ul} n_{\rm atom} \left[\frac{(g_u/g_l) e^{-E_{ul}/k_{\rm B}T}}{1 + (g_u/g_l) e^{-E_{ul}/k_{\rm B}T}}\right] . \tag{44}$$

The term in square brackets is simply the fraction of all of the atoms that are in the upper level u when the system is in LTE, a quantity that we will refer to as $f_{u,\,\rm LTE}$. In this limit, we write

$$\Lambda_{ul,\,\rm LTE} = A_{ul} E_{ul} f_{u,\,\rm LTE} n_{\rm atom} \ . \tag{45}$$

This is known as the LTE limit. In this limit, the *mean* cooling rate per atom depends only on the temperature, and not on the collisional excitation rate. Consequently, the cooling rate scales linearly with the density,

$$\Lambda_{ul,\,\rm LTE} \propto n \ . \tag{46}$$

3.2 Effects of Line Opacity

So far, we have assumed that the strength of the local radiation field at the frequency of the atomic transition is negligible, allowing us to ignore the effects of absorption and stimulated emission. This is a reasonable approximation when the gas is optically thin, provided that the ISRF is not too strong, but it becomes a poor approximation once the gas becomes optically thick. Therefore, we now generalize our analysis to handle the effects of absorption and stimulated emission.

Consider once again our two-level atom, with level populations that are in statistical equilibrium. In this case, we have

$$(C_{lu} + B_{lu} J_{ul}) n_l = (A_{ul} + B_{ul} J_{ul} + C_{ul}) n_u, \tag{47}$$

where J_{ul} is the mean specific intensity,

$$J_{ul} = \frac{1}{4\pi} \oint I_{ul}(\mathbf{n}) d\Omega, \tag{48}$$

where the integral is over all directions $\mathbf{n}$ and $d\Omega$ is the solid angle element.

In general, to solve this equation throughout our medium, we need to know J_{ul} at every point, and since J_{ul} depends on the level populations, we end up with a tightly coupled problem that is difficult to solve even for highly symmetric systems, and that in general requires a numerical treatment. A detailed discussion of the different numerical methods that can be used to solve this optically-thick line transfer problem is outside the scope of our lecture notes. Instead, we refer the reader to the paper by van Zadelhoff et al. (2002) and the references therein.

Here, we restrict our attention to a simple but important limiting case. We start by assuming that any incident radiation field is negligible, and hence that the only important contribution to J_{ul} comes from the emission of the atoms themselves. We also assume that there are only three possible fates for the emitted photons:

(1) Local absorption, followed by collisional de-excitation of the atom[2]
(2) Local absorption, followed by re-emission (i.e. scattering)
(3) Escape from the gas.

Photons which scatter may do so once or many times, before either escaping from the gas, or being removed by absorption followed by collisional de-excitation. The probability that the photon eventually escapes from the gas is termed the escape probability. In its most general form, this can be written as

$$\beta(\mathbf{x}) = \frac{1}{4\pi} \oint \int e^{-\tau_\nu(\mathbf{x},\mathbf{n})} \phi(\nu) d\nu d\Omega, \tag{49}$$

where $\beta(\mathbf{x})$ is the escape probability at a position $\mathbf{x}$, $\phi(\nu)$ is the line profile function, a function normalized to unity that describes the shape of the line, and $\tau_\nu(\mathbf{x}, \mathbf{n})$ is the optical depth at frequency ν at position $\mathbf{x}$ in the direction $\mathbf{n}$.

We note that the net number of absorptions (i.e. the number of photons absorbed minus the number produced by stimulated emission) must equal the number of photons emitted that do not escape from the gas, i.e.

$$(n_l B_{lu} - n_u B_{ul}) J_{ul} = n_u (1 - \beta) A_{ul}. \tag{50}$$

Using this, we can rewrite Eq. (47) for the statistical equilibrium level populations as

$$C_{lu} n_l = (C_{ul} + \beta A_{ul}) n_u. \tag{51}$$

[2]By local, we generally mean within a small volume around the emission site, within which we can assume that physical conditions such as density and temperature do not vary appreciably.

Local absorptions reduce the effective radiative de-excitation rate by a factor determined by the escape probability β, i.e. we go from A_{ul} in the optically thin case to $A'_{ul} = \beta A_{ul}$ in the optically thick case. Therefore, all of our previously derived results still hold provided that we make the substitution $A_{ul} \rightarrow A'_{ul}$. One important consequence of this is that the critical density decreases. Since $n_{\rm cr} \propto A_{ul}$, we see that when the gas is optically thick, $n_{\rm cr} \propto \beta$. This means that the effect of local absorption (also known as photon trapping) is to lower the density at which LTE is reached. The higher the optical depth, the more pronounced this effect becomes.

In order for the escape probability approach to be useful in practice, we need to be able to calculate β in a computationally efficient fashion. Unfortunately, the expression for β given in Eq. (49) is not well suited for this. The reason for this is the dependence of β on the direction-dependent optical depth $\tau_{\nu}(\mathbf{x}, \mathbf{n})$. This can be written in terms of the absorption coefficient α_{ν} as

$$\tau_{\nu}(\mathbf{x}, \mathbf{n}) = \int_{0}^{\infty} \alpha_{\nu}(\mathbf{x} + s\mathbf{n}, \mathbf{n}) ds, \tag{52}$$

where $\alpha_{\nu}(\mathbf{x} + s\mathbf{n}, \mathbf{n})$ is the absorption coefficient at position $\mathbf{x} + s\mathbf{n}$ for photons propagating in the direction $\mathbf{n}$. To compute the integral over solid angle in Eq. (49), we need to integrate for τ_{ν} along many rays between the point of interest and the edge of the cloud. This can be done, but if we want to properly sample the spatial distribution of the gas, then the computational cost of performing these integrals will typically scale as $N^{2/3}$, where N is the number of fluid elements in our model cloud. If we then need to repeat this calculation for every fluid element (e.g. in order to calculate the cooling rate at every point in the cloud), the result is a calculation that scales as $N^{5/3}$. For comparison, modeling the hydrodynamical or chemical evolution of the cloud has a cost that scales as N.

Because of the high computational cost involved in computing β accurately, most applications of the escape probability formalism make further simplifications to allow β to be estimated more easily. One common approach is to simplify the geometry of the cloud model under consideration. For example, if we adopt a spherically symmetric or slab-symmetric geometry, the inherent dimensionality of the problem can be reduced from three to one, greatly speeding up our calculation of β. This approach can work very well in objects such as prestellar cores that are quasi-spherical, but becomes less applicable as we move to larger scales in the ISM, since real molecular clouds exhibit complex and highly inhomogeneous density and velocity structure (Sect. 4.1.2) and are not particularly well described as either slabs or spheres.

A more useful approximation for treating cooling in interstellar clouds is the Large Velocity Gradient (LVG) approximation, also known as the Sobolev approximation (Sobolev 1957). The basic idea here is that when there are large differences in the velocities of adjacent fluid elements, photons can more easily escape from the gas. Suppose a photon is emitted at position $\mathbf{x}$ from gas moving with velocity $\mathbf{v}$, and propagates a distance $\Delta\mathbf{x}$, to a point in the gas where the velocity is $\mathbf{v} + \Delta\mathbf{v}$. The probability of the photon being absorbed at this point depends on the frequency of the photon in the rest frame of the gas at that point. If the change in velocity is sufficient to have Doppler-shifted the photon out of the core of the line, the probability of it being absorbed is small. For lines dominated by thermal broadening, the required change in velocity is roughly equal to the thermal velocity of the absorber, $v_{\rm th}$. If the photon can successfully propagate a distance

$$L_{\rm s} \equiv \frac{v_{\rm th}}{|dv/dx|}, \tag{53}$$

where dv/dx is the velocity gradient, then it is extremely likely that it will escape from the gas. The length-scale defined by Eq. (53) is known as the Sobolev length, and the LVG approximation can be used successfully when this length-scale is significantly shorter than the length-scales corresponding to variations in the density, temperature or velocity of the gas, or in the fractional abundance of the absorbing species.

More quantitatively, when the Sobolev approximation applies, the integral over frequency can be solved analytically, yielding

$$\int e^{-\tau_\nu(\mathbf{x},\mathbf{n})}\phi(\nu)d\nu = \frac{1 - e^{-\tau_{\rm LVG}(\mathbf{x},\mathbf{n})}}{\tau_{\rm LVG}(\mathbf{x},\mathbf{n})}, \tag{54}$$

where $\tau_{\rm LVG}(\mathbf{x},\mathbf{n})$ is the direction-dependent LVG optical depth,

$$\tau_{\rm LVG}(\mathbf{x},\mathbf{n}) = \frac{A_{ul}c^3}{8\pi\nu_{ul}^3}\frac{1}{|\mathbf{n}\cdot\nabla v|}\left(\frac{g_l}{g_u}n_u - n_l\right). \tag{55}$$

In this case, β is given by the expression

$$\beta(\mathbf{x}) = \frac{1}{4\pi}\oint \frac{1 - e^{-\tau_{\rm LVG}(\mathbf{x},\mathbf{n})}}{\tau_{\rm LVG}(\mathbf{x},\mathbf{n})}d\Omega. \tag{56}$$

The validity of the Sobolev approximation for line transfer in turbulent molecular clouds was examined by Ossenkopf (1997), who showed that most of the fluid elements contributing significantly to the ^{13}CO line emission produced by a typical turbulent cloud had short Sobolev lengths, justifying the use of the LVG approximation for modeling the emission (see also Ossenkopf 2002).

3.3 Multi-level Systems

So far, we have restricted our discussion to the case of a simple two-level system. However, most of the important coolants in the ISM have more than two energy levels that need to be taken into account when computing the cooling rate, and so in this section, we briefly look at how we can generalize our analysis to the case of multiple levels.

When we are dealing with more than two levels, and hence more than a single transition contributing to the cooling rate, then the net cooling rate can be written in terms of the level populations as

$$\Lambda = \sum_{u} \sum_{l<u} E_{ul} \left[(A_{ul} + B_{ul} J_{ul}) n_u - B_{lu} J_{ul} n_l \right], \tag{57}$$

where the second sum is over all states l that have energies $E_l < E_u$. A major difficulty here comes from the need to compute the level populations n_u. If the levels are in statistical equilibrium, then the level populations satisfy the equation

$$\sum_{j>i} \left[n_j A_{ji} + (n_j B_{ji} - n_i B_{ij}) J_{ij} \right] - \sum_{j<i} \left[n_i A_{ij} + (n_i B_{ij} - n_j B_{ji}) J_{ij} \right] + \sum_{j \neq i} \left[n_j C_{ji} - n_i C_{ij} \right] = 0. \tag{58}$$

If we have N different levels, then this equation can also be written in the form of N coupled linear equations. These are straightforward to solve numerically if the mean specific intensities J_{ij} are known, but just as in the two-level case, these specific intensities will in general depend on the level populations at every point in our gas, meaning that the general form of the non-LTE statistical equilibrium equation is very challenging to solve numerically in an efficient manner.

Consequently, when computing cooling rates for complicated multi-level systems, we often make use of simplifying assumptions similar to those we have already discussed in the case of the two-level system. For example, when the gas density is very low, it is reasonable to assume that essentially all of our coolant atoms or molecules will be in the ground state, and that every collisional excitation from the ground state will be followed by the loss of a photon from the gas (possible after one or more scattering events, if the gas is optically thick). In this limit, the cooling rate simplifies to

$$\Lambda_{n \to 0} = \sum_{u} E_{u0} C_{u0} n_0, \tag{59}$$

where n_0 is the number density of coolant atoms/molecules in the ground state, C_{u0} is the collisional excitation rate from the ground state to state u, and E_{u0} is the difference in energy between state u and the ground state.

In the LTE limit, the cooling rate is also easy to calculate, as the level populations will simply have the values implied by the Boltzmann distribution,

$$n_u = \frac{(g_u/g_0) e^{-E_{u0}/k_B T}}{Z(T)} n_{\rm atom}, \tag{60}$$

where $n_{\rm atom}$ is the total number density of the coolant of interest, g_0 is the statistical weight of the ground state, g_u is the statistical weight of level u, E_{u0} is the energy difference between level u and the ground state, and Z is the partition function. It is given by the expression

$$Z(T) = \sum_i \frac{g_i}{g_0} e^{-E_{i0}/k_{\rm B}T}, \tag{61}$$

where we sum over all of the states of the coolant, including the ground state. In this limit, we still need to know the mean specific intensities of the various lines in order to calculate the total cooling rate. In principle, these are straightforward to compute when the level populations are fixed, although for reasons of computational efficiency, a further simplifying assumption such as the LVG approximation is often adopted.

In the case of our simple two-level system, we have already seen that at densities $n \ll n_{\rm crit}$, it is safe to use the low-density limit of the cooling rate, while at densities $n \gg n_{\rm crit}$, the LTE limit applies. In the multi-level case, the situation is slightly more complicated, as in principle there is a critical density $n_{{\rm crit},ul}$ associated with every possible transition that can occur. Moreover, since there can be large differences in the value of A_{ul} between one transition and another, these individual critical densities can differ by orders of magnitude. We can therefore often be in the situation where some of the energy levels of our coolant are in LTE, while others are not. In practice, what is often done if we are interested in the total cooling rate and not in the strengths of the individual lines is to define an effective critical density (Hollenbach and McKee 1979)

$$n_{\rm crit,eff} = \frac{\Lambda_{\rm LTE}}{\Lambda_{n\to 0}} n , \tag{62}$$

and write the density-dependent cooling rate as

$$\Lambda = \frac{\Lambda_{\rm LTE}}{1 + n_{\rm crit,eff}/n}. \tag{63}$$

This expression can be somewhat approximate at densities close to $n_{\rm crit,eff}$, but becomes very accurate in the limit of low density or high density.

3.4 Atomic and Molecular Coolants in the ISM

Having briefly outlined the basic physical principles of line cooling, we now go on to examine which of the many possible forms of line emission are most important for the cooling of interstellar gas.

3.4.1 Permitted Transitions

At high temperatures, in regions dominated by atomic or ionized gas, the cooling of the ISM takes place largely via the permitted (i.e. dipole-allowed) electronic transitions of various atoms and ions. At temperatures close to 10^4 K, excitation of the Lyman series lines of atomic hydrogen is the dominant process,[3] giving rise to a cooling rate per unit volume (Black 1981; Cen 1992) of

$$\Lambda_{\rm H} = 7.5 \times 10^{-19} \frac{1}{1 + (T/10^5)^{1/2}} \exp\left(\frac{-118348}{T}\right) n_{\rm e} n_{\rm H}, \tag{64}$$

where $n_{\rm e}$ and $n_{\rm H}$ are the number densities of free electrons and atomic hydrogen, respectively, and temperature T are in kelvin. At temperatures $T \sim 3 \times 10^4$ K and above, however, the abundance of atomic hydrogen generally becomes very small, and other elements, particularly C, O, Ne and Fe, start to dominate the cooling (see e.g. Gnat and Ferland 2012).

In conditions where collisional ionization equilibrium (CIE) applies, and where the fractional abundance of each ion or neutral atom is set by the balance between collisional ionization and radiative recombination, the total cooling function is relatively straightforward to compute and only depends on the temperature and metallicity. As an example, we show in Fig. 3 the CIE cooling efficiency, $\mathcal{L}_{\rm CIE}$, as a function of temperature, computed for a solar metallicity gas using the data given in Gnat and Ferland (2012). The cooling efficiency plotted in the figure has units of erg cm^3 s^{-1} and is related to the cooling rate per unit volume by the expression

$$\Lambda_{\rm CIE} = \mathcal{L}_{\rm CIE} n_{\rm e} n, \tag{65}$$

where $n_{\rm e}$ is the number density of free electrons and n is the number density of hydrogen nuclei.

Unfortunately, there are many cases in the real ISM in which the CIE assumption does not apply. As an obvious example, consider gas in the H II regions around massive stars, where the ionization state of the various elements is determined primarily by photoionization rather than collisional ionization. The CIE assumption also breaks down whenever the gas cools rapidly. When the cooling time becomes shorter than the recombination time the gas cannot adjust its ionization state rapidly enough to remain in equilibrium. Instead, it becomes over-ionized compared to the CIE case (see e.g. Kafatos 1973; Sutherland and Dopita 1993; Gnat and Sternberg 2007). Similarly, if gas is heated more rapidly than it can collisionally ionize itself, such as in a very strong shock, then it can be under-ionized compared to the CIE case. These non-equilibrium effects are particularly important around 10^4 K (see e.g. Micic et al. 2013; Richings et al. 2014), but can also significantly affect the cooling rate at higher temperatures.

---

[3]Note that this is often referred to in the literature simply as "Lyman-α" cooling.

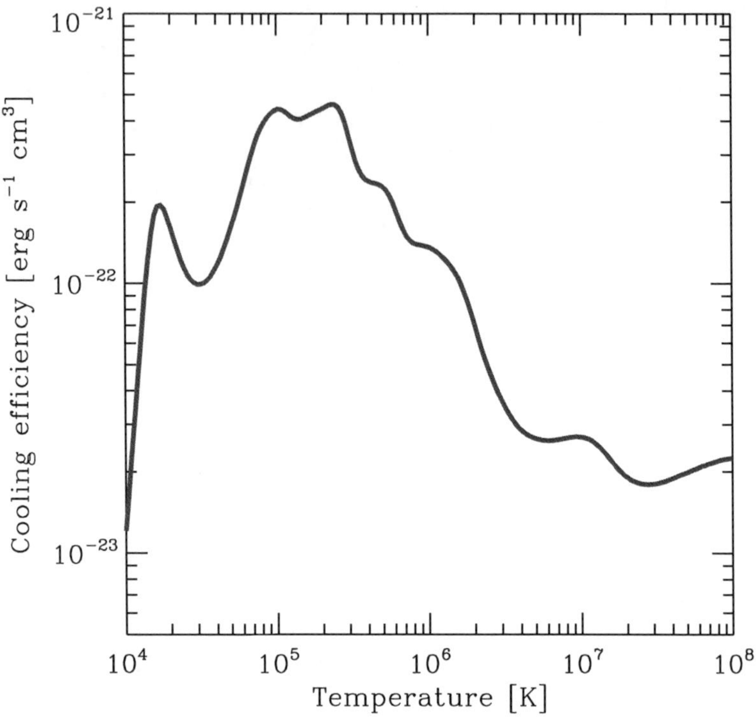

Fig. 3 Cooling efficiency of solar-metallicity gas in collisional ionization equilibrium, plotted as a function of temperature. This plot is based on data taken from Gnat and Ferland (2012)

Efforts have been made to account for these non-equilibrium effects, either by explicitly solving for the non-equilibrium ionization state of the main elements contributing to the high temperature cooling (see e.g. Cen and Fang 2006; de Avillez and Breitschwerdt 2012; Oppenheimer and Schaye 2013; Richings et al. 2014), or by pre-computing and tabulating rates appropriate for gas cooling at constant pressure or constant density (Gnat and Sternberg 2007), or with an ionization state dominated by photoionization rather than collisional ionization (e.g. Wiersma et al. 2009; Gnedin and Hollon 2012). In any case, there is inevitably a trade-off between accuracy and speed—full non-equilibrium calculations best represent the behavior of the real ISM but have a considerably larger computational cost than simple CIE-based calculations.

3.4.2 Fine Structure Lines

At temperatures below around 10^4 K, it becomes extremely difficult for the gas to cool via radiation from permitted atomic transitions, such as the Lyman series lines of atomic hydrogen, as the number of electrons available with sufficient energy to excite these transitions declines exponentially with decreasing temperature. Atomic cooling continues to play a role in this low temperature regime, but the focus now shifts from permitted transitions between atomic energy levels with different principal quantum numbers to forbidden transitions between different fine structure energy levels.

Fine structure splitting is a phenomenon caused by the interaction between the orbital and spin angular momenta of the electrons in an atom, an effect known as spin-orbit coupling (see e.g. Atkins and Friedman 2010). Each electron within an atom has a magnetic moment due to its orbital motion and also an intrinsic magnetic moment due to its spin. States where these magnetic moments are parallel have higher energy than states where they are anti-parallel, which in the right conditions can lead to a splitting of energy levels that would otherwise remain degenerate. In order for an atom or ion to display fine structure splitting in its ground state, the electrons in the outermost shell must have both non-zero total orbital angular momentum (i.e. $L > 0$) and non-zero total spin angular momentum (i.e. $S > 0$), or else the spin-orbit coupling term in the Hamiltonian, which is proportional to $\mathbf{L} \cdot \mathbf{S}$, will vanish. For example, the ground state of the hydrogen atom has $S = 1/2$ but $L = 0$, and hence has no fine structure. On the other hand, the ground state of neutral atomic carbon has $L = 1$ and $S = 1$ and hence does have fine structure.

As the size of the spin-orbit term in the Hamiltonian is typically quite small compared to the other terms, the resulting splitting of the energy levels is also small, with energy separations of the order of 10^{-2} eV. This corresponds to a temperature of about 100 K, meaning that it is possible to excite these transitions even at relatively low gas temperatures.

The radiative transition rates associated with these fine structure transitions are very small in comparison to the rates of the permitted atomic transitions discussed above, for a couple of reasons. First, they are typically magnetic dipole transitions, with transition matrix elements that are of the order of $\alpha^2 \approx 5 \times 10^{-5}$ times smaller than for electric dipole transitions, where α is the fine structure constant. Second, it is easy to show that transitions with similar transition matrix elements but different frequencies have spontaneous transition rates that scale as $A_{ij} \propto \nu_{ij}^3$. Since the frequencies associated with the fine structure transitions are of the order of a thousand times smaller than those associated with the most important permitted electronic transitions, such as Lyman-α, one expects the spontaneous transition rates to be a factor of 10^9 smaller.

Together, these two effects mean that we expect the size of the spontaneous transition rates associated with the fine structure transitions to be of the order of 10^{14} or more times smaller than those associated with the most important permitted atomic transitions. Consequently, the critical densities associated with many of the important fine structure transitions are relatively low: $n_{\rm crit} \sim 10^2$–10^6 cm^{-3} in conditions when collisions with H or H_2 dominate, and up to two to three orders of magnitude smaller when collisions with electrons dominate (Hollenbach and McKee 1989). We therefore would expect cooling from fine structure emission to be effective at moderate densities, e.g. in the WNM or CNM, but to become much less effective at the much higher densities found in gravitationally collapsing regions within molecular clouds.

As hydrogen and helium have no fine structure in their ground states, fine structure cooling in the ISM is dominated by the contribution made by the next most abundant elements, carbon and oxygen (Wolfire et al. 1995). In the diffuse ISM, carbon is found mainly in the form of C^+, as neutral atomic carbon can be photoionized by photons with energies $E > 11.26$ eV, below the Lyman limit. Singly ionized carbon has two fine structure levels in its ground state, an upper level with total angular momentum $J = 3/2$ and a lower level with total angular momentum $J = 1/2$. The energy separation between these two levels is approximately $E/k_B = 92$ K, and so this transition remains easy to excite down to temperatures of around 20 K (see Fig. 4).

In denser regions of the ISM, where dust provides some shielding from the effects of the ISRF, C^+ recombines to yield C. Atomic carbon has three fine structure levels, with total angular momenta $J = 0, 1, 2$ and energies relative to the ground state $\Delta E/k_B = 0.0, 23.6, 62.4$ K, respectively. The small energy separations of

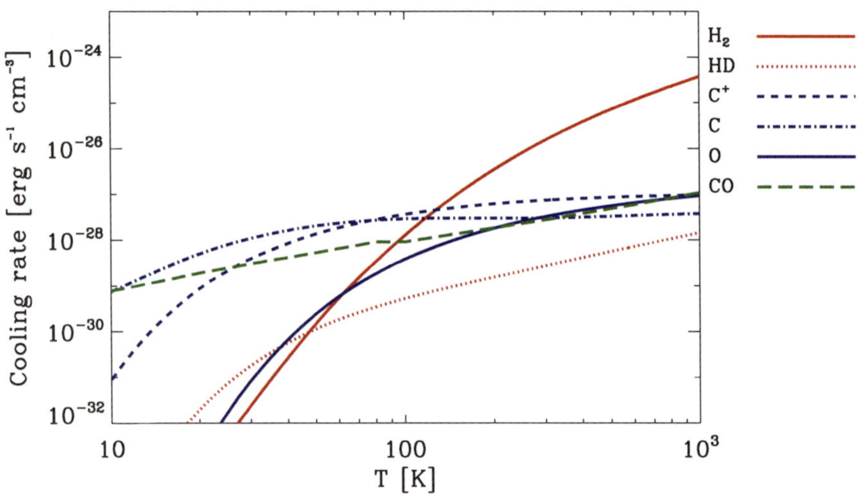

Fig. 4 Cooling rates for selected ISM coolants. The values plotted were computed assuming that $n = 1\,\mathrm{cm}^{-3}$ and are weighted by the fractional abundance (relative to the total number of hydrogen nuclei) of the coolant in question. For H_2 and HD, we assume that the gas is fully molecular, so that $x_{H_2} = 0.5$ and $x_{HD} = 2.5 \times 10^{-5}$. For the metals, we adopt total C and O abundances from Sembach et al. (2000), and show the cooling rate that we would have if all of the relevant element were in the form of the indicated species. In the case of CO, the adopted abundance is the total C abundance, since this is smaller than the total abundance of oxygen (Sembach et al. 2000). For cooling from H_2 and HD, we account for collisions with both H_2 and He, while for the other species, we account only for collisions with H_2, owing to a lack of data on the collision rates with He. The error introduced by omitting He is unlikely to exceed 10–20 %. The data on which these cooling rates are based is taken from Flower and Roueff (1998, 1999a), and Flower et al. (1998) for H_2, Flower and Roueff (1999b) and Roueff and Zeippen (1999) for HD, Wiesenfeld and Goldsmith (2014) for C^+, Schroder et al. (1991) and Warin et al. (1996) for C, Jaquet et al. (1992) for O, and Neufeld and Kaufman (1993); Neufeld et al. (1995); Flower (2001) and Wernli et al. (2006) for CO

these levels mean that neutral atomic carbon remains an effective coolant down to very low temperatures. Indeed, in the low density limit, neutral atomic carbon is a more effective coolant than CO (Fig. 4), although it becomes ineffective at densities $n \gg 100\,\mathrm{cm}^{-3}$, owing to its low critical density.

The ionization energy of neutral oxygen is very similar to that of hydrogen, and so in the WNM and CNM, oxygen is present largely in neutral atomic form. Neutral oxygen also has fine structure in its ground state. In this case, there are three fine structure levels, with total angular momenta $J = 0, 1, 2$ and energies relative to the ground state $\Delta E/k_{\rm B} = 0.0, 227.7, 326.6$ K, respectively. The larger energy separation of these levels compared to the C^{+} fine structure levels means that in the CNM, C^{+} cooling is considerably more effective than cooling from oxygen, despite the larger abundance of oxygen relative to carbon (see e.g. Wolfire et al. 1995). In warmer gas, however, carbon and oxygen are both important coolants.

In gas with standard solar metallicity, other metals such as N, Ne, Si, Fe and S also have relatively high abundances, but in practice they do not contribute significantly to the cooling of the ISM. Nitrogen and neon are present in the WNM and CNM primarily in neutral form, and have no fine structure in their ground state in this form. Silicon and iron do have ground state fine structure, but are strongly depleted in the ISM, particularly in the colder and denser phases (Jenkins 2009). Finally, sulfur is present primarily in the form of S^{+}, which has no fine structure in its ground state.

Data on the collisional excitation rates of the fine structure transitions of C^{+}, C and O can be found in a number of places in the literature. Compilations of excitation rate data are given in Hollenbach and McKee (1989); Glover and Jappsen (2007) and Maio et al. (2007), as well as in the LAMDA database (Schöier et al. 2005).

3.4.3 Molecular Hydrogen

Molecular hydrogen is the dominant molecular species in the ISM and can have an abundance that is orders of magnitude larger than that of any other molecule or that of any of the elements responsible for fine structure cooling. Because of this, it is natural to expect H_2 to play an important role in the cooling of the ISM. In practice, however, at metallicities close to solar, H_2 cooling is important only in shocks (see e.g. Hollenbach and McKee 1979, 1989) and not in more quiescent regions of the diffuse ISM (Glover and Clark 2014). H_2 is not a particularly effective coolant at low temperatures.

There are several reasons for that. To a first approximation, we can treat H_2 as a linear rigid rotor, with rotational energy levels separated by energies

$$\Delta E = 2BJ, \tag{66}$$

where J is the rotational quantum number of the upper level and B is the rotational constant,

$$B \equiv \frac{\hbar^2}{2I_m}, \tag{67}$$

and I_m is the moment of inertia of the molecule (see e.g. Atkins and Friedman 2010). Since H_2 is a light molecule, it has a small moment of inertia, and hence a large rotational constant, leading to widely spaced energy levels: $\Delta E/k_B \approx 170J$ K when the rotational quantum number J is small. In addition, radiative transitions between states with odd J and even J are strongly forbidden. The reason for this is that the hydrogen molecule has two distinct forms, distinguished by the value of the nuclear spin quantum number I. If the two protons have anti-parallel spins, so that $I = 0$, then the total wave-function is anti-symmetric with respect to exchange of the two protons (as required by the Pauli exclusion principle) if and only if the rotational quantum number J is even. Molecular hydrogen in this form is known as para-hydrogen. On the other hand, if the protons have parallel spins, then the Pauli principle requires that J be odd. H_2 in this form is known as ortho-hydrogen. Radiative transitions between the ortho and para states (or vice versa) therefore require a change in the nuclear spin, which is highly unlikely, and hence the associated transition rates are very small. The first accessible rotational transition is therefore the $J = 2 \rightarrow 0$ transition, which has an associated energy separation of around 510 K. At low temperatures, it becomes extremely hard to excite this transition, and therefore H_2 cooling becomes extremely ineffective.

This is illustrated in Fig. 4, where we compare the cooling rate due to H_2 with the cooling rates of a number of other potentially important coolants, discussed in more detail below. All of the cooling rates are computed in the low density limit and assume that the hydrogen is fully molecular and that the fractional ionization is zero.

From the figure, we see that in these conditions, H_2 cooling can be important at temperatures $T > 100$ K, but becomes insignificant in comparison to fine structure line cooling or CO rotational emission at $T < 100$ K, owing to the exponential fall-off in the H_2 cooling rate. Changing the composition of the gas will change the relative contributions of the different coolants, but in practice will typically make H_2 cooling less important. For example, reducing the fractional abundance of H_2 causes the H_2 cooling rate to drop significantly, because not only does one have fewer H_2 molecules to provide the cooling, but their collisional excitation rates also decrease, since collisions with H atoms are much less effective at exciting the rotational levels of H_2 than collisions with other H_2 molecules (see e.g. Glover and Abel 2008 and references therein). We therefore see that at solar metallicity, H_2 cooling is important only in gas with a high H_2 fraction *and* a temperature $T > 100$ K. In practice, it is difficult to satisfy both of these conditions at the same time in quiescent gas. Temperatures of 100 K or more are easy to reach in low density CNM clouds, but the equilibrium H_2 abundance in these clouds is small. Increasing the density and/or column density of the clouds increases the equilibrium H_2 abundance, but at the same time decreases the typical gas temperature. For this reason, high H_2 fractions tend to be found only in cold gas (Krumholz et al. 2011), and therefore in conditions where H_2 cooling is ineffective.

The combination of high H_2 fraction and a gas temperature $T > 100$ K can occur in shocked molecular gas, provided that the shock is not so strong as to completely dissociate the H_2, and H_2 has long been known to be a significant coolant in these conditions (Hollenbach and McKee 1979, 1989; Pon et al. 2012).

3.4.4 Hydrogen Deuteride

Although H_2 is an ineffective low temperature coolant, the same is not true for its deuterated analogue, HD. Unlike H_2, HD does not have distinct ortho and para forms, and hence for HD molecules in the $J = 0$ ground state, the first accessible excited level is the $J = 1$ rotational level. In addition, HD is 50 % heavier than H_2, and hence has a smaller rotational constant and more narrowly spaced energy levels. The energy separation of its $J = 0$ and $J = 1$ rotational levels is $\Delta E_{10}/k_B = 128$ K, around a factor of four smaller than the separation of the $J = 0$ and $J = 2$ levels of H_2. We would therefore expect HD cooling to remain effective down to much lower temperatures than H_2.

One important factor working against HD is the fact that the deuterium abundance is only a small fraction of the hydrogen abundance, meaning that in general H_2 is orders of magnitude more abundant than HD. However, in cold gas that is not yet fully molecular, the HD abundance can be significantly enhanced by a process known as chemical fractionation. HD is formed from H_2 by the reaction

$$H_2 + D^+ \rightarrow HD + H^+ \tag{68}$$

and is destroyed by the inverse reaction

$$HD + H^+ \rightarrow H_2 + D^+. \tag{69}$$

The formation of HD via reaction (68) is exothermic and can take place at all temperatures, but the destruction of HD via reaction (69) is mildly endothermic and becomes very slow at low temperatures. As a result, the equilibrium HD/H_2 ratio is enhanced by a factor (Galli and Palla 2002)

$$f_{en} = 2 \exp\left(\frac{462}{T}\right) \tag{70}$$

over the elemental D/H ratio. At temperatures $T < 100$ K, characteristic of the CNM, this corresponds to an enhancement in the equilibrium abundance by a factor of hundreds to thousands. This fractionation effect helps HD to be a more effective low temperature coolant than one might initially suspect. Nevertheless, there is a limit to how effective HD can become, since the HD abundance obviously cannot exceed the total deuterium abundance. The total abundance of deuterium relative to hydrogen in primordial gas is (Cooke et al. 2014)

$$(D/H) = (2.53 \pm 0.04) \times 10^{-5}. \tag{71}$$

In the local ISM, the ratio of D/H is even smaller (see e.g. Linsky et al. 2006; Prodanović et al. 2010), as some of the primordial deuterium has been destroyed by stellar processing.

From the comparison of cooling rates in Fig. 4, we see that in fully molecular gas, HD becomes a more effective coolant than H_2 once the temperature drops below 50 K, despite the fact that in these conditions, the abundance of HD is more than 10^4 times smaller than that of H_2. However, we also see that at these low temperatures, the amount of cooling provided by HD is a factor of a hundred or more smaller than the cooling provided by C^+, C or CO. It is therefore safe to conclude that HD cooling is negligible in low density, solar metallicity gas. At higher densities, HD cooling could potentially become more important, as HD has a larger critical density than C or C^+, but at the relevant densities ($n \sim 10^6$ cm^{-3}), dust cooling generally dominates.

3.4.5 Carbon Monoxide

Heavier molecules can also contribute significantly to the cooling of interstellar gas. In particular, carbon monoxide (CO), the second most abundant molecular species in the local ISM, can play an important role in regulating the temperature within giant molecular clouds (GMCs). As we can see from Fig. 4, CO is a particularly important coolant at very low gas temperatures, $T < 20$ K, owing to the very small energy separations between its excited rotational levels. However, we also see from the figure that at low densities, fine structure cooling from neutral atomic carbon is more effective than CO cooling, and that at $T \sim 20$ K and above, the contribution from C^+ also becomes significant. The overall importance of CO therefore depends strongly on the chemical state of the gas. If the gas-phase carbon is primarily in the form of C or C^+, then fine structure emission from these species will dominate, implying that CO becomes important only once the fraction of carbon in CO becomes large. As we will discuss in more detail later, this only occurs in dense, well-shielded gas, and so in typical GMCs, CO cooling only dominates once the gas density exceeds $n \sim 1000$ cm^{-3}.

In practice, CO is able to dominate the cooling only over a restricted range in densities, as it becomes ineffective at densities $n \gg 1000$ cm^{-3}. In part, this is because CO has only a small dipole moment and hence the CO rotational transitions have low critical densities. For example, in optically thin gas, the relative populations of the $J = 0$ and $J = 1$ rotational levels reach their LTE values at a density $n_{\rm crit} \sim 2200$ cm^{-3}, while the $J = 2$ level reaches LTE at $n_{\rm crit} \sim 23000$ cm^{-3}. In addition, the low J transitions of ^{12}CO rapidly become optically thick in these conditions, further lowering their effective critical densities and significantly limiting their contribution to the cooling rate of the gas. This behavior has a couple of interesting implications. First, it means that cooling from isotopic variants of CO, such as ^{13}CO or C^{18}O can become important, despite the low abundances of these species relative to ^{12}CO (e.g. Szűcs et al. 2014), since they will often remain optically thin even if ^{12}CO is optically thick. Second, it means that the freeze-out of CO onto the surface of dust grains, which is thought to occur in the cold, dense gas at the center of many prestellar cores, has very little effect on the overall CO cooling rate. This was demonstrated in striking fashion by Goldsmith (2001), who showed that at densities of order 10^4–

10^5 cm^{-3} within a typical prestellar core, reducing the CO abundance by a factor of a hundred reduces the CO cooling rate by only a factor of a few.

3.4.6 Other Heavy Molecules

Other molecular species can become important coolants in comparison to H_2 and CO in the right conditions. An interesting example is water. H_2O molecules have a very large number of accessible rotational and vibrational transitions and also have high critical densities. Therefore, over a wide range of temperatures and densities, the amount of cooling that one gets per water molecule can be much larger than the amount that one gets per CO molecule (see e.g. the comparison in Neufeld and Kaufman 1993). Despite this, water does not contribute significantly to the thermal balance of cold gas in molecular clouds, because the fractional abundance of water in these regions is very small (see e.g. Snell et al. 2000). This is, because most of the water molecules that form rapidly freeze out onto the surface of dust grains, forming a significant part of the ice mantles that surround these grains (Bergin et al. 2000; Hollenbach et al. 2009). On the other hand, in warm regions, such as the shocked gas in molecular outflows, H_2O can be a very important coolant (Nisini et al. 2010).

The other molecules and molecular ions present in interstellar gas also provide some cooling, but at low gas densities, their total contribution is relatively small compared to that of CO, since the latter generally has a much larger abundance. In very dense gas, however, their contributions become much more important, owing to the high optical depth of the CO rotational lines. Of particular importance in this high density regime are species that have large dipole moments, such as HCN or N_2H^+, as these species have high critical densities and hence remain effective coolants up to very high densities. That said, in typical molecular cloud conditions, dust cooling takes over from molecular line cooling as the main energy loss route well before these species start to dominate the line cooling, and so their overall influence on the thermal balance of the cloud remains small.

3.5 *Gas-Grain Energy Transfer*

Dust can also play an important role in the cooling of the ISM (Goldreich and Kwan 1974; Leung 1975). Individual dust grains are extremely efficient radiators, and so the mean temperature of the population of dust grains very quickly relaxes to an equilibrium value given by the balance between radiative heating caused by the absorption of photons from the ISRF and radiative cooling via the thermal emission from the grains.[4] If the resulting dust temperature, T_d, differs from the gas temperature, T_K,

---

[4]The chemical energy released when H_2 molecules form on grain surfaces and the direct interaction between dust grains and cosmic rays also affect the grain temperature, but their influence on the mean grain temperature is relatively minor (Leger et al. 1985).

then collisions between gas particles and dust grains lead to a net flow of energy from one component to the other, potentially changing both $T_{\rm K}$ and $T_{\rm d}$.

The mean energy transferred from the gas to the dust by a single collision is given by

$$\Delta E = \frac{1}{2} k_{\rm B} (T_{\rm K} - T_{\rm d}) \alpha, \tag{72}$$

where α is the thermal energy accommodation coefficient, which describes how efficiently energy is shared between the dust and the gas (Burke and Hollenbach 1983). This efficiency typically varies stochastically from collision to collision, but since we are always dealing with a large number of collisions, it is common to work in terms of the mean value of α, which we denote as $\bar{\alpha}$. However, even then, the treatment of the accommodation coefficient can be complicated, as $\bar{\alpha}$ depends in a complicated fashion on the nature of the dust grain, the nature of the collider (e.g. whether it is a proton, a hydrogen atom or an H_2 molecule), and the gas and grain temperatures (Burke and Hollenbach 1983).

The total rate at which energy flows from the gas to the dust is the product of the mean energy per collision and the total collision rate. The latter can be written as

$$R_{\rm coll} = 4\pi \sigma_{\rm d} \bar{v} n_{\rm tot} n_{\rm d}, \tag{73}$$

where $\sigma_{\rm d}$ is the mean cross-sectional area of a dust grain, $n_{\rm d}$ is the number density of dust grains, $n_{\rm tot}$ is the number density of particles, and $\bar{v}$ is the mean thermal velocity of the particles in the gas. Note that both $n_{\rm tot}$ and $\bar{v}$ are functions of the composition of the gas—in a fully atomic gas, $n_{\rm tot}$ and $\bar{v}$ are both larger than in a fully molecular gas.

Combining Eqs. (72) and (73), we can write the cooling rate per unit volume due to energy transfer from the gas to the dust as

$$\Lambda_{\rm gd} = \pi \sigma_{\rm d} \bar{v} \bar{\alpha} (2kT_{\rm K} - 2kT_{\rm d}) n_{\rm tot} n_{\rm d}. \tag{74}$$

Note that although it is common to talk about this in terms of cooling, if $T_{\rm d} > T_{\rm K}$ then energy will flow from the dust to the gas, i.e. this will become a heating rate.

Expressions given in the astrophysical literature for $\Lambda_{\rm gd}$ are typically written in the form

$$\Lambda_{\rm gd} = C_{\rm gd} T_{\rm K}^{1/2} \,(T_{\rm K} - T_{\rm d})\, n^2 \ {\rm erg\ s^{-1}\ cm^{-3}}, \tag{75}$$

where n is the number density of hydrogen nuclei and $C_{\rm gd}$ is a cooling rate coefficient given by

$$C_{\rm gd} = 2\pi k \sigma_{\rm d} \left(\frac{\bar{v}}{T_{\rm K}^{1/2}} \right) \bar{\alpha} \frac{n_{\rm tot} n_{\rm d}}{n^2}. \tag{76}$$

The value of $C_{\rm gd}$ is largely determined by the assumptions that we make regarding the chemical state of the gas and the nature of the dust grain population, but in principle it also depends on temperature, through the temperature dependence of the mean accommodation coefficient, $\bar{\alpha}$.

Different authors introduce different assumptions about various of these issues, leading to a wide spread of values for $C_{\rm gd}$ being quoted in the literature for Milky Way dust. For example, Hollenbach and McKee (1989) write $C_{\rm gd}$ as

$$C_{\rm gd} = 3.8 \times 10^{-33} \left[1 - 0.8 \exp\left(-\frac{75}{T_{\rm K}}\right)\right] \left(\frac{10\,{\rm nm}}{a_{\rm min}}\right)^{1/2} {\rm erg\,s^{-1}\,cm^{3}\,K^{-3/2}}, \quad (77)$$

where $a_{\rm min}$ is the minimum radius of a dust grain, often taken to be simply $a_{\rm min} = 10\,{\rm nm}$. However, Tielens and Hollenbach (1985) quote a value for the same process that is almost an order of magnitude smaller

$$C_{\rm gd} = 3.5 \times 10^{-34}\,{\rm erg\,s^{-1}\,cm^{3}\,K^{-3/2}}, \quad (78)$$

while Goldsmith (2001) quotes a value that is smaller still,

$$C_{\rm gd} = 1.6 \times 10^{-34}\,{\rm erg\,s^{-1}\,cm^{3}\,K^{-3/2}}. \quad (79)$$

Finally, Evans (private communication) argues for a rate

$$C_{\rm gd} = 1.8 \times 10^{-33} \left[1 - 0.8 \exp\left(-\frac{75}{T_{\rm K}}\right)\right] {\rm erg\,s^{-1}\,cm^{3}\,K^{-3/2}}, \quad (80)$$

close to the Hollenbach and McKee (1989) rate. Although it is not always clearly stated, all of these rates seem to be intended for use in H_2-dominated regions. In regions dominated by atomic hydrogen, one would expect the cooling rate to vary, owing to the difference in the value of $\bar{\alpha}$ appropriate for H atoms and that appropriate for H_2 molecules (Burke and Hollenbach 1983). In a recent study, Krumholz et al. (2011) attempted to distinguish between the molecular-dominated and atomic-dominated cases, using

$$C_{\rm gd} = 3.8 \times 10^{-33}\,{\rm erg\,s^{-1}\,cm^{3}\,K^{-3/2}} \quad (81)$$

for molecular gas and

$$C_{\rm gd} = 1.0 \times 10^{-33}\,{\rm erg\,s^{-1}\,cm^{3}\,K^{-3/2}} \quad (82)$$

for atomic gas.

The uncertainty in $C_{\rm gd}$ becomes even greater as we move to lower metallicity, as less is known about the properties of the dust. It is often assumed that $C_{\rm gd}$ scales linearly with metallicity (e.g. Glover and Clark 2012c), but this is at best a crude approximation, particularly as the dust abundance appears to scale non-linearly with metallicity in metal-poor galaxies (Galametz et al. 2011; Herrera-Camus et al. 2012).

The importance of gas-grain energy transfer as a cooling mechanism depends strongly on the gas density, regardless of which value of C_{gd} we adopt. At low densities, the cooling rate is low in comparison to that provided by atomic fine structure lines or molecular transitions, as can be seen by comparing the values quoted above with the atomic and molecular cooling rates plotted in Fig. 4. This remains true for as long as we remain in the low-density line cooling regime, where the cooling rate due to line emission scales as n^2. However, as the density increases, we will eventually pass the critical densities of our main atomic and molecular coolants. Once we do so, the line cooling rate will begin to scale linearly with the density, while the gas-grain rate will continue to scale as n^2. We therefore see that eventually gas-grain energy transfer will come to dominate the cooling rate of the gas. The considerable optical depths that can build up in the main coolant lines simply act to hasten this process.

Once gas-grain energy transfer dominates the cooling rate, the gas temperature is quickly driven towards the dust temperature. This is illustrated in Fig. 5, which shows the evolution of dust and gas temperature as function of number density in the turbulent ISM for different combinations of metallicity and strength of the ISRF (based on numerical simulations similar to those described in Glover and Clark 2012a, c). The gas density at which the two temperatures become strongly coupled depends on the value of C_{gd}. In a quiescent pre-stellar core, coupling occurs once the cooling due to dust becomes larger than the cosmic ray heating rate of the gas. In solar metallicity gas, this takes place at a density $n \sim 10^5$ cm^{-3} if one uses the

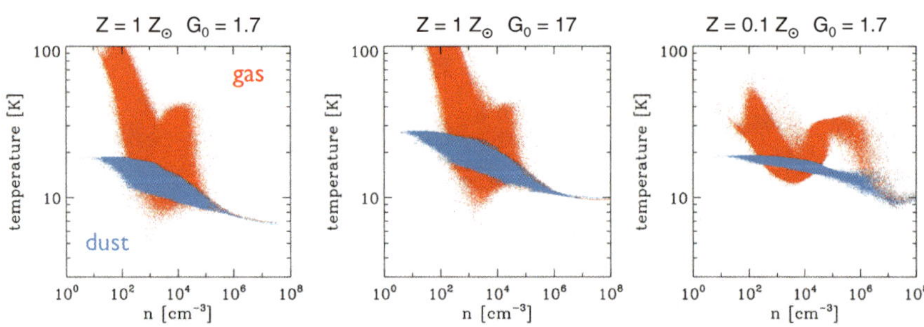

Fig. 5 Gas and dust temperatures, T_K and T_d, as a function of the hydrogen nuclei number density n in the turbulent ISM for two different metallicities (the solar value, $Z = 1\,Z_{\odot}$, and a much smaller value typical of metal-poor dwarf galaxies, $Z = 0.1\,Z_{\odot}$), and for two different strength of the ISRF (the solar neighborhood value with $G_0 = 1.7$, and a value ten times larger with $G_0 = 17$). We assume here that the grain size distribution is the same in each case, and that the dust-to-gas ratio scales linearly with the metallicity. The distribution of dust temperatures varies only weakly with changes in Z or G_0. For $Z = 1\,Z_{\odot}$, the gas becomes thermally coupled to the gas at densities larger than $n \sim 10^5$ cm^{-3}, so that $T_K \approx T_d$. For $Z = 0.1\,Z_{\odot}$, this happens instead at densities above $n \sim 10^6$ cm^{-3}. In these models, which started with atomic initial conditions, H_2 formation on dust grains releases latent heat and leads to a bump in the gas temperature at densities $n \sim 10^4$ cm^{-3} for $Z = 1\,Z_{\odot}$, and at $n \sim 10^5$ cm^{-3} for $Z = 0.1\,Z_{\odot}$. This feature is absent in clouds that have already converted all of their hydrogen to molecular form

Hollenbach and McKee (1989) prescription for $C_{\rm gd}$, but not until $n \sim 10^6\ {\rm cm}^{-3}$ if one uses the Goldsmith (2001) prescription, provided that we assume a standard value for the cosmic ray heating rate. In regions where the cosmic ray flux is highly enhanced or where the metallicity is low, however, the coupling between gas and dust can be delayed until much higher densities (see e.g. Papadopoulos 2010; Clark et al. 2013).

3.6 Computing the Dust Temperature

As we saw in the previous subsection, energy transfer between gas and grains acts to couple the gas temperature to the dust temperature in dense gas. It is therefore important to understand the physics responsible for determining $T_{\rm d}$. Because dust grains are extremely efficient radiators, it is usually a good approximation to treat them as being in thermal equilibrium, with a temperature set by the balance between three processes: heating by photon absorption and by collisions with warmer gas particles, and cooling by photon emission.

The dust temperature is set by the following equation:

$$\Gamma_{\rm ext} - \Lambda_{\rm dust} + \Lambda_{\rm gd} = 0\,. \tag{83}$$

Here $\Gamma_{\rm ext}$ is the dust heating rate per unit volume due to the absorption of radiation, $\Lambda_{\rm dust}$ is the radiative cooling rate of the dust, and $\Lambda_{\rm gd}$, as we have already seen, is the net rate at which energy is transferred from the gas to the dust by collisions.

In the simple case in which the main contribution to $\Gamma_{\rm ext}$ comes from the interstellar radiation field, we can write this term as the product of a optically thin heating rate, $\Gamma_{\rm ext,0}$, and a dimensionless factor, χ, that represents the attenuation of the interstellar radiation field by dust absorption (Goldsmith 2001),

$$\Gamma_{\rm ext} = \chi \Gamma_{\rm ext,0}. \tag{84}$$

The optically thin heating rate is given by

$$\Gamma_{\rm ext,0} = 4\pi \mathcal{D} \rho \int_0^\infty J_\nu \kappa_\nu \, d\nu, \tag{85}$$

where $\mathcal{D}$ is the dust-to-gas ratio, ρ is the gas density, J_ν is the mean specific intensity of the incident interstellar radiation field, and κ_ν is the dust opacity in units of ${\rm cm}^2\ {\rm g}^{-1}$. To determine the attenuation factor χ at a specified point in the cloud, we can use the following expression:

$$\chi = \frac{\oint \int_0^\infty J_\nu \kappa_\nu \exp\left[-\kappa_\nu \Sigma(\mathbf{n})\right] \, d\nu \, d\Omega}{4\pi \int_0^\infty J_\nu \kappa_\nu \, d\nu}\,, \tag{86}$$

where $\Sigma(\mathbf{n})$ is the column density of the gas between the point in question and the edge of the cloud in the direction $\mathbf{n}$.

The values of both $\Gamma_{\rm ext,0}$ and χ depend on the parameterization we use for the ISRF and on our choice of dust opacities. For example, Goldsmith (2001) uses values for the radiation field from Mathis et al. (1983) plus a highly simplified dust grain model and derives an optically thin heating rate

$$\Gamma_{\rm ext,0} = 1.95 \times 10^{-24} n \, {\rm erg \, s^{-1} \, cm^{-3}}. \tag{87}$$

On the other hand, Glover and Clark (2012b) make use of a more complicated model, involving values for the ISRF taken from Black (1994) and dust opacities taken from Ossenkopf and Henning (1994) at long wavelengths and Mathis et al. (1983) at short wavelengths, but their resulting value for $\Gamma_{\rm ext,0}$ is relatively similar:

$$\Gamma_{\rm ext,0} = 5.6 \times 10^{-24} n \, {\rm erg \, s^{-1} \, cm^{-3}}. \tag{88}$$

The dust cooling rate, $\Lambda_{\rm dust}$, is given by

$$\Lambda_{\rm dust}(T_{\rm d}) = 4\pi \mathcal{D} \rho \int_0^{\infty} B_{\nu}(T_{\rm d}) \kappa_{\nu} \, d\nu, \tag{89}$$

where $B_{\nu}(T_{\rm d})$ is the Planck function for a temperature $T_{\rm d}$. Again, the resulting rate is sensitive to our choice of opacities. Using values from Ossenkopf and Henning (1994) yields a cooling rate that is well fit by the expression (Glover and Clark 2012b)

$$\Lambda_{\rm dust}(T_{\rm d}) = 4.68 \times 10^{-31} T_{\rm d}^{6} n \, {\rm erg \, s^{-1} \, cm^{-3}} \tag{90}$$

for dust temperatures $5 < T_{\rm d} < 100$ K.

Comparing the expressions given above for $\Gamma_{\rm ext,0}$ and $\Lambda_{\rm dust}(T_{\rm d})$, we see that in optically thin, quiescent gas illuminated by a standard ISRF, the equilibrium dust temperature is $T_{\rm d} \sim 15$ K.[5] Moreover, $T_{\rm d}$ decreases only very slowly as χ increases, since $T_{\rm d} \propto \chi^{1/6}$, and so substantial attenuation of the ISRF is required in order to significantly alter the dust temperature. Note also, however, that once the attenuation becomes very large, this simple prescription for computing $T_{\rm d}$ breaks down, as the re-emitted far infrared radiation from the grains themselves starts to make a significant contribution to the overall heating rate (see e.g. Mathis et al. 1983).

Comparison of $\Gamma_{\rm ext}$ and $\Lambda_{\rm gd}$ allows us to explore the role played by the gas in heating the grains. In optically thin gas, $\Gamma_{\rm ext} \gg \Lambda_{\rm gd}$ even when $T_{\rm K} \gg T_{\rm d}$ unless the gas density is very high, of the order of $10^5 \, {\rm cm^{-3}}$ or higher, and so in these conditions, dust is heated primarily by the ISRF, with energy transfer from the gas becoming important only in extreme conditions, such as in supernova blast waves. In dense

[5]This is somewhat smaller than the mean value of ~20 K that we quote in Sect. 2.3.2, but this discrepancy is most likely due to our use of the Ossenkopf and Henning (1994) opacities here, as these are intended to represent the behavior of dust in dense molecular clouds and not in the diffuse WNM and CNM.

cores, where n is large and χ is small, the importance of gas-grain energy transfer for heating the dust depends on the difference between the gas temperature and the dust temperature, which in these conditions will generally be small. However, if we assume that we are at densities where the gas and dust temperatures are strongly coupled, we can get a good idea of the importance of the $\Lambda_{\rm gd}$ term by comparing the heating rate of the gas (e.g. by cosmic rays or compressional heating) with $\Gamma_{\rm ext}$, since most of the energy deposited in the gas will be quickly transferred to the dust grains.

If cosmic ray heating of the gas dominates, and we adopt the prescription for cosmic ray heating given in Goldsmith and Langer (1978), then gas-grain energy transfer and heating from the ISRF become comparable once $\chi \sim 10^{-4}\zeta_{17}$, where ζ_{17} is the cosmic ray ionization rate of atomic hydrogen in units of $10^{-17}\,{\rm s}^{-1}$, and where we have adopted the Hollenbach and McKee (1989) form for $\Lambda_{\rm gd}$. In dense cores, $\zeta_{17} \sim 1$ (van der Tak and van Dishoeck 2000), and so in this scenario, gas-grain energy transfer only becomes important for heating the grains once $\chi \sim 10^{-4}$, corresponding to an extremely high dust extinction (see e.g. Fig. A1 in Glover and Clark 2012b). On the other hand, if compressional heating or turbulent dissipation dominate, as appears to be the case in gravitationally collapsing cores (Glover and Clark 2012a), then the heating rate can be considerably larger. One important consequence of this is that once dynamical effects dominate the heating of the dust grains, the dust (and hence the gas) will start to heat up with increasing density, evolving with an effective adiabatic index $\gamma_{\rm eff} \approx 1.1$ (Larson 2005; Banerjee et al. 2006).

3.7 Photoelectric Heating

One of the most important forms of radiative heating in the diffuse ISM is the photoelectric heating caused by the interaction between dust grains and UV photons. If a dust grain is hit by a suitably energetic photon, it can emit a photo-electron. The energy of this photo-electron is equal to the difference between the energy of the photon and the energy barrier that needs to be overcome in order to detach the electron from the grain, a quantity often known as the work function. This difference in energies can often be substantial (of the order of an eV or more), and this energy is rapidly redistributed amongst the other particles in the gas in the form of heat.

For a dust grain with radius a, photon absorption cross-section $\sigma_{\rm d}(a, \nu)$, and charge $Z_{\rm d}e$, the rate at which photo-electrons are ejected can be written as

$$R_{\rm pe}(a, Z_{\rm d}) = 4\pi \int_{\nu_{Z_{\rm d}}}^{\nu_{\rm H}} \frac{J_{\nu}}{h\nu} \sigma_{\rm d}(a, \nu) Y_{\rm ion}(Z_{\rm d}, a, \nu) d\nu. \quad (91)$$

Here, J_{ν} is the mean specific intensity of the ISRF, $h\nu_{Z_{\rm d}}$ is the ionization potential of the grain (i.e. the energy required to remove a single electron), and $h\nu_{\rm H} = 13.6$ eV is the ionization potential of atomic hydrogen; we assume that photons with $\nu > \nu_{\rm H}$ are absorbed by the neutral atomic hydrogen in the ISM and do not reach the grains.

The $Y_{\rm ion}$ term is the photo-ionization yield. It can be written as

$$Y_{\rm ion} = Y_{\infty} \left(1 - \frac{\nu_{\rm Z_d}}{\nu}\right) f_y(a), \tag{92}$$

where Y_{∞} is the yield for very large grains in the limit $\nu \gg \nu_{\rm Z_d}$, and $f_y(a)$ is a yield correction factor. This correction factor accounts for the fact that in large grains, the photon attenuation depth, l_a, i.e. the distance that a photon can penetrate into the material before it is absorbed, can be larger than the electron mean free path l_e. We typically normalize Y_{∞} such that $f_y(a) = 1$ for large grains, in which case our values for small grains are enhanced by a factor $f_y = (l_e + l_a)/l_e$. Typical values for these yield-related parameters are $Y_{\infty} = 0.15$, $l_e = 1$ nm and $l_a = 10$ nm, respectively.

We therefore see that there are three main parameters that influence the size of the photoelectric heating rate. These are (1) the strength of the ISRF at the relevant frequencies, as quantified by the mean specific intensity J_{ν}, (2) the size distribution of the grains, often taken to be given by the simple MRN distribution (Eq. 1), and (3) the charge of the grains. The charge is important because it influences how easy it is to eject electrons from the grains. When the grains are highly negatively charged, electron ejection is easy, the work function is small, and the photo-ionization yield is high. As the grains become more neutral or even positively charged, it becomes much harder to detach electrons from the grains: the work function increases and the photo-ionization yield drops.

The photoelectric heating rate is therefore much larger in conditions when most grains are neutral, or negatively charged, than when most grains are positively charged. The main processes determining the charge on a typical grain are photo-ionization—i.e. the same process that gives us the photoelectric heating—together with the accretion of free electrons and the recombination of gas-phase ions with surface electrons. Schematically, we can write these reactions as

$$\mathrm{GR}^{+n} + \gamma \rightarrow \mathrm{GR}^{+n+1} + \mathrm{e}^-, \tag{93}$$

$$\mathrm{GR}^{+n} + \mathrm{e}^- \rightarrow \mathrm{GR}^{+n-1}, \tag{94}$$

$$\mathrm{GR}^{+n} + \mathrm{A}^+ \rightarrow \mathrm{GR}^{+n+1} + \mathrm{A}. \tag{95}$$

In general, collisions with electrons are more important than collisions with positive ions, since the electron thermal velocity is much larger than the thermal velocity of the ions. The level of charge on the grains is therefore set primarily by the balance between photo-ionization and recombination with free electrons.

Although a detailed analysis of grain charging is rather complex, and beyond the scope of these lecture notes, in practice one finds that the dependence of the photoelectric heating rate on the physical conditions in the gas is fairly accurately described as a function of a single parameter, the combination

$$\psi \equiv \frac{G_0 T^{1/2}}{n_e}, \tag{96}$$

where G_0 is the strength of the ISRF, n_e is the number density of gas-phase electrons and T is the gas temperature (Draine and Sutin 1987; Bakes and Tielens 1994; Weingartner and Draine 2001b). Physically, this behavior makes sense: a strong ISRF or a paucity of free electrons will tend to lead to the grains being more positively charged, while the converse will lead to grains being more negatively charged. The $T^{1/2}$ dependence simply reflects the temperature dependence of the rate coefficient for electron recombination with the grains.

For standard interstellar dust, the photoelectric heating rate has been parameterized as a function of this ψ parameter by Bakes and Tielens (1994). Their prescription for the heating rate per unit volume can be written as

$$\Gamma_{pe} = 1.3 \times 10^{-24} \epsilon G_0 n \text{ erg s}^{-1} \text{ cm}^{-3}, \tag{97}$$

where ϵ is the photoelectric heating efficiency, given by

$$\epsilon = \frac{0.049}{1 + (\psi/1925)^{0.73}} + \frac{0.037(T/10000)^{0.7}}{1 + (\psi/5000)}, \tag{98}$$

and G_0 is the strength of the interstellar radiation field in units of the Habing (1968) field (see Eq. 8). In the limit of small ψ, we have $\epsilon \approx 0.05$ when the temperature is low, and $\epsilon \approx 0.09$ when the temperature is high (see Fig. 6). A more recent treatment by Weingartner and Draine (2001b) gives similar values for ϵ for small ψ, but predicts a more rapid fall-off in ϵ and Γ_{pe} with increasing ψ for $\psi > 10^4$.

A final important point to note is that because the photons required to eject photoelectrons must be energetic, with minimum energies typically around 6 eV, the photoelectric heating rate is highly sensitive to the dust extinction. This sensitivity can be approximately represented by a scaling factor $f_{thick} = \exp(-2.5 A_V)$ (Bergin et al. 2004). From this, we see that photoelectric heating will become ineffective once the visual extinction of the gas exceeds $A_V \sim 1$–2.

3.8 Other Processes Responsible for Heating

Ultraviolet radiation

As well as heating the gas via the photoelectric effect, the ultraviolet component of the ISRF also heats the gas in two other important ways. First, the photodissociation of H_2 by UV photons results in a transfer of energy to the gas, as the hydrogen atoms produced in this process have kinetic energies that on average are greater than the mean kinetic energy of the gas particles. The amount of energy released varies depending upon which rovibrational level of the excited electronic state was involved

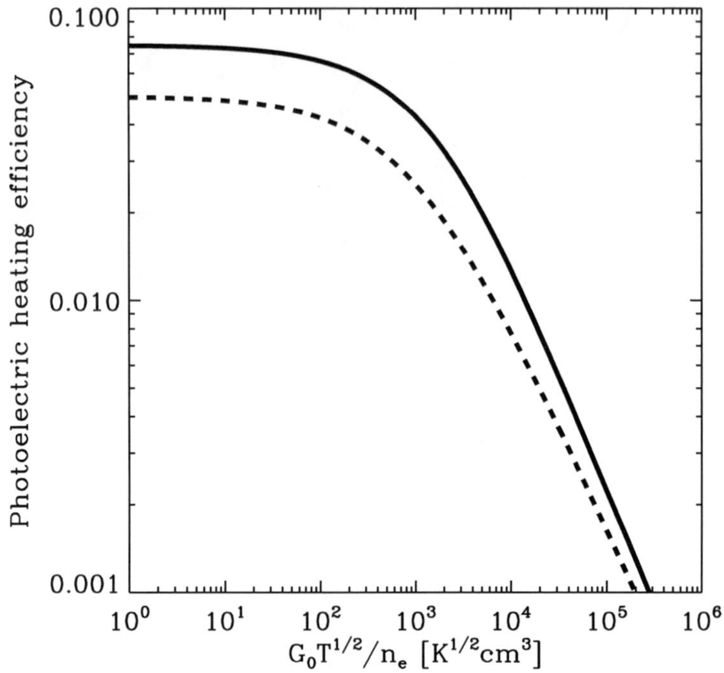

Fig. 6 Photoelectric heating efficiency ϵ as a function of $\psi \equiv G_0 T^{1/2} n_e^{-1}$. Values are plotted for gas temperatures of 6000 K (*solid line*), characteristic of the WNM, and 60 K (*dashed line*), characteristic of the CNM. The values shown here are based on the work of Bakes and Tielens (1994), as modified by Wolfire et al. (2003). In the local ISM, $\psi \sim 2 \times 10^4$ in the WNM and ~2000 in the CNM (Wolfire et al. 2003)

in the dissociation (Stephens and Dalgarno 1973; Abgrall et al. 2000). Averaged over all the available dissociative transitions, the total heating rate is around 0.4 eV per dissociation (Black and Dalgarno 1977).

Second, UV irradiation of molecular hydrogen can lead to heating via a process known as UV pumping. The absorption of a UV photon by H_2 leads to photodissociation only around 15 % of the time (Draine and Bertoldi 1996). The rest of the time, the H_2 molecule decays from its electronically excited state back into a bound rovibrational level in the electronic ground state. Although the molecule will occasionally decay directly back into the $v = 0$ vibrational ground state, it is far more likely to end up in a vibrationally excited level. In low density gas, it then radiatively decays back to the rovibrational ground state, producing a number of near infrared photons in the process. In high density gas, on the other hand, collisional de-excitation occurs more rapidly than radiative de-excitation, and so most of the excitation energy is converted into heat. In this case, the resulting heating rate is around 2 eV per pumping event, corresponding to around 10–11 eV per photodissociation (see e.g. Burton et al. 1990). The density at which this process becomes important is simply the critical density of H_2, $n_{\rm crit} \sim 10^4\ {\rm cm}^{-3}$. This process is therefore not a major heat source at typical molecular cloud densities, but can become important in dense cores exposed to strong UV radiation fields.

Cosmic rays

In gas that is well shielded from the ISRF, both of these processes become unimportant, as does photoelectric heating. In this regime, cosmic rays provide one of the main sources of heat. When a cosmic ray proton ionizes a hydrogen or helium atom, or an H_2 molecule, the energy lost by the cosmic ray is typically considerably larger than the ionization energy of the atom or molecule (Glassgold and Langer 1973). The excess energy is shared between the resulting ion and electron as kinetic energy, and the collisions of these particles with other atoms or molecules can lead to further ionizations, known as secondary ionizations. Alternatively, the excess kinetic energy can be redistributed in collisions as heat. The amount of heat transferred to the gas per cosmic ray ionization depends upon the composition of the gas (Dalgarno et al. 1999; Glassgold et al. 2012), but is typically around 10–20 eV. Most models of thermal balance in dark clouds adopt a heating rate that is a fixed multiple of the cosmic ray ionization rate, rather than trying to account for the dependence of the heating rate on the local composition of the gas (see e.g. Goldsmith and Langer 1978; Goldsmith 2001; Glover et al. 2010; Krumholz et al. 2011). A commonly adopted parameterization is

$$\Gamma_{\rm cr} \sim 3.2 \times 10^{-28} (\zeta_{\rm H}/10^{-17}\,{\rm s}^{-1})\, n \quad {\rm erg\,cm^{-3} s^{-1}}\,, \tag{99}$$

where the cosmic ray ionization rate of atomic hydrogen $\zeta_{\rm H}$ is scaled by its typical value of $10^{-17}\,{\rm s}^{-1}$, and where n is the number density of hydrogen nuclei. Note that the uncertainty introduced by averaging procedure is typically much smaller than the current uncertainty in the actual cosmic ray ionization rate in the considered region (see Sect. 2.4).

X-rays

X-rays can also heat interstellar gas, and indeed in this case the chain of events is very similar to that in the case of cosmic ray heating: X-ray ionization produces an energetic electron that can cause a significant number of secondary ionizations, with some fraction of the excess energy also going into heat. Unlike cosmic rays, X-rays are rather more sensitive to the effects of absorption, since their mean free paths are typically much smaller. Therefore, although X-ray heating can be important in the diffuse ISM (see e.g. Wolfire et al. 1995), it is generally not important in the dense gas inside molecular clouds, unless these clouds are located close to a strong X-ray source such as an AGN (see e.g. Hocuk and Spaans 2010).

Chemical reactions

Another way in which the gas can gain energy is through changes in its chemical composition. The formation of a new chemical bond, such as that between the two hydrogen nuclei in an H_2 molecule, leads to a release of energy. Much of this energy will be in the form of rotational and/or vibrational excitation of the newly-formed molecule, and in low density environments, this will rapidly be radiated away. At high densities, however, collisional de-excitation can convert this energy into heat before it can be lost via radiation. Some of the energy released in a reaction may also be in the form of translational energy of the newly-formed molecule, and this will also be rapidly converted into heat via collisions. Many of the reactions occurring in interstellar gas lead to heating in this way, but for the most part, their effects are unimportant, as the quantities of the reactants involved are too small to do much. The one case in which this process can become significant, however, is the formation of H_2. In the local ISM, the H_2 formation rate is approximately (Jura 1975)

$$R_{H_2} \sim 3 \times 10^{-17} n n_H \, \mathrm{cm}^{-3} \mathrm{s}^{-1}, \tag{100}$$

and a total of 4.48 eV of energy is released for each H_2 molecule that is formed. If this energy is converted to heat with an efficiency ϵ_{H_2}, then the resulting heating rate is

$$\Gamma_{H_2 \mathrm{form}} \sim 2 \times 10^{-28} \epsilon_{H_2} n n_H \, \mathrm{erg\,cm}^{-3} \mathrm{s}^{-1}. \tag{101}$$

Comparing this with the heating rate (99) due to cosmic ray ionization, we see that H_2 formation heating will dominate whenever $\epsilon_{H_2} n_H > (\zeta_H / 10^{-17}\ \mathrm{s}^{-1})$. In principle, therefore, H_2 formation heating can be an important process, provided that the efficiency factor ϵ_{H_2} is not too small. Unfortunately, the value of ϵ_{H_2} remains a matter of debate within the astrochemical community. Some studies (see e.g. Le Bourlot et al. 2012 and references therein) indicate that a significant fraction of the H_2 binding energy should be available for heating the gas, while others (e.g. Roser et al. 2003; Congiu et al. 2009) predict that ϵ_{H_2} should be small.

Dynamical processes

Finally, hydrodynamical and magnetohydrodynamical effects can also lead to significant heating. In subsonic, gravitationally collapsing regions, such as low mass prestellar cores, adiabatic compression (PdV heating) can be a major source of heat and can actually be more important in the overall thermal balance of the core than cosmic ray heating. In less quiescent regions, where the gas flow is supersonic, turbulent dissipation in shocks is another major heat source. Figure 7 provides an overview of the most important heating and cooling processes for the solar neighborhood ISM. Unlike in Fig. 4, the rates are here plotted as a function of the hydrogen nuclei number density n. The figure shows that initially atomic gas exhibits three different regimes. At densities $n < 2000\,\mathrm{cm}^{-3}$, the gas heating is dominated by photoelectric emission from dust grains (Sect. 3.7), while cooling is provided by fine structure emission from C^+. In the density regime $2000 < n < 10^5\ \mathrm{cm}^{-3}$, rotational line emission from CO becomes the main coolant. Photoelectric heating remains the main heat source initially, but steadily becomes less effective, owing to the larger visual extinction of the cloud at these densities, and other processes—adiabatic compression of the gas, dissipation of turbulent kinetic energy in shocks and cosmic ray ionization heating—become more important at $n \sim 6000\ \mathrm{cm}^{-3}$ and above. Finally, at densities above about $10^5\ \mathrm{cm}^{-3}$, the gas couples to the dust (Sect. 3.5), which acts as a thermostat

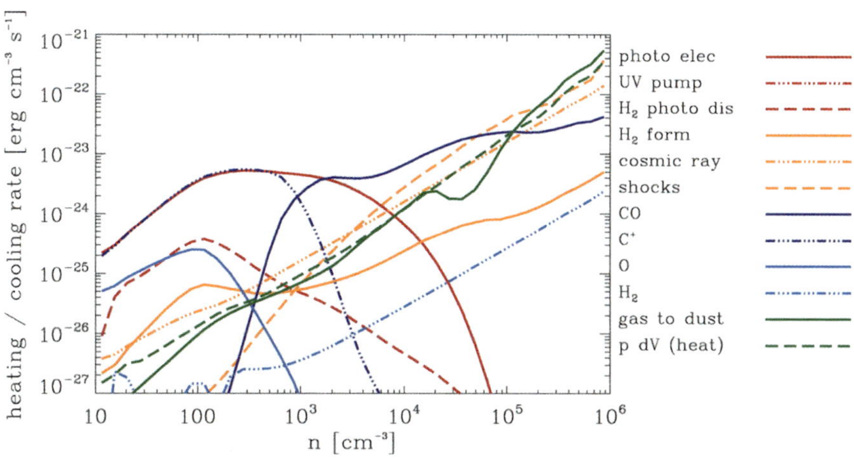

Fig. 7 Overview of the main heating and cooling processes plotted as a function of the hydrogen nuclei number density n calculated from a simulation of molecular cloud formation from initially atomic gas in the solar neighborhood. Adopted from Glover and Clark (2012a)

and provides most of the cooling power. Weak shocks and adiabatic compressions together dominate the heating of the gas in this regime, each contributing close to half of the total heating rate (for a more detailed discussion, see Glover and Clark 2012a).

The rate at which turbulent kinetic energy is dissipated in regions where the turbulence is supersonic is well established (Mac Low et al. 1998; Stone et al. 1998; Mac Low 1999). The energy dissipation rate within a cloud of mass M and velocity dispersion σ can be written to within a factor of order unity as (Mac Low 1999)

$$\dot{E}_{\rm kin} \sim -M k_{\rm d} \sigma^3, \tag{102}$$

where $k_{\rm d}$ is the wavenumber on which energy is injected into the system. If we assume that this is comparable to the size of the cloud (see e.g. Brunt et al. 2009), and adopt Larson's relations between the size of the cloud and its velocity dispersion and number density (Larson 1981), then we arrive at an average turbulent heating rate (Pan and Padoan 2009)

$$\Gamma_{\rm turb} = 3 \times 10^{-27} \left(\frac{L}{1\,{\rm pc}}\right)^{0.2} n \,{\rm erg\,s^{-1}\,cm^{-3}}. \tag{103}$$

This heating rate is of a similar order of magnitude to the cosmic ray heating rate. Unlike cosmic ray heating, however, turbulent heating is highly intermittent (Pan and Padoan 2009). This means that in much of the cloud, the influence of the turbulent dissipation is small, while in small, localized regions, very high heating rates can be produced (see e.g. Falgarone et al. 1995, Godard et al. 2009). We provide a more detailed account of ISM turbulence in the next Section.

Finally, note that the physical nature of the heating process depends upon the strength of the magnetic field within the gas. If the field is weak, energy dissipation occurs mostly through shocks, whereas if the field is strong, a substantial amount of energy is dissipated via ambipolar diffusion (Padoan et al. 2000; Li et al. 2012).

4 ISM Turbulence

The dynamical evolution of the ISM and many of its observational parameters cannot be understood without acknowledging the importance of supersonic turbulence. Here, we summarize some of the key measurements that point towards the presence of strong supersonic turbulent motions in the various phases of the ISM on a wide range of spatial scales. We introduce the most important theoretical concepts behind our current understanding of ISM turbulence, and discuss some statistical properties of compressible turbulent flows. Finally, we speculate about the physical origin of the observed turbulence in the ISM. For an overview of ISM turbulence we refer the reader to the review articles by Elmegreen and Scalo (2004) and Scalo and Elmegreen (2004), and for a discussion of the relation between turbulence and star formation on local as well as galactic scales, we point to the reviews by Mac Low and Klessen (2004) and Ballesteros-Paredes et al. (2007). More recent discussions on the topic of ISM dynamics can be found in Hennebelle and Falgarone (2012) as well as in *Protostars and Planets VI*, in particular in the chapters by Padoan et al. (2014) or Dobbs et al. (2014).

4.1 Observations

4.1.1 Observational Tracers of ISM Dynamics

The best approach to learn more about the dynamical and kinematic state of the ISM is to look for the line emission (or sometimes absorption) of various atomic and molecular species. We take a spectrum, and once we have identified the line, we can compare the observed frequency with the rest-frame frequency in order to obtain information about the velocity distribution of gas along the line of sight (LOS). Ideally, we obtain spectra at multiple positions and fully cover the projected area of the object of interest on the sky. By doing so, we obtain a three-dimensional data cube containing the line intensity at different positions on the sky and different LOS velocities. Such position-position-velocity (PPV) cubes form the basis of most kinematic studies of ISM dynamics.

For the warm neutral medium (Sect. 2.1), most studies focus on the 21 cm hyperfine structure line of atomic hydrogen (HI). It occurs with a spin flip from the excited state ($S = 1$), where the spins of proton and electron are parallel, to the ground state ($S = 0$), where the two spins are anti-parallel. The energy difference is $\Delta E = 5.87 \times 10^{-6}$ eV or $\Delta E/k_B = 0.06816$ K, corresponding to the emission of a photon with the wavelength $\lambda = 21.106$ cm or the frequency $\nu = 1.4204$ GHz (for further details, see e.g. Chap. 8 in Draine 2011).

Molecular hydrogen is much more difficult to observe. It is a homonuclear molecule, and as a consequence its dipole moment vanishes. The quadrupole radiation requires high excitation temperatures and is extremely weak under normal molecular cloud conditions (Sect. 3.4.3). Direct detection of cold interstellar H_2 requires ultraviolet absorption studies. However, due to the atmospheric absorption properties, this is only possible from space and limited to pencil-beam measurements of the absorption of light from bright stars or from AGN.[6] Studies of the molecular ISM therefore typically rely on measuring the radio and sub-millimeter emission either from dust grains or from other molecules that tend to be found in the same locations as H_2.

The most prominent of these tracer molecules is CO and its various isotopologues. As previously mentioned, the most abundant of these isotopologues is $^{12}C^{16}O$, often referred to just as ^{12}CO or simply CO. However, the high abundance of this tracer can actually become problematic, as it is often optically thick, and hence we cannot use it to trace the properties of the turbulence in the whole of the cloud. For example, numerical studies have shown that many of the smaller-scale structures identified in PPV cubes of ^{12}CO emission are actually blends of multiple unrelated features along the LOS (Ballesteros-Paredes and Mac Low 2002; Beaumont et al. 2013), and that the statistical properties of the velocity field that can be derived using ^{12}CO emission are not the same as those derived using the ^{12}CO number density (Bertram et al. 2014). For this reason, studies of the properties of the turbulence within molecular clouds often focus on less abundant isotopologues, such as $^{13}C^{16}O$ (usually written simply as ^{13}CO) or $^{12}C^{18}O$ (often written just as $C^{18}O$). The optical depths of these tracers are much lower, and we therefore expect them to provide a less biased view of the properties of the turbulent velocity field. Nevertheless, problems still remain. The lowest rotational transition of CO, the $J = 1$–0 transition, has a critical density of only $n_{cr} = 1.1 \times 10^3$ particles per cm^3, only a factor of a few larger than the typical mean density of a molecular cloud. Observations of this transition are therefore useful at providing us with information on the properties of the cloud at densities close to the mean density, but provide little information on highly underdense or highly overdense regions. This is exacerbated by the chemical inhomogeneity of the CO distribution within molecular clouds. In low density, low extinction regions, much of the CO is photodissociated (see Sect. 5.1.2), and most of the available carbon is found instead in the form of C^+, while in high density cores, CO freezes out onto the surface of dust grains.

[6]Note that rotational and ro-vibrational emission lines from H_2 have also been detected in the infrared, both in the Milky Way and in other galaxies. However, this emission comes from gas that has been strongly heated by shocks or radiation, and it traces only a small fraction of the total H_2 mass (e.g. van der Werf 2000).

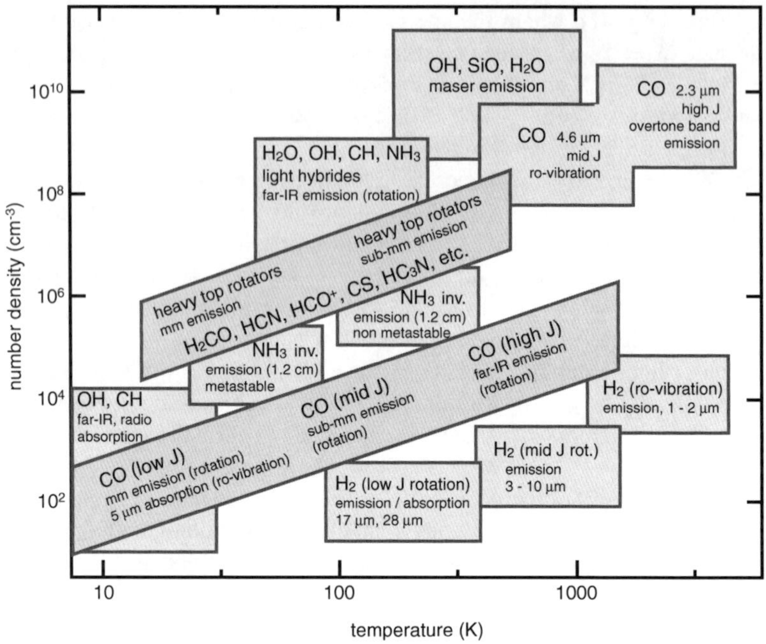

Fig. 8 Temperature and density range of various observational tracers of molecular cloud structure and dynamics. Adapted from Genzel (1991)

To trace the properties of the turbulence in these regions, different observational tracers are required. In low density regions, this is difficult, as C^+ emits at a wavelength of $158\,\mu$m which cannot be observed from the ground owing to the

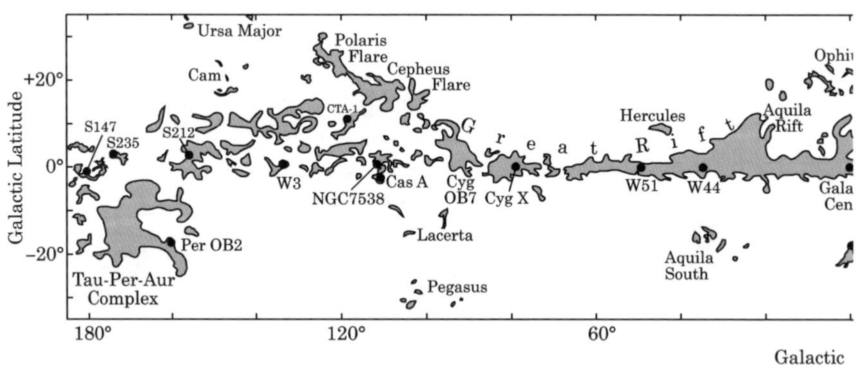

Fig. 9 Schematic distribution of molecular cloud complexes in the disk of the Milky Way. Data from Dame et al. (2001)

effects of atmospheric attenuation. It has been observed from the stratosphere by the Kuiper Airborne Observatory (see e.g. Chokshi et al. 1988) and more recently by the Stratospheric Observatory for Infrared Astronomy (SOFIA; see e.g. Simon et al. 2012), and from space by ISO and by the Herschel space telescope (e.g. Pineda et al. 2013), but efforts to map the large-scale distribution of C^+ emission within molecular clouds are still in their infancy. In addition, they are hampered by the fact that the energy separation of the ground state and first excited state of C^+ corresponds to a temperature of around 92 K, higher than one expects to find in the low density regions of most molecular clouds, making the properties of the observed C^+ emission highly sensitive to the temperature distribution of the gas in the cloud. In high density regions, the situation is much simpler, as a number of different observational tracers are readily available, with the most popular ones being HCN, NH_3, HCO^+ and N_2H^+. A summary of the most relevant tracers, together with the range of temperature and density they are most suitable for, is depicted in Fig. 8.

For studying the properties of HII regions, atomic recombination lines are the best available tool. These are electronic transitions that occur when the recombination event leaves the electron in an excited state, which consequently decays down towards the ground state by emitting photons. The classic example is line emission from the hydrogen atom itself in the Lyman, Balmer, Paschen, etc. series (e.g. Spitzer 1978; Osterbrock 1989). For low quantum numbers these photons typically have UV or optical wavelengths, but if highly excited Rydberg states are involved, the emission can be detected at radio or sub-mm wavelengths. Besides hydrogen (and in part helium) recombination lines, HII regions also show a large number of metal lines, both at optical wavelengths, where they result from the recombination of (multiply) ionized atoms (such as O^{++} or N^+), and in the infrared, where they result from fine structure transitions of ions or atoms with high ionization potentials.

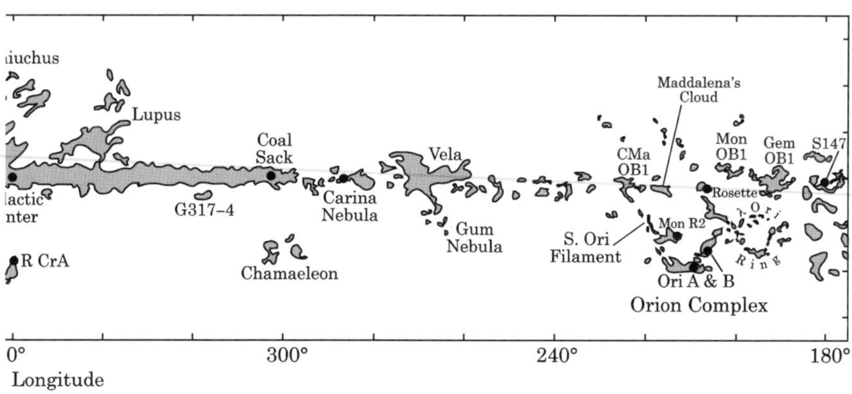

Fig. 9 (continued)

4.1.2 Properties of Molecular Clouds

In the following, we focus on the properties of the molecular component of the ISM. The all-sky survey conducted by Dame et al. (2001) shows that molecular gas is mostly confined to a thin layer in the Galactic disk, and that the gas in this layer is organized into cloud complexes of various sizes and masses (see also Combes 1991). Figure 9 illustrates the distribution of Galactic H_2 as traced by the $J = 1-0$ line of ^{12}CO with some of the most prominent molecular cloud complexes indicated by name. One of the best studied complexes in the northern sky contains the two giant molecular clouds Orion A and B, lying between a Galactic longitude $200° < \ell < 220°$ and a latitude $-20° < b < -10°$. A detailed map of the total CO intensity from this region is shown in Fig. 10, taken from the study of Wilson et al. (2005). Their observations reveal a complex hierarchy of filaments and clumps on all resolved scales.

Studies of the structure of the molecular gas show that molecular clouds appear to display self-similar behavior over a wide distribution of spatial scales (see e.g. the review by Williams et al. 2000), ranging from scales comparable to the disk thickness down to the size of individual prestellar cores, where thermal pressure starts to dominate the dynamics. The molecular cloud mass spectrum is well described by a power law of the form

$$\frac{dN}{dm} \propto m^{-\alpha} , \qquad (104)$$

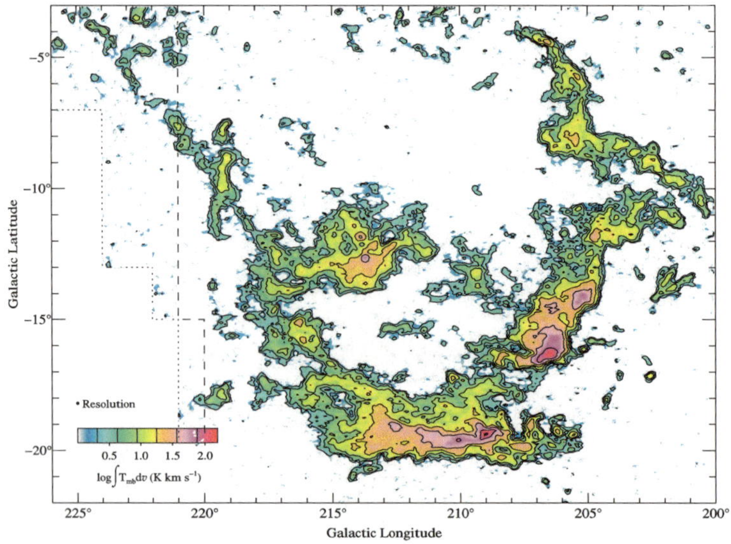

Fig. 10 Map of the velocity-integrated J=1–0 rotational line emission of the ^{12}CO molecule as tracer of the total molecular hydrogen gas in the Orion/Monoceros region. The image shows the complex spatial structure of H_2 gas in a typical molecular cloud complex. The figure is taken from Wilson et al. (2005)

Table 3 Physical properties of molecular clouds, clumps, and cores

	Molecular clouds	Cluster-forming clumps	Protostellar cores
Size (pc)	2–20	0.1–2	$\lesssim 0.1$
Mean density (H_2 cm^{-3})	10^2–10^3	10^3–10^5	$>10^5$
Mass ($M_\odot$)	10^2–10^6	10–10^3	0.1–10
Temperature (K)	10–30	10–20	7–12
Line width (km s^{-1})	1–10	0.5–3	0.2–0.5
Turbulent Mach number	5–50	2–15	0–2
Column density (g cm^{-2})	0.03	0.03–1.0	0.3–3
Crossing time (Myr)	2–10	$\lesssim 1$	0.1–0.5
Free-fall time (Myr)	0.3–3	0.1–1	$\lesssim 0.1$
Examples	Orion, Perseus	L1641, L1709	B68, L1544

Adapted from Cernicharo (1991) and Bergin and Tafalla (2007)

with the exponent being somewhere in the range $3/2 < \alpha < 2$. Consequently there is no natural mass or size scale for molecular clouds between the observed lower and upper limits. The largest molecular structures are giant molecular clouds (GMCs). They have masses of typically 10^5–10^6 $M_\odot$ and extend over a few tens of parsecs. On the other hand, the smallest observed entities are protostellar cores with masses of a few solar masses or less and sizes of $\lesssim 10^{-2}$ pc. The volume filling factor of dense clumps, even denser subclumps and so on, is very low. It is of the order of 10 % or less. In the following, we distinguish between molecular cloud complexes, cluster-forming clumps (often called infrared dark clouds, IRDCs, in the phases prior to the onset of massive star formation), and protostellar cores (which give rise to individual stars or binary systems). Table 3 summarizes their basic parameters.

The fact that all studies obtain a similar power law is remarkable, and we argue below that it is the result of turbulent motions acting on self-gravitating gas (see also Mac Low and Klessen 2004; Ballesteros-Paredes et al. 2007). This result holds for clouds over a wide range of masses and densities, and is based on data obtained with different reduction and analysis techniques. Furthermore, the result seems to be independent of whether it was derived for very actively star-forming clouds or very cold and quiescent ones. Given the uncertainties in determining the slope, it appears reasonable to conclude that there is a universal mass spectrum, and it appears plausible that the physical processes at work are rather similar from cloud to cloud. And vice versa, clouds that show significant deviation from this universal distribution most likely have different dynamical histories or live in different environments (for a discussion of molecular cloud properties in the spiral galaxy M51 based on probability distribution functions of ^{12}CO integrated intensity, see Hughes et al. 2013).

Temperatures within molecular clouds are generally lower than in other phases of the ISM. Simulations suggest that the low density portions of molecular clouds that are CO-dark (i.e. that are H_2-rich, but CO-poor; see Sect. 5.2) have temperatures ranging from around 20 K up to the temperatures of 50–100 K that are typical of the CNM (see e.g. Glover and Clark 2012c; Glover et al. 2015). Unfortunately, observational verification of this prediction is difficult, as the main observational tracer of the gas in these regions, C^+, has only a single fine structure emission line and hence does not directly constrain the temperature of the gas. In the denser, CO-emitting gas, both observations and simulations find temperatures of around 10–20 K for dark, quiescent clouds, and somewhat higher values in clouds close to sites of ongoing high-mass star formation. It is notable that within this dense, well-shielded gas, the temperature remains remarkably constant over several orders of magnitude in density (see e.g. Goldsmith 1988; Glover and Clark 2012a; Glover et al. 2015). This has important consequences for theoretical and numerical models of molecular cloud dynamics and evolution, because to a good approximation the gas can be described by a simple isothermal equation of state, where pressure P and density ρ are linearly related,

$$P = c_{\rm s}^2\,\rho, \tag{105}$$

with the sound speed $c_{\rm s}$ being the proportionality factor. The assumption of isothermality breaks down when the gas becomes optically thick and heat can no longer be radiated away efficiently. In the local ISM, this occurs when the number density exceeds values of $n(\mathrm{H_2}) \approx 10^{10}\,\mathrm{cm}^{-3}$.

The masses of molecular clouds are orders of magnitude larger than the critical mass for gravitational collapse computed from the average density and temperature (see Sect. 6). If we assume that only thermal pressure opposes gravitational attraction they should collapse and quickly form stars on timescales comparable to the free-fall time. However, this is *not* observed (for an early discussion, see Zuckerman and Evans 1974; for more recent discussions, consult Kennicutt and Evans 2012). The typical lifetime of giant molecular clouds is about 10^7 years (Blitz et al. 2007; Dobbs et al. 2014), and the average star formation efficiency is low, with values ranging between 1 and 10 % (Blitz and Shu 1980; Krumholz and Tan 2007). This tells us that there must be additional physical agents that provide stability against large-scale cloud collapse.

For a long time, magnetic fields have been proposed as the main agent responsible for preventing collapse (e.g. Shu et al. 1987). However, it appears that the typical field strengths observed in molecular clouds are not sufficient to stabilize the clouds as a whole (Verschuur 1995a, b; Troland et al. 1996; Padoan and Nordlund 1999; Lunttila et al. 2009; Crutcher et al. 2009a; Crutcher 2010; Bertram et al. 2012). This is the point at which ISM turbulence comes into play (Elmegreen and Scalo 2004; Scalo and Elmegreen 2004). Virtually all observations of molecular cloud dynamics reveal highly supersonic gas motions on scales above a few tenths of a parsec. The observed linewidths are always wider than what is implied by the excitation temperature of the molecules. This is illustrated in Fig. 11, which shows the ^{12}CO $J = 1$–0 integrated intensity from the Orion A cloud in the top panel together with the distribution of

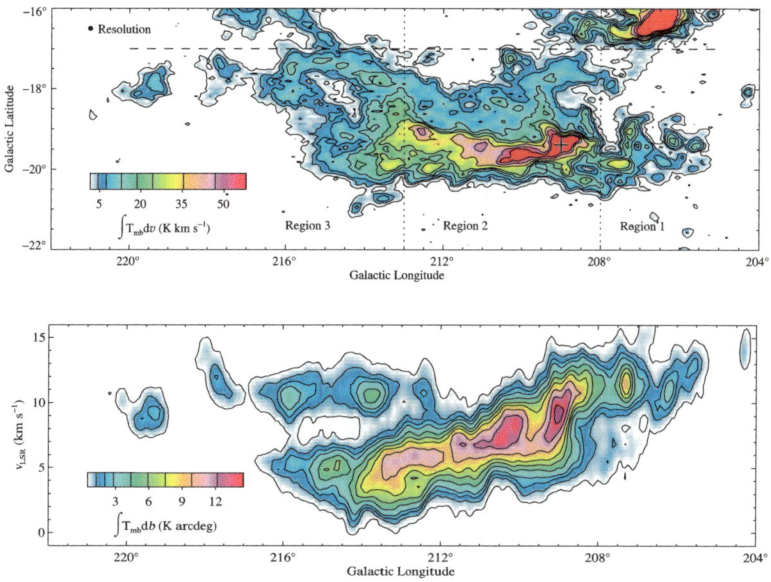

Fig. 11 *Top* Integrated intensities of the $J = 1$–0 transition of ^{12}CO of the Orion A cloud (enlargement of the lower central parts of Fig. 10). *Bottom* Distribution of line-of-sight velocities as a function of Galactic longitude for the same region, based on data integrated in stripes of Galactic latitude from $-22° < b < -17°$. The width of the velocity distribution gives a good indication of the turbulent velocity dispersion in the region. (For more information on both panels, see Wilson et al. 2005)

Doppler velocities of the line peak as a function of the cloud's major axis in the bottom panel, each entry sampled in strips across the face of the cloud parallel to the minor axis. The width of the velocity distribution along the ordinate is a good indicator of the one-dimensional velocity dispersion $\sigma_{\rm 1D}$ of the cloud. We see that $\sigma_{\rm 1D}$ reaches values of a few km s^{-1}, about an order of magnitude larger than the sound speed of the dense molecular gas, $c_{\rm s} \approx 0.2\,$km s^{-1}.

More detailed analysis reveals that the observed velocity dispersion $\sigma_{\rm 1D}$ is related to the size L of the cloud by

$$\sigma_{\rm 1D} \approx 0.5 \left(\frac{L}{1.0\,{\rm pc}}\right)^{1/2} \,{\rm km\,s^{-1}}. \tag{106}$$

This goes back to the seminal work by Larson (1981), who compared measurements of different clouds available at that time, and it has been confirmed by many follow-up studies both in our Milky Way as well as neighboring satellite galaxies (e.g. Solomon et al. 1987; Heyer and Brunt 2004; Bolatto et al. 2008; Falgarone et al. 2009; Roman-Duval et al. 2011; Caldú-Primo et al. 2013). There is still some debate about the normalization and about slight variations in the slope (Heyer et al. 2009; Shetty et al. 2012; Hughes et al. 2013), but in general the relation (106) is thought to reflect a more

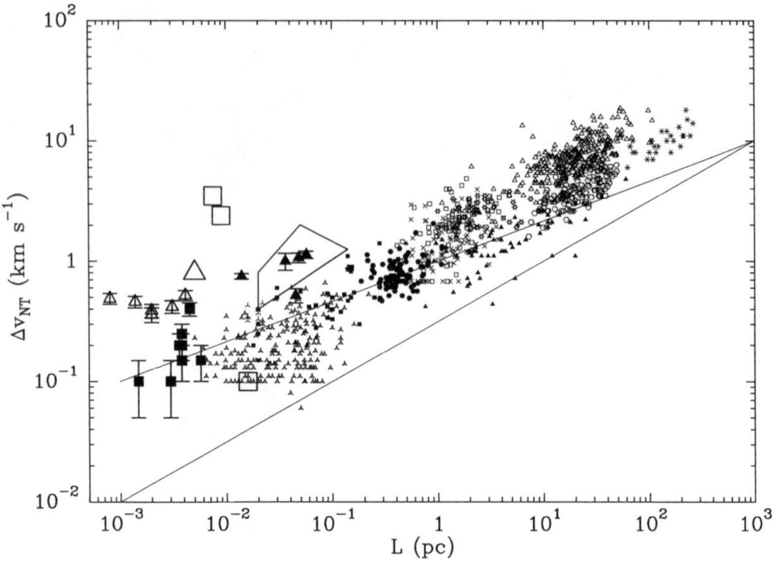

Fig. 12 Relation between velocity dispersion (as measured by the width of the J =1–0 rotational line transition of ^{12}CO) and spatial scale in the Galactic ISM. The data points come from a wide range of observations that trace different structures and physical conditions (in terms of density, temperature, excitation parameter, etc.). This in part contributes to the large scatter in the data. However, altogether the observations reveal a clear power-law relation of the form $\Delta v_{\rm NT} \propto L^{\alpha}$. To guide the eyes, the *solid lines* illustrate the slopes $\alpha = 1/3$ and $\alpha = 1/2$. The lower limit of $\Delta v_{\rm NT} \approx 0.1$ km s^{-1} is due to the spectral resolution in the data and corresponding noise level. The figure is taken from Falgarone et al. (2009), where further details and a full list of references can be found

or less universal property of the ISM (see Fig. 12). The most common interpretation is the presence of turbulent gas motions. On scales above ~0.1 pc (which corresponds to the typical sizes of prestellar cores; see Sect. 6.1), the velocities inferred from Eq. (106) exceed values of the thermal line broadening (where the one-dimensional velocity dispersion $\sigma_{\rm 1D}$ is comparable to the sound speed $c_{\rm s}$). On scales of molecular cloud complexes, we measure root mean square Mach numbers of 10 or larger, clearly indicative of highly supersonic turbulence. We also note that these motions seem to exceed the typical Alfvén velocities in molecular clouds,

$$v_{\rm A} = \left(\frac{B^2}{4\pi\rho}\right)^{1/2}, \tag{107}$$

with B and ρ being the magnetic field strength and the mass density of the gas, respectively. The observed turbulence is not only supersonic but also super-Alfvénic (e.g. Padoan and Nordlund 1999; Heyer and Brunt 2012). In essence, this means that the energy density associated with turbulent gas motions dominates over both the thermal energy density as well as the magnetic energy density. We also note that the

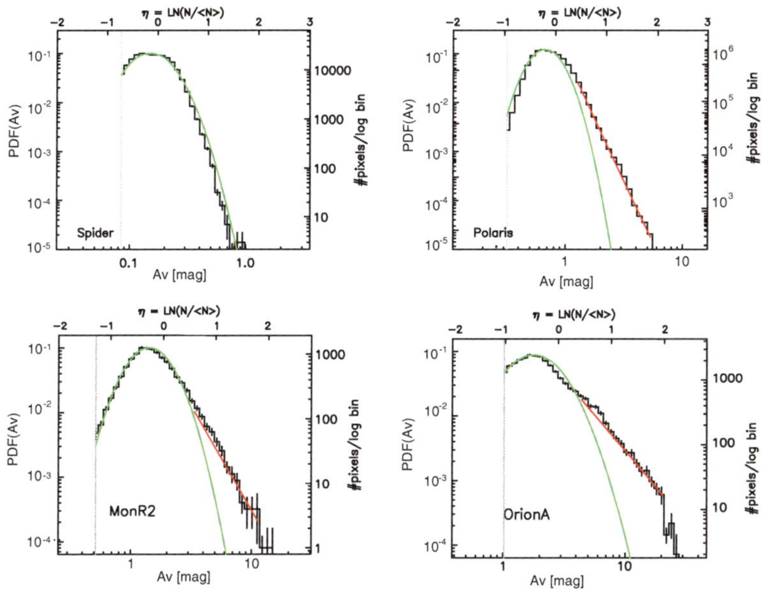

Fig. 13 Column density PDFs derived from *Herschel* observations (see e.g. Schneider et al. 2013) for four different nearby molecular clouds: two tenuous high-latitude clouds, Spider (*top left*) and Polaris (*top right*), and two star dense star-forming regions, Monoceros R2 (*bottom left*) and Orion A (*bottom right*). For a map of the latter two clouds, see Fig. 10. The *lower* abscissa gives the visual extinction, A_V, which we take as a proxy of the column density N. The *upper axis* indicates the natural logarithm of the column density normalized to the mean value, $\eta = \ln(N/\langle N\rangle)$. The *left* ordinate is the PDF of the extinction, and to the *right*, we provide the corresponding total number of pixels to indicate the statistical significance of the observation. The *green curve* indicates the fitted PDF, and the *red line* shows a possible power-law fit to the high A_V tail. The plots are adopted from Schneider et al. (2015)

observed linewidths generally are not due to large-scale collapse as inferred from the generally rather low star formation rates and the absence of inverse P-Cygni line profiles.

The analysis of extinction or dust emission maps in nearby molecular clouds reveals a roughly log-normal distribution of column densities in tenuous cirrus-like clouds with no or little star formation, and they show the development of a power-law tail at high column density that becomes more pronounced for more massive and more vigorously star-forming clouds (Lada et al. 2010; Kainulainen et al. 2011; 2013; Schneider et al. 2012, 2015; Alves et al. 2014). Typical examples are provided in Fig. 13, where we take the visual extinction, A_V, as a proxy for the column density. Spider (top left) and Polaris (top right) are high latitude clouds located in the North Celestial Loop (Meyerdierks et al. 1991). Spider shows no signs of star formation and is a prototypical example of a cloud with a log-normal PDF, while Polaris seems to be forming some low-mass stars and exhibits a weak power-law tail. Monoceros and Orion A (see also Fig. 10) have much higher average densities and are forming clusters containing intermediate to high-mass stars. They exhibit a clear power-law tail at high extinctions. These observations are essential, because the characteristics of the (column) density distribution function are important input to our current theoretical star formation models. This is discussed in detail in Sect. 6.4.4.

4.2 Simple Theoretical Considerations

At this point, we need to digress from our discussion of molecular cloud properties and turn our attention to the theoretical models introduced to describe turbulent flows. We begin by introducing the classical picture of incompressible turbulence. This is a good description for turbulent flows with velocities that are significantly smaller than the speed of sound, such as those we typically encounter on Earth. For very subsonic flows we can infer from the continuity equation (108) that density fluctuations are negligible. We note that for typical ISM conditions, however, the turbulence is highly supersonic, and we need to go beyond this simple picture as the compressibility of the medium becomes important, for instance when we want to understand the formation of stars as discussed in Sect. 6.3. In any case, we assume that kinetic energy is inserted into the system on some well-defined, large scale $L_{\rm D}$, and that it cascades down through a sequence of eddies of decreasing size until the size of the eddies becomes comparable to the mean free path λ. The kinetic energy associated with eddy motion turns into heat (random thermal motion) and is dissipated away. This picture of the turbulent cascade goes back to Richardson (1920).

We first follow Kolmogorov (1941) and derive the corresponding power spectrum of the turbulent kinetic energy, which describes terrestrial flows, such as the motion of air in the Earth's atmosphere or the flow of water in rivers and oceans. Then we turn to supersonic motions and focus on additional aspects that are characteristic of ISM turbulence. For the level of our discussion, it is sufficient to think of turbulence as the gas flow resulting from random motions on many scales, consistent with the simple scaling relations discussed above. For a more detailed discussion of the complex statistical characteristics of turbulence, we refer the reader to the excellent textbooks by Frisch (1996); Lesieur (1997), or Pope (2000). For a thorough account of ISM turbulence, we point again to the reviews by Elmegreen and Scalo (2004) and Scalo and Elmegreen (2004), and for the relation to star formation to Mac Low and Klessen (2004) and Ballesteros-Paredes et al. (2007).

4.2.1 Energy Cascade in Stationary Subsonic Turbulence

Hydrodynamical flows exhibit two fundamentally different states. For small velocities, they tend to be laminar and smooth. If the velocity increases, however, the flow becomes unstable. It becomes turbulent and highly chaotic. This transition occurs when advection strongly dominates over dissipation. To see this, let us consider the equations describing the motion of a fluid element. From the continuity equation,

$$\frac{\partial \rho}{\partial t} + \mathbf{v} \cdot \nabla \rho = -\rho \nabla \cdot \mathbf{v} \;, \tag{108}$$

we can infer for incompressible flows ($\rho =$ const.) that $\nabla \cdot \mathbf{v} = 0$. The equation of motion, also called the Navier-Stokes equation, then simplifies to

$$\frac{\partial \mathbf{v}}{\partial t} + (\mathbf{v} \cdot \nabla)\mathbf{v} - \nu \nabla^2 \mathbf{v} = -\frac{1}{\rho} \nabla P \;, \tag{109}$$

where $\mathbf{v}$ is the fluid velocity, P and ρ are pressure and density, and ν is the kinetic viscosity with the units $\mathrm{cm^2\,s^{-1}}$. At any spatial scale ℓ, we can compare the advection term $(\mathbf{v} \cdot \nabla)\mathbf{v}$ with the dissipation term $\nu \nabla^2 \mathbf{v}$. To get an estimate of their relative importance for the flow dynamics we approximate ∇ by $1/\ell$ and obtain

$$(\mathbf{v} \cdot \nabla)\mathbf{v} \approx \frac{v_\ell^2}{\ell} \text{ and } \nu \nabla^2 \mathbf{v} \approx \frac{\nu v_\ell}{\ell^2} \;, \tag{110}$$

with v_ℓ being the typical velocity on scale ℓ. This ratio defines the dimensionless Reynolds number on that scale,

$$\mathrm{Re}_\ell = \frac{v_\ell \ell}{\nu} \;. \tag{111}$$

It turns out that a flow becomes unstable and changes from being laminar to turbulent if the Reynolds number exceeds a critical value $\mathrm{Re_{cr}} \approx \text{few} \times 10^3$. The exact value depends on the flow characteristics. For example, pipe flows with $\mathrm{Re} < 2 \times 10^3$ are usually laminar, while flows with $\mathrm{Re} > 4 \times 10^3$ are most certainly turbulent. For intermediate Reynolds numbers both laminar and turbulent flows are possible, depending on other factors, such as pipe roughness and flow uniformity. These flows are often called transition flows. In the ISM, the Reynolds number can easily exceed values of 10^9 or more, indicating that the ISM is highly turbulent.

A full analysis of the turbulent instability is very difficult and in general an unsolved problem. In the classical picture turbulence causes the formation of eddies on a wide range of different length scales. In this picture, some driving mechanism creates eddies on some large scale. These live for about one crossing time, and then fragment into smaller eddies, which again break up into even smaller eddies, and so forth. Most of the kinetic energy of the turbulent motion is contained on large scales. It cascades down to smaller and smaller ones in an inertial and essentially inviscid way. This holds as long as the advection term dominates over dissipation, i.e. as long as $\mathrm{Re} \gg \mathrm{Re_{cr}}$. Eventually this hierarchy creates structures on scales that are small enough so that molecular diffusion or other forms of dissipation become important. The turbulent eddies become so tiny that they essentially turn into random thermal motion, the kinetic energy they carry becomes heat, and may be radiated away.

To obtain an estimate of the scaling behavior of turbulent flows let us look at the specific kinetic energy ϵ_ℓ carried by eddies of size ℓ. With v_ℓ being the typical rotational velocity across the eddy and with $t_\ell \approx \ell/v_\ell$ being the typical eddy turnover time, we can estimate the energy flow rate through eddies of size ℓ as

$$\dot{\epsilon} \approx \frac{\epsilon_\ell}{t_\ell} \approx \left(\frac{v_\ell^2}{2}\right)\left(\frac{\ell}{v_\ell}\right)^{-1} \approx \frac{v_\ell^3}{\ell} . \tag{112}$$

As long as $\mathrm{Re}_\ell \gg 1$, dissipation is negligible. The rate $\dot{\epsilon}$ is conserved, and the kinetic energy simply flows across ℓ from larger scales down to smaller ones. This defines the inertial range of the turbulent cascade. It ends when Re_ℓ approaches unity at the dissipation scale λ_ν. Assume now that kinetic energy is inserted into the system on some large scale L with a typical velocity v_L. Then, the inertial range covers the scales

$$L > \ell > \lambda_\nu . \tag{113}$$

In this regime, the energy flow $\dot{\epsilon}$ is independent of scale, as kinetic energy cannot be accumulated along the turbulent cascade. This implies that the typical eddy velocity v_ℓ changes with eddy scale ℓ as $v_\ell = \epsilon_\ell \ell^{1/3}$. As a consequence, the largest eddies carry the highest velocities,

$$v_\ell \approx v_L \left(\frac{\ell}{L}\right)^{1/3} , \tag{114}$$

but the smallest ones have the highest vorticity,

$$\Omega_\ell = \nabla \times \mathbf{v}_\ell \approx \frac{v_\ell}{\ell} \approx \frac{v_L}{(\ell^2 L)^{1/3}} \approx \left(\frac{L}{\ell}\right)^{2/3} \Omega_L . \tag{115}$$

Indeed, in agreement with this picture of the turbulent cascade, observations of nearby molecular clouds reveal that the energy is carried by large-scale modes, indicating that the turbulent velocity field in these clouds is driven by external sources (see Sect. 4.5).

We can obtain an estimate for the size of the inertial range (113) based on the requirement that $\mathrm{Re} \approx 1$ on the dissipation scale λ_ν. In combination with (114), this leads to

$$\frac{L}{\lambda_\nu} \approx \mathrm{Re}^{3/4} . \tag{116}$$

With Reynolds numbers $\mathrm{Re} \approx 10^9$ and above, the turbulent cascade in the ISM extends over more than six orders of magnitude in spatial scale.

We now look at the autocorrelation function of the velocity fluctuations on the scale ℓ defined as

$$\xi_v(\boldsymbol{\ell}) = \langle [\mathbf{v}(\mathbf{x}+\boldsymbol{\ell}) - \mathbf{v}(\mathbf{x})]^2 \rangle , \tag{117}$$

which is the average of the square of all velocity differences between any two points in space separated by a lag $\boldsymbol{\ell}$. We have assumed that the system has zero net velocity,

$\langle \mathbf{v}(\mathbf{x}) \rangle = 0$. If turbulence is isotropic, the autocorrelation function depends only on the separation $\ell = |\boldsymbol{\ell}|$ and not on the direction, and so $\xi_v(\boldsymbol{\ell}) = \xi_v(\ell)$. From (112) we obtain

$$\xi_v(\ell) \propto \langle v_\ell^2 \rangle \propto (\dot{\epsilon}\ell)^{2/3} . \tag{118}$$

In particular, we are interested in the Fourier transform of $\xi_v(\boldsymbol{\ell})$, the power spectrum $P(k)$ of the velocity fluctuations. For random Gaussian fluctuations, both are related via

$$\frac{1}{2}\int_0^\infty \langle v_\ell'^2 \rangle d^3\ell' = \int_\infty^0 P_v(k) d^3k , \tag{119}$$

which simply is the specific kinetic energy in the system. On each scale $\ell = 2\pi/k$ Eq. (119) can be approximated by

$$P_v(k) \propto \ell^3 \xi_v \propto k^{-3} \left(\dot{\epsilon} k^{-1}\right)^{2/3} \propto \dot{\epsilon}^{2/3} k^{-11/3} . \tag{120}$$

If we consider isotropic turbulence with $d^3k \to 4\pi k^2 dk$, then the power in modes in the wave number range k to $k + dk$ is

$$P_v\, k^2 dk \propto \dot{\epsilon}^{2/3} k^{-5/3} dk . \tag{121}$$

This is the Kolmogorov spectrum of isotropic incompressible turbulence in the inertial range.

4.2.2 Energy Cascade in Stationary Supersonic Turbulence

We now turn to the opposite limit of highly compressible turbulence, where the flow can be described as a network of interacting shock fronts. To simplify our discussion, we neglect the effects of pressure forces. This leads to the so-called Burgers (1939) turbulence. He introduced a simplified non-linear partial differential equation of the form

$$\frac{\partial v}{\partial t} + v \frac{\partial v}{\partial x} = \nu \frac{\partial^2 v}{\partial x^2} , \tag{122}$$

as approximation to the full Navier-Stokes equation, in order to study the mathematical properties of turbulent flows. Indeed, in the high Mach number regime, where the velocities v are much larger than the sound speed c_s, we can neglect the pressure term, since $P \propto c_\mathrm{s}^2$, and Burgers' equation (122) is identical to Eq. (109) in one dimension. If we consider the width of the shock transition to be infinitely thin, then the density or the velocity jump in the shock can be mathematically described by a Heaviside step function. For isotropic turbulence, there is always a shock that runs parallel to the considered line-of-sight, and we can naively estimate the power in the wavenumber range k to $k + dk$ as

$$P_v dk \propto v_k^2 dk \propto k^{-2} dk \, . \tag{123}$$

In a more general sense, this follows from the fact that the energy spectrum of a field is determined by its strongest singularity. If a function $u(x)$ has a discontinuity in the $(m-1)$th order derivative, then its energy spectrum has the form of a power-law $P(k) \propto k^{-2m}$. In a shock the velocity itself is discontiuous, and with $m = 1$ we obtain Eq. (123). A less handwaving derivation is very involved (see e.g. Boldyrev 1998; Verma 2000; Boldyrev et al. 2004; Bec and Khanin 2007), but in general leads to a similar conclusion.

4.2.3 Decay Rate of Turbulent Energy

So far, we have considered the case of stationary turbulence. This requires an energy source that continuously excites large-scales modes to compensate for the loss of energy at the dissipation scale. For a long time, it was thought that the rate of energy dissipation in a magnetized gas differs significantly from purely hydrodynamic turbulence. Arons and Max (1975), for example, suggested that presence of strong magnetic fields could explain the supersonic motions ubiquitously observed in molecular clouds (Sect. 4.1). In their view, interstellar turbulence is a superposition of Alfvén waves, propagating in many different directions with different wavelengths and amplitudes. These are transverse waves traveling with the Alfvén velocity (Eq. 107), and they are dissipationless in the linear regime under the assumption of ideal magnetohydrodynamics. However, if ambipolar diffusion is taken into account, i.e. the drift between charged and neutral particles in the partially ionized ISM, these waves can be dissipated away at a rate substantial enough to require energy input from a driving source to maintain the observed motions (e.g. Zweibel and Josafatsson 1983; Zweibel 2002). Furthermore, if one includes higher order effects, then the dissipation becomes comparable to the purely hydrodynamic case (e.g. Cho and Lazarian 2003).

This analysis is supported by numerical simulations. One-dimensional calculations of non-driven, compressible, isothermal, magnetized turbulence by Gammie and Ostriker (1996) indicate a very efficient dissipation of kinetic energy. They also found that the decay rate depends strongly on the adopted initial and boundary conditions. Mac Low et al. (1998); Stone et al. (1998), and Padoan and Nordlund (1999) determined the decay rate in direct numerical simulations in three dimensions, using a range of different numerical methods. They uniformly report very rapid decay rates and propose a power-law behavior for the decay of the specific kinetic energy of the form $\dot{\epsilon} \propto t^{-\eta}$, with $0.85 < \eta < 1.1$. A typical result is shown in Fig. 14. Magnetic fields with strengths ranging up to equipartition with the turbulent motions seem to reduce η to the lower end of this range, while unmagnetized supersonic turbulence shows values closer to $\eta \lesssim 1.1$.

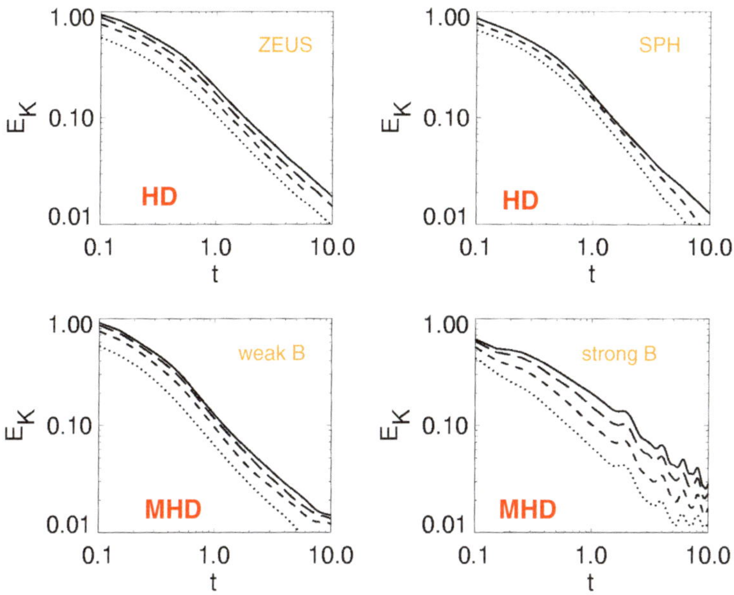

Fig. 14 Decay of supersonic turbulence. The plots show the time evolution of the total kinetic energy E_K in a variety of three-dimensional numerical calculations of decaying supersonic turbulence in isothermal ideal gas with initial root mean square Mach number of 5, calculated with two different numerical codes. ZEUS is a Eulerian grid code that solves the equations of magnetohydrodynamics (Stone and Norman 1992a, b), while SPH follows a particle-based approach (Benz 1990; Springel 2010). The *top panels* depict the decay of purely hydrodynamic turbulence, the *bottom panels* show the decay properties with weak and strong magnetic fields. For more details see Mac Low et al. (1998)

Besides directly measuring the decay of the kinetic energy in the absence of driving sources in a closed system, we can also continuously insert energy and determine the resulting velocity dispersion. Mac Low (1999) and Elmegreen (2000b) argue that the dissipation time, $t_\mathrm{d} = \epsilon/\dot{\epsilon}$, is comparable to the turbulent crossing time in the system,

$$t_\mathrm{d} \approx \frac{L}{\sigma}\,, \tag{124}$$

where L is again the driving scale and σ the velocity dispersion. This holds regardless of whether the gas is magnetized or not and also extends into the subsonic regime. The loss of the specific turbulent kinetic energy, $\epsilon = 1/2\sigma^2$ is then,

$$\dot{\epsilon} = \frac{\epsilon}{t_\mathrm{d}} = \xi \frac{1}{2}\frac{\sigma^3}{L}\,. \tag{125}$$

We have recovered Kolmogorov's formula for the energy decay rate (112), modulo an efficiency coefficient $\xi/2$ that depends on the physical parameters of the system or on the details of the numerical method employed.

4.3 Scales of ISM Turbulence

As introduced in Sect. 4.2.3, interstellar turbulence decays on roughly a crossing time. It needs to be continuously driven in order to maintain a steady state. Here we compare the various astrophysical processes that have been proposed as the origin of ISM turbulence, and mostly follow the discussion in Mac Low and Klessen (2004) and Klessen and Hennebelle (2010).

We begin with an analysis of the typical scales of ISM turbulence in our Galaxy, then calculate the corresponding turbulent energy loss, and finally turn our attention to the various astrophysical driving mechanisms that have been proposed to compensate for the decay of turbulent energy. We point out that a key assumption is that the ISM in the Milky Way evolves in a quasi steady state, so that energy input and energy loss rates roughly balance when being averaged over secular timescales and over large enough volumes of the Galactic disk. We caution the reader that this need not necessarily be the case.

The self-similar behavior of turbulent flows only hold in the inertial range, i.e. on scales between the driving and dissipation scales. We now want to address the question of what these scales are in the Galactic ISM. The answer clearly depends on the different physical processes that stir the turbulence and that provide dissipation. There is a wide variety of driving mechanisms proposed in the literature, ranging from stellar feedback acting only on very local scales up to accretion onto the Galaxy as a whole, inserting energy on very large scales. As we discuss below, we favor the latter idea.

Regardless of the detailed driving process, a firm outer limit to the turbulent cascade in disk galaxies is given by the disk scale height. If molecular clouds are created at least in part by converging large-scale flows triggered by accretion, or by spiral shocks, or by the collective influence of recurring supernovae explosions, then the extent of the Galactic disk is indeed the true upper scale of turbulence in the Milky Way. For individual molecular clouds this means that turbulent energy is fed in at scales well above the size of the cloud itself. This picture is supported by the observation that the clouds' density and velocity structure exhibits a power-law scaling behavior extending all the way up to the largest scales observed in today's surveys (Ossenkopf and Mac Low 2002; Brunt 2003; Brunt et al. 2009).

One could argue that the outer scale of the ISM turbulence actually corresponds to the diameter of the Galaxy as a whole (rather than the disk scale height) and that the largest turbulent eddy is the rotational motion of the disk itself. However, because the disk scale height H is typically less than 10 % of the disk radius R, this motion is intrinsically two-dimensional and if we restrict our discussion to three-dimensional turbulence then H is the maximum outer scale. In addition, we note that the decay time (124) is comparable for both approaches. At the solar radius, $R = 8.5\,\mathrm{kpc}$, the rotational speed is $v_{\rm rot} = 220\,\mathrm{km\ s^{-1}}$, leading to $t_{\rm d} \approx R/v_{\rm rot} \approx 38\,\mathrm{Myr}$. If we adopt an average HI disk scale height of $H = 0.5\,\mathrm{kpc}$ and a typical velocity dispersion of $\sigma = 12\,\mathrm{km\ s^{-1}}$ (Ferriére 2001; Kalberla 2003), then our estimate of the turbulent decay time, $t_{\rm d} \approx L/\sigma \approx 40\,\mathrm{Myr}$, is essentially the same. In conclusion, for the estimate of the energy decay rate in typical disk galaxies, and by the same token, for the calculation of the required turbulent driving rate, it does not really matter what we assume for the outer scale of the turbulence (for further discussions, see e.g. Klessen and Hennebelle 2010). This follows, because for typical disk galaxies, the disk scale height and radius, as well as the velocity dispersion and rotational velocity scale similarly, that is $H/R \approx 0.1$ and $\sigma/v_{\rm rot} \approx 0.1$.

The estimate of the dissipation scale is also difficult. For purely hydrodynamic turbulence, dissipation sets in when molecular viscosity becomes important. The corresponding spatial scales are tiny. In the ISM the situation is more complex because we are dealing with a magnetized, partially ionized, dusty plasma. Zweibel and Josafatsson (1983) argue that ambipolar diffusion (i.e. the drift between charged and neutral species in this plasma) is the most important dissipation mechanism in typical molecular clouds with very low ionization fractions $x = \rho_i/\rho_n$, where ρ_i and ρ_n are is the densities of ions and neutrals, respectively, with $\rho = \rho_i + \rho_n$ being the total density. We can then replace the kinetic viscosity in the Navier-Stokes equation (109) by the ambipolar diffusion coefficient

$$\nu_{\rm AD} = v_{\rm A}^2/\zeta_{ni}, \tag{126}$$

where $v_{\rm A}^2 = B^2/4\pi\rho_n$ approximates the effective Alfvén speed for the coupled neutrals and ions if $\rho_n \gg \rho_i$, and $\zeta_{ni} = \alpha\rho_i$ is the rate at which each neutral is hit by ions. The coupling constant α is given by

$$\alpha = \langle\sigma v\rangle/(m_i + m_n) \approx 9.2 \times 10^{13}\,{\rm cm^3\,s^{-1}\,g^{-1}}\,, \tag{127}$$

with m_i and m_n being the mean mass per particle for the ions and neutrals, respectively. Typical values in molecular clouds are $m_i = 10\,m_{\rm H}$ and $m_n = 2.35m_{\rm H}$. It turns out that α is roughly independent of the mean velocity, as the ion-neutral cross-section σ scales inversely with velocity in the regime of interest. For further details on the microphysics, consult the excellent textbooks by Osterbrock (1989); Tielens (2010), or Draine (2011).

We can define an ambipolar diffusion Reynolds number in analogy to Eq. (111) as

$$\mathrm{Re}_{\mathrm{AD},\ell} = \ell v_\ell/\nu_{\rm AD} = \mathcal{M}_{\rm A}\ell\,\zeta_{ni}/v_{\rm A}, \tag{128}$$

which must fall below unity on scales where ambipolar diffusion becomes important. As before, v_ℓ is the characteristic velocity at scale ℓ, and we define $\mathcal{M}_{\rm A} = v_\ell/v_{\rm A}$ as the characteristic Alfvén Mach number at that scale. From the condition $\mathrm{Re}_{\mathrm{AD},\lambda} = 1$, we derive the dissipation scale due to ambipolar diffusion as

$$\lambda_{\rm AD} = \frac{v_{\rm A}}{\mathcal{M}_{\rm A}\zeta_{ni}} \approx 0.041\ {\rm pc}\left(\frac{B}{10\,\mu{\rm G}}\right)\mathcal{M}_{\rm A}^{-1}\left(\frac{x}{10^{-6}}\right)^{-1}\left(\frac{n_n}{10^3\,{\rm cm^{-3}}}\right)^{-3/2}, \tag{129}$$

with the magnetic field strength B, the ionization fraction x, the neutral number density n_n, and where we have taken $\rho_n = \mu n_n$, with a mean particle mass $\mu =$

$2.35\, m_{\rm H} = 3.92 \times 10^{-24}$ g typical for molecular clouds. It is interesting to note, that the resulting value for $\lambda_{\rm AD}$ is comparable to the typical sizes of protostellar cores (e.g. Bergin and Tafalla 2007). Indeed, the velocity dispersion in these objects is dominated by thermal motions (Goodman et al. 1998).

We note that there are wave families that can survive below $\lambda_{\rm AD}$ that resemble gas dynamic sound waves. Consequently, even on scales where the magnetic field becomes uniform, the gas dynamical turbulent cascade could continue. This is determined by the dimensionless magnetic Prandtl number,

$$\mathrm{Pr}_{\rm mag} = \frac{\mathrm{Re}_{\rm AD}}{\mathrm{Re}} = \frac{\nu}{\nu_{\rm AD}} , \qquad (130)$$

which compares the relative importance of viscous and magnetic diffusion processes. For small magnetic Prandtl numbers the hydrodynamic inertial range extends beyond the magnetic one, and vice versa, for $\mathrm{Pr}_{\rm mag} \gg 1$ fluctuations in the magnetic field can occur on scales much smaller than the hydrodynamic diffusion limit (for further discussions see e.g. Schekochihin et al. 2004a, b; Schober et al. 2012a, b).

Combining these findings with the molecular cloud properties discussed in Sect. 4.1.2 we arrive at the following picture, as illustrated in Fig. 15. On the scales of individual molecular clouds and large molecular cloud complexes, the observed turbulence is highly supersonic. We know that the density contrast created by shocks in isothermal gas scales with the Mach number $\mathcal{M}$ to the second power, $\Delta\rho/\rho \propto \mathcal{M}^2$ (e.g. Landau and Lifshitz 1959). Consequently, for $\mathcal{M} \approx 10$ we expect density contrasts of roughy 100. This is indeed observed in molecular clouds, where the mean density is around 100 particles per cubic centimeter and where the high-density

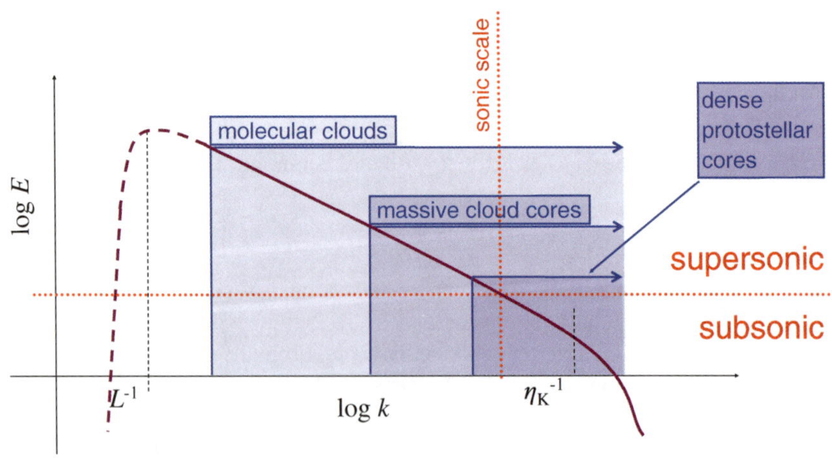

Fig. 15 Simple cartoon picture of the turbulent energy spectrum, i.e. of the kinetic energy carried by modes of different wave numbers k, and their relation to different cloud structures (see also Table 3). Turbulence is driven on large scales comparable to the size L of the cloud and is dissipated on very small scales $\eta_{\rm K}$. Adopted from Klessen (2011)

cores exceed values of $10^4\,\mathrm{cm}^{-3}$ and more (see Table 3). When we focus on cluster-forming cloud cores (or their not-yet-star-forming counterparts, the so-called infrared dark clouds) we still measure $\mathcal{M} \approx 5$ leading to localized density fluctuations of $\Delta\rho/\rho \approx 25$, on average. As we discuss further in Sect. 6, some of these fluctuations may exceed the critical mass for gravitational collapse to set in. The presence of turbulence thus leads to the break-up into smaller units. The core fragments to build up a cluster of stars with a wide range of masses rather than forming one single high-mass star. We call this process gravoturbulent fragmentation, because turbulence generates the distribution of clumps in the first place, and then gravity selects a subset of them for subsequent star formation (see also Sect. 6.3). Finally, when focusing on low-mass cores, the velocity field becomes more coherent and the turbulence subsonic (see Sect. 6.1). This defines the sonic scale at around 0.1 pc. Such structures are no longer subject to gravoturbulent fragmentation and are the direct progenitors of individual stars or binary systems. We note, however, that gravitational fragmentation may still occur within the protostellar accretion disk that builds up in the center of the core due to angular momentum conservation (Sect. 6.6). This process is likely to produce close binaries (see e.g. Bodenheimer 1995; Machida 2008). The fact that the observed velocity dispersion approaches the thermal value as one zooms in on smaller and smaller scales is a direct consequence of the turbulent cascade, as expressed in the observed Larson relation (Eq. 106).

4.4 Decay of ISM Turbulence

With the above considerations, we are now in a position to calculate the rate of energy loss in the Galactic ISM due to the decay of turbulence. Our Milky Way is a typical $L_\star$ galaxy with a total mass including dark matter of about $10^{12}\,\mathrm{M}_\odot$ out to the virial radius at $\sim$250 kpc (e.g. Xue et al. 2008). The resulting rotation curve is $220\,\mathrm{km\,s}^{-1}$ at the solar radius, $R_\odot \approx 8.5$ kpc, and it declines to values slightly below 200 km s^{-1} at a radius of 60 kpc (Xue et al. 2008). Star formation occurs out to a radius of about 15 kpc (Rix and Bovy 2013). The total mass in the disk in stars is about $2.7 \times 10^{10}\,\mathrm{M}_\odot$, and in gas it is about $8 \times 10^9\,\mathrm{M}_\odot$ (see Table 4, as well as Naab and Ostriker 2006). Assuming a global baryon fraction of 17 %, this corresponds to 40 % of all the baryonic mass within the virial radius and implies that roughly the same amount of baryons is in an extended halo in form of hot and tenuous gas. The gaseous disk of the Milky Way can be decomposed into a number of different phases. We follow Ferriére (2001) and Kalberla (2003), and consider molecular gas (H_2 as traced e.g. via CO emission) as well as atomic hydrogen gas (as observed, e.g. by the HI 21 cm emission). The HI component can be separated into a cold ($T \approx$ few $\times\, 10^2$ K) and a hot ($T \approx$ few $\times\, 10^3$ K) component. Because they have similar overall distributions we consider them together. The scale height of HI ranges from $\sim$230 pc within 4 kpc up to values of $\sim$3 kpc at the outer Galactic boundaries. The HI disk therefore is strongly flared. We take 500 pc as a reasonable mean value, but note that this introduces a high degree of uncertainty. Also, we neglect the warm

Table 4 Properties of gas components of the Milky Way

Component	M (10^9 M$_\odot$)[a]	n (cm^{-3})[b]	L (pc)[c]	σ (km s^{-1})[d]	$E_{\rm kin}$ (10^{55} erg)[e]
Molecular gas	2	0.7	150	5	0.5
Atomic gas	6	0.4	1000	12	8.6

[a]Total mass of the component. Values from Ferriére (2001) and Kalberla (2003)
[b]Estimate of volume-averaged midplane number density. Values from Ferriére (2001) and Kalberla (2003). Note that the value for H_2 gas is so low compared to Table 3, because the volume filling factor of molecular clouds in the Galactic disk is very small
[c]We consider the disk thickness as being twice the observed scale height
[d]The parameter σ is the three-dimensional velocity dispersion
[e]Total kinetic energy of the component, $E_{\rm kin} = 1/2\, M\sigma^2$

and the hot ionized medium in our analysis, since the ionized gas within HII regions or supernova remnants carries little of the turbulent kinetic energy compared to the other components. Indeed, roughly 95 % of the turbulent kinetic energy is carried by the atomic component. The adopted values are summarized in Table 4.

One of the remarkable features of spiral galaxies is the nearly constant velocity dispersion σ, e.g. as measured using the HI 21 cm emission line, seemingly independent of galaxy mass and type (Dickey and Lockman 1990; van Zee and Bryant 1999; Tamburro et al. 2009). The inferred values of σ typically fall in a range between 10 and 20 km s^{-1} (Bigiel et al. 2008; Walter et al. 2008) and extend well beyond the optical radius of the galaxy with only moderate fall-off as one goes outwards. Quite similar behavior is found in the molecular gas, when increasing the sensitivity in the outer regions by stacking the data (Caldú-Primo et al. 2013). It is interesting in this context that the transition from the star-forming parts of the galaxy to the non-star-forming outer disk seems not to cause significant changes in the velocity dispersion. This approximate indifference to the presence of stellar feedback sources sets severe constraints on the physical processes that can drive the observed level of turbulence (see Sects. 4.5 and 4.6).

Using Eq. (125), we can calculate the average loss of kinetic energy density e (erg/cm^3) per unit time in the ISM. With $e = \rho\epsilon$, where ϵ is the specific energy (in units of erg/g) and $\rho = \mu n$ is the mass density, we obtain

$$\begin{aligned} \dot{e} &= -\frac{1}{2}\frac{\mu n \sigma^3}{L} \\ &\approx -3.5\times 10^{-27}\,\mathrm{erg\,cm^{-3}\,s^{-1}} \left(\frac{n}{1\,\mathrm{cm^{-3}}}\right)\left(\frac{\sigma}{10\,\mathrm{km\,s^{-1}}}\right)^3\left(\frac{L}{100\,\mathrm{pc}}\right)^{-1}, \end{aligned} \tag{131}$$

where we have set the efficiency parameter ξ in Eq. (125) to unity, and where again n, σ, and L are the number density, the velocity dispersion, and the turbulent driving scale, respectively. For simplicity, we have assumed a mean mass per particle $\mu = 1.26\,m_{\rm H} = 2.11\times 10^{-24}$ g typical for purely atomic gas. If we consider different ISM phases, this number needs to be adapted. If we plug in the values from Table 4, we obtain the following estimate for our Galaxy:

$$\dot{e}_{\rm ISM} = \dot{e}_{\rm H_2} + \dot{e}_{\rm HI} \approx -4.5 \times 10^{-28}\,{\rm erg\,cm^{-3}\,s^{-1}}\,. \tag{132}$$

According to (124) the decay timescale can be computed as

$$t_{\rm d} = \frac{e}{\dot{e}} = \frac{L}{\sigma} \approx 10\,{\rm Myr}\left(\frac{L}{100\,{\rm pc}}\right)\left(\frac{\sigma}{10\,{\rm km\,s^{-1}}}\right)^{-1}\,, \tag{133}$$

which is simply the turbulent crossing time on the driving scale.

4.5 Sources of ISM Turbulence: Gravity and Rotation

There is a wide range of physical processes that could potentially drive the observed turbulent flows in the ISM. We will introduce and discuss the most important ones that have been proposed in the literature. We identify two main categories of sources. In this Section we focus on processes that convert a fraction of the potential energy available in the galaxy into turbulent gas motions. We first look at the process of accretion-driven turbulence, and then turn to rotation. As the rotational motion of the Galaxy is ultimately driven by gravity, we include all mechanisms that can tap the rotational energy here as well. In Sect. 4.6 we then assess the influence of stellar feedback on the large-scale dynamics of the ISM. Our list is sorted in such a way that the processes which seem most important to us are introduced first.

4.5.1 Accretion onto the Galaxy

We argue that it is the accretion process that inevitably accompanies any astrophysical structure formation, whether it is the formation of galaxies or the birth of stars, that drives the observed turbulent motions. We propose that this process is universal and makes significant contributions to the turbulent energy on all scales (see also Field et al. 2008). When cosmic structures grow, they gain mass via accretion. This transport of matter is associated with kinetic energy and provides an ubiquitous source for the internal turbulence on smaller scales. We follow the analysis of Klessen and Hennebelle (2010) and ask whether the accretion flow onto galaxies provides enough energy to account for the observed ISM turbulence.

We begin with a summary of what we know about accretion onto spiral galaxies like our Milky Way. Our Galaxy forms new stars at a rate of $\dot{M}_{\rm SF} \sim 2 - 4\,{\rm M_\odot\,yr^{-1}}$ (e.g. Naab and Ostriker 2006; Adams et al. 2013). Its gas mass is about $M_{\rm gas} \approx 8 \times 10^9\,{\rm M_\odot}$ (see Table 4, and also Ferriére 2001 and Kalberla 2003). If we assume a constant star formation rate, then the remaining gas should be converted into stars within about 2–4 Gyr. Similar gas depletion timescales of the order of a few billion years are reported for many nearby spiral galaxies (Bigiel et al. 2008). We note, however, that there is a debate in the community as to whether the depletion timescale is constant or whether it varies with the surface gas density. While Leroy et al. (2008, 2012, 2013) argue in favor of a more or less universal gas depletion time of ~2 Gyr, Shetty et al. (2013) and Shetty et al. (2014) find that the depletion time varies from galaxy to galaxy, and as a general trend, increases with surface density. Nevertheless, it is fair to say that the inferred overall gas depletion times are shorter by a factor of a few than the ages of these galaxies of ~10 Gyr. If we discard the possibility that most spiral galaxies are observed at an evolutionary phase close to running out of gas, and instead assume that they evolve in quasi steady state, then this requires a supply of fresh gas at a rate roughly equal to the star formation rate.

There is additional support for this picture. Dekel et al. (2009) and Ceverino et al. (2010), for example, argue that massive galaxies are continuously fed by steady, narrow, cold gas streams that penetrate through the accretion shock associated with the dark matter halo down to the central galaxy. This is a natural outcome of cosmological structure formation calculations if baryonic physics is considered consistently. In this case, roughly three quarters of all galaxies forming stars at a given rate are fed by smooth streams (see also Agertz et al. 2009). The details of this process, however, seem to depend on the numerical method employed and on the way gas cooling is implemented (e.g. Bird et al. 2013). Further evidence for accretion onto galaxies comes from the observation that the total amount of atomic gas in the universe appears to be roughly constant for redshifts $z \lesssim 3$. This holds despite the continuous transformation of gas into stars, and it suggests that HI is continuously replenished (Hopkins et al. 2008; Prochaska and Wolfe 2009). In our Galaxy, the presence of deuterium at the solar neighborhood (Linsky 2003) as well as in the Galactic Center (Lubowich et al. 2000) also points towards a continuous inflow of low-metallicity material. As deuterium is destroyed in stars and as there is no other known source of deuterium in the Milky Way, it must be of cosmological and extragalactic origin (Ostriker and Tinsley 1975; Chiappini et al. 2002).

In order to calculate the energy input rate from gas accretion we need to know the velocity $v_{\rm in}$ with which this gas falls onto the disk of the Galaxy and the efficiency with which the kinetic energy of the infalling gas is converted into ISM turbulence. As the cold accretion flow originates from the outer reaches of the halo and beyond, and because it lies in the nature of these cold streams that gas comes in almost in free fall, $v_{\rm in}$ can in principle be as high as the escape velocity $v_{\rm esc}$ of the halo. For the Milky Way in the solar neighborhood we find $v_{\rm esc} \sim 550\,{\rm km\ s^{-1}}$ (Fich and Tremaine 1991; Smith et al. 2007). However, numerical experiments indicate that the inflow velocity of cold streams is of order of the virialization velocity of the halo (Dekel et al. 2009), which typically is $\sim 200\,{\rm km\ s^{-1}}$. The actual impact velocity with which this gas interacts with disk material will also depend on the sense of rotation. Streams which come in co-rotating with the disk will have smaller impact velocities than material that comes in counter-rotating. To relate to quantities that are easily observable and to within the limits of our approximations, we adopt $v_{\rm in} = v_{\rm rot}$ as our fiducial value, but note that considerable deviations are possible. We also note that even gas that shocks at the virial radius and thus heats up to 10^5–10^6 K, may cool down again and some fraction of it may be available for disk accretion. This gas can condense into higher-density clumps that sink towards the center and replenish the disk (Peek 2009). Again, $v_{\rm in} \approx v_{\rm rot}$ is a reasonable estimate.

Putting everything together, we can now calculate the energy input rate associated with gas accretion onto the Milky Way as

$$\dot{e} = \rho\dot{\epsilon} = \frac{1}{2}\rho\frac{\dot{M}_{\rm in}}{M_{\rm gas}}v_{\rm in}^2 \tag{134}$$

$$= 6.2 \times 10^{-27}\,{\rm erg\,cm^{-3}\,s^{-1}} \left(\frac{n}{1\,{\rm cm^{-3}}}\right)\left(\frac{\dot{M}_{\rm in}}{3\,{\rm M_\odot\,yr^{-1}}}\right)\left(\frac{v_{\rm in}}{220\,{\rm km\,s^{-1}}}\right)^2 ,$$

where again $\rho = \mu n$ with the mean particle mass $\mu = 2.11 \times 10^{-24}$ g suitable for atomic gas. We take an average number density n in the Galactic disk from Table 4, set the mass infall rate $\dot{M}_{\rm in}$ to the average star formation rate of $3\,{\rm M_\odot\,yr^{-1}}$, and approximate the infall velocity $v_{\rm in}$ by the circular velocity of $220\,{\rm km\,s^{-1}}$.

If we compare the input rate (134) with the decay rate (132), we note that only about 7 % of the infall energy is needed to explain the observed ISM turbulence. However, the fraction of the infall energy that actually is converted into turbulent motions is very difficult to estimate. Some fraction will turn into heat and is radiated away. In addition, if the system is highly inhomogeneous with most of the mass residing in high density clumps with a low volume filling factor, most of the incoming flux will feed the tenuous interclump medium rather than the dense clumps, and again, will not contribute directly to driving turbulence in the dense ISM. Numerical experiments indicate that the efficiency of converting infall energy into turbulence scales linearly with the density contrast between the infalling gas and the ISM (Klessen and Hennebelle 2010). For the Milky Way, this means that the infalling material should have average densities of $\lesssim 0.1\,{\rm cm^{-3}}$.

It seems attractive to speculate that the population of high-velocity clouds observed around the Milky Way is the visible signpost for high-density peaks in this accretion flow. Indeed the inferred infall rates of high-velocity clouds are in the range 0.5–5 $\rm M_\odot yr^{-1}$ (Wakker et al. 1999; Blitz et al. 1999; Braun and Thilker 2004; Putman 2006), in good agreement with the Galactic star formation rate or with chemical enrichment models (see e.g. Casuso and Beckman 2004 and references therein). An important question in this context is where and in what form the gas reaches the Galaxy. This is not known well. Recent numerical simulations indicate that small clouds (with masses less than a few $10^4\,{\rm M_\odot}$) most likely will dissolve, heat up and merge with the hot halo gas, while larger complexes will be able to deliver cold atomic gas even to the inner disk (Heitsch and Putman 2009). In any case, it is likely that the gas is predominantly accreted onto the outer disk of the Milky Way. However, it is consumed by star formation mostly in the inner regions. To keep the Galaxy in a steady state there must be an inwards gas motion of the order of $v_{\rm R} \approx \dot{M}_{\rm in}/(2\pi R\Sigma) \approx 3\,{\rm km\,s^{-1}}$, where we adopt a gas surface density at the solar radius $R_\odot = 8.5$ kpc of $\Sigma = 15\,{\rm M_\odot\,pc^{-1}}$ (Naab and Ostriker 2006). Whether this net inward flow exists is not known, given our viewpoint from within the Galaxy and given a typical velocity dispersion of $\sim$10 km s^{-1}, which exceeds the strength of the signal we are interested in. In other galaxies, where we have an outside view onto the disk, we could in principle try to decompose the observed line-of-sight motions and find signs of the proposed inward mass transport.

4.5.2 Spiral Arms

Spiral galaxies such as our Milky Way are rotationally supported. This means that the gas is prevented from freely flowing towards the Galactic Center by angular momentum conservation, which forces the gas into circular orbits about a common origin. The process is very similar to the formation of protostellar accretion disks during the collapse of rotating cloud cores, which are a natural part of the process of stellar birth (see Sect. 6.6). Parcels of gas can only change their radial distance from the center of the disk by exchanging angular momentum with neighboring gas. Fluid elements that lose angular momentum move inwards, while those that gain angular momentum move outwards. Exchange of angular momentum between fluid elements may be due to dynamical friction or to the influence of some effective viscosity. In the ISM, molecular viscosity is far too small to explain the observed gas motions. However, we can resort to either spiral arms (which reflect the onset of gravitational instability) or to magnetic fields in the disk. In both cases, some of the energy stored in Galactic rotation can be converted into turbulent kinetic energy as the gas moves inwards.

Indeed, the spiral structure that is almost ubiquitously observed in disk galaxies has long been proposed as an important source of ISM turbulence. Roberts (1969) argued that the gas that flows through spiral arms forming in marginally stable disks (Toomre 1964; Lin and Shu 1964; Lin et al. 1969) may shock and so distribute energy throughout different scales. Gómez and Cox (2002) and Martos and Cox (1998), for example, found that some fraction of the gas will be lifted up in a sudden vertical jump at the position of the shock. Some portion of this flow will contribute to interstellar turbulence. However, we note that the observed presence of interstellar turbulence in irregular galaxies without spiral arms as well as in the outer regions of disk galaxies beyond the extent of the spiral arm structure suggests that there must be additional physical mechanisms driving turbulence.

For purely hydrodynamic turbulence in the absence of magnetic fields or for flows with weak Maxwell stresses (see Eq. 136 below), purely gravitational stress terms may become important. Wada et al. (2002) estimated the energy input resulting from these Newton stress terms. They result from correlations in the different components of the flow velocity $\mathbf{v}$ as $T_{R\Phi} = \langle \rho v_R v_\Phi \rangle$ (Lynden-Bell and Kalnajs 1972), and will only add energy for a positive correlation between radial and azimuthal gravitational forces. It is not clear, however, whether this is always the case. Despite this fact, we can use this approach to get an upper limit to the energy input. As an order of magnitude estimate, we obtain

$$\dot{e} \approx G(\Sigma_g/H)^2 L^2 \Omega \tag{135}$$
$$\approx 4 \times 10^{-29} \text{erg cm}^{-3}\text{s}^{-1} \times \left(\frac{\Sigma_g}{10\,\text{M}_\odot\text{pc}^{-2}}\right)^2 \left(\frac{H}{100\,\text{pc}}\right)^{-2} \left(\frac{L}{100\,\text{pc}}\right)^2 \left(\frac{\Omega}{(220\,\text{Myr})^{-1}}\right),$$

where G is the gravitational constant, Σ_g is the density of gas, H is the scale height of the disk, L is the length scale of turbulent perturbations, and Ω is the angular velocity. The normalization is appropriate for the Milky Way. This is about an order of magnitude less than the value required to maintain the observed ISM turbulence (Eq. 132).

We note that the fact that spiral arms are curved adds another pathway to driving ISM turbulence. Curved shocks are able to generate vortex motions as the gas flows through the discontinuity. Kevlahan and Pudritz (2009) argue that this process is able to produce a Kolmogorov-type energy spectrum in successive shock passages (see also Wada 2008). However, further investigations are needed to determine whether this process is able to produce the observed energy density in the Galactic ISM.

4.5.3 Magnetorotational Instabilities

Sellwood and Balbus (1999) proposed that the magnetorotational instability (Balbus and Hawley 1998) could efficiently couple large scale rotation with small-scale turbulence. The instability generates Maxwell stresses, which lead to a positive correlation between radial B_R and azimuthal B_Φ components of the magnetic field, transfering energy from shear into turbulent motions at a rate

$$\dot{e} = -T_{R\Phi}(d\Omega/d\ln R) = T_{R\Phi}\Omega\,, \tag{136}$$

where the last equality holds for a flat rotation curve. Typical values are $T_{R\Phi} \approx 0.6\,B^2/(8\pi)$ (Hawley et al. 1995). At the radius of the Sun, $R_\odot = 8.5$ kpc and a circular velocity of $v_{\rm rot} = 220\,\text{km s}^{-1}$, we obtain an angular velocity of

$$\Omega = \frac{v_{\rm rot}}{2\pi R_\odot} = \frac{1}{220\,\text{Myr}} \approx 1.4 \times 10^{-16}\,\text{rad s}^{-1}\,. \tag{137}$$

If we put both together, we conclude that the magnetorotational instability could contribute energy at a rate

$$\dot{e} = 3 \times 10^{-29}\,\text{erg cm}^{-3}\,\text{s}^{-1} \left(\frac{B}{3\mu\text{G}}\right)^2 \left(\frac{\Omega}{(220\,\text{Myr})^{-1}}\right). \tag{138}$$

Sellwood and Balbus (1999) tested this hypothesis for the small galaxy NGC 1058 and concluded that the magnetic field required to produce the observed velocity dispersion of 6 Km s^{-1} is roughly 3 μG which is reasonable value for such a galaxy.

Whether the process is efficient enough to explain ISM turbulence in large spiral galaxies such as our Milky Way remains an open question. The typical values derived from Eq. (138) are considerably lower than the energy required to compensate for the loss of turbulent energy (Eq. 132). Numerical simulations geared towards the Galactic disk (e.g. Dziourkevitch et al. 2004; Piontek and Ostriker 2004; Piontek and Ostriker 2005) are not fully conclusive and in general deliver values of $\dot{e}$ that are too small. Overall, the magnetorotational instability may provide a base value for the velocity dispersion below which no galaxy will fall, but it seems likely that additional processes are needed to explain the observations.

4.6 Sources of ISM Turbulence: Stellar Feedback

There are various stellar feedback processes that could also act as potential sources of ISM turbulence. In general, we can distinguish between mechanical and radiative energy input. Supernova explosions that accompany the death of massive stars, line-driven winds in the late phases of stellar evolution, as well as the protostellar jets and outflows that are associated with stellar birth belong to the first category. The ionizing and non-ionizing radiation that stars emit during all of their life belongs to the latter one. As before, we discuss these various feedback processes in decreasing order of importance.

4.6.1 Supernovae

The largest contribution from massive stars to interstellar turbulence most likely comes from supernova explosions. In order to understand their impact on ISM dynamics in our Galaxy, we first need to determine the supernova rate $\sigma_{\rm SN}$. The exact number is quite uncertain, but typical estimates fall in the range of 2–5 supernovae per century (e.g. McKee 1989; McKee and Williams 1997; Adams et al. 2013). Note that we do not distinguish between core collapse supernovae from massive stars and type Ia explosions which are triggered by accretion onto white dwarfs. In addition, we assume for simplicity that each event releases the same energy of $E_{\rm SN} = 10^{51}$ erg. Next, we need to obtain an estimate for the volume of the star forming disk of the Galaxy. Following the values discussed in Sect. 4.4, we take the star forming radius to be $R = 15$ kpc and the disk thickness to be $H = 100$ pc. The corresponding energy input rate normalized to Milky Way values becomes

$$\begin{aligned} \dot{e} &= \frac{\sigma_{\rm SN}\xi_{\rm SN}E_{\rm SN}}{\pi R_{\rm sf}^2 H} \\ &= 3\times 10^{-26}\ {\rm erg\ s^{-1}cm^{-3}} \\ &\quad \times \left(\frac{\xi_{\rm SN}}{0.1}\right)\left(\frac{\sigma_{\rm SN}}{(100\,{\rm yr})^{-1}}\right)\left(\frac{H}{100\ {\rm pc}}\right)^{-1}\left(\frac{R}{15\ {\rm kpc}}\right)^{-2}\left(\frac{E_{\rm SN}}{10^{51}{\rm erg}}\right). \end{aligned} \tag{139}$$

The efficiency of energy transfer from supernova blast waves to the interstellar gas ξ_{SN} depends on many factors, including the strength of radiative cooling in the initial shock, or whether the explosion occurs within a hot and tenuous HII region or in dense gas. Substantial amounts of energy can escape in the vertical direction in galactic fountain flows. The scaling factor $\xi_{SN} \approx 0.1$ used here was derived by Thornton et al. (1998) from detailed one-dimensional numerical simulations of supernovae expanding in a uniform medium. The efficiency can also be estimated analytically (Norman and Ferrara 1996), mostly easily by assuming momentum conservation, comparing the typical expansion velocity of 100 Km s^{-1} to the typical velocity of ISM turbulence of 10 Km s^{-1}. Clearly, fully three-dimensional models, describing the interaction of multiple supernovae in the multi-phase ISM are needed to better constrain the efficiency factor ξ_{SN}.

Supernova driving appears to be powerful enough to maintain ISM turbulence at the observed levels and to compensate for the energy loss estimated in Eq. (125). In the star-forming parts of the Galactic disk, it provides a large-scale self-regulation mechanism. As the disk becomes more unstable, the star formation rate goes up. Consequently, the number of OB stars increases which leads to a higher supernova rate. As the velocity dispersion increases, the disk becomes more stable again and the star formation rate goes down again. However, this process does not explain the large velocity dispersion observed in the outer parts of disk galaxies, which show little signs of star formation, and hence, will not have much energy input from supernovae. Here other processes, such as those described in Sect. 4.5, appear to be required.

4.6.2 Stellar Winds

The total energy input from a line-driven stellar wind over the main-sequence lifetime of an early O star can equal the energy from its supernova explosion, and the Wolf-Rayet wind can be even more powerful (Nugis and Lamers 2000). The wind mass-loss rate scales somewhat less than quadratically with the stellar luminosity (e.g. Pauldrach and Puls 1990; Puls et al. 1996; Vink et al. 2000, 2001), and as the luminosity L itself is a very steep function of stellar mass M, with $L \propto M^{3.5}$ providing a reasonable approximation (e.g. Kippenhahn et al. 2012), only the most massive stars contribute substantial energy input (for a review, see Lamers and Cassinelli 1999). We also note that stellar rotation can dramatically change the derived stellar mass loss rates and the energy and momentum inserted by line-driven winds (for recent reviews, see Meynet 2009 or Maeder and Meynet 2012, or for a grid of evolutionary tracks, see Ekström et al. 2012 and Georgy et al. 2012). Krumholz et al. (2014) concluded that even the most optimistic wind models lead to momentum and energy input rates comparable to the radiation field (see below, Sect. 4.6.4). In comparison, the energy from supernova explosions remains nearly constant down to the least massive star that can explode. Because there are far more low-mass stars than massive stars in the Milky Way and other nearby galaxies (for a discussion of the stellar initial mass function, see Sect. 6.2.3), supernova explosions inevitably dominate over stellar winds after the first few million years of the lifetime of an

OB association. Nevertheless, realistic three-dimensional numerical models of the momentum and kinetic energy input into the ISM and its effects on molecular cloud evolution and on interstellar turbulence are needed. At the moment too little is known about this process (see also Krumholz et al. 2006; Yeh and Matzner 2012).

4.6.3 Protostellar Jets and Outflows

Protostellar jets and outflows are another very popular potential energy source for the observed ISM turbulence. They propagate with velocities of about 300 km s^{-1} as seen in the radial velocity shift of forbidden emission lines, but also in proper motion of jet knots. Many of these jets remain highly collimated with opening angles less than 5° over a distance up to several parsec (e.g. Mundt et al. 1990, 1991).

Protostellar jets and outflows are launched by magnetic forces (for a summary, see Pudritz et al. 2007). The scenario of magnetohydrodynamic jet formation has been studied with stationary models (Camenzind 1990; Shu et al. 1994; Fendt and Camenzind 1996; Ferreira 1997) as well as by time-dependent MHD simulations (e.g. Ouyed and Pudritz 1997; Ouyed et al. 2003; Krasnopolsky et al. 1999).

Essentially, the MHD jet formation process works by transferring magnetic energy (Poynting flux) into kinetic energy. As a consequence, the asymptotic, collimated jet flow is in energy equipartition between magnetic and kinetic energy. The general characteristics of jet propagation can be summarized as follows. Along the interface between the propagating jet and the surrounding material at rest, Kelvin-Helmholtz instabilities develop and lead to the entrainment of matter from this region into the jet. This slows down the outward propagation while roughly conserving the overall momentum of the flow. At the front of the jet two leading shocks build up, a bow shock at the interface between the jet and the ambient medium and a Mach shock where the propagating matter is decelerated to low velocities. It is diverted into a cocoon of back-flowing material which is highly turbulent and heated up to high temperatures, leading to emission from the fine structure lines of carbon, nitrogen, oxygen, or sulfur atoms and to some degree from their ions (Sect. 3.4.2). Eventually the outflow dissolves as it reaches a speed that is comparable to the typical velocity dispersion in the ISM.

Norman and Silk (1980) estimated the amount of energy injected into the ISM by protostellar outflows, and showed that they could be an important energy source for turbulent motions in molecular clouds. They suggested that this in turn may influence the structure of the clouds and regulate the rate of gravitational collapse and star formation (see also Li and Nakamura 2006; Banerjee et al. 2007; Nakamura and Li 2008; Wang et al. 2010). The existence of a kinematic interrelation between outflows and their ambient medium has been inferred from high resolution CO observations, e.g. of the PV Cephei outflow HH 315 (Arce and Goodman 2002a, b). Optical observations surveying nearby molecular clouds furthermore indicate a similar influence of the outflows on the ionization state and energetics of the inter-cloud medium that surrounds low-mass star forming regions (for Perseus, see Bally et al. 1997 or Arce et al. 2010; for Orion A, see Stanke et al. 2002).

We begin with an estimate of the protostellar jet kinetic luminosity. It can be described as

$$L_{\rm jet} = \frac{1}{2}\dot{M}_{\rm jet}\, v_{\rm jet}^2 = 1.3\times 10^{32}\,{\rm erg\,s^{-1}}\left(\frac{\dot{M}_{\rm jet}}{10^{-8}\,{\rm M_\odot yr^{-1}}}\right)\left(\frac{v_{\rm jet}}{200\,{\rm km\,s^{-1}}}\right)^2 , \quad (140)$$

with $\dot{M}_{\rm jet} \approx 10^{-8}\,{\rm M_\odot yr^{-1}}$ being the mass loss associated with the jet material that departs from the protostellar disk system at typical velocities of $v_{\rm jet} \approx 200\,{\rm km\,s^{-1}}$. This outflow rate is closely coupled to the accretion rate $\dot{M}_{\rm acc}$ onto the central star by $\dot{M}_{\rm jet} = f_{\rm jet}\dot{M}_{\rm acc}$, with the efficiency factor typically being in the range $0.1 \lesssim f_{\rm jet} \lesssim 0.4$ (see e.g. Shu et al. 2000; Ouyed and Pudritz 1997; Bontemps et al. 1996; or consult the reviews by Bally et al. 2007; Pudritz et al. 2007; Frank et al. 2014; or Li et al. 2014 for further details).

A simple estimate of the jet lifetime in this phase is $t_{\rm jet} \approx 2\,{\rm pc}/200\,{\rm km\,s^{-1}} \approx 10^4\,{\rm yr}$. This coincides to within factors of a few with the typical duration of the class 0 and early class 1 phases of protostellar evolution (see Sect. 6.6). During these phases, we expect the strongest outflow activity (see the review by André et al. 2000). The total amount of energy provided by the jet is therefore

$$E_{\rm jet} = L_{\rm jet}\, t_{\rm jet} \approx 8\times 10^{43}{\rm erg}\ . \quad (141)$$

This kinetic luminosity is smaller than but comparable to the radiative luminosity of protostars. The outflow-ISM coupling is more direct and, thus, supposedly more efficient than the energy exchange between the protostellar radiation and the ISM. However, determinations of the coupling strength are controversial and require further investigation (Banerjee et al. 2007; Nakamura and Li 2008; Cunningham et al. 2008; Wang et al. 2010; Carroll et al. 2010; Federrath et al. 2014).

The total energy input from protostellar winds will substantially exceed the amount that can be transferred to the turbulence, because of radiative cooling at the wind termination shock. This introduces another efficiency factor $\xi_{\rm jet}$. A reasonable upper limit to the energy loss can be obtained by assuming that this cooling process is very efficient so that only momentum conservation holds,

$$\xi_{\rm jet} \lesssim \frac{\sigma}{v_{\rm jet}} = 0.05\left(\frac{\sigma}{10\,{\rm km\,s^{-1}}}\right)\left(\frac{v_{\rm jet}}{200\,{\rm km\,s^{-1}}}\right)^{-1} , \quad (142)$$

where σ as before is the velocity dispersion of ISM turbulence. If we assumed that most of the energy went into driving dense gas, the efficiency would be lower, as typical velocities for CO outflows are only 1–2 km s^{-1}. The energy injection rate per unit volume then follows as

$$\dot{e} = \frac{1}{2}\xi_{\rm jet} f_{\rm jet} \frac{\dot{M}_{\rm SF} v_{\rm jet}^2}{\pi R^2 H} = \frac{1}{2} f_{\rm jet} \frac{\dot{M}_{\rm SF} v_{\rm jet} \sigma}{\pi R^2 H} \tag{143}$$
$$= 1.4 \times 10^{-28}\,{\rm erg\,cm^{-3}\,s^{-1}}$$
$$\times \left(\frac{f_{\rm jet}}{0.2}\right)\left(\frac{\dot{M}_{\rm SF}}{3\,{\rm M_\odot\,yr^{-1}}}\right)\left(\frac{v_{\rm jet}}{200\,{\rm km\,s^{-1}}}\right)\left(\frac{\sigma}{10\,{\rm km\,s^{-1}}}\right)\left(\frac{H}{100\,{\rm pc}}\right)^{-1}\left(\frac{R}{15\,{\rm kpc}}\right)^{-2},$$

where we again normalize to the Galactic star formation rate $\dot{M}_{\rm SF} = 3\,{\rm M_\odot\,yr^{-1}}$ and take the volume of the star forming disk as $V = \pi R^2 H$, with radius $R = 15$ kpc and disk thickness $H = 100$ pc.

Although protostellar jets and outflows are very energetic, they are likely to deposit most of their energy into low density gas (Henning 1989), as is shown by the observation of multi-parsec long jets extending completely out of molecular clouds (Bally and Devine 1994). Furthermore, observed motions of molecular gas show increasing power on scales all the way up to and perhaps beyond the largest scale of molecular cloud complexes (Ossenkopf and Mac Low 2002). It is hard to see how such large scales could be driven by protostars embedded in the clouds.

4.6.4 Radiation

Next, we consider the radiation from massive stars. We focus our attention on ionizing radiation, because HII regions can affect large volumes of interstellar gas and their expansion converts thermal energy into kinetic energy. To a much lesser degree, the same holds for radiation in the spectral bands that can lead to the dissociation of molecular hydrogen into atomic gas. The thermal radiation mostly from low-mass stars will not be able to trigger large gas motions in the ISM, and we will not concern ourselves with it here (but see Sect. 2.3).

The total energy density carried by photons at frequencies high enough to ionize hydrogen is very large. We use the information provided, e.g. by Tielens (2010) or Draine (2011), and estimate the integrated luminosity of ionizing radiation in the disk of the Milky Way to be

$$\dot{e} = 1.5 \times 10^{-24}\,{\rm erg\,s^{-1}\,cm^{-3}}\,. \tag{144}$$

(See also earlier work by Abbott 1982). We note, however, that only a small fraction of this energy is converted into turbulent gas motions. There are two main pathways for this to happen. First, ionizing radiation will produce free electrons with relatively large velocities. This process heats up the resulting plasma to 7000–10000 K. The ionized regions are over pressured compared to the ambient gas and start to expand. They cool adiabatically and convert thermal energy into kinetic energy. Second, the medium can also cool radiatively and possibly contract. The ISM in this regime is thermally unstable (Field 1965; McKee and Ostriker 1977). This instability can excite turbulent motions (e.g. Vázquez-Semadeni et al. 2000; Kritsuk and Norman 2002a;

Piontek and Ostriker 2005; Hennebelle and Audit 2007) with typical conversion factors from thermal to kinetic energy of less than 10 %.

We begin with the first process, and look at the supersonic expansion of HII regions after photoionization heating raises their pressures above that of the surrounding neutral gas. By integrating over the HII region luminosity function derived by McKee and Williams (1997), Matzner (2002) estimates the average momentum input from expanding HII regions as

$$p_{\rm HII} \approx 260\,{\rm km\,s^{-1}} \left(\frac{N}{1.5\times10^{22}\,{\rm cm^{-2}}}\right)^{-3/14} \left(\frac{M_{\rm cloud}}{10^6\,{\rm M_\odot}}\right)^{1/14} M_* \,, \tag{145}$$

where the column density N is scaled to the mean value for Galactic molecular clouds (Solomon et al. 1987), $M_{\rm cloud}$ is a typical molecular cloud mass, and $M_* = 440\,{\rm M_\odot}$ is the mean stellar mass per cluster in the Galaxy (Matzner 2002; Lada and Lada 2003).

We focus our attention on clusters and OB associations producing more than 10^{49} ionizing photons per second, because these are responsible for most of the available ionizing photons. From the luminosity function presented by McKee and Williams (1997), we estimate that there are about $N_{49} = 650$ such clusters in the Milky Way. To derive an energy input rate per unit volume from the mean momentum input per cluster (Eq. 145), we need to obtain an estimate for the typical expansion velocity $v_{\rm HII}$ of the HII regions as well as for the duration of this process. While expansion is supersonic with respect to the ambient gas, it is by definition subsonic with respect to the hot interior. The age spread in massive star clusters and OB associations can be several million years (Preibisch and Zinnecker 1999; Portegies Zwart et al. 2010; Longmore et al. 2014). We take a value of $t_* = 10$ Myr. With the stellar mass—luminosity relation on the main sequence being $L/{\rm L_\odot} \approx 1.5\,(M/{\rm M_\odot})^{3.5}$ for stars with masses up to $M \approx 20\,{\rm M_\odot}$ and $L/{\rm L_\odot} \approx 3200\,(M/{\rm M_\odot})$ for stars with $M \gtrsim 20\,{\rm M_\odot}$, the energy output in a cluster is dominated by the most massive stars. The main sequence lifetime can be estimated as $t_{\rm MS} \approx 10^{10}$ yr $(M/{\rm M_\odot})(L/{\rm L_\odot})^{-1} \approx 10^{10}$ yr $(M/{\rm M_\odot})^{-2.5}$ for stars with $M < 20\,{\rm M_\odot}$, asymptoting to a value of around $t_{\rm MS} \approx 3$ Myr for more massive stars (e.g. Hansen and Kawaler 1994). For typical O-type stars, $t_{\rm MS} < t_*$, and as a consequence we can take t_* as a good order of magnitude estimate for the duration of strong ionizing feedback from the clusters of interest. Once again, we estimate the volume of the star forming disk of the Galaxy as $V = \pi R^2 H$, with radius $R = 15$ kpc and disk thickness $H = 100$ pc. Putting this all together, the estimated energy input rate from expanding HII regions is then

$$\begin{aligned} \dot{e} &= \frac{N_{49}\,p_{\rm HII}\,v_{\rm HII}}{\pi R^2 H\, t_*} \\ &= 3\times10^{-30}\ {\rm erg\ s^{-1}\ cm^{-3}} \left(\frac{N}{1.5\times10^{22}\ {\rm cm^{-2}}}\right)^{-3/14} \left(\frac{M_{\rm cloud}}{10^6\ {\rm M_\odot}}\right)^{1/14} \\ &\quad \times \left(\frac{M_*}{440\ {\rm M_\odot}}\right) \left(\frac{N_{49}}{650}\right) \left(\frac{v_{\rm HII}}{10\ {\rm km\ s^{-1}}}\right) \left(\frac{H}{100\ {\rm pc}}\right)^{-1} \left(\frac{R}{15\ {\rm kpc}}\right)^{-2} \left(\frac{t_*}{10\ {\rm Myr}}\right)^{-1}. \end{aligned} \tag{146}$$

Nearly all of the energy in ionizing radiation goes towards maintaining the ionization and temperature of the diffuse medium, and hardly any towards driving turbulence. Flows of ionized gas may be important very close to young clusters and may terminate star formation locally (for the difference between two- and three-dimensional simulations, see Yorke and Sonnhalter 2002; Krumholz et al. 2007, 2009, Peters et al. 2010b, 2011; Kuiper et al. 2011). They can also influence the molecular cloud material that surrounds the young star cluster (e.g. Dale et al. 2005; Dale and Bonnell 2011; Walch et al. 2012, 2013). However, they appear not to contribute significantly on a global scale.

Now we turn our attention to the second process, to the thermal instability. Kritsuk and Norman (2002b) find that the thermal energy released can be converted into turbulent kinetic energy, $e_{\rm kin} = \xi_{\rm ion} e_{\rm th}$, with an efficiency $\xi_{\rm ion} \approx 0.07$. Parravano et al. (2003) study the time dependence of the local UV radiation field. They find that the corresponding photoelectric heating rate increases by a factor of 2–3 due to the formation of a nearby OB association every 100–200 Myr. Note, however, that substantial motions only last about 1 Myr after a heating event (Kritsuk and Norman 2002b; de Avillez and Breitschwerdt 2004, 2005, 2007). We follow this line of reasoning and estimate the resulting average energy input by taking the kinetic energy input from the heating event and dividing by the typical time $t_{\rm OB}$ between heating events. We determine the thermal energy for gas at a number density of $n = 1\,{\rm cm}^{-3}$ at a temperature of $T = 10^4$ K and find that

$$\begin{aligned} \dot{e} &= \frac{3}{2}\frac{nkT\xi_{\rm ion}}{t_{\rm OB}} \qquad (147) \\ &= 5 \times 10^{-29}\,{\rm erg\,cm^{-3}\,s^{-1}} \left(\frac{n}{1\,{\rm cm}^{-3}}\right)\left(\frac{T}{10^4\,{\rm K}}\right)\left(\frac{\xi_{\rm ion}}{0.07}\right)\left(\frac{t_{\rm OB}}{100\,{\rm Myr}}\right)^{-1} . \end{aligned}$$

In comparison to some other proposed energy sources discussed here, this mechanism appears unlikely to be as important as the supernova explosions from the same OB stars discussed before.

5 Formation of Molecular Clouds

5.1 Transition from Atomic to Molecular Gas

Our starting point for considering the physics of molecular cloud formation is the chemistry of the gas. After all, molecular clouds are, by definition, dominated by molecular gas, while the gas in the more diffuse neutral phases of the ISM is almost entirely atomic. Cloud formation must therefore involve, at some stage of the process, a chemical transition from gas which is mainly atomic to gas which is mainly molecular.

There are two main chemical transitions, occurring at different points in the assembly of a molecular cloud, that we could use to identify the point at which our assembling cloud becomes "molecular". The first and most obvious of these is the transition between atomic and molecular hydrogen: once most of the hydrogen in the cloud is in the form of H_2 rather than H, it is obviously reasonable to talk of the cloud as being molecular. However, this transition has the disadvantage that it is extremely difficult to observe, since H_2 does not emit radiation at typical molecular cloud temperatures. Therefore, it is common to use a different, observationally-motivated definition of the point when a cloud becomes molecular, which is the moment it becomes visible in CO emission. Understanding when this occurs requires understanding the chemical transition from C^+ to C to CO that occurs within the assembling cloud.

Below, we discuss the chemistry involved in both of these transitions in more detail, and then examine some of the approximations used to model the atomic-to-molecular transition in numerical studies of molecular cloud formation.

5.1.1 Transition from H to H_2

The simplest way to form H_2 in the ISM is via the radiative association of two hydrogen atoms, i.e.

$$\mathrm{H} + \mathrm{H} \rightarrow \mathrm{H_2} + \gamma. \tag{148}$$

However, in practice the rate coefficient for this reaction is so small that only a very small amount of H_2 forms in this way. Somewhat more can form via the ion-neutral reaction pathways

$$\mathrm{H} + \mathrm{e^-} \rightarrow \mathrm{H^-} + \gamma, \tag{149}$$

$$\mathrm{H^-} + \mathrm{H} \rightarrow \mathrm{H_2} + \mathrm{e^-}, \tag{150}$$

and

$$\mathrm{H} + \mathrm{H^+} \rightarrow \mathrm{H_2^+} + \gamma, \tag{151}$$

$$\mathrm{H_2^+} + \mathrm{H} \rightarrow \mathrm{H_2} + \mathrm{H^+}, \tag{152}$$

but it is difficult to produce H_2 fractional abundances larger than around $f_{\mathrm{H_2}} \sim 10^{-2}$ with these reactions, even in the most optimal conditions (see e.g. Tegmark et al. 1997). Moreover, in the local ISM, photodetachment of H^- and photodissociation of H_2^+ by the ISRF render these pathways considerably less effective (Glover 2003). We are therefore forced to conclude that gas-phase formation of H_2 is extremely inefficient in typical ISM conditions. Nevertheless, we do observe large quantities of H_2 in Galactic molecular clouds.

The resolution to this apparent puzzle comes when we realize that most of the H_2 in the ISM does not form in the gas-phase, but instead forms on the surface of dust grains (Gould and Salpeter 1963). Association reactions between adsorbed hydrogen atoms occur readily on grain surfaces, and the rate at which H_2 forms there is limited primarily by the rate at which H atoms are adsorbed onto the surface. For typical Milky Way conditions, the resulting H_2 formation rate is approximately (Jura 1975)

$$R_{H_2} \sim 3 \times 10^{-17} n n_{H} \, cm^{-3} s^{-1}. \tag{153}$$

Here, n is the total number density of gas particles, while n_H is the number density of atomic hydrogen. For atomic hydrogen gas, both quantities are identical if we neglect contributions from helium and possibly metals. Note that n_H goes down as the molecular fraction increases, while n remains the same in the absence of compression or expansion. The H_2 formation timescale corresponding to the formation rate (153) is approximately

$$t_{form} = \frac{n_H}{R_{H_2}} \sim 10^9 n^{-1} \, yr. \tag{154}$$

When the gas density is low, this timescale can be considerably longer than the most important dynamical timescales, such as the turbulent crossing time or the gravitational free-fall time. Accounting for the effects of the small-scale transient density structures produced by supersonic turbulence does shorten the timescale somewhat (Glover and Mac Low 2007b; Micic et al. 2012), but typically not by more than an order of magnitude.

Molecular hydrogen in the ISM can be collisionally dissociated by

$$H_2 + H \rightarrow H + H + H, \tag{155}$$

$$H_2 + H_2 \rightarrow H + H + H_2. \tag{156}$$

However, these reactions are effective at destroying H_2 only in warm, dense gas, and so although they are important in certain circumstances, such as in molecular outflows (see e.g. Flower et al. 2003), they do not play a major role in regulating the molecular content of the ISM. Instead, the dominant process responsible for destroying H_2 in the local ISM is photodissociation.

Photodissociation of H_2 occurs via a process known as spontaneous radiative dissociation (Stecher and Williams 1967; van Dishoeck 1987). The H_2 molecule first absorbs a UV photon with energy $E > 11.2$ eV, placing it in an excited electronic state. The excited H_2 molecule then undergoes a radiative transition back to the electronic ground state. This transition can occur either into a bound ro-vibrational level in the ground state, in which case the molecule survives, or into the vibrational continuum, in which case it dissociates. The dissociation probability depends strongly on the rotational and vibrational quantum numbers that the molecule has while in the excited electronic state, but on average, it is around 15 % (Draine and Bertoldi 1996). The discrete set of UV absorption lines produced by this process are known as the Lyman and Werner bands, and hence it has become common to refer to the energetic photons responsible for destroying H_2 as Lyman-Werner photons.

Because H_2 photodissociation is line-based, rather than continuum-based, the H_2 photodissociation rate in the ISM is highly sensitive to an effect known as self-shielding. This term refers to the fact that in a region with a high H_2 column density, the Lyman-Werner photons with energies corresponding to the main absorption lines are mostly absorbed by H_2 on the outskirts of the region, with only a few surviving to reach the center. Consequently, the H_2 photodissociation rate in the gas at the center of the region is reduced by a large factor compared to the rate in the unshielded, optically thin gas. Detailed studies of this process show that it starts to significantly affect the H_2 photodissociation rate once the H_2 column density exceeds $N_{H_2} \sim 10^{14}\,\mathrm{cm}^{-2}$ (Draine and Bertoldi 1996). The corresponding total column density of hydrogen depends on the strength of the ISRF and the density of the gas. In unshielded gas illuminated by an ISRF with a strength G_0 in Habing units (see Sect. 2.3), the equilibrium number density of H_2 is given approximately by

$$n_{H_2} \sim \frac{3 \times 10^{-17} n n_H}{3 \times 10^{-10} G_0} = 10^{-6} n n_H G_0^{-1}. \tag{157}$$

The resulting H_2 column density, N_{H_2}, is therefore related to the total hydrogen column density N by

$$N_{H_2} = 10^{-6} n_H G_0^{-1} N. \tag{158}$$

From this, we see that in order to produce an H_2 column density of $10^{14}\,\mathrm{cm}^{-2}$, we need a total column density

$$N = 10^{20} G_0 n^{-1}\,\mathrm{cm}^{-2}. \tag{159}$$

For comparison, the visual extinction required to reduce the H_2 photodissociation rate by a factor of ten is approximately $A_V \approx 0.65$, which in the diffuse ISM corresponds to a total hydrogen column density $N \sim 10^{21}\,\mathrm{cm}^{-2}$. Therefore, H_2 self-shielding becomes important earlier, at lower total column densities, than dust shielding in conditions when G_0/n is small, such as in CNM clouds far from regions of massive star formation. On the other hand, if G_0/n is large, such as can be the case in photodissociation regions close to massive stars, then dust extinction typically dominates.

5.1.2 Transition from C^+ to C to CO

The chemistry involved in the transition from C^+ to C is very simple: atomic carbon forms via the radiative recombination of C^+,

$$C^+ + e^- \rightarrow C + \gamma, \tag{160}$$

and is destroyed by photoionization,

$$C + \rightarrow C^+ + e^-. \tag{161}$$

However, the formation of CO is considerably more complicated, as in this case there is not a single dominant process responsible for CO formation, but rather a variety of different pathways that one can follow to get to CO. In this section, we give a very brief introduction to the basics of CO formation chemistry, but we refer readers in search of a more detailed and comprehensive treatment to the classic papers by Glassgold and Langer (1975), Langer (1976), Dalgarno and Black (1976), Tielens and Hollenbach (1985) and Sternberg and Dalgarno (1995).

CO Formation

The majority of the CO found in molecular clouds forms via one or the other of two main sets of chemical intermediates. One set of intermediates includes hydroxyl (OH), its positive ion (OH^+) and their products, while the other set includes the simple hydrocarbons CH and CH_2 and their positive ions.

The formation of CO from OH occurs rapidly via the neutral-neutral reaction

$$\mathrm{C} + \mathrm{OH} \rightarrow \mathrm{CO} + \mathrm{H}. \tag{162}$$

Unlike many neutral-neutral gas-phase reactions, this reaction has no activation energy and hence remains effective even at the very low temperatures found within molecular clouds. In addition, in gas with a high C^+ to C ratio, CO^+ ions are produced by

$$\mathrm{C}^+ + \mathrm{OH} \rightarrow \mathrm{CO}^+ + \mathrm{H}, \tag{163}$$

which then form CO either directly,

$$\mathrm{CO}^+ + \mathrm{H} \rightarrow \mathrm{CO} + \mathrm{H}^+, \tag{164}$$

or indirectly, via HCO^+ in the reactions

$$\mathrm{CO}^+ + \mathrm{H}_2 \rightarrow \mathrm{HCO}^+ + \mathrm{H}, \tag{165}$$
$$\mathrm{HCO}^+ + \mathrm{e}^- \rightarrow \mathrm{CO} + \mathrm{H}. \tag{166}$$

We therefore see that once OH forms, CO follows rapidly. However, forming the necessary OH radical is not so straightforward. One obvious pathway to OH involves the reaction of atomic oxygen with H_2,

$$\mathrm{O} + \mathrm{H}_2 \rightarrow \mathrm{OH} + \mathrm{H}. \tag{167}$$

However, this reaction has an activation energy of 0.26 eV, and so although it is an important source of OH in hot gas (see e.g. Hollenbach and McKee 1979), in the cold gas found in CNM clouds and molecular clouds, other less-direct routes to OH dominate. One of these involves the reaction of atomic oxygen with H_3^+:

$$O + H_3^+ \rightarrow OH^+ + H_2. \tag{168}$$

The OH^+ ions react rapidly with H_2, forming H_2O^+ and H_3O^+ via

$$OH^+ + H_2 \rightarrow H_2O^+ + H, \tag{169}$$
$$H_2O^+ + H_2 \rightarrow H_3O^+ + H. \tag{170}$$

H_3O^+ does not readily react further with H_2, but instead is removed from the gas via dissociative recombination, yielding a variety of products that include OH and water (see e.g. Jensen et al. 2000),

$$H_3O^+ + e^- \rightarrow H_2O + H, \tag{171}$$
$$H_3O^+ + e^- \rightarrow OH + H_2, \tag{172}$$
$$H_3O^+ + e^- \rightarrow OH + H + H, \tag{173}$$
$$H_3O^+ + e^- \rightarrow O + H_2 + H. \tag{174}$$

The other main route to OH involves O^+. This can be produced by cosmic ray ionization of neutral oxygen, or by charge transfer from H^+, and can react with H_2 to yield OH^+,

$$O^+ + H_2 \rightarrow OH^+ + H. \tag{175}$$

The OH^+ ions produced in this reaction then follow the same chain of reactions as outlined above.

An important point to note here is that in every case, the rate-limiting step is the formation of the initial OH^+ ion. Although the reactions between O and H_3^+ and between O^+ and H_2 are rapid, the fractional abundances of O^+ and H_3^+ are small, and so the overall rate of OH^+ formation is relatively small. Once the OH^+ ions have formed, however, the remainder of the reactions in the chain leading to CO are rapid. Since all of the reactions involved in the formation of OH^+ depend on H_2, either directly or as a source for the H_3^+ ions, one consequence of this is that CO formation via the OH pathway is sensitive to the molecular hydrogen abundance.

The other main route to CO involves the simple hydrocarbons CH and CH_2 and their ions. In gas with a high C^+ fraction, CH^+ can be formed via the reaction with H_2,

$$C^+ + H_2 \rightarrow CH^+ + H, \tag{176}$$

or by radiative association with atomic hydrogen,

$$C^+ + H \rightarrow CH^+ + \gamma. \tag{177}$$

As radiative association is a slow process, one might expect that the reaction with H_2 would dominate. However, this suffers from the same problem as reaction (167). It has a substantial energy barrier, in this case 0.4 eV, and therefore proceeds at a very slow rate at the temperatures typical of the CNM or of molecular clouds. Indeed, this presents something of a problem, as the resulting CH^+ formation rate is too slow to explain the observed CH^+ abundance in the diffuse atomic ISM, possibly indicating that some form of non-thermal chemistry is active there (see e.g. Sheffer et al. 2008; Godard et al. 2009).

In gas with significant fractions of C^+ and H_2, the CH_2^+ ion can be formed by radiative association,

$$C^+ + H_2 \rightarrow CH_2^+ + \gamma. \tag{178}$$

The rate coefficient for this reaction is significantly larger than the rate coefficient for reaction (177) as discussed by McElroy et al. (2013), and so in regions with $n_{H_2} \geq n_H$, this reaction is usually the main starting point for the formation of CO via the hydrocarbon pathway.

Once CH^+ or CH_2^+ has formed via one of the above reactions, it quickly reacts further with H_2,

$$CH^+ + H_2 \rightarrow CH_2^+ + H, \tag{179}$$
$$CH_2^+ + H_2 \rightarrow CH_3^+ + H. \tag{180}$$

Although the CH_3^+ ions can react further with H_2, they do so via a slow radiative association reaction,

$$CH_3^+ + H_2 \rightarrow CH_5^+ + \gamma. \tag{181}$$

Therefore, most of the CH_3^+ is destroyed instead by dissociative recombination,

$$\begin{aligned} CH_3^+ + e^- &\rightarrow CH + H_2, \\ CH_3^+ + e^- &\rightarrow CH + H + H, \\ CH_3^+ + e^- &\rightarrow CH_2 + H. \end{aligned} \tag{182}$$

The CH and CH_2 radicals produced by this process react readily with atomic oxygen, forming CO via

$$CH + O \rightarrow CO + H, \tag{183}$$
$$CH_2 + O \rightarrow CO + H_2, \tag{184}$$
$$CH_2 + O \rightarrow CO + H + H. \tag{185}$$

The CH_2 radicals are also destroyed rapidly in a reaction with atomic hydrogen,

$$CH_2 + H \rightarrow CH + H_2, \tag{186}$$

but in the case of CH, the analogous reaction has a significant energy barrier and hence is negligible at the temperatures of interest. The end result is therefore that a large fraction of the carbon incorporated into CH^+ or CH_2^+ by the reactions described above ultimately ends up in the form of CO.

In gas with a high abundance of neutral atomic carbon, a few other processes contribute significantly to the formation rate of the CH and CH_2 radicals that are the precursor of CO. They can be formed directly via radiative association of C with H or H_2,

$$C + H \rightarrow CH + \gamma, \tag{187}$$

$$C + H_2 \rightarrow CH_2 + \gamma, \tag{188}$$

although these reactions are relatively slow. Alternatively, atomic carbon can react with H_3^+,

$$C + H_3^+ \rightarrow CH^+ + H_2, \tag{189}$$

forming a CH^+ ion that the reacts further as described above

Looking at the hydrocarbon pathway as a whole, we see that it shares some common features with the OH pathway. In each case, the rate-limiting step is the initiating reaction, whether this is the formation of CH, CH_2, CH^+ or CH_2^+ by radiative association, or the formation of H_3^+ as a consequence of the cosmic ray ionization of H_2. Once the initial molecular ion or radical has formed, the remainder of the reactions that lead to CO proceed relatively quickly. This behavior forms the basis of several simplified methods for treating CO formation that we discuss in Sect. 5.1.3. In addition, we also see that all of the different ways that we can proceed from C^+ or C to CO rely on the presence of molecular hydrogen. This is important, as it implies that substantial quantities of CO will form only in regions that already have high H_2 fractions. Therefore, although the characteristic timescales of the chemical reactions involved in CO formation are generally shorter than the H_2 formation time, non-equilibrium, time-dependent behavior can nevertheless still be important, owing to the dependence on the H_2 fraction.

CO Destruction

In gas with a low visual extinction, the destruction of CO is dominated by photodissociation:

$$CO + \gamma \rightarrow C + O. \tag{190}$$

The photodissociation of the CO molecule occurs via a process known as predissociation (van Dishoeck 1987). The molecule first absorbs a UV photon with energy $E > 11.09$ eV, placing it in an excited electronic state. From here, it can either return to the ground state via radiative decay, or it can undergo a transition to a repulsive electronic state via a radiationless process. In the latter case, the molecule very rapidly dissociates. In the case of CO, dissociation is typically far more likely than decay back to the ground state (van Dishoeck and Black 1988). Consequently, the lifetimes of the excited electronic states are very short. This is important, as Heisenberg's uncertainty principle then implies that their energy is comparatively uncertain. The UV absorption lines associated with the photodissociation of CO are therefore much broader than the lines associated with H_2 photodissociation. As a result, CO self-shielding is less effective than the analogous process for H_2.

The classic work on the photodissociation of CO in an astrophysical context is the paper by van Dishoeck and Black (1988). However, in the two decades since this paper was published, improved experimental data on the properties of the CO molecule has become available, and a revised treatment of CO photodissociation was recently given by Visser et al. (2009).

Once the visual extinction of the gas becomes large, CO photodissociation becomes unimportant. In these circumstances, two other processes take over as the main routes by which CO is destroyed. First, cosmic ray ionization of hydrogen molecules or hydrogen atoms produces energetic photo-electrons. If these collide with other hydrogen molecules before dissipating their energy, they can excite the H_2 molecules into excited electronic states. The subsequent radiative decay of the molecules back to the ground state produces UV photons that can produce localized photodissociation of CO and other molecules (Prasad and Tarafdar 1983; Gredel et al. 1987, 1989). Second, CO is also destroyed via dissociative charge transfer with He^+ ions,

$$CO + He^+ \rightarrow C^+ + O + He. \tag{191}$$

The He^+ ions required by this reaction are produced by cosmic ray ionization of neutral helium. We therefore see that the rate at which CO molecules are destroyed in high A_V gas is controlled by the cosmic ray ionization rate. In local molecular clouds, this is relatively small (van der Tak and van Dishoeck 2000), and so almost all of the carbon in these high A_V regions is found in the form of CO. In clouds illuminated by a much higher cosmic ray flux, however, such as those in the Central Molecular Zone of the Galaxy, the destruction of CO by these processes in high extinction gas is considerably more important, and the CO fraction can be significantly suppressed even in well-shielded gas (see e.g. Clark et al. 2013).

5.1.3 Modeling the Atomic-to-Molecular Transition

There are a number of different approaches that one can use in order to numerically model the transition from atomic to molecular gas that occurs as one builds up a molecular cloud, each with their own strengths and weaknesses.

One of the most obvious approaches is to build a model that incorporates all of the main chemical reactions occurring in the gas. The forty or so reactions discussed in Sects. 5.1.1 and 5.1.2 above represent only a small fraction of the full range of possible reactions that can occur, particular once one accounts for the role played by additional chemical elements such as nitrogen or sulfur. An example of the degree of chemical complexity that is possible is given by the UMIST Database for Astrochemistry (McElroy et al. 2013). The latest release of this database contains details of 6173 gas-phase reactions of astrophysical interest, involving 467 different chemical species. If one also attempts to account for the full range of possible grain-surface chemistry and also for important isotopic variants of the main chemical species (e.g. molecules with one or more deuterium atoms in place of a hydrogen atom), then the size of the resulting chemical network can easily be an order of magnitude larger still (see e.g. Albertsson et al. 2013 for a recent example). By coupling a comprehensive chemical model such as this to a detailed model for the penetration of UV radiation through the gas and in addition a treatment of its magnetohydrodynamical and thermal evolution, we can in principle model the chemical evolution of the gas with a very high degree of accuracy.

Unfortunately, the computational requirements of such an approach are currently prohibitive. The chemistry of the ISM evolves on a wide range of different timescales, and hence the set of coupled ordinary differential equations (ODEs) that describe the chemical evolution of the gas are what is known as "stiff". To ensure stability, these equations must be solved implicitly, and the cost of doing so scales as the cube of the number of ODEs. Consequently, solving for the chemical evolution of the gas using a comprehensive chemical model is rather time-consuming, owing to the large number of ODEs involved. This is not necessarily a problem if one is interested in solving for the chemical evolution of only a small number of fluid elements, but becomes a major difficulty once one tries to solve for the chemical evolution of the gas within a high-resolution three-dimensional simulation, when one is dealing with tens or hundreds of millions of fluid elements. Chemical networks involving $\sim$10 to 20 different species can be used within such models (see e.g. Glover et al. 2010), although this is already computationally demanding, but scaling up to $\sim$400 to 500 species requires approximately 10^4 times more computational power, rendering it completely impractical at the present time.

Because of this, any attempt to model the atomic-to-molecular transition numerically must make some simplifications. If one is interested in a time-dependent, non-equilibrium description of the transition, then there are two main strategies that can be used to make the problem simpler. First, we can simplify the chemistry while continuing to use a detailed model of the hydrodynamical evolution of the gas. The basic idea here is to strip the chemical model down to its bare essentials, i.e. only those reactions that most directly affect the abundances of H, H_2, C^+, C and CO. In the case of H and H_2, the simplicity of the chemistry makes this relatively straightforward, and a number of different implementations of H_2 formation chemistry within large hydrodynamical simulations are now available (see e.g. Anninos et al. 1997; Glover and Mac Low 2007a; Dobbs et al. 2008; Gnedin et al. 2009; Christensen et al. 2012). The only real difficulty in this case is how to handle the effects of H_2 self-shielding and dust shielding. Several different approaches have been used in the literature, ranging from simple Sobolev-like approximations (Gnedin et al. 2009), to more sophisticated approximations based on computing the column density of dust and H_2 along a limited number of sight-lines (see e.g. Clark et al. 2012a, b; Hartwig et al. 2015).

Modeling the chemistry involved in the transition from C^+ to C to CO is rather harder, owing to the significantly greater complexity of the required chemical network. Nevertheless, several different possibilities have been put forward in the literature (Nelson and Langer 1997, 1999; Keto and Caselli 2008, 2010; Glover et al. 2010). Typically, these approximate treatments ignore any reactions involving elements other than H, He, C and O, ignore those parts of the carbon chemistry not directly involved in the formation or destruction of CO, and greatly simplify the treatment of the main pathways from C^+ to CO. As we have seen, the rate limiting step in these pathways is typically the initiating reaction, and so a decent estimate of the CO formation rate can be arrived at by computing how rapidly carbon is incorporated into any one of CH, CH_2, CH^+ and CH_2^+, and how rapidly oxygen is incorporated into OH^+, without the need to follow all of the details of the subsequent chemistry. A number of these approximate treatments were compared with each other by Glover and Clark (2012b), who showed that although very simple treatments such as that of Nelson and Langer (1997) tend to over-produce CO, more detailed models such as those of Nelson and Langer (1999) and Glover et al. (2010) produced results that agreed well with each other.

The other main non-equilibrium approach retains far more of the chemical complexity of the full network, choosing to simplify instead the treatment of the gas dynamics and often also the geometry of the gas. This is the strategy used, for example, in most PDR codes.[7] For a long time, the standard approach has been to ignore the effects of dynamics completely, and to adopt either spherical symmetry or one-dimensional slab symmetry in order to model the clouds. Neglecting the hydrodynamical evolution of the gas is often justified, if the chemical species one is interested in have characteristic evolutionary timescales that are much shorter than a representative dynamical timescale such as the turbulent crossing time or gravitational free-fall time. However, it is probably not a good approximation for treating species such as H or H_2 that have long chemical timescales and whose abundances at any given time in the evolution of a molecular cloud are therefore sensitive to the previous dynamical history of the gas (see e.g. Bergin et al. 2004; Glover and Mac Low 2007b). The assumption of one-dimensional symmetry, although computationally convenient, is less easy to justify, as the resulting models are unable to explain some notable features of real molecular clouds such as the widespread distribution of atomic carbon (Frerking et al. 1989; Little et al. 1994; Schilke et al. 1995). Accounting for clumping within the cloud greatly alleviates this issue (see e.g. Kramer et al. 2008), and although it is possible to model a clumpy cloud using a one-dimensional PDR code by representing the cloud as an ensemble of spherically-symmetric clumps (see e.g. Stutzki et al. 1988), ideally one would use a full three-dimensional approach. Recently, three-dimensional PDR codes are starting to become available (see e.g. Levrier et al. 2012; Offner et al. 2013), although they are not yet as fully-featured as their one-dimensional cousins.

An even simpler approach to modeling the atomic-to-molecular transition involves relaxing the assumption that the chemistry is out of equilibrium. If we assume that the gas is in chemical equilibrium, then instead of solving a set of coupled ordinary differential equations in order to obtain the current values of the chemical abun-

[7] The acronym PDR stands for photodissociation region or photon dominated region.

dances, we have the simpler task of solving a set of linear equations. In particular, note the studies by Krumholz et al. (2008, 2009) and McKee and Krumholz (2010) in which they solve for the equilibrium H and H_2 abundances as a function of the gas surface density, the UV field strength, and the metallicity. They also derive simple analytical approximations to their numerical results, suitable for implementing in large-scale numerical simulations that do not have the resolution to model the structure of individual molecular clouds (see e.g. Kuhlen et al. 2012; Thompson et al. 2014).

The validity of the equilibrium approach depends upon the extent to which the equilibrium abundances reflect the true chemical abundances in the ISM, and hence on the relative sizes of the H_2 formation timescale, $t_{\rm form}$, and the dynamical time, $t_{\rm dyn}$. At solar metallicity, the two timescales are roughly equal in dense GMCs, and so it is reasonable to expect equilibrium models to be a good guide to the behavior of the H_2 fraction in these clouds. Indeed, recent observations of the H and H_2 content of the Perseus molecular cloud made by Lee et al. (2012) yield results that are well fit by the Krumholz et al. (2008) model. However, as $t_{\rm form} \propto n^{-1}$, the formation time can be significantly longer than the dynamical time in lower density clouds, such as the diffuse H_2 clouds observed in UV absorption line studies (Snow and McCall 2006), and it is therefore unclear whether the H_2 content of these clouds has yet reached equilibrium (see also Mac Low and Glover 2012). Furthermore, since H_2 formation timescale scales inversely with the dust-to-gas ratio, the equilibrium approximation can fail badly in very low metallicity, dust-poor systems. In these conditions, cloud formation, gravitational collapse and star formation can all take place before the gas has had a chance to reach chemical equilibrium (Glover and Clark 2012c; Krumholz 2012). At very low metallicities, this can even lead to star formation occurring in regions that are primarily atomic rather than molecular.

5.2 Importance of Dust Shielding

In our discussion of the atomic to molecular transition in the previous Section, we saw that the column density of the gas in the ISM plays an important role in regulating its chemical state. Regions with high column densities have large visual extinctions, and hence can shield themselves effectively from the UV portion of the ISRF (see Sect. 2.3). In these regions, the gas is primarily molecular once it reaches chemical equilibrium, although the approach to equilibrium can take a long time when the volume density of the gas is small. On the other hand, regions with a low column density have low visual extinctions and so are unable to resist the dissociating effects of the ISRF. These regions are generally dominated by atomic gas.

The precise value of the visual extinction corresponding to the transition between mostly-atomic and mostly-molecular gas depends on a number of factors: the strength of the ISRF, the volume density of the gas, and the effectiveness of self-shielding. However, the equilibrium molecular fraction typically depends only linearly on these quantities, but exponentially on the visual extinction. In conditions typical of local

GMCs, the transition from atomic to molecular hydrogen occurs at a visual extinction $A_V \sim 0.1$–0.2 (Draine and Bertoldi 1996; Krumholz et al. 2008) and that from C^+ to CO occurs at $A_V \sim 1$ (Wolfire et al. 2010). Large differences in either the density of the gas or the strength of the ISRF are required in order to significantly alter these values. In solar metallicity gas, the total hydrogen column densities corresponding to the two transitions are $N_{H,tot} \approx 2 \times 10^{20}$ cm^{-2} for the H–H_2 transition and $N_{H,tot} \approx 2 \times 10^{21}$ cm^{-2} for the C^+–CO transition. However, in lower metallicity environments, such as the Magellanic Clouds, the lower dust-to-gas ratio means that a higher column density is required.

The transition from unshielded gas to gas with a significant visual extinction also has an important influence on the thermal state of the gas. As we have already discussed, photoelectric heating is the dominant form of radiative heating in the diffuse ISM (Sect. 3.7), but its effectiveness falls off rapidly with increasing extinction for $A_V > 1$. Consequently, the equilibrium gas temperature drops significantly as we move from unshielded to shielded gas, as illustrated in Fig. 16.

The chemical and thermal changes that occur as we move from unshielded to shielded regions have important implications for the gravitational stability of the gas. The Jeans mass—the critical mass scale above which quasi-spherical overdensities become unstable to their own self-gravity—is related to the gas density and the temperature by

$$M_J \approx 60 \, M_\odot \, \mu^{-2} T^{3/2} n_{tot}^{-1/2} , \tag{192}$$

where μ is the mean particle mass and n_{tot} is the total particle number density. The factor of six decrease in T that we find as we move from low extinction to high extinction gas therefore results in a drop in M_J of roughly a factor of fifteen. The associated chemical transition from gas which is primarily atomic to gas which is primarily molecular results in a further decrease in M_J by roughly a factor of 2.5, and so the overall effect of increasing the extinction is to decrease the Jeans mass by more than an order of magnitude.

The decrease in T and increase in μ that occur as we move from low A_V to high A_V gas are also responsible for a drop in the sound speed of the gas. If the turbulent kinetic energy remains fixed, the result is an increase in the Mach number of the turbulence. This makes it easier for the turbulence to create high density regions (see also Sect. 6.4.4). The high density gas is more likely to be gravitationally bound, since $M_J \propto n^{-1/2}$, and so the result of an increase in the Mach number of the turbulence will typically be an increase in the star formation rate of the gas (Krumholz and McKee 2005; Padoan and Nordlund 2011; Hennebelle and Chabrier 2011; Federrath and Klessen 2012).

We therefore see that it is much easier to form stars in gas clouds with high visual extinctions than in those with low visual extinctions. In high A_V clouds, the gas temperature is lower and the gas is more likely to be molecular, and both of these effects make star formation more likely. We therefore expect to find a correlation between high A_V clouds and star formation. Moreover, since these high A_V clouds

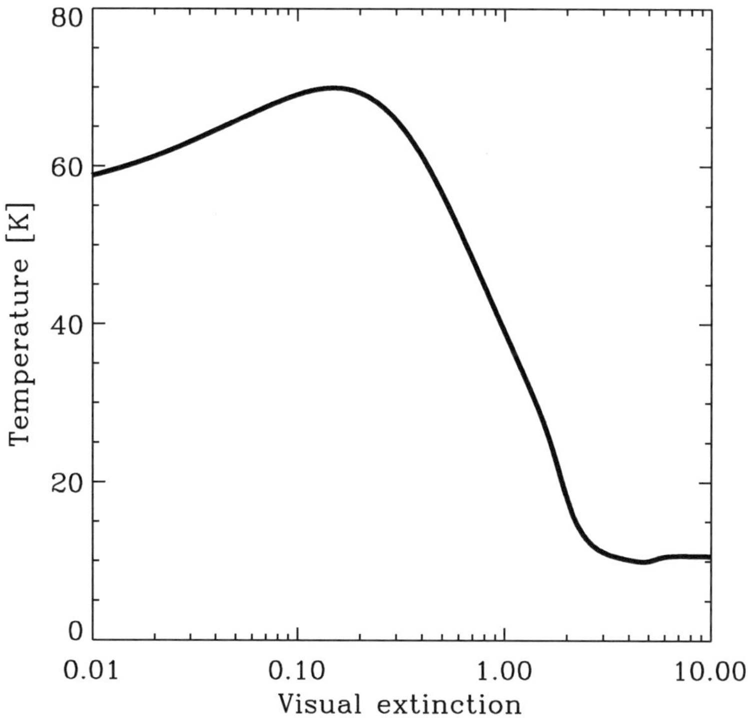

Fig. 16 Equilibrium temperature as a function of visual extinction within a uniform density, semi-infinite slab of solar metallicity gas. The slab had a fixed hydrogen nuclei number density $n = 300\ \mathrm{cm}^{-3}$, a velocity gradient of $3\ \mathrm{km\,s^{-1}\,pc^{-1}}$, and was illuminated with a model of the ISRF based on Draine (1978) at UV wavelengths and Mathis et al. (1983) at longer wavelengths. The cosmic ray ionization rate of atomic hydrogen was taken to be $\zeta_{\rm H} = 3 \times 10^{-17}\ \mathrm{s}^{-1}$. The rise in the temperature at very low $A_{\rm V}$ is a consequence of the transition from H to H_2: the cooling in this regime is dominated by C^+ fine structure cooling, and at these densities and at fixed temperature, the C^+ cooling rate in fully atomic gas is two to three times larger than the C^+ cooling rate in fully molecular gas

are dominated by molecular gas, we also expect there to be a correlation between molecular gas and star formation (Krumholz et al. 2011; Federrath 2013).

A correlation of just this kind is seen when we examine how stars form in our own Milky Way or in nearby spiral galaxies. Work by a number of groups has shown that on large scales, there is a tight correlation between the surface density of molecular gas and the surface density of star formation in spiral galaxies (Wong and Blitz 2002; Leroy et al. 2008; Bigiel et al. 2008, 2011; Schruba et al. 2011). This correlation is close to linear, although arguments continue as to whether it is truly linear (Leroy et al. 2013), or is actually slightly sub-linear (Shetty et al. 2013, 2014; Federrath 2013).

This correlation is often interpreted as being a consequence of molecular cooling. It is argued that only CO cooling can lower the gas temperature to the value of ~10 K characteristic of prestellar cores within molecular clouds (Bergin and Tafalla 2007), and that these low temperatures are required for star formation. Although it is true that gas cannot reach 10 K purely due to C^+ or O fine structure cooling, it can quite easily reach temperatures as low as 20 K in well-shielded gas, as demonstrated in the detailed numerical simulations of Glover and Clark (2012a). These simulations also show that the star formation rate of the gas is relatively insensitive to whether the gas is dominated by atomic or molecular cooling (see also Glover and Clark 2012c; Krumholz 2012), further supporting the idea that the observed correlation between molecular gas and star formation is a consequence of the fact that both are associated with regions of cold, dense gas. In short, molecular gas is a tracer of star formation but not its cause.

Dust shielding may also play an important role in determining which sub-regions within molecular clouds can successfully form stars. We know from observations of local star-forming GMCs that stellar birth is not a completely random process. Instead, there is a clear relationship between the observed column density of the gas and the star formation rate. The process occurs predominantly in regions with column densities $N_{H_2} > 7.5 \times 10^{21}\ cm^{-2}$, corresponding to visual extinctions $A_V > 8$ (see e.g. the discussion in Molinari et al. 2014). It remains an open question as to whether this relationship is best described in terms of a column density *threshold* (Onishi et al. 1998; Johnstone et al. 2004; Lada et al. 2010) or simply a steep dependence of the star formation rate on the column density (Hatchell et al. 2005; Enoch et al. 2008; Heiderman et al. 2010; Gutermuth et al. 2011; Burkert and Hartmann 2013; Lada et al. 2013; Evans et al. 2014).

A complete theoretical understanding of why this correlation exists remains lacking, but Clark and Glover (2014) argue that it is a further consequence of dust shielding. They point out that in a turbulent cloud, the angle-averaged extinction seen by an arbitrarily chosen point along a high extinction line of sight will frequently be much lower than the extinction along that line of sight. In their simulations, the dense structures traced by line of sight extinctions $A_V > 8$ typically have much smaller mean extinctions, $\langle A_V \rangle \sim 1$–2. This roughly corresponds to the point at which dust shielding renders photoelectric heating of the gas ineffective, and so Clark and Glover (2014) argue that the observed column density threshold merely reflects the extinction required for clouds to shield themselves effectively from their environment.

5.3 Molecular Cloud Formation in a Galactic Context

As we have seen above, the transition between regions of the ISM that are dominated by atomic gas and regions that are dominated by molecular gas is primarily driven by changes in the column density. In regions with low column densities, photodissociation of H_2 and CO is very efficient and the equilibrium molecular fraction is small. On the other hand, in regions with high column densities, molecular self-shielding and dust shielding dramatically reduce the photodissociation rates of H_2 and CO, allowing the equilibrium molecular fraction to become large. From one point of view, then, the question of how molecular clouds form in the ISM has a simple answer. This happens whenever sufficient gas is brought together in one place to raise the column density above the value needed to provide effective shielding against the ISRF, as long as the gas remains in this configuration for longer than the H_2 formation timescale, $t_{\rm form}$. The real question, therefore, is what are the most important processes responsible for gathering together sufficient gas out of the diffuse ISM to make a dense molecular cloud. This is a highly complex topic, and in these lecture notes we will do little more than to give a brief outline of the main models that have been proposed to explain molecular cloud formation. We refer readers in search of a more in-depth treatment of these issues to the recent reviews by Hennebelle and Falgarone (2012), Dobbs et al. (2014) and Molinari et al. (2014).

One of the simplest models for molecular cloud formation is the coagulation model, originally proposed by Oort (1954) and subsequently elaborated by many other authors (see e.g. Field 1965; Kwan 1979; Tomisaka 1984; Tasker and Tan 2009). This model is based on a picture of the ISM in which the cold atomic and molecular gas is organized into a series of discrete clouds with a range of different masses. Small atomic clouds are formed directly from warmer atomic gas by thermal instability (Field 1965). Collisions between these small clouds efficiently dissipate energy, and so colliding clouds tend to coagulate, forming successively larger clouds. Once the clouds have grown large enough, they become able to shield themselves from the effects of the ISRF, at which point they become dominated by molecular gas. Even once they have become molecular, however, the clouds continue to undergo regular collisions, and can potentially grow to very large masses. This process is terminated for a particular cloud once the feedback from the stars forming within it becomes strong enough to disrupt the cloud.

This model has a number of appealing features. The stochasticity of the process of cloud-cloud collisions is thought to naturally lead to a power law cloud mass function (e.g. Field 1965), and the fact that collisions occur more frequently in denser regions of the galactic disk also provides a simple explanation for the enhanced concentrations of molecular gas and ongoing star formation found within most spiral arms. In addition, the coagulation model also can easily produce clouds that are counter-rotating compared to the galactic disk (e.g. Dobbs 2008; Tasker and Tan 2009), explaining why clouds with retrograde rotation appear to be common within the ISM (see e.g. Phillips 1999; Imara and Blitz 2011).

Unfortunately, this model also suffers from a major problem. Small molecular clouds can be built by coagulation relatively rapidly, but large molecular clouds with masses of 10^5–10^6 $M_\odot$ require of the order of 100 Myr or more to form by this method (Blitz and Shu 1980). Since this is an order of magnitude larger than most estimates for typical GMC lifetimes (Blitz et al. 2007), it seems to be impossible to form massive GMCs in low density environments in this model. In the dense environments of spiral arms, the much higher cloud collision rate alleviates this problem to a large extent (Casoli and Combes 1982; Kwan and Valdes 1983; Dobbs 2008), but this does not provide an explanation for the existence of very massive clouds in inter-arm regions, as observed in galaxies such as M51 (Hughes et al. 2013).

A more fundamental issue with the coagulation model is that it is not clear that the picture of the ISM on which it is based, in which GMCs are discrete objects that evolve in equilibrium between collisions and that have well-defined masses and clear edges, is a good description of the real ISM. As we discuss in Sect. 4.1, observations show that GMCs are ubiquitously surrounded by extended envelopes of atomic gas (see e.g. Wannier et al. 1983; Elmegreen and Elmegreen 1987; Lee et al. 2012; Heiner and Vázquez-Semadeni 2013; Motte et al. 2014) and so the observational "edge" of a GMC—the point at which we cease to be able to detect CO emission—more likely represents a chemical transition in the gas (see Sect. 5.1), rather than any sudden change in the density. In addition, a considerable fraction of the molecular gas of a galaxy seems to be in an extended diffuse component, rather than in discrete clouds (see e.g. Pety et al. 2013; Shetty et al. 2014; Smith et al. 2014). This finding casts further doubts on any astrophysical conclusion derived from the cloud collision picture.

Altogether, it is highly plausible that rather than being discrete objects with identities that persist over long periods of time, molecular clouds are instead merely the highest density regions within a far more extended turbulent flow of gas. This picture motivates an alternative way of thinking about molecular cloud formation, known as the converging (or colliding) flow model for cloud formation. The basic idea in this case is that molecular clouds form in dense, post-shock regions formed when converging flows of lower density gas collide and interact. If the flows initially consist of warm atomic hydrogen, then their collision can trigger a thermal instability, leading to the rapid production of a cloud of much denser, cooler gas (see e.g. Hennebelle and Pérault 1999, 2000; Koyama and Inutsuka 2002; Audit and Hennebelle 2005; Heitsch et al. 2005, 2006; Vázquez-Semadeni et al. 2006; Hennebelle and Audit 2007; Heitsch and Hartmann 2008; Banerjee et al. 2009). The mean density of the cold gas clouds produced in this way is typically of the order of $100\ \mathrm{cm}^{-3}$, high enough to allow H_2 formation to occur on a timescale shorter than the duration of the collision (see e.g. Clark et al. 2012b). CO will also form in these cold clouds in regions where the column density is high enough to provide effective shielding from the ISRF, although simulations have shown that the production of these high column density regions generally requires at least some part of the cold cloud to undergo gravitational collapse (Heitsch and Hartmann 2008; Clark et al. 2012b).

The converging flow model for GMC formation naturally explains why we see so few molecular clouds that are not associated with ongoing star formation. CO observations are blind to the inflow during its early evolution, since at this stage, the molecular abundance in the gas is very small (Hartmann et al. 2001). High molecular abundances and detectable CO luminosities are produced only during relatively late evolutionary phases, and work by Clark et al. (2012b) has shown that the time lag between the appearance of detectable CO emission and the onset of star formation is typically only 1–2 Myr. This picture is supported by growing observational evidence that molecular clouds (as traced by CO) are continuously gaining mass during their evolution. For example, Fukui et al. (2009) and Kawamura et al. (2009) report in their analysis of GMCs in the Large Magellanic Cloud mass growth rates of several $10^{-2}\,\mathrm{M_\odot\,yr^{-1}}$. It is a very appealing feature that this continuous accretion process provides a simple explanation for the presence of turbulence within GMCs. The kinetic energy associated with the convergent flow that forms the cloud in the first place is also able to drive its internal turbulence and explain many of its internal properties (see Klessen and Hennebelle 2010; Goldbaum et al. 2011). As a consequence, the turbulent cascade extends from global galactic scales all the way down to the dissipation regime on sub-parsec scales (Sect. 4.3).

Much of the work that has been done to model the formation of molecular clouds in converging flows has focused on the case where the flow is essentially one-dimensional, with two streams of gas colliding head-on. However, in this scenario, it is difficult to form very massive clouds, as a simple calculation demonstrates. Suppose we have two flows of convergent gas, each of which has a cross-sectional area A, an initial number density n_0, and a length $L_{\rm flow}/2$. The total mass of the cloud that can be formed by the collision of these flows is given approximately by

$$M_{\rm cloud} \sim \mu n_0 A L_{\rm flow}, \tag{193}$$

where we have assumed that all of the gas in the flows becomes part of the cold cloud, and $\mu = 1.26\,m_{\rm H} = 2.11 \times 10^{-24}$ g typical for atomic gas. If the gas in the flows is initially part of the warm neutral medium, then the number density is $n_0 \sim 0.5\,\mathrm{cm^{-3}}$ (see Table 1), and

$$M_{\rm cloud} \sim 2300\,\mathrm{M_\odot} \left(\frac{A}{1000\,\mathrm{pc^2}}\right) \left(\frac{L_{\rm flow}}{150\,\mathrm{pc}}\right). \tag{194}$$

If the flows together have a total length $L_{\rm flow} \sim 150$ pc that is comparable to the molecular gas scale height of the Galactic disk (see Table 4), and a cross-sectional area typical of a reasonably large GMC (Solomon et al. 1987), then the total mass of the resulting cloud is only a few thousand solar masses, much smaller than the mass of most GMCs.

There are several ways in which we might try to avoid this problem. First, we can make $L_{\rm flow}$ larger. However, even if we make it comparable to the atomic gas scale height, so that $L_{\rm flow} \sim 1000$ pc, the resulting cloud mass is still small, $M_{\rm cloud} \sim 15000\,\mathrm{M_\odot}$. Second, we can make n_0 larger. The value that we have adopted above is typical of the stable WNM, but thermally unstable diffuse atomic gas could have a density that is an order of magnitude higher (see e.g. Dobbs et al. 2012). However, once again this does not increase $M_{\rm cloud}$ by a large enough amount to explain how the most massive GMCs form. Finally, we could make A larger. Simulations show that clouds formed in one-dimensional flows tend to collapse gravitationally in the directions perpendicular to the flow (see e.g. Burkert and Hartmann 2004; Heitsch et al. 2008; Vázquez-Semadeni et al. 2009). Therefore, it is reasonable to suppose that the cross-sectional area of the flows involved in forming the cloud may be much larger than the cross-sectional area of the final GMC. However, even if we increase A by a factor of 20, so that the width and height of the flow are comparable to its length, we again only increase $M_{\rm cloud}$ by an order of magnitude. In addition, if all of the dimensions of the flow are similar, it is unclear whether we should really think of it as a one-dimensional flow any longer. In the end, what is needed in order to explain the formation of the most massive GMCs in this model is a combination of these points. The flow must consist of gas that is denser than is typical for the WNM, that has a coherent velocity over a relatively large distance, and that either has a large cross-sectional area or is actually inflowing from multiple directions simultaneously. How often these conditions are realized in the real ISM remains an open question.

Another issue that is not yet completely settled is which of several different possible physical processes is primarily responsible for driving these convergent flows of gas. One obvious possibility is that these flows are driven by large-scale gravitational instability. Analysis of the behavior of small perturbations in a thin rotating gas disk shows that the key parameter that determines whether or not they grow exponentially is the so-called Toomre parameter (Toomre 1964),

$$Q = \frac{c_{\rm s,eff}\,\kappa}{\pi G \Sigma}. \tag{195}$$

Here, $c_{\rm s,eff}$ is the effective sound-speed of the gas, which accounts not only for the thermal sound speed, but also for the influence of the small-scale turbulent velocity dispersion, κ is the epicyclic frequency of the disk, and Σ is the surface density of the gas. A pure gas disk is unstable whenever $Q < 1$. In the case of a disk that contains a mix of gas and stars, the analysis is more complex (see e.g. Rafikov 2001; Elmegreen 2011), but the required value of Q remains close to unity. Measurements of Q in nearby spirals and dwarf galaxies suggest that in most of these systems, the gas is marginally Toomre stable, even when the gravity of the stellar component is taken into account (Leroy et al. 2008). However, this does not mean that gravitational instability is unimportant in these systems, as simulations show that star formation in disk galaxies tends to self-regulate so that $Q \sim 1$ (e.g. Krumholz and Burkert 2010; Faucher-Giguère et al. 2013). Briefly, the reason for this is that if $Q \ll 1$, the disk will be highly unstable and will form stars rapidly (see e.g. Li et al. 2005, 2006). This will both deplete the gas surface density, and also increase $c_{\rm s,eff}$, due to the injection of thermal and turbulent energy into the gas by the various stellar feedback processes discussed in Sect. 4.6. These effects combine to increase Q until the disk becomes marginally stable.

Another mechanism that can drive large-scale convergent flows of gas in spiral galaxies is the Parker instability (Parker 1966). This is a magnetic instability which causes a field that is stratified horizontally in the disk to buckle due to the influence of magnetic buoyancy. Gas then flows down the buckled magnetic field lines, accumulating near the midplane of the disk. The characteristic length scale associated with this instability is a factor of a few larger than the disk scale height. It therefore allows gas to be accumulated from within a large volume, and is hence capable of producing even the most massive GMCs (Mouschovias 1974; Mouschovias et al. 1974). However, the density contrasts produced by the Parker instability are relatively small (see e.g. Kim et al. 1998, 2001, 2002) and so, although this instability may play a role in triggering thermal instability in the galactic midplane (Mouschovias et al. 2009), it seems unlikely to be the main mechanism responsible for GMC formation.

Finally, stellar feedback in the form of expanding HII regions, stellar wind bubbles, supernova remnants and super-bubbles may also drive converging flows of gas in the ISM (see e.g. Ntormousi et al. 2011; Dobbs et al. 2012; Hennebelle and Iffrig 2014 for some recent examples). The idea that stellar feedback may trigger cloud formation, and hence also star formation, has a long history (see e.g. Elmegreen and Lada 1977 for a seminal early study). At first sight it has considerable observational support, since examples of spatial associations between molecular clouds and feedback-driven bubbles are widespread (e.g. Beaumont and Williams 2010; Deharveng et al. 2010; Hou and Gao 2014). However, this is a case in which the observations are somewhat misleading. The fact that a molecular cloud is associated with the edge of a feedback-driven bubble does not necessarily imply that the bubble is responsible for creating the cloud, since the expanding bubble may simply have swept up some dense, pre-existing structure (e.g. Pringle et al. 2001). Models of cloud formation in a supernova-driven turbulent ISM without self-gravity find that although some cold, dense clouds are formed in the expanding shells bounding the supernova remnants, the total star formation rate expected for these regions is only $\sim$10% of the rate required to produce the assumed supernova driving (Joung and Mac Low 2006). Recent efforts to quantify the effectiveness of triggering in the LMC also find that no more than about 5–10 % of the total molecular gas mass budget can be ascribed to the direct effect of stellar feedback (Dawson et al. 2013). Therefore, although stellar feedback clearly plays an important role in structuring the ISM on small scales and contributes significantly to the energy budget of interstellar turbulence (Sect. 4.6), it does not appear to be the main process responsible for the formation of molecular clouds.

6 Star Formation

6.1 Molecular Cloud Cores as Sites of Star Formation

In this Section, we focus on the small-scale characteristics of molecular clouds and discuss the properties of the low-mass cores that are the immediate progenitors of individual stars or binary systems. We begin with a discussion of the core mass spectrum, and then turn our attention to the density, thermal, chemical, kinematic, and magnetic field structure of individual cores. We distinguish between prestellar cores, which are dense cloud cores that are about to form stars in their interior, but have not yet done so (or at least show no detectable sign of stellar activity), and protostellar cores, for which we can infer the presence of embedded protostars in the main accretion phase.

6.1.1 Mass Spectrum of Molecular Cloud Cores

In Sect. 4.1.2, we discussed the global statistical properties of molecular clouds. While a complete structural decomposition of an entire cloud leads to a power-law mass spectrum (Eq. 104), focusing on the densest parts of the clouds, on the pre- and protostellar cores, yields a different picture. As one probes smaller and smaller scales and more strongly bound objects, the inferred mass distribution becomes closer to the stellar IMF. The first large study of this kind was published by Motte et al. (1998), for a population of submillimeter cores in ρ Oph. Using data obtained with the IRAM 30m-telescope,[8] they discovered a total of 58 starless clumps, ranging in mass from 0.05 $M_{\odot}$ to 3 $M_{\odot}$. Similar results have been obtained for the Serpens cloud (Testi and Sargent 1998), for Orion B North (Johnstone et al. 2001) and Orion B South (Johnstone et al. 2006), and for the Pipe Nebula (Lada et al. 2006). Currently all observational data (e.g. Motte et al. 1998; Testi and Sargent 1998; Johnstone et al. 2000, 2001, 2006; Nutter and Ward-Thompson 2007; Alves et al. 2007; Di Francesco et al. 2007; Ward-Thompson et al. 2007; Lada et al. 2008; Könyves et al. 2010) reveal a striking similarity to the IMF. To reach complete overlap one is required to introduce a mass scaling or efficiency factor of 0.2–0.5, depending on the considered region. An exciting interpretation of these observations is that we are witnessing the direct formation of the IMF via fragmentation of the parent cloud. However, we note that the observational data also indicate that a considerable fraction of the prestellar cores do not exceed the critical mass for gravitational collapse, much like the clumps on larger scales. The evidence for a one-to-one mapping between prestellar cores and the stellar mass, thus, is by no means conclusive. For an extended discussion of potential caveats, see Clark et al. (2007) or consult the *Protostars and Planets VI* review by Offner et al. (2014).

6.1.2 Density Structure

The density structure of prestellar cores is typically inferred through the analysis of dust emission or absorption using near-infrared extinction mapping of background starlight, millimeter/submillimeter dust continuum emission, or dust absorption against the bright mid-infrared background emission (Bergin and Tafalla 2007). A main characteristic of the density profiles derived with the above techniques is that they require a central flattening within radii smaller than 2500–5000 AU, with typical central densities of 10^5–10^6 cm^{-3} (Motte et al. 1998; Ward-Thompson et al. 1999). A popular approach is to describe these cores as truncated isothermal (Bonnor-Ebert) spheres (Ebert 1955; Bonnor 1956) that often (but not always) provide a good fit to the data (Bacmann et al. 2001; Alves et al. 2001; Kandori et al. 2005). These are equilibrium solutions for the density structure of self-gravitating gas spheres bounded by external pressure. However, this density structure is not unique. Numerical calculations of the dynamical evolution of supersonically turbulent clouds show that the

[8] http://www.iram-institute.org/EN/30-meter-telescope.php.

transient cores forming at the stagnation points of convergent flows exhibit similar morphology despite not being in dynamical equilibrium (Ballesteros-Paredes et al. 2003).

6.1.3 Kinematic Structure

In contrast to the supersonic velocity fields observed in molecular clouds, dense cores have low internal velocities. Starless cores in clouds like Taurus, Perseus, and Ophiuchus systematically exhibit spectra with close to thermal linewidths, even when observed at low angular resolution (Myers 1983; Jijina et al. 1999). This indicates that the gas motions inside the cores are subsonic or at best transsonic, with Mach numbers less than $\sim$2 (Kirk et al. 2007; André et al. 2007; Rosolowsky et al. 2008). In addition, in some cores, inward motions have also been detected. They are inferred from the observation of optically thick, self-absorbed lines of species like CS, H_2CO, or HCO^+, in which low-excitation foreground gas absorbs part of the background emission. Typical inflow velocities are of order of 0.05–0.1 km s^{-1} and are observed on scales of 0.05–0.15 pc, comparable to the total size of the cores (Lee et al. 1999). The overall velocity structure of starless cores appears broadly consistent with the structure predicted by models in which protostellar cores form at the stagnation points of convergent flows, but the agreement is not perfect (Klessen et al. 2005; Offner et al. 2008). Clearly more theoretical and numerical work is needed. In particular, the comparison should be based on synthetic line emission maps, requiring one to account for the chemical evolution of the gas in the core and the effects of radiative transfer (e.g. Smith et al. 2012, 2013; Chira et al. 2014). In addition, it is also plausible that the discrepancy occurs because the simulations do not include all the necessary physics such as radiative feedback and magnetic fields.

Subsonic turbulence contributes less to the energy budget of the cloud than thermal pressure and so cannot provide sufficient support against gravitational collapse (Myers 1983; Goodman et al. 1998; Tafalla et al. 2006). If cores are longer lasting entities there must be other mechanisms to provide stability. Obvious candidates are magnetic fields (Shu et al. 1987). However, they are usually not strong enough to provide sufficient support (Crutcher et al. 1999, 2009a, 2010b; Crutcher and Troland 2000; Bourke et al. 2001). It seem reasonable to conclude that most observed cores are continuously evolving transient objects rather than long-lived equilibrium structures.

6.1.4 Thermal Structure

The kinetic temperature of dust and gas in a core is regulated by the interplay between various heating and cooling processes. At densities above 10^5 cm^{-3} in the inner part of the cores, gas and dust are coupled thermally via collisions (Goldsmith and Langer 1978; Burke and Hollenbach 1983; Goldsmith 2001, and see also Sect. 3.5). At lower densities, corresponding to the outer parts of the cores, the two temperatures are not necessarily expected to be the same. Thus, the dust and gas temperature distributions need to be independently inferred from the observations. Large-scale studies of the dust temperature show that the grains in starless cores are colder than in the surrounding lower-density medium. Far-infrared observations toward the vicinity of a number of dense cores provide evidence for flat or decreasing temperature gradients with cloud temperatures of 15–20 K and core values of 8–12 K (Ward-Thompson et al. 2002; Tóth et al. 2004; Launhardt et al. 2013). These observations are consistent with dust radiative transfer modeling in cores illuminated by interstellar radiation field (Langer et al. 2005; Keto and Field 2005; Stamatellos et al. 2007). The gas temperature in molecular clouds and cores is commonly infered from the level excitation of simple molecules like CO and NH_3 (Evans 1999; Walmsley and Ungerechts 1983). One finds gas temperatures of 10–15 K, with a possible increase toward the lower density gas near the cloud edges. However, these measurements are difficult, since as the density drops, the molecular emission can become sub-thermal, in which case its excitation temperature no longer traces the kinetic temperature of the gas (see the discussion in Sect. 3.1). In static prestellar cores (if such things exist), the main heat source is cosmic ray ionization, while in gravitationally collapsing cores, compressional heating and the dissipation of turbulence can also make significant contributions to the total heating rate (Glover and Clark 2012a). Cooling in dense cores is dominated by molecular line emission, particularly from CO, and by heat transfer from the gas to the grains (Goldsmith and Langer 1978).

6.1.5 Chemical Structure

Maps of integrated line intensity can look very different for different molecular tracers. This is illustrated in Fig. 17. It shows that the emission from nitrogen-bearing species, such as N_2H^+, more closely follows the dust emission, while emission from carbon-bearing molecules, such as $C^{18}O$ or CS, often appears as a "ring-like" structure around the dust emission peak (Bergin et al. 2002; Tafalla et al. 2002; Lada et al. 2003; Maret et al. 2007). The common theoretical interpretation of these data is that carbon-bearing species freeze-out on the surfaces of cold dust grains in dense portions of the cloud, while nitrogen-bearing molecules largely remain in the gas phase. At the same time, chemical models of prestellar cores predict that molecules in the envelope of the core are destroyed by the interstellar UV field (Pavlyuchenkov et al. 2006; Aikawa et al. 2008). The resulting chemical stratification significantly complicates the interpretation of molecular line observations, and again requires the use of sophisticated chemical models which have to be coupled to the dynamical evolution (e.g. Aikawa et al. 2008; van Weeren et al. 2009; Furuya et al. 2012). From the observational side, the freeze-out of many molecules makes it difficult to use their emission lines for probing the physical conditions in the inner regions of the cores. Nevertheless, modeling of the chemical evolution of the gas can provide us with important information on the cores. For example, the level of CS depletion can be used to constrain the age of the prestellar cores, while the deficit of CS in the envelope can indicate the strength of the external UV field (Bergin and Tafalla 2007).

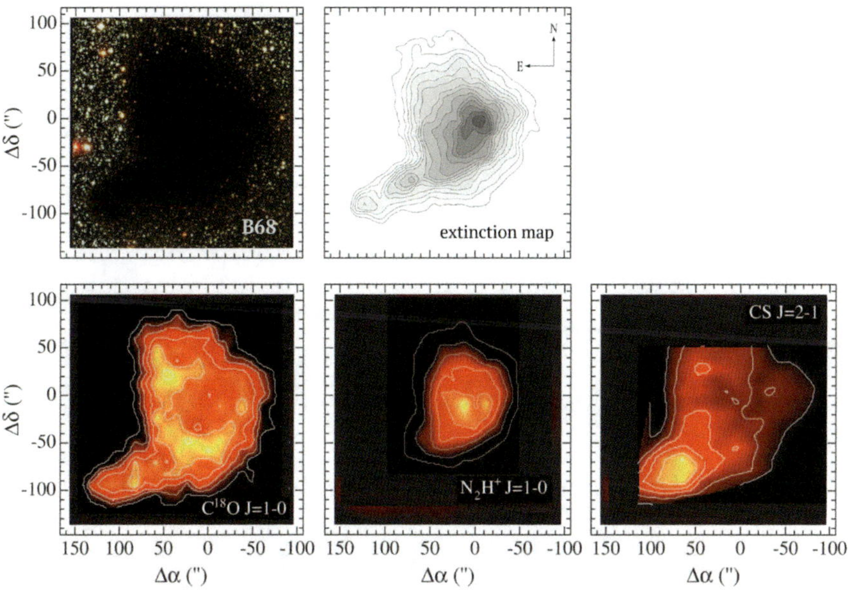

Fig. 17 Isolated prestellar molecular cloud core Barnard 68. *Upper panel* Optical image and extinction map of the object (see Alves et al. 2001; images from ESO website). *Lower panel* Maps of molecular line emission from $C^{18}O$, N_2H^+, and CS (images adopted from Lada et al. 2003; see also see Bergin et al. 2002). The lower images illustrate the effects of depletion onto grains in the high-density central region of the core. $C^{18}O$ is clearly underabundant in the central, high-density regions of Barnard 68, while N_2H^+ traces this region very well. CS is brightest in the tail structure in the south-east corner and is highly depleted in the core center (see also Bergin and Tafalla 2007)

In any case, any physical interpretation of the molecular lines in prestellar cores has to be based on chemical models and should do justice to the underlying density and velocity pattern of the gas.

6.1.6 Magnetic Field Structure

Magnetic fields are ubiquitously observed in the interstellar gas on all scales (Crutcher et al. 2003; Heiles and Troland 2005). However, their importance for star formation and for the morphology and evolution of molecular cloud cores remains a source of considerable controversy. A crucial parameter in this debate is the ratio between core mass and magnetic flux. In supercritical cores, this ratio exceeds a threshold value and collapse can proceed. In subcritical ones, magnetic fields provide stability (Spitzer 1978; Mouschovias 1991a, b). Measurements of the Zeeman splitting of molecular lines in nearby cloud cores indicate mass-to-flux ratios that lie above the critical value, in some cases only by a small margin, but very often by factors of many if non-detections are included (Crutcher 1999; Bourke et al. 2001; Crutcher et al. 2009a, 2010a). The polarization of dust emission offers an alternative path-

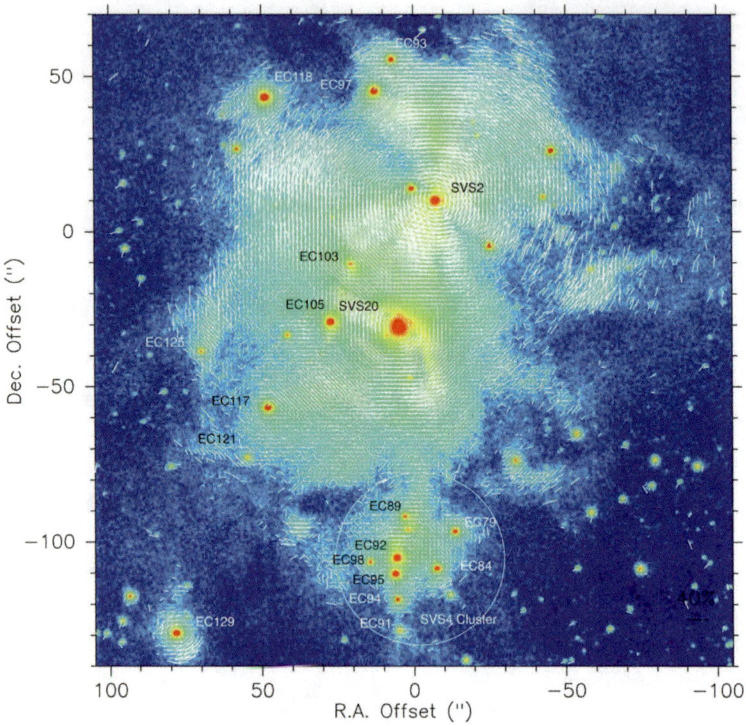

Fig. 18 Polarization vector map of the central region of the Serpens cloud core, superimposed on the total intensity images in logarithmic scaling. The area of the image is 220" × 220". From Sugitani et al. (2010)

way to studying the magnetic field structure of molecular cloud cores (as illustrated in Fig. 18). Magnetohydrodynamic (MHD) simulations of turbulent clouds predict degrees of polarization between 1 and 10 %, regardless of whether turbulent energy dominates over the magnetic energy (i.e. the turbulence is super-Alfvénic) or not (Padoan and Nordlund 1999; Padoan et al. 2001). However, converting polarization into magnetic field strength is very difficult (Heitsch et al. 2001b). Altogether, the current observational findings imply that magnetic fields must be considered when studying stellar birth, but also that they are not the dominant agent that determines when and where star formation sets in within a cloud. It seems fair to conclude that magnetic fields appear to be too weak to prevent gravitational collapse from occurring.

This means that in many cases and to a reasonable approximation purely hydrodynamic simulations are sufficient to model ISM dynamics and stellar birth. However, when more precise and quantitative predictions are desired, e.g. when attempting to predict star formation timescales or binary properties, it is necessary to perform magnetohydrodynamic simulations or even to consider non-ideal MHD. The latter means to take ambipolar diffusion (drift between charged and neutral particles) or

Ohmic dissipation into account. Recent numerical calculations have shown that even a weak magnetic field can have noticeable dynamical effects. It can alter how cores fragment (Price and Bate 2007b, 2008; Hennebelle and Fromang 2008; Hennebelle and Teyssier 2008; Hennebelle et al. 2011; Peters et al. 2011), change the coupling between stellar feedback processes and their parent clouds (Nakamura and Li 2007; Krumholz et al. 2007b), influence the properties of protostellar disks due to magnetic braking (Price and Bate 2007a; Mellon and Li 2009; Hennebelle and Ciardi 2009; Seifried et al. 2011, 2012a, b, 2013), or slow down the overall evolution (Heitsch et al. 2001a).

6.2 Statistical Properties of Stars and Star Clusters

In order to better understand how gas turns into stars, we also need to introduce some of the key properties of young stellar systems. We restrict ourselves to a discussion of the star formation timescale, the spatial distribution of young stars, and the stellar initial mass function (IMF). We note, however, that other statistical characteristics, such as the binary fraction, its relation to the stellar mass, and the orbital parameters of binary stars are equally important for distinguishing between different star formation models. As the study of stars and star clusters is central to many areas of astronomy and astrophysics, there are a large number of excellent reviews that cover various aspects of this wide field. For further reading on embedded star clusters, we refer to Lada and Lada (2003). For the early evolution of young star clusters, we point to Krumholz et al. (2014) and Longmore et al. (2014), as well as to Kroupa (2005) and Portegies Zwart et al. (2010). More information on the stellar IMF can be found in the seminal papers by Scalo (1986), Kroupa (2002), and Chabrier (2003a), as well as in the reviews by Kroupa et al. (2013) and Offner et al. (2014). General reviews of star formation are provided by Mac Low and Klessen (2004), Ballesteros-Paredes et al. (2007), McKee and Ostriker (2007), Krumholz (2014), or Zinnecker and Yorke (2007), with the latter focusing specifically on the formation of high-mass stars.

6.2.1 Star Formation Timescales

The star formation process in molecular clouds appears to be fast (Hartmann et al. 2001; Elmegreen 2007). Once the collapse of a cloud region sets in, it rapidly forms an entire cluster of stars within 10^6 years or less. This is indicated by the young stars associated with star-forming regions, typically T Tauri stars with ages less than 10^6 years (Gomez et al. 1992; Greene and Meyer 1995; Carpenter et al. 1997), and by the small age spread in more evolved stellar clusters (Hillenbrand 1997; Palla and Stahler 1999). Star clusters in the Milky Way also exhibit an amazing degree of chemical homogeneity (in the case of the Pleiades, see Wilden et al. 2002), implying that the gas out of which these stars formed must have been chemically well-mixed initially, which could provide interesting pathways to better understand turbulent mixing in the ISM (see also de Avillez and Mac Low 2002; Klessen and Lin 2003; Feng and Krumholz 2014).

6.2.2 Spatial Distribution

The advent of sensitive infrared detectors in the last decade or so allowed us to conduct wide-area surveys. These have revealed that most stars form in clusters and aggregates of various size and mass scales, and that isolated or widely distributed star formation is the exception rather than the rule (Lada and Lada 2003). The complex hierarchical structure of molecular clouds (see e.g. Fig. 10) provides a natural explanation for this finding.

Star-forming molecular cloud cores can vary strongly in size and mass. In small, low-density clouds, stars form with low efficiency, more or less in isolation or scattered around in small groups of up to a few dozen members. Denser and more massive clouds may build up stars in associations and clusters of a few hundred members. This appears to be the most common mode of star formation in the solar neighborhood (Adams and Myers 2001). Examples of star formation in small groups and associations are found in the Taurus-Aurigae molecular cloud (Hartmann 2002). Young stellar groups with a few hundred members form in the Chamaeleon I (Persi et al. 2000) or ρ-Ophiuchi (Bontemps et al. 2001) dark clouds. Each of these clouds is at a distance of about 130–160 pc from the Sun. Like most of the nearby young star forming regions they appear to be associated with a ring-like structure in the Galactic disk called Gould's Belt (Poppel 1997).

The formation of dense rich clusters with thousands of stars is rare. The closest region where this happens is the Orion Nebula Cluster (Hillenbrand 1997; Hillenbrand and Hartmann 1998). It lies at a distance of 410 pc (Sandstrom et al. 2007; Menten et al. 2007; Hirota et al. 2007; Caballero 2008). A rich cluster somewhat further away is associated with the Monoceros R2 cloud (Carpenter et al. 1997) at a distance of $\sim$830 pc. The cluster NGC 3603 is roughly ten times more massive than the Orion Nebula Cluster. It lies in the Carina region, at about 7 kpc distance. It contains about a dozen O stars, and is the nearest object analogous to a starburst knot (Brandl et al. 1999; Moffat et al. 2002). To find star-forming regions building up hundreds of O stars one has to look towards giant extragalactic HII regions, the nearest of which is 30 Doradus in the Large Magellanic Cloud, a satellite galaxy of our Milky Way at a distance at 55 kpc. The giant star-forming region 30 Doradus is thought to contain up to a hundred thousand young stars, including more than 400 O stars (Hunter et al. 1995; Walborn and Blades 1997; Townsley et al. 2006). Figure 19 shows that the star formation process spans many orders of magnitude in spatial scale and mass, ranging from stellar groups with no or only a few high-mass stars to massive clusters with several tens of thousands of stars and dozens if not hundreds of O stars. This variety of star-forming regions appears to be controlled by the competition between self-gravity and opposing agents such as the turbulence in the parental gas clouds, its gas pressure and magnetic field content.

Fig. 19 Comparison of clusters of different masses scaled to same relative distance. The cluster in the *upper left corner* is the Orion Nebula Cluster (McCaughrean 2001) and the one at the *lower left* is NGC 3603 (Brandl et al. 1999), both observed with the Very Large Telescope at infrared wavelength. The large cluster in the center is 30 Doradus in the LMC observed with the Hubble Space Telescope (courtesy of M.J. McCaughrean). The total mass increases roughly by a factor of ten from one cluster to the other. Image from Zinnecker and Yorke (2007)

6.2.3 Observations of the Stellar IMF

Mass is the most important parameter determining the evolution of individual stars. The luminosity L of a star scales as a very steep function of the mass M. The relation $L \propto M^{3.5}$ provides a reasonable estimate except for very low-mass stars and very massive ones (Kippenhahn et al. 2012). Stars on the main sequence generate energy by nuclear fusion. The total energy available is some fraction of Mc^2, with c being the speed of light. Consequently, we can estimate the main sequence lifetime as $t \propto M/L$ or $t \propto M^{-2.5}$. Massive stars with high pressures and temperatures in their centers

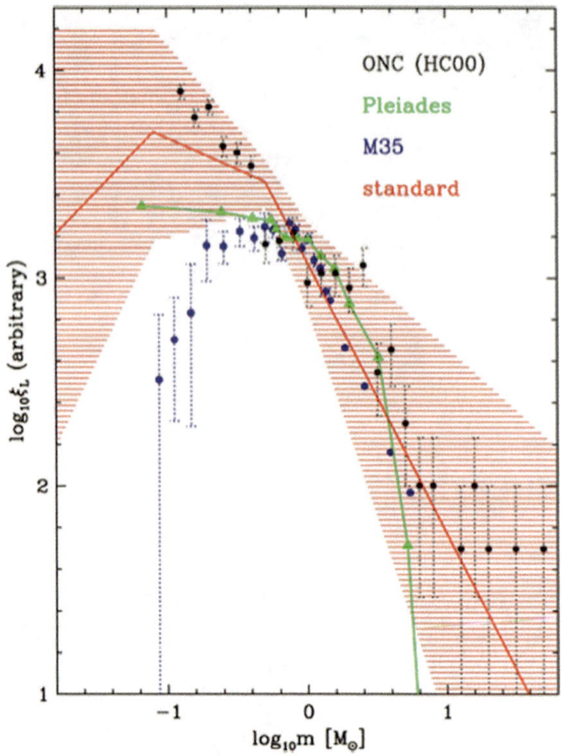

Fig. 20 Stellar mass spectrum in different nearby clusters (*black symbols* Orion Nebula Cluster, *green* Pleiades, *blue* M35) and its description by a three-component power law (*red lines* with overall uncertainties indicated by the hatched region). From Kroupa (2002)

convert hydrogen into helium very efficiently. This makes them very luminous but also short-lived. Conversely, low-mass stars are much fainter but long-lived.

Explaining the distribution of stellar masses at birth, the so-called initial mass function (IMF), is a key prerequisite to any theory of star formation. The IMF has three properties that appear to be relatively robust in diverse environments (see Fig. 20). These are the power law behavior $dN/dM \propto M^{-\alpha}$ with slope $\alpha \approx 2.3$ for masses M above about 1 $M_\odot$, originally determined by Salpeter (1955), the lower mass limit for the power law and the broad plateau below it before the brown dwarf regime (Miller and Scalo 1979; Scalo 1986), and the maximum mass of stars at around 100 $M_\odot$ (Weidner and Kroupa 2004, 2006; Oey and Clarke 2005). Comprehensive reviews of cluster and field IMFs may be found in Scalo (1986), Kroupa (2002), Chabrier (2003a), Bastian et al. (2010), Kroupa et al. (2013), and Offner et al. (2014).

There are two widely accepted functional parameterizations of the IMF. The first one is based on the continuous combination of multiple power-law segments. It was proposed by Kroupa (2001, 2002), and introducing the dimensionless mass $m = M/1M_\odot$, it reads

$$f(m) = \begin{cases} Ak_0\, m^{-0.3} & \text{for } 0.01 < m < 0.08\,, \\ Ak_1\, m^{-1.3} & \text{for } 0.08 < m < 0.5\,, \\ Ak_2\, m^{-2.3} & \text{for } 0.5 < m\,, \end{cases} \tag{196}$$

where A is a global normalization factor, and $k_0 = 1$, $k_1 = k_0 m_1^{-0.3+1.3}$, and $k_2 = k_1 m_2^{-1.3+2.3}$ are chosen to provide a continuous transition between the power-law segments at $m_1 = 0.08$ and $m_2 = 0.5$. The quantity $f(m)dm$ denotes the number of stars in the mass interval $[m, m + dm]$. A method to calculate k_i is provided by Pflamm-Altenburg and Kroupa (2006); see also Maschberger (2013a).

Another parameterization is suggested by Chabrier (2003a). It combines a log-normal with a power-law,

$$f(m) = \begin{cases} Ak_1 m^{-1} \exp\left[-\frac{1}{2}\left(\frac{\log_{10} m - \log_{10} 0.079}{0.69}\right)^2\right] & \text{for } m < 1\,, \\ Ak_2 m^{-2.3} & \text{for } m > 1\,. \end{cases} \tag{197}$$

Again A is a global normalization factor, and $k_1 = 0.158$ and $k_2 = 0.0443$ provide a continuous connection at $m = 1$ (Chabrier 2003b, 2005; Maschberger 2013a). Equation (196) is easier to integrate than (197), as this does not involve special functions. On the other hand it has several kinks. Both converge to the Salpeter (1955) power law with a slope of -2.3 for large masses. They differ by about a factor of 2 at low masses. However, within the observational errors, both functional forms are more or less equivalent.

We need to point out that the observational knowledge of the IMF is quite limited at the extreme ends of the stellar mass spectrum. Because massive stars are very rare and short-lived, only very few are sufficiently near to study them in detail and with very high spatial resolution, for example to determine multiplicity (Zinnecker and Yorke 2007). We do not even know what is the upper mass limit for stability, both in terms of observations as well as theoretical models (Massey 2003; Vink et al. 2015). In addition, there is evidence that the upper mass end of the IMF depends on the properties of the cluster where it is measured. The upper mass limit in more massive clusters seems to be higher than in lower-mass clusters, an effect that goes beyond the statistical fluctuations expected for purely random sampling from a universal distribution (see e.g. Weidner and Kroupa 2004, 2006; Weidner et al. 2010).

At the other end of the IMF, low-mass stars and brown dwarfs are faint, so they too are difficult to study in detail (Burrows et al. 2001). Such studies, however, are in great demand, because secondary indicators such as the fraction of binaries and higher-order multiples as a function of mass, or the distribution of disks around very young stars and possible signatures of accretion during their formation are probably better suited to distinguish between different star formation models than just looking at the IMF (e.g. Goodwin and Kroupa 2005; Marks and Kroupa 2012).

6.3 Gravoturbulent Star Formation

The past decade has seen a paradigm shift in low-mass star formation theory (Mac Low and Klessen 2004; McKee and Ostriker 2007; Offner et al. 2014). The general belief since the 1980s was that prestellar cores in low-mass star-forming regions evolve quasi-statically in magnetically subcritical clouds (Shu et al. 1987). In this picture, gravitational contraction is mediated by ambipolar diffusion (Mouschovias 1976, 1979, 1991a; Mouschovias and Paleologou 1981) causing a redistribution of magnetic flux until the inner regions of the core become supercritical and go into dynamical collapse. This process was originally thought to be slow, because in highly subcritical clouds the ambipolar diffusion timescale is about 10 times larger than the dynamical one. However, for cores close to the critical value, as is suggested by observations, both timescales are comparable. Numerical simulations furthermore indicate that the ambipolar diffusion timescale becomes significantly shorter for turbulent velocities similar to the values observed in nearby star-forming region (Fatuzzo and Adams 2002; Heitsch et al. 2004; Li and Nakamura 2004). The fact that ambipolar diffusion may not be a slow process under realistic cloud conditions, as well as the fact that most cloud cores are magnetically supercritical (Crutcher et al. 1999, 2009a; Crutcher and Troland 2000; Bourke et al. 2001) has cast significant doubts on any magnetically-dominated quasi-static models of stellar birth. For a more detailed account of the shortcomings of the quasi-static star formation model, see Mac Low and Klessen (2004).

For this reason, star formation research has turned to considering supersonic turbulence as being one of the primary physical agents regulating stellar birth. The presence of turbulence, in particular of supersonic turbulence, has important consequences for molecular cloud evolution (see e.g. Padoan et al. 2014; Dobbs et al. 2014). On large scales it can support clouds against contraction, while on small scales it can provoke localized collapse. Turbulence establishes a complex network of interacting shocks, where dense cores form at the stagnation points of convergent flows. The density can be large enough for gravitational collapse to set in. However, the fluctuations in turbulent velocity fields are highly transient. The random flow that creates local density enhancements can disperse them again. For local collapse to actually result in the formation of stars, high density fluctuations must collapse on timescales shorter than the typical time interval between two successive shock passages. Only then are they able to decouple from the ambient flow and survive subsequent shock interactions. The shorter the time between shock passages, the less likely these fluctuations are to survive. Hence, the timescale and efficiency of protostellar core formation depend strongly on the wavelength and strength of the driving source (Klessen et al. 2000; Heitsch et al. 2001a; Vázquez-Semadeni et al. 2003; Mac Low and Klessen 2004; Krumholz and McKee 2005; Ballesteros-Paredes et al. 2007; McKee and Ostriker 2007), and accretion histories of individual protostars are strongly time-varying (Klessen 2001a; Schmeja and Klessen 2004).

Altogether, we propose an evolutionary sequence as outlined in Fig. 21. Star cluster formation takes place in massive cloud cores of several 10^2–10^3 solar masses with

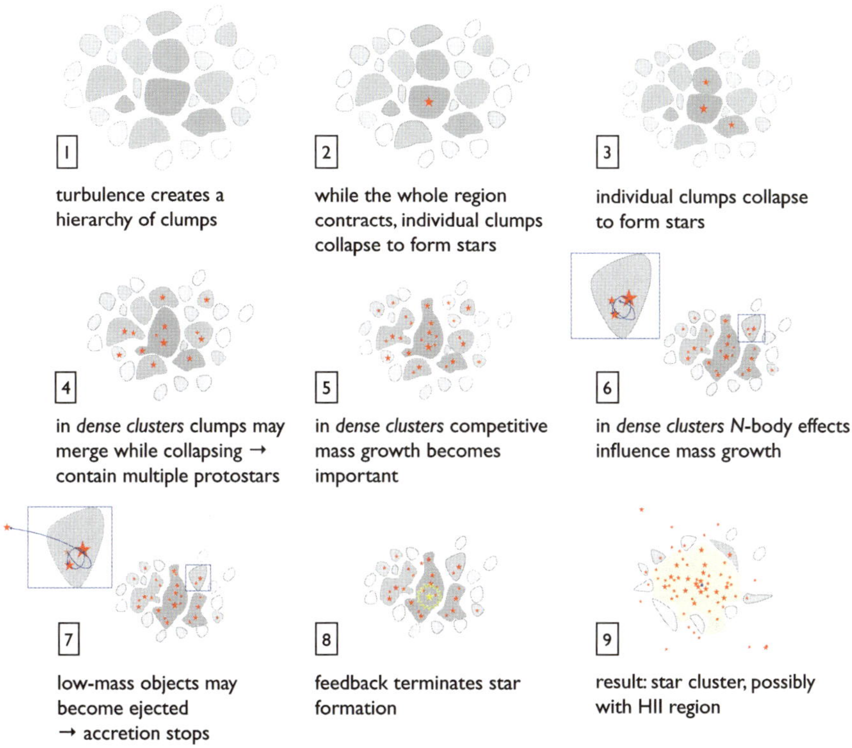

Fig. 21 Cartoon picture of star cluster formation in a molecular cloud core. From Klessen (2011)

sizes of a few parsecs and a velocity dispersion of about 1 km s^{-1} (see also Table 3). In order to form a bound cluster, the potential energy must dominate the energy budget, meaning that the entire region is contracting. The cluster-forming massive cloud cores are still in the supersonic range of the turbulent cascade (see Sect. 4.3), and as a consequence they exhibit a high degree of internal substructure with large density contrasts. Some of these density fluctuations are gravitationally unstable in their own right and begin to collapse on timescales much shorter than the global contraction time, as the free-fall time $\tau_{\rm ff}$ scales with the density ρ as $\tau_{\rm ff} \propto \rho^{-1/2}$.

Typically, the most massive fluctuations have the highest density and form a protostar in their center first. This nascent star can accrete from the immediate environment, but because it is located in a minimum of the cloud core's gravitational potential more gas flows towards it, and it can maintain a high accretion rate for a longer time. In contrast, stars that form in lower-mass gas clumps typically can only accrete material from their immediate surrounding and not much beyond that (see e.g. Klessen and Burkert 2000, 2001; Klessen et al. 2000; Bonnell et al. 2004). Because this preferentially happens in the cluster outskirts, these processes naturally lead to mass segregation as we often observe in young clusters (see e.g. Hillenbrand 1997; Hillenbrand and Hartmann 1998 for the Orion Nebula Cluster). In very dense clusters, there is the possibility that clumps merge while still accreting onto their central protostars. These protostars now compete for further mass growth from a common gas reservoir. This gives rise to collective phenomena which can strongly modify the accretion behavior and hence influence the resulting mass spectrum (see Sect. 6.4).

Once a star has reached a mass of about 10 $M_{\odot}$, it begins to ionize its environment. It carves out a bubble of hot and tenuous gas, which eventually will expand and enclose the entire stellar cluster. At this point no new stars can form and stellar birth has come to an end. We can observe the young cluster at infrared or even optical wavelengths, as illustrated in Fig. 19.

6.4 Theoretical Models for the Origin of the IMF

There are three principal pathways towards better understanding the origin of the IMF, depending on which aspects of gravitational collapse in the turbulent ISM one decides to focus on. We begin by introducing the underlying physical concepts behind the three different models in a qualitative way. Then we follow a more rigorous approach and discuss the most popular theoretical models for the IMF in some mathematical detail. We point out that the boundaries between these approaches are not clearly defined and that numerous hybrid models have been proposed in the current literature.

It turns out that essentially all theoretical models that are able to reproduce the IMF rely on two basic ingredients. They propose that the stellar mass spectrum is determined by a sequence of stochastic processes (such as turbulence or the probabilistic nature of stellar encounters in dense clusters) combined with scale-free physics (again, as provided by the power-law nature of the turbulent energy cascade or by the simple distance dependence of gravitational interactions). The former leads to a log-normal mass spectrum, the latter to a power-law contribution. Put together, they constitute one of the most popular parameterizations of the IMF (e.g. Chabrier 2003a).

6.4.1 Basic Concepts and Caveats

Here we introduce the three basic physical concepts behind the IMF.

Core Accretion

This model takes as its starting point the striking similarity between the shape of the observed core mass distribution and the IMF. It is based on the assumption of a one-to-one relation between the two distributions, such that the observed cores are the direct progenitors of individual stars or binary systems. The factor of ~3 decrease in mass between cores and stars is thought to be the result of feedback processes, mostly protostellar outflows, that eject a fixed fraction of the mass in a core rather than letting it accrete onto the star (Matzner and McKee 2000). This model reduces the problem of the origin of the IMF to understanding the mass spectrum of bound cores. Arguments to explain the core mass distribution generally rely on the statistical properties of turbulence. Its scale-free nature leads to a power-law behavior for high masses. The thermal Jeans mass in the cloud then imposes the flattening and turn-down in the observed mass spectrum. For further discussion, see Sect. 6.4.4.

Collective Models

A second line of reasoning accounts for the fact that stars almost always form in clusters, where the interaction between protostars, as well as between a protostellar population and the gas cloud around it may become important. In these collective models, the origin of the peak in the IMF is much the same as in the core accretion model: it is set by the Jeans mass in the prestellar gas cloud. However, rather than fragmentation in the gas phase producing a spectrum of core masses, each of which collapses down to a single star or star system, the final stellar mass spectrum in the collective accretion model is the result of the mutual interaction between the protostars in a cluster during their main accretion phase.

In the original competitive accretion picture (Bonnell et al. 2001a, b; Bonnell and Bate 2002; Bate et al. 2003) protostars start out with roughly the same small mass close to the opacity limit of fragmentation (Rees 1976). These protostars then compete with each other for mass accretion from the same reservoir of gas. In a simple Bondi-Hoyle-Lyttleton accretion scenario (e.g. Bondi 1952), the accretion rate scales as the square of the protostellar mass ($dM/dt \propto M^2$). That means small initial differences in mass quickly amplify and leads to a run-away growth of a few selected objects. The original idea of putting roughly equal-mass protostellar seeds into an pre-existing gas reservoir was later extended by taking the original cloud fragmentation process into account (Klessen et al. 1998; Klessen and Burkert 2000, 2001; Bate and Bonnell 2005; Bonnell et al. 2006, 2008). The fragmentation down to the local Jeans scale creates a mass function that lacks the power law tail at high masses that we observe in the stellar mass function. This part of the distribution forms via a second phase in which protostars with initial masses close to the Jeans mass compete for gas in the center of a dense cluster. The cluster potential channels mass toward the center, so stars that remain in the center grow to large masses, while those that are ejected from the cluster center by N-body interactions remain low mass (Bonnell et al. 2004).

The fact that fragmentation and the formation of multiple protostars strongly influence the subsequent accretion flow in the entire cluster also allows for a different interpretation. Peters et al. (2010b) and Girichidis et al. (2012b) point out that the processes described above limit the accretion of gas onto the central protostars in a cluster. They find that the gas flowing towards the potential minimum at the center of the cluster is efficiently accreted by protostars that are located at larger radii. As a consequence, the central region is effectively shielded from further accretion and none of the central objects can sustain its initially high accretion rate for a very long time. The fact that the gas fragments into a cluster of stars limits the mass growth onto the central object, which would otherwise have the available gas reservoir all for itself (as in the core accretion model described before). The gas flow towards the cluster center is reduced due to the efficient shielding by secondary protostars. Consequently, this process has been termed fragmentation-induced starvation (Peters et al. 2010a; Girichidis et al. 2011, 2012a). In these collective models, the apparent similarity between the core and stellar mass functions is an illusion, because the observed cores do not map the gas reservoir that is accreted by the stars (Clark and Bonnell 2006; Smith et al. 2008).

Importance of the Thermodynamic Behavior of the Gas

One potential drawback to both the core accretion and collective models is that they rely on the Jeans mass to determine the peak of the IMF, but do not answer the question of how to compute it. This is subtle, because molecular clouds are nearly isothermal but at the same time contain a very wide range of densities. At a fixed temperature, the Jeans mass scales as $M_{\rm J} \propto \rho^{-1/2}$, and it is not obvious what value of the density should be used to calculate $M_{\rm J}$. A promising idea to resolve this problem forms the basis for a third model of the IMF. It focuses on the thermodynamic properties of the gas. The amount of fragmentation occurring during gravitational collapse depends on the compressibility of the gas (Li et al. 2003). For a polytropic equation of state (215) with an index $\gamma < 1$, the gas reacts very strongly to pressure gradients. Turbulent compression can thus lead to large density contrasts. The local Jeans mass (Eq. 192) drops rapidly and many high-density fluctuations in the turbulent flow become gravitationally unstable and collapse. On the other hand, when $\gamma > 1$, compression leads to heating and turbulence can only induce small density variations. As the gas heats up, the decrease in the Jeans mass in the compressed gas is much smaller. Indeed, for $\gamma > 4/3$, compression actually results in an increasing Jeans mass. In addition, Larson (2005) argues that $\gamma = 1$ is a critical value, because filaments in which $\gamma < 1$ are unstable to continued gravitational collapse, while those with $\gamma > 1$ are stabilized against collapse and hence cannot decrease their Jeans mass to very small values. In real molecular clouds, the effective polytropic index varies significantly as the gas density increases. At low densities, $\gamma \approx 0.7$ (Larson 1985, 2005; Glover and Clark 2012a), but once the gas and dust temperatures become thermally coupled at $n_{\rm crit} \approx 10^5\,{\rm cm}^{-3}$ (see Fig. 5 and the discussion in Sect. 3.5), one expects this value to increase, reaching $\gamma \approx 1.1$ at densities $n \gg n_{\rm crit}$ (Banerjee et al. 2006). This suggests that fragmentation will tend to occur at densities $n \approx n_{\rm crit}$, and that the Jeans mass evaluated at this point sets the mass scale for the peak of the IMF. In this model, the apparent universality of the IMF in the Milky Way and nearby galaxies is then a result of the insensitivity of the dust temperature to the intensity of the interstellar radiation field (Elmegreen et al. 2008). Not only does this mechanism set the peak mass, but it also appears to produce a power-law distribution of masses at the high-mass end comparable to the observed distribution (Jappsen et al. 2005).

Caveats

Each of these models has potential problems. In the core accretion picture, hydrodynamic simulations seem to indicate that massive cores should fragment into many stars rather than collapsing monolithically (Dobbs et al. 2005; Clark and Bonnell 2006; Bonnell and Bate 2006). The hydrodynamic simulations often suffer from over-fragmentation because they do not include radiative feedback from embedded massive stars (Krumholz 2006; Krumholz et al. 2007; Krumholz and McKee 2008). The suggestion of a one-to-one mapping between the observed clumps and the final IMF is subject to strong debate, too. Many of the prestellar cores discussed in Sect. 6.1 appear to be stable entities (Johnstone et al. 2000, 2001, 2006; Lada et al. 2008), and thus are unlikely to be in a state of active star formation. In addition, the simple interpretation that one core forms on average one star, and that all cores contain the same number of thermal Jeans masses, leads to a timescale problem (Clark et al. 2007; see also the discussion in the last paragraph of Sect. 6.4.4). Its solution actually requires a difference between the core mass function and the stellar IMF. We also note that the problems associated with neglecting radiative feedback effects also apply to the gas thermodynamic idea. The assumed cooling curves typically ignore the influence of protostellar radiation on the temperature of the gas, which simulations show can reduce fragmentation (Krumholz et al. 2007; Commerçon et al. 2011; Peters et al. 2010b, 2011). The collective accretion picture has also been challenged, on the grounds that the kinematic structure observed in star-forming regions sometimes appears inconsistent with the idea that protostars have time to interact with one another strongly before they completely accrete their parent cores (André et al. 2007). For a comprehensive overview of the big open questions in star formation theory, see Krumholz (2014).

6.4.2 IMF from Simple Statistical Theory

In the previous section, we discussed various models for the origin of the stellar mass function based on a range of different physical processes. Here we approach the problem from a purely statistical point of view without specifying up front which of these processes will become dominant. We consider the distribution of stellar masses as the result of a sequence of independent stochastic processes. Invoking the central limit theorem then naturally leads to a log-normal IMF (for early discussions, see Zinnecker 1984; Adams and Fatuzzo 1996). The key assumption is that the mass M of a star can be expressed as the product of N independent variables x_j. At this point it is not necessary to specify these variables, as long as they are statistically independent and their values are determined by stochastic processes. We introduce again the dimensionless mass variable $m = M/(1\,\mathrm{M}_\odot)$ and write

$$m = \prod_{j=1}^{N} x_j \ . \tag{198}$$

Taking the logarithm of this equation, the logarithm of the mass is a sum of the random variables,

$$\ln m = \sum_{j=1}^{N} \ln x_j + \text{constant} \;, \tag{199}$$

where the constant term includes all quantities that are truly constant, e.g. the gravitational constant G or Boltzmann's constant k_{B} or others. The central limit theorem shows that the distribution of the composite variable $\ln m$ always approaches a normal distribution as the number N of variables approaches infinity (Bronstein and Semendjajew 1987). For the application of the theorem, a transformation into normalized variables ξ_j is useful, which are given by

$$\xi_j \equiv \ln x_j - \langle \ln x_j \rangle \equiv \ln \left(\frac{x_j}{\bar{x}_j} \right) \;. \tag{200}$$

The angle brackets denote averages taken over the logarithm of the variables,

$$\ln \bar{x}_j = \langle \ln x_j \rangle = \int_{-\infty}^{\infty} \ln x_j \, f_j(\ln x_j) d \ln x_j \;. \tag{201}$$

Here, f_j is the distribution function of the variable x_j. The normalized variables ξ_j have zero mean and their dispersions σ_j are given by

$$\sigma_j^2 = \int_{-\infty}^{\infty} \xi_j^2 f_j(\xi_j) d\xi_j \;. \tag{202}$$

We can define the new composite variable Ξ as

$$\Xi \equiv \sum_{j=1}^{N} \xi_j = \sum_{j=1}^{N} \ln \left(\frac{x_j}{\bar{x}_j} \right) \;. \tag{203}$$

It also has zero mean and, since the variables are assumed to be independent, it follows that

$$\Sigma^2 = \sum_{j=1}^{N} \sigma_j^2 \;. \tag{204}$$

For $N \to \infty$, the central limit theorem describes its distribution function as being Gaussian with

$$f(\Xi) = (2\pi\Sigma^2)^{-1/2} \exp\left(-\frac{1}{2} \frac{\Xi^2}{\Sigma^2} \right) \;, \tag{205}$$

independent of the distribution f_j of the individual variables x_j. The mass function (198) then can be expressed as

$$\ln m = \ln m_0 + \Xi \ , \tag{206}$$

with m_0 being a characteristic mass scale defined by

$$\ln m_0 \equiv \sum_{j=1}^{N} \langle \ln \bar{x}_j \rangle \ . \tag{207}$$

Combining the two Eqs. (205) and (206), we can write the distribution f of stellar masses in the form

$$\ln f(\ln m) = A - \frac{1}{2\Sigma^2}\left[\ln\left(\frac{m}{m_0}\right)\right]^2 \ , \tag{208}$$

where A is a constant. This is the log-normal form of the IMF first introduced by Miller and Scalo (1979). It fits very well the mass distribution of multiple stellar systems in the solar vicinity with masses less than few solar masses (e.g. Kroupa et al. 1990, 1991) and it is often used to describe the peak of the single star IMF (Sect. 6.2.3).

6.4.3 IMF from Stochastically Varying Accretion Rates

To obtain the observed power-law behavior at the high-mass end of the IMF, we need to add complexity to the model and extend this simple statistical approach. As a highly illustrative example, we follow the discussion provided by Maschberger (2013b) and consider the case where the stellar mass is determined by accretion in a stochastically fluctuating medium. If we disregard the fluctuating part for the time being, and if we assume that the accretion rate depends on the mass to some power of α, then the growth of an individual star can be described by the simple differential equation

$$dm = m^{\alpha} A dt \ , \tag{209}$$

where the constant $A > 0$ and the exponent α account for all physical processes involved. For example, if the protostars move with constant velocity v through isothermal gas with temperature T and sound speed $c_s = (k_B T/\mu)^{1/2}$ with Boltzmann constant k_B and mean particle mass μ (in grams), we can apply the Bondi-Hoyle-Lyttleton accretion formula (e.g. Bondi 1952) to obtain

$$A = \frac{2\pi G^2 \rho}{(v^2 + c_s^2)^{3/2}} \ ,$$
$$\alpha = 2 \ .$$

Note that this is an approximate formula. Replacing the factor 2π by 4π gives a better fit to the Hoyle-Lyttleton rate (Hoyle and Lyttleton 1939), where the object moves highly supersonically and we can neglect the contribution of the sound speed. Detailed numerical simulations yield a more complex parameterization of A, depending on the physical parameters of the system (e.g. Ruffert and Arnett 1994; Krumholz et al. 2006). What remains, however, is the quadratic power-law dependence of the accretion rate on the mass.

Now assume that the star grows with a statistically fluctuating mass accretion rate. This could be due to the stochastic nature of gas flows in turbulent media, and/or due to N-body dynamics in dense embedded clusters (e.g. Bonnell et al. 2001a, b; Klessen 2001a), or due to other processes that lead to stochastic protostellar mass growth. In this case, Eq. (209) turns into a stochastic differential equation,

$$dm = m^{\alpha}(Adt + BdW) , \tag{210}$$

where Adt describes the mean growth rate and BdW the fluctuations around this mean. Depending on the statistical properties of BdW the sum $Adt + BdW$ could become negative, which would imply mass loss and could potentially lead to negative masses. In order to avoid that, it is often sensible to restrict the stochastic variable BdW to positive values or to very small amplitudes. For $\alpha \neq 1$ (as well as for $\alpha \neq 0$) we obtain the formal solution

$$m(t) = \left[(1-\alpha)\left(\frac{m_0^{1-\alpha}}{1-\alpha} + At + BW(t)\right)\right]^{\frac{1}{1-\alpha}} , \tag{211}$$

where the integration constant m_0 is the initial mass and $W(t) = \int_0^t dW$ is the integral of the stochastic variable. For Gaussian fluctuations, its distribution has zero mean and variance t, as is well known from the random walk problem. For a more detailed discussion, see e.g. (Øksendal 2000). To get a mass spectrum, many realizations of the random variable need to be considered. For $B = 0$, Eq. (211) reduces to the solution of the deterministic growth problem, which for $\alpha > 1$ reaches infinite mass in the finite time

$$t_{\infty} = \frac{m_0^{1-\alpha}}{A(\alpha - 1)} . \tag{212}$$

For $B \neq 0$, the time t_{∞} no longer takes on a single value, but instead depends on the stochastic path $W(t)$. In reality, this solution is not desired, because the mass of a cloud core is limited. In addition, once feedback from massive stars sets in, the local reservoir of gas available for star formation is reduced even further. Consequently, the solution (211) makes sense only for $t \ll t_{\infty}$.

If we know the statistical properties of the random process $W(t)$, we can calculate the mass spectrum for an ensemble of stars. For Gaussian fluctuations with zero mean and variance t, we obtain

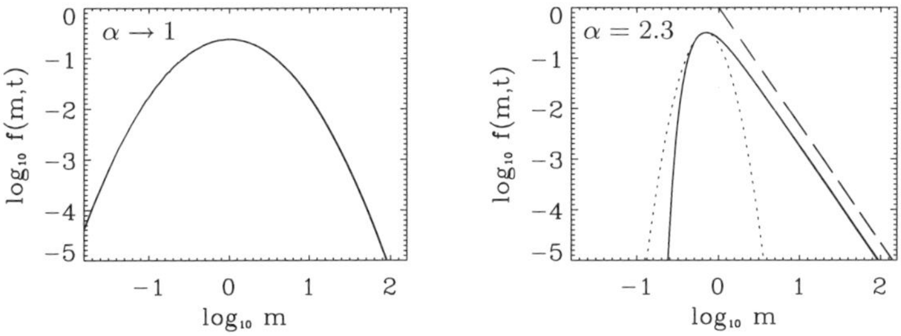

Fig. 22 Mass distribution function (213) from stochastic accretion for $\alpha \to 1$ (*left*) and $\alpha = 2.3$ (*right*). The parameters A, B, and m_0 are all set to unity. As α approaches unity, the mass function becomes log-normal. The case $\alpha = 2.3$ develops a power law tail at large m with slope $\alpha = 2.3$ (as indicted by the long *dashed line*) and is very similar to the observed IMF (Sect. 6.2.3). The *dotted curves* indicates a log-normal fit to the peak of the mass distribution. Adopted from Maschberger (2013b)

$$f(m,t) = \frac{1}{(2\pi)^{1/2}} \frac{1}{m^{\alpha}} \frac{n_{\infty}(t)}{Bt^{1/2}} \exp\left[-\frac{1}{2B^2 t} \left(\frac{m^{1-\alpha} - m_0^{1-\alpha}}{1-\alpha} - At \right)^2 \right], \quad (213)$$

where $f(m,t)dm$ gives the fraction of stars in the mass range m to $m + dm$. The factor $n_{\infty}(t)$ corrects for the possible contributions of stars with masses approaching infinity for $\alpha > 1$. It needs to be introduced to ensure the normalization of $f(m,t)dm$ as a probability distribution function at any time t (for further details, see Maschberger 2013b). For $\alpha < 1$, we set $n_{\infty}(t)$ to unity. For $\alpha \to 1$, i.e. for average exponential growth, we obtain the log-normal distribution function motivated before (see Eq. 208),

$$f(m,t) = \frac{1}{(2\pi)^{1/2}} \frac{1}{m} \frac{1}{Bt^{1/2}} \exp\left[-\frac{1}{2B^2 t} (\log m - \log m_0 - At)^2 \right], \quad (214)$$

where the factor $1/m$ is due to the conversion from $\log m$ to m, as $d \log m = dm/m$. We plot the two cases $\alpha = 1$ and $\alpha = 2.3$ in Fig. 22. The log-normal distribution function (left) peaks at the mass m_0. For all values $\alpha > 1$ the function $f(m,t)dm$ (at the right) develops a power-law tail at large masses m with slope α, reaches a maximum slightly below m_0, and exhibits a sharp decline for small masses. The case $\alpha = 2.3$ is therefore very similar to the observed stellar IMF, as discussed in Sect. 6.2.3. In short, the power-law tail traces the accretion behavior, while the log-normal part of the spectrum comes from the intrinsic stochasticity of the process.

6.4.4 IMF from Turbulence Statistics

Besides leading to stochastic variations in the protostellar accretion rate (as discussed before in Sect. 6.4.3), interstellar turbulence can influence the IMF by producing the clump structure within molecular clouds. These clumps or cores define the mass reservoir available for the formation of individual stars and small-multiple stellar systems. As a start, let us assume—most likely wrongly (see Sect. 6.4.1)—that each core forms exactly one star with some fixed efficiency factor. If we furthermore assume that there are no other stochastic processes at play, such as competitive accretion or fragmentation-induced starvation (Sect. 6.4.1), then understanding the origin of the stellar IMF boils down to identifying the physical processes that determine the clump mass function (CMF) in star-forming molecular clouds.

ISM turbulence is intrinsically a scale-free process as long as one stays within the inertial range (Sect. 4.3). It is therefore conceivable that it could play a key role in producing the power-law tail at the high-mass end of the stellar mass distribution. Most analytic models that attempt to do so involve the following four steps. First, they come up with a model that relates key parameters of the turbulent ISM to the probability distribution function (PDF) of gas density. Second, they relate the density PDF to the clump mass spectrum. Third, they identify a set of criteria by which some of these clumps go into gravitational collapse and begin to form stars. Typically, these involve some kind of Jeans argument and give preference to the most massive and densest clumps in the cloud. Fourth, they involve a mapping procedure, which converts a certain fraction of the clump mass into the final stellar mass.

Density Distribution Function

As discussed at the end of Sect. 4.1.2, the PDF of column densities in tenuous, non star-forming clouds is well approximated by a log-normal function. However, it develops a power-law tail at high column densities in more massive and star-forming cloud complexes. This is a signpost of gravitational contraction. A typical example for this case is the Orion A cloud. A map of its integrated CO emission is shown in the top panel of Fig. 11, and the corresponding distribution function of column densities (derived from dust emission measurements) is plotted at the bottom right of Fig. 13.

In order to obtain an estimate of the three-dimensional density distribution we need to convert the projected column density PDFs. Numerical simulations show that the column density PDFs have a smaller width than the density PDFs and can exhibit different shapes in the high- and low-density regimes (Ostriker et al. 2001; Federrath et al. 2010). However, both generally show very similar statistical properties (Federrath and Klessen 2012). This can be used to derive an estimate of the three-dimensional density PDF from the two-dimensional column density PDF (for further details, see Brunt et al. 2010). The shape and width of the density PDF are governed by the presence of compressive motions in the turbulent ISM. The medium is highly compressive and locally convergent flows lead to spatially and temporally confined regions of increased density. By the same token, expansion creates lower-density voids. Consequently, the overall distribution of density in the ISM is a sensitive function of the statistical properties of the underlying turbulent flow, with key parameters being the effective Mach number, the turbulent forcing scheme (i.e.

the ratio between compressional and rotational modes), the magnetic field strength, and the thermodynamic properties of the gas. Magnetic field lines resist compression and distortion and therefore reduce the compressibility of the gas. The competition between heating and cooling processes in the ISM (see Sect. 3) can act both ways. This is best seen when adopting a effective polytropic equation of state of the form

$$P \propto \rho^{\gamma} \,. \tag{215}$$

If the gas heats up when being compressed (for $\gamma > 1$), then pressure differences lead only to moderate density increase. However, when the gas gets colder when compressed (in the case $\gamma < 1$), the same pressure gradient can result in large density excursions.

Analytical theory as well as numerical simulations show that the distribution of the gas density in isothermal ($\gamma = 1$), non self-gravitating, and well sampled turbulent media follows a log-normal distribution,

$$\mathrm{PDF}(s) = \frac{1}{\sqrt{2\pi\sigma_s^2}} \exp\left(-\frac{(s - s_0)^2}{2\sigma_s^2} \right) . \tag{216}$$

Here, we introduce the logarithmic density,

$$s = \ln\left(\rho/\rho_0\right) , \tag{217}$$

and $\rho_0 = \langle \rho \rangle$ as well as $s_0 = \langle s \rangle$ denote the corresponding mean values. For a purely Gaussian distribution, the mean s_0 is related to the variance σ_s^2 of the logarithmic density s via the equation

$$s_0 = -\frac{1}{2}\sigma_s^2 \,. \tag{218}$$

This results from the normalization and mass-conservation constraints of the PDF (Vázquez-Semadeni 1994; Federrath et al. 2008). In turn, we can relate σ_s to the Mach number $\mathcal{M}$, to the forcing parameter b, and to the ratio of the thermal energy density to the magnetic energy density β,

$$\sigma_s^2 = \ln\left(1 + b^2\mathcal{M}^2 \frac{\beta}{\beta + 1}\right) . \tag{219}$$

For further discussions, see Padoan and Nordlund (2011) or Molina et al. (2012). The forcing parameter b varies from a value of approximately 0.3 for turbulence that is purely driven by solenoidal (divergence-free) modes to $b \approx 1$ for purely compressive (curl-free) schemes. A natural mix of forcing modes results in $b \approx 0.4$ (see e.g. Federrath et al. 2008, 2010; Schmidt et al. 2009; Seifried et al. 2011; Konstandin et al. 2012). The parameter β describes the ratio between thermal energy density and magnetic energy density,

$$\beta = \frac{c_s^2}{B^2/8\pi\rho} = 2\frac{c_s^2}{v_A^2}\,, \tag{220}$$

and can be expressed as the ratio between sound speed c_s and Alfvén velocity $v_A = B/\sqrt{4\pi\rho}$ (Eq. 107).

As indicated by the column density PDF in nearby molecular clouds depicted in Fig. 13, deviations from the pure log-normal behavior occur when parts of the gas undergo gravitational collapse and form stars. The velocity field, and as a consequence the density distribution, are no longer solely governed by turbulence statistics, but are also influenced by varying degrees of self-gravity (Klessen 2000; Dib and Burkert 2005; Collins et al. 2011; Kritsuk et al. 2011). Furthermore, Passot and Vázquez-Semadeni (1998) found deviations of the log-normal behavior in simulations of non-isothermal gas. Depending on the polytropic exponent γ in the equation of state (215), the PDF develops a power-law tail at low densities for $\gamma > 1$ and at high densities for $\gamma < 1$ (see e.g. Li et al. 2003). The latter effect is very similar to gravitational collapse. In addition, the PDFs from hydrodynamic simulations typically change with time as the overall cloud evolution progresses (Ballesteros-Paredes et al. 2011; Cho and Kim 2011). For example, Federrath and Klessen (2013) quantify how the slope of the high-density tail of the PDF in their numerical models flattens with increasing star-formation efficiency. Girichidis et al. (2014) demonstrate analytically that free-fall contraction of a single core or an ensemble of collapsing spheres forms a power- law tail similar to the observed PDFs.

Clump Mass Function

The next step in this sequence is to relate the density PDF to the clump mass function (CMF). A first attempt to analytically derive the CMF from turbulence properties goes back to Elmegreen (1993) and was then refined by Padoan et al. (1997) and Padoan and Nordlund (2002). They argue that high-density clumps are simply the shock-compressed regions that are the natural outcome of supersonic turbulence. They then invoke the shock jump conditions to calculate the achievable density contrast from the distribution of Mach numbers in the flow. Because they consider magnetized media, they base their considerations on the Alfvénic Mach number ($\mathcal{M}_A = v_A/c_s$), but similar conclusions follow for purely hydrodynamic flows (Elmegreen 2002b). To get the number of cores at a given density and length scale they argue that the flow is self-similar and that this quantity is simply determined by the available volume compressed to the density under consideration. This method has a number of shortcomings and we refer the reader to Krumholz (2014) for a more detailed account.

More refined statistical approaches are based on the Press-Schechter (1974) and excursion set (Bond et al. 1991) formalisms. These were originally introduced to describe the stochastic properties of cosmological fluctuations and quantify the behavior of random fields with structure over a wide range of scales. In particular, they can be used to count the number of objects above a certain density threshold for a given distribution function. This is exactly what is needed to determine the CMF. The first to realize this and to employ the Press-Schechter formalism to construct a model of the mass distribution of clumps from ISM turbulence was Inutsuka (2001). This was later extended by Hennebelle and Chabrier (2008, 2009, 2013). The Press-Schechter method, however, gives rise to the so-called 'cloud-in-cloud' problem, which occurs because the same object may be counted several times at different spatial scales. It can be resolved by introducing an appropriate correction factor (Jedamzik 1995), but a better a approach is to resort to the excursion set formalism. In essence, one computes a large number of Monte Carlo realizations of the stochastic variable by performing a random walk in the available parameter space. This allows one to determine the expected number of cores for a given length scale and density (and thus mass) with high precision (for a more detailed discussion, see Hopkins 2012a, b, 2013b).

We would also like to mention that alternative statistical models have been proposed that are based on stochastic sampling in fractal media (see Elmegreen 1996, 1997a, b, 1999, 2000a, 2002a).

Collapsing Cores

Once the CMF is obtained, the next step towards the stellar mass function is to select the subset of clumps that are gravitationally unstable and that begin to collapse in order to form stars. The most simple approach is to base the selection of bound clumps on a thermal Jeans argument. Jeans (1902) studied the stability of isothermal gas spheres. He found that the competition between thermal pressure gradients and potential gradients introduces a critical mass, $M_{\rm J} \propto \rho^{-1/2}(T/\mu)^{3/2}$, as expressed by Eq. (192). The Jeans mass only depends on the density ρ and the temperature T, as well as on the chemical composition of the gas through the mean particle mass μ. If a clump is more massive than $M_{\rm J}$, then it will collapse; otherwise, it will expand.

This approach can be extended by including the effects of micro-turbulence (von Weizsäcker 1951; Chandrasekhar 1951) and by considering the presence of magnetic fields (Mestel and Spitzer 1956; Mouschovias and Spitzer 1976; Shu et al. 1987). For a more detailed account of the historic development of star formation criteria, see Mac Low and Klessen (2004). Probably the most intuitive approach to assess the stability of molecular cloud clumps is based on the virial theorem, which relates the time evolution of the moment of inertia tensor of an object to its volumetric energy densities and surface terms, and allows us to take all physical processes into account that influence the dynamical evolution of the system (e.g. McKee and Zweibel 1992; Ballesteros-Paredes 2006). For an application to star formation, see for example Krumholz and McKee (2005).

The problem that arises from the turbulent compression (Padoan et al. 1997; Padoan and Nordlund 2002) or Press-Schechter approach is that it provides an estimate of the mass of a clump, but not of the density and of other physical properties that allow us to calculate the stability of the clump. To solve this problem, Padoan and Nordlund (2002), for example, take a typical cloud temperature T (see Table 3) and pick a random density ρ from the assumed density PDF (see above) for each clump M in the CMF. With these values they calculate the Jeans mass (Eq. 192). If the clump mass exceeds the Jeans mass, $M > M_J$, the clump is considered to be gravitationally bound and forming stars. If it is less massive than the Jeans mass, it is disregarded. The probability for massive clumps to be unstable for randomly picked ρ and T values is very high, and the mass spectrum of bound clumps is similar to the CMF at high masses. However, for lower-mass clumps, the likelihood of picking a combination of ρ and T such that the clump mass exceeds the Jeans mass gets smaller and smaller. As a consequence, the mass spectrum of bound clumps turns over towards smaller masses. This calculation becomes somewhat easier in the excursion set approach. This is because the random walk through parameter space provides both the length scale ℓ and the density ρ for each clump. If one picks a temperature T, one can calculate at each step in the process whether the mass $M \sim \rho\ell^3$ of the clump under consideration exceeds the Jeans mass. Objects on the largest scales ℓ with $M > M_J$ are identified as giant molecular clouds and objects on the smallest scales ℓ with $M > M_J$ as star-forming clumps or prestellar cores (Hopkins 2012a, b). This approach can readily be extended to include the stabilizing effects of turbulence and magnetic fields (Krumholz and McKee 2005; Padoan et al. 2007; Hennebelle and Chabrier 2009, 2013; Chabrier and Hennebelle 2010; Hopkins 2013a; Federrath and Klessen 2012) or the influence of changes in the equation of state (Guszejnov and Hopkins 2015). The overall peak of the mass spectrum is most likely determined by the balance between heating and cooling processes in the star-forming gas which sets a characteristic range of values for M_J and its variants (Sect. 6.4.1).

Stellar IMF

Once an ensemble of collapsing cloud clumps is selected, as outlined above, the stellar IMF is often determined by simply mapping the clump mass to the stellar mass with some given fixed efficiency. Typical values are around 30 % (see Sect. 6.4.1). As a result the IMF has the same functional form as the mass function of bound cores (see Sect. 6.1.1). However, as outlined in Sect. 6.4.1, this simple approach has its problems. If indeed each core only forms one star (or maybe a binary system), then it needs to have about one Jeans mass. Otherwise the core is likely to fragment (Goodwin et al. 2004a, b, 2006; Dobbs and Bonnell 2008; Holman et al. 2013). Because $M_J \propto \rho^{-1/2}$ for a given temperature T, high-mass clumps should be less dense than low mass ones. This immediately leads to a timescale problem (Clark et al. 2007). The free-fall time $\tau_{ff} \propto \rho^{-1/2}$, and so the collapse time scales linearly with the clump mass. The time it takes to build up a star with $10\,M_\odot$ is sufficient to form ten stars with $1\,M_\odot$. As a consequence the resulting stellar IMF should be considerably steeper than the CMF (e.g. Veltchev et al. 2011). In addition, high-mass clumps are not observed to be less dense than low-mass ones. If anything, they tend to be denser, and they are typically highly Jeans-unstable (Battersby et al. 2010; Ragan et al. 2012, 2013; Marsh et al. 2014).

A potential way out of this dilemma is to assume that high-mass clumps are hotter. High mass stars can indeed heat up their surroundings quite considerably (Sect. 6.5), since their luminosity L scales with stellar mass M as $L \propto M^{3.5}$ (e.g. Hansen and Kawaler 1994). However, there are many low-mass star-forming regions which show no signs of massive star formation (e.g. Taurus or ρ-Ophiuchi) and where the temperatures inferred for prestellar cores are uniformly low (Bergin and Tafalla 2007).

In general there is thus no good reason to believe in a one-to-one mapping between the core mass function and the stellar IMF. None of the current analytic models for the IMF includes processes such as stellar feedback in form of radiation or outflows, or fragmentation during core collapse and during the accretion disk phase, in a realistic and consistent way. The same holds for numerical simulations of star cluster formation. Altogether, it is likely that the transition from core to stars follows a complicated and stochastic pathway that may change with varying environmental conditions. For a simple cartoon picture, see Sect. 6.6.

6.5 Massive Star Formation

Because their formation time is short, of the order of 10^5 yr, and because they grow while deeply embedded in massive cloud cores, very little is known about the initial and environmental conditions of high-mass stellar birth. In general, regions forming high-mass stars are characterized by more extreme physical conditions than regions forming only low-mass stars, containing cores of size, mass, and velocity dispersion roughly an order of magnitude larger than those of cores in regions without high-mass star formation (e.g. Beltrán et al. 2006; Beuther et al. 2002, 2007; Motte et al. 2008; Krumholz et al. 2014). Typical sizes of cluster-forming clumps are about 1 pc. They have mean densities of $n \approx 10^5$ cm^{-3}, masses of $\sim 10^3$ M$_\odot$ and above, and velocity dispersions ranging between 1.5 and 4 km s^{-1}. Whenever observed with high resolution, these clumps break up into even denser cores that are believed to be the immediate precursors of single or gravitationally bound multiple massive protostars.

Massive stars usually form as members of multiple stellar systems (Ho and Haschick 1981; Lada 2006; Zinnecker and Yorke 2007; Reipurth et al. 2014) which themselves are parts of larger clusters (Lada and Lada 2003; de Wit et al. 2004; Testi et al. 1997; Longmore et al. 2014). This fact adds additional challenges to the interpretation of observational data from high-mass star forming regions as it is difficult to disentangle mutual dynamical interactions from the influence of individual stars (e.g. Goto et al. 2006; Linz et al. 2005). Furthermore, high-mass stars reach the main sequence while still accreting. Their Kelvin-Helmholtz pre-main sequence contraction time is considerably shorter than their accretion time. Once a star has reached a mass of about 10 M$_\odot$, its spectrum becomes UV-dominated and it begins to ionize its environment. This means that accretion as well as ionizing and non-ionizing radiation needs to be considered in concert (Keto 2002, 2003, 2007; Keto and Wood 2006; Peters et al. 2010a, b). It was realized decades ago that in simple one-dimensional collapse models, the outward radiation force on the accreting material should be significantly stronger than the inward pull of gravity (Larson and Starrfield 1971; Kahn 1974), in particular if one accounts for dust opacity. Since we see stars with $100\,M_{\odot}$ or even more (Bonanos et al. 2004; Figer 2005; Rauw et al. 2005; Bestenlehner et al. 2011; Borissova et al. 2012; Doran et al. 2013), a simple spherically symmetric approach to high-mass star formation must fail.

Consequently, two different models for massive star formation have been proposed. The first one takes advantage of the fact that high-mass stars always form as members of stellar clusters. If the central density in the cluster is high enough, there is a chance that low-mass protostars collide and so successively build up more massive objects (Bonnell et al. 1998). As the radii of protostars usually are considerably larger than the radii of main sequence stars in the same mass range (Hosokawa and Omukai 2009), this could be a viable option. However, the stellar densities required to produce massive stars by collisions are extremely high (Baumgardt and Klessen 2011). They seem inconsistent with the observed stellar densities of most Galactic star clusters (e.g. Portegies Zwart et al. 2010 and references therein), but could be reached in the central regions of the most extreme and massive clusters in the Local Group (such as 30 Doradus in the LMC as shown in Fig. 19; see e.g. Banerjee et al. 2013).

An alternative approach is to argue that high-mass stars build up like low-mass stars by accretion of ambient gas that goes through a rotationally supported disk formed by angular momentum conservation. Indeed, such disk structures are observed around a number of high-mass protostars (Chini et al. 2004, 2006; Jiang et al. 2008; Davies et al. 2010). Their presence breaks any spherical symmetry that might have been present in the initial cloud and thus solves the opacity problem. Radiation tends to escape along the polar axis, while matter is transported inwards through parts of the equatorial plane shielded by the disk. Hydrodynamic simulations in two and three dimensions focusing on the transport of non-ionizing radiation strongly support this picture (Yorke and Sonnhalter 2002; Krumholz et al. 2009; Kuiper et al. 2010, 2011). The same holds when taking the effects of ionizing radiation into account (Peters et al. 2010a, b, 2011; Commerçon et al. 2011). Once the disk becomes gravitationally unstable, material flows along dense, opaque filaments, whereas the radiation escapes through optically thin channels in and above the disk. Even ionized material can be accreted, if the accretion flow is strong enough (Keto 2003, 2007). HII regions are gravitationally trapped at this stage, but soon begin to rapidly fluctuate between trapped and extended states, as seen in some Galactic massive star-formation regions (Peters et al. 2010a; Galván-Madrid et al. 2011; De Pree et al. 2014). Over time, the same ultracompact HII region can expand anisotropically, contract again, and take on any of the observed morphological classes (Wood and Churchwell 1989; Kurtz et al. 1994; Peters et al. 2010c). In their extended phases, expanding HII regions drive bipolar neutral outflows characteristic of high-mass star formation (Peters et al. 2010a).

Another key fact that any theory of massive star formation must account for is the apparent presence of an upper mass limit at around 100–$150\,M_{\odot}$ (Massey 2003).

It holds for the Galactic field, but in dense clusters, apparently higher-mass stars have been reported (e.g. Crowther et al. 2010; Doran et al. 2013). If this mass limit holds, then purely random sampling of the initial mass function (IMF) (Kroupa 2002; Chabrier 2003a) without an upper mass limit should have yielded stars above $150\,M_{\odot}$ (Weidner and Kroupa 2004; Figer 2005; Oey and Clarke 2005; Weidner et al. 2010; see however, Selman and Melnick 2008). Altogether, the situation is not fully conclusive. If indeed there is an upper mass limit, it raises the question of its physical origin. It has been speculated before that radiative stellar feedback might be responsible for this limit (for a detailed discussion see e.g. Zinnecker and Yorke 2007) or alternatively that the internal stability limit of stars with non-zero metallicity lies in this mass regime (Appenzeller 1970a, b, 1987; Baraffe et al. 2001). However, fragmentation could also limit protostellar mass growth, as suggested by the numerical simulations of Peters et al. (2010b). The likelihood of fragmentation to occur and the number of fragments to form depends sensitively on the physical conditions in the star-forming cloud and its initial and environmental parameters (see e.g. Girichidis et al. 2012b). Understanding the build-up of massive stars therefore requires detailed knowledge about the physical processes that initiate and regulate the formation and dynamical evolution of the molecular clouds these stars form in (Vázquez-Semadeni et al. 2009).

Peters et al. (2010b, 2011), Kuiper et al. (2011), and Commerçon et al. (2011) argue that ionizing radiation (see also Krumholz et al. 2014), just like its non-ionizing, lower-energy counterpart, cannot shut off the accretion flow onto massive stars. Instead it is the dynamical processes in the gravitationally unstable accretion flow that inevitably occurs during the collapse of high-mass cloud cores that control the mass growth of individual protostars. Accretion onto the central star is shut off by the fragmentation of the disk and the formation of lower-mass companions which intercept inward-moving material. Peters et al. (2010b, 2011) call this process fragmentation-induced starvation and show that it occurs unavoidably in regions of high-mass star formation where the mass flow onto the disk exceeds the inward transport of matter due to viscosity only and thus renders the disk unstable to fragmentation (see also Sect. 6.4.1).

As a side note, it is interesting to speculate that fragmentation-induced starvation is important not only for present-day star formation but also in the primordial universe during the formation of metal-free Population III stars. Consequently, we expect these stars to be in binary or small number multiple systems and to be of lower mass than usually inferred (Abel et al. 2002; Bromm et al. 2009). Indeed, current numerical simulations provide the first hints that this might be the case (e.g. Clark et al. 2011; Greif et al. 2011; Stacy and Bromm 2013).

6.6 Final Stages of Star and Planet Formation

Here, we summarize again the main phases of the star and planet formation process. The entire sequence is illustrated in Fig. 23. It begins with the formation of molecular cloud complexes in the turbulent multi-phase ISM of the Galaxy (as we discuss in Sect. 4.1.2), and continues (a) with supersonic turbulence generating high density clumps with a wide range of densities and sizes (Sect. 6.1). (b) Some of these density fluctuations may become gravitationally unstable and begin to collapse. The central density increases until the compressional heat generated by the contraction can no longer be radiated away efficiently. A quasi-hydrostatic object then forms in the center of the core. This protostar continues to grows in mass by accreting material from the infalling envelope. In the class 0 phase of protostellar evolution, the mass budget is still dominated by the enclosing envelope. It is optically thin and absorbs the accretion luminosity generated as the infalling material comes to a halt at the protostellar surface. The spectral energy distribution (SED) of the system is thus dominated by the reprocessed emission from the cold envelope radiating mainly at sub-mm wavelengths.

Due to the conservation of angular momentum, most of the infalling matter will not directly end up in the central protostar, (c) but instead it will build up a rotationally supported accretion disk. If the mass load onto this disk during the main accretion phase exceeds its capability to transport material inwards by gravitoviscous processes, then the disk becomes unstable and will fragment into a binary or higher-order multiple stellar system. This is very likely to happen for high-mass stars, but occurs less frequently for low-mass objects, as indicated by the strong mass dependence of the stellar multiplicity fraction (Lada 2006). Molecular cloud cores are magnetized. The magnetic field is compressed and amplified by dynamo processes during the contraction, and eventually the accretion disk is able to launch a magnetically driven outflow along the rotational axis of the system (for a summary, see Pudritz et al. 2007). The outflow begins to disperse the remaining envelope material. First this happens along the polar axis, but later on larger and larger volumes are affected. This influences the observed SED. As more and more of the inner regions of the disk become visible, the peak of the emission moves towards infrared wavelengths. Favorable viewing angles even permit us to look down onto the protostellar photosphere, which contributes to the emission at near infrared and optical wavelengths. (d) Clearing the infalling envelope and the corresponding changes in SED mark the transition from class 1 to class 2 objects. In this T Tauri phase of protostellar evolution, most of the mass is already assembled in the central star with the remaining accretion disk contributing only a few percent to the overall matter budget. Nevertheless, this is where planets begin to form.

At some point, (e) the envelope is completely removed (or accreted by the disk), and the central star becomes fully visible. Only a debris disk remains in which planet formation continues. The central protostar is expected to be fully convective and the energy loss due to the emission of radiation at its surface is compensated by the release of gravitational energy. It slowly contracts and by doing so becomes hotter and denser. This is the classical Kelvin-Helmholtz contraction phase of pre-main sequence evolution. For solar-type stars it lasts about 20–30 million years. Finally, the central conditions are right for nuclear fusion to set in. The star enters the main sequence and settles into a quasi-equillibrium state, where its radiative energy losses are compensated by nuclear burning processes converting hydrogen into helium. (f)

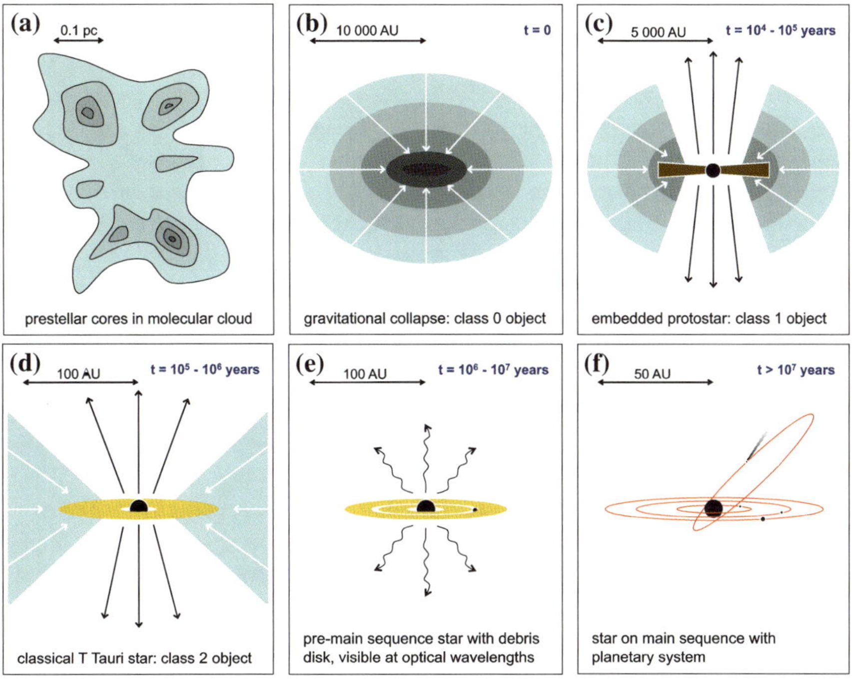

Fig. 23 Schematic view of the star and planet formation process. **a** Prestellar cores form by turbulent compression inside larger molecular clouds. **b** Some become gravitational unstable and begin to collapse. During the main accretion phase, the young protostar in the center experiences rapid mass growth. This is the class 0 phase of protostellar evolution. **c** Because of angular momentum conservation, the infalling material settles into a protostellar/protoplanetary disk. Magnetically launched outflows begin to disperse the infalling envelope. This is the class 1 phase. **d** The central protostar becomes visible as more and more of the envelope is removed. This is the class 2 phase. **e** The envelope is removed, and the central star becomes fully visible. Only a remnant disk remains in which planet formation continues. Low mass stars are still on the Kelvin-Helmholtz pre-main sequence contraction phase. **f** Finally, the original gas and dust disk is cleared and what remains is a central star with a planetary system, such as we observe in our solar system. Note that the cartoon picture describes the situation for isolated low-mass stars. For high-mass stars, the situation is more complicated, because the disk is likely to fragment into a binary or higher-order stellar system during the main accretion phase

After some time, the remaining debris disk is also cleared away and we are left with the central star, most likely being surrounded by a planetary system, such as we see in our own solar system or as we observe around other nearby stars.[9]

[9]The latest updates and findings of the research activities on extrasolar planets can be found at the following websites: www.exoplanet.eu and www.exoplanets.org.

7 Summary

In these lecture notes, we have made an attempt to identify and characterize the key astrophysical processes that provide the link between the dynamical behavior of the interstellar medium and the formation of stars. We hope that we have made it clear that one part cannot be understood without solid knowledge of the other. Both are connected via a number of competing feedback loops. We have argued that the evolution of the galactic ISM on large scales depends on the detailed microphysics in very complicated and often counter-intuitive ways. Conversely, global dynamical processes set the initial and boundary conditions for the formation of dense clouds on small scales and the birth of stars in their interior. Altogether, ISM dynamics spans an enormous dynamic range, from the extent of the galaxy as a whole down to the local blobs of gas that collapse to form individual stars or binary systems. Similarly, it covers many orders of magnitude in time, from the hundreds of millions of years it takes to complete one galactic rotation down to the hundreds of years it takes an ionization front to travel through a star-forming cloud. Improving our understanding of the interstellar medium and its ability to give birth to new stars is a complex multi-scale and multi-physics problem. It requires insights from quantum physics and chemistry, and knowledge of magnetohydrodynamics, plasma physics as well as gravitational dynamics. It also asks for a deep understanding of the coupling between matter and radiation, together with input from high-resolution multi-frequency and multi-messenger astronomical observations.

After a brief introduction to the field in Sect. 1, we began our discussion in Sect. 2 with a detailed account of the main constituents of the interstellar medium. These are gas, dust, the interstellar radiation field, and cosmic rays. Next, we turned our attention to the various heating and cooling processes that govern the thermodynamic behavior of the ISM. In Sect. 3, we introduced the microphysical processes that regulate the coupling between matter and radiation as well as between the different matter components. We identified the observed interstellar turbulence as a key agent driving the dynamical evolution of the Galactic ISM. These turbulent flows play a dual role. As we discussed in Sect. 4, turbulence can prevent or delay collapse on large scales, but on small scales it may actually trigger local compression and star formation. In addition, we showed that ISM turbulence dissipates quickly and needs to be continuously replenished for a galaxy to reach an approximate steady state. This led to a critical comparison of the various astrophysical processes that have been proposed to drive interstellar turbulence in galaxies such as our Milky Way. Because star formation is always found to be associated with molecular clouds, we discussed the physical (and chemical) processes that govern the formation of these densest and coldest components of the ISM in Sect. 5. We paid special attention to the chemical reactions that lead to the formation of H_2 and to its most important tracer CO. We found that dust attenuation plays a key role in this process, and we discussed molecular cloud formation in the context of global ISM dynamics on galactic scales. Finally, in Sect. 6 we zoomed in on smaller and smaller scales, and summarized the properties of molecular cloud cores as the direct progenitors of individual stars and stellar systems. Furthermore, we motivated and described the current statistical and theoretical models of stellar birth and tried to explain the seemingly universal observed distribution of stellar masses at birth, the initial mass function (IMF), as the result of a sequence of stochastic events mostly governed by the interplay between turbulence and self-gravity in the star-forming gas.

We hope that we have illustrated in these lecture notes that the question of stellar birth in the multi-phase ISM of our Milky Way and elsewhere in the universe is far from being solved. On the contrary, the field of ISM dynamics and star formation is rapidly evolving and has gone through a significant transformation in recent years. We acknowledge that scientific progress in this area requires the concerted and combined efforts of theory, observations, as well as laboratory experiments. We notice a general trend away from only taking isolated processes and phenomena into account and towards a more integrated multi-scale and multi-physics approach in today's theoretical models and computer simulations. Observational studies now regularly attempt to accumulate and combine information from as many different wavebands as possible, and to cover as large an area on the sky with as much detail and resolution as possible. New large facilities such as ALMA on the ground or Gaia in space have the potential for real scientific breakthroughs.[10] All our theoretical and observational efforts would be in vain without complementary laboratory studies that provide fundamental information and cross sections for molecular and ionic reactions as well as transition frequencies and data on dust physics, that constitute the physical and chemical basis of our understanding of the ISM.

We end this summary with a list of questions, which we think are amongst the most important open problems in the field of ISM dynamics and star formation studies. We note that these questions are closely related to each other, and that the answer to one question may hold the key to resolving another.

What drives interstellar turbulence? Observations show that turbulence in molecular clouds is ubiquitous. With the exception of the dense cores discussed in Sect. 6.1, it seems to follow a universal relationship between velocity dispersion and size (Larson's relation, see Sect. 4.1). Even extragalactic molecular clouds exhibit similar behavior. In addition, there are few variations in the turbulent properties between molecular cloud regions with ongoing star formation and those without. This seems to argue in favor of a galaxy-scale driving process (Sect. 4.5). On the other hand, there are also no systematic variations in GMC properties within a galaxy or between galaxies, which would seem to argue that internal processes must be important as well (Sect. 4.6). What is the relative importance of internal and external forcing mechanisms in driving ISM turbulence? Does the answer depend on the length scales that one examines, or on the place where one looks? So far, the 'smoking gun' to answer these questions, both observationally or theoretically, remains to be found.

How is the star formation process correlated with galaxy properties? And how can we best study that problem? On large scales, star formation appears to follow a fairly universal scaling behavior. This holds for galaxies that range from being mildly

[10]Information about the Atacama Large Millimeter/Submillimeter Array (ALMA) and about the Gaia satellite can be found at http://www.almaobservatory.org and http://sci.esa.int/gaia/.

dominated by atomic hydrogen (such as the Milky Way) to those that are strongly dominated by molecular hydrogen (such as local starbursts). Does the presence or absence of a significant atomic phase play an important role in regulating star formation, either directly (e.g. by limiting the amount of molecular gas available for star formation) or indirectly (e.g. by driving turbulent motions via thermal instability)? How does the star formation process change, if at all, in galaxies such as dwarfs that contain very little molecular gas? On the observational side, one of the key questions is whether and to what extent commonly used observational tracers of the star formation rate (SFR), such as Hα, 24 μm dust emission, or [CII] fine structure emission, can reliably recover the true rates? Accurate measurements of the SFR in galaxies are of great importance for many different fields of astrophysical research, and yet remain difficult to carry out. In nearby molecular clouds, counts of young stellar objects can give a direct measurement of the SFR, but this technique cannot be used in extragalactic systems where individual objects cannot be detected and resolved. Instead, indirect indicators of the SFR must be used, such as the Hα luminosity or the total far-infrared emission. A central assumption underpinning these methods is that the energy radiated by these tracers comes primarily from newly-formed massive stars. If this is not the case, then these tracers will give a misleading view of the SFR. Answering these questions requires both dedicated observations and numerical models that allow us to explore the conditions in which the different tracers of the SFR can be used safely, and to understand when and why they fail. These can then also address the question of the scale over which the above correlations hold. Do we still see a good correlation between the tracers and the SFR on small scales (tens of parsecs or less), or only when we average on scales of hundreds of parsecs?

What are the best observational tracers to study ISM dynamics and molecular cloud assembly? We know that dense molecular clouds must be assembled from gas that is initially in a more diffuse state (Sect. 5), but whether this process is driven primarily by turbulence or by gravity is unclear. At the present time, we are not even sure what we should observe in order to best distinguish between these two models. It seems likely that CO forms in significant quantities only once a large fraction of the cloud mass is already assembled, since it resists photodissociation only in regions with relatively high extinctions (Sect. 5.2), and so CO observations are unlikely to provide strong constraints on the assembly process. HI 21 cm observations may be better suited for this purpose, but only if the inflowing gas is primarily atomic. If, instead, it is largely composed of H_2, then chemical tracers of this phase (e.g. HD or HF) may be more useful. Fine structure emission from [CII] or [OI] may also trace the inflowing gas, but only if it is warm enough to excite the lines. Addressing this issue requires us to perform dedicated numerical simulations coupled with a time-dependent chemical network and to produce synthetic observations in the tracers of interest in a post-processing step, which then can be compared one-to-one with real observational data.

Which observational diagnostics are best suited to recover the true physical properties of the ISM? Our knowledge of physical cloud properties (such as mass or spatial extent) often relies on indirect measurements, particularly in extragalactic systems. For example, it is often assumed that all molecular clouds are in virial equilibrium, which allows us to estimate their masses from the observed CO linewidths. In the Milky Way, we can hope to benchmark this approach by using more direct measurements of cloud masses, as derived from e.g. dust extinction. However, this can be done easily only for nearby clouds, meaning that the range of environments in which these estimates can be directly tested is quite limited. Besides better and more detailed observations, progress will require us to examine the performance of the different available measures of cloud mass in dedicated numerical simulations for a wide variety of different physical environments. Only these calculations provide full access to the six-dimensional phase space, and by doing so enable us to figure out which measures are the least biased and potentially also to derive correction factors to improve the observational estimates.

What physical processes determine the distribution of stellar masses at birth? How reliable are observations that suggest that the stellar IMF and binary distribution at the present day are similar in different galactic environments? In particular, in rich (and more distant) clusters, our observational basis needs to be extended to lower masses. The same holds for variations with metallicity as can be traced in the Local Group. Is the IMF in the Large Magellanic Cloud (with a metal abundance of about half of the solar value) and the Small Magellanic Cloud (with a metal abundance of about one fifth solar or less) really similar to the Milky Way? On the theoretical side, what processes are responsible for the (non-)variation of the IMF? The critical mass for gravitational collapse can vary enormously between different environments. Yet the IMF in globular clusters, for example, appears to be the same as in regions of distributed star formation such as Taurus. How can the statistical theoretical models introduced in Sect. 6.4 be extended to address these questions? Better understanding the physical origin of the IMF will remain a key driver of star formation research for a long time to come.

Acknowledgments Writing these lecture notes would have been impossible without the help and input from many collaborators and colleagues. In particular, we want to thank Christian Baczynski, Javier Ballesteros-Paredes, Robi Banerjee, Erik Bertram, Henrik Beuther, Frank Bigiel, Peter Bodenheimer, Ian A. Bonnell, Andreas Burkert, Paul C. Clark, Cornelis P. Dullemond, Edith Falgarone, Christoph Federrath, Philipp Girichidis, Alyssa Goodman, Dimitrios Gouliermis, Fabian Heitsch, Patrick Hennebelle, Thomas Henning, Mark H. Heyer, Philip F. Hopkins, Juan Ibañez Mejia, Eric R. Keto, Lukas Konstandin, Pavel Kroupa, Mark R. Krumholz, Mordecai-Mark Mac Low, Faviola Molina, Volker Ossenkopf, Thomas Peters, Ralph E. Pudritz, Sarah Ragan, Julia Roman-Duval, Daniel Seifried, Dominik R.G. Schleicher, Wolfram Schmidt, Nicola Schneider, Jennifer Schober, Rahul Shetty, Rowan J. Smith, Jürgen Stutzki, Làszlò Szűcs, Enrique Vazquez-Semadeni, Antony P. Whitworth, and Hans Zinnecker for many stimulating and encouraging discussions.
We acknowledge support from the Deutsche Forschungsgemeinschaft (DFG) via the SFB 881 *The Milky Way System* (subprojects B1, B2, B5 and B8), and the SPP (priority program) 1573 *Physics of the ISM*. We also acknowledge substantial support from the European Research Council under the European Community's Seventh Framework Program (FP7/2007-2013) via the ERC Advanced Grant *STARLIGHT: Formation of the First Stars* (project number 339177).

References

Abbott, D. C., The return of mass and energy to the interstellar medium by winds from early-type stars, *ApJ*, **263**, 723 (1982)

Abel, T., Bryan, G. L., Norman, M. L., The Formation of the First Star in the Universe, *Science*, **295**, 93 (2002)

Abgrall, H., Roueff, E., Drira, I., Total transition probability and spontaneous radiative dissociation of B, C, B′ and D states of molecular hydrogen, *A&AS*, **141**, 297 (2000)

Adams, F. C., Fatuzzo, M., A Theory of the Initial Mass Function for Star Formation in Molecular Clouds, *ApJ*, **464**, 256 (1996)

Adams, F. C., Myers, P. C., Modes of Multiple Star Formation, *ApJ*, **553**, 744 (2001)

Adams, S. M., Kochanek, C. S., Beacom, J. F., Vagins, M. R., Stanek, K. Z., Observing the next galactic supernova, *ApJ*, **778**, 164 (2013)

Agertz, O., Teyssier, R., Moore, B., Disc formation and the origin of clumpy galaxies at high redshift, *MNRAS*, **397**, L64 (2009)

Aikawa, Y., Wakelam, V., Garrod, R. T., Herbst, E., Molecular Evolution and Star Formation: From Prestellar Cores to Protostellar Cores, *ApJ*, **674**, 984 (2008)

Albertsson, T., Semenov, D. A., Vasyunin, A. I., Henning, Th., Herbst, E., New Extended Deuterium Fractionation Model: Assessment at Dense ISM Conditions and Sensitivity Analysis, *ApJS*, **207**, 27 (2013)

Alves, J., Lombardi, M., Lada, C. J., The mass function of dense molecular cores and the origin of the IMF, *A&A*, **462**, L17 (2007)

Alves, J. F., Lada, C. J., Lada, E. A., Internal structure of a cold dark molecular cloud inferred from the extinction of background starlight, *Nature*, **409**, 159 (2001)

Alves J., Lombardi M., Lada C. J., 2MASS wide-field extinction maps. V. Corona Australis, *A&A*, **565**, A18 (2014)

Amenomori, M., et al., Anisotropy and Corotation of Galactic Cosmic Rays, *Science*, **314**, 439 (2006)

André, P., Ward-Thompson, D., Barsony, M., From Prestellar Cores to Protostars: the Initial Conditions of Star Formation, in: Protostars and Planets IV, edited by V. Mannings, A. P. Boss, S. S. Russell, p. 59 (2000)

André, P., Belloche, A., Motte, F., Peretto, N., The initial conditions of star formation in the Ophiuchus main cloud: Kinematics of the protocluster condensations, *A&A*, **472**, 519 (2007)

Anninos, P., Zhang, Y., Abel, T., Norman, M. L., Cosmological hydrodynamics with multi-species chemistry and nonequilibrium ionization and cooling, *New. Astron.* **2**, 209 (1997)

Appenzeller, I., Mass Loss Rates for Vibrationally Unstable Very Massive Main-sequence Stars, *A&A*, **9**, 216 (1970a)

Appenzeller, I., The Evolution of a Vibrationally Unstable Main-sequence Star of 130 $M_{\odot}$, *A&A*, **5**, 355 (1970b)

Appenzeller, I., Theory of vibrational instabilities in luminous early type stars, in: Instabilities in Luminous Early Type Stars, vol. 136 of Astrophysics and Space Science Library, edited by H. J. G. L. M. Lamers C. W. H. de Loore, p. 55 (1987)

Arce, H. G., Borkin, M. A., Goodman, A. A., Pineda, J. E., Halle, M. W., The COMPLETE survey of outflows in Perseus, *ApJ*, **715**, 1170 (2010)

Arce, H. G., Goodman, A. A., Bow shocks, wiggling jets, and wide-angle winds: A high-resolution study of the entrainment mechanism of the PV Cephei molecular (CO) outflow, *ApJ*, **575**, 928 (2002a)

Arce, H. G., Goodman, A. A., The great PV Cephei outflow: A case study in outflow-cloud interaction, *ApJ*, **575**, 911 (2002b)

Arons, J., Max, C. E., Hydromagnetic Waves in Molecular Clouds, *ApJ*, **196**, L77 (1975)

Atkins, P., Friedman, R., Molecular Quantum Mechanics, 5th Edition. OUP, Oxford (2010)

Audit, E., Hennebelle, P., Thermal condensation in a turbulent atomic hydrogen flow, *A&A*, **433**, 1 (2005)

Bacmann, A., André, P., Ward-Thompson, D., The Structure of Prestellar Cores as Derived from ISO Observations, in: From Darkness to Light: Origin and Evolution of Young Stellar Clusters, vol. 243 of Astronomical Society of the Pacific Conference Series, edited by T. Montmerle P. André, p. 113 (2001)

Bakes, E. L. O., Tielens, A. G. G. M., The photoelectric heating mechanism for very small graphitic grains and polycyclic aromatic hydrocarbons, *ApJ*, **427**, 822 (1994)

Balbus, S. A., Hawley, J. F., Instability, turbulence, and enhanced transport in accretion disks, *Rev. Mod. Phys.*, **70**, 1 (1998)

Ballesteros-Paredes J., Six myths on the virial theorem for interstellar clouds, *MNRAS*, **372**, 443 (2006)

Ballesteros-Paredes, J., Vázquez-Semadeni, E., Gazol, A., Hartmann, L. W., Heitsch, F., Colín, P., Gravity or turbulence? - II. Evolving column density probability distribution functions in molecular clouds, *MNRAS*, **416**, 1436 (2011)

Ballesteros-Paredes, J., Klessen, R. S., Mac Low, M.-M., Vazquez-Semadeni, E., Molecular Cloud Turbulence and Star Formation, in: Protostars and Planets V, edited by B. Reipurth, D. Jewitt, K. Keil, p. 63 (2007)

Ballesteros-Paredes, J., Klessen, R. S., Vázquez-Semadeni, E., Dynamic Cores in Hydrostatic Disguise, *ApJ*, **592**, 188 (2003)

Ballesteros-Paredes, J., Mac Low, M.-M., Physical versus observational properties of clouds in turbulent molecular cloud models, *ApJ*, **570**, 734 (2002)

Bally, J., Devine, D., A parsec-scale 'superjet' and quasi-periodic structure in the HH 34 outflow?, *ApJ*, **428**, L65 (1994)

Bally, J., Devine, D., Alten, V., Sutherland, R. S., New Herbig-Haro flows in I1448 and I1455, *ApJ*, **478**, 603 (1997)

Bally, J., Reipurth, B., Davis, C. J., Observations of Jets and Outflows from Young Stars, in: Protostars and Planets V, edited by B. Reipurth, D. Jewitt, K. Keil, p. 215 (2007)

Banerjee, R., Klessen, R. S., Fendt, C., Can protostellar jets drive supersonic turbulence in molecular clouds? *ApJ*, **668**, 1028 (2007)

Banerjee, S., Kroupa, P., Oh, S., The emergence of super-canonical stars in R136-type starburst clusters, *MNRAS*, **426**, 1416 (2013)

Banerjee, R., Pudritz, R. E., Anderson, D. W., Supersonic turbulence, filamentary accretion and the rapid assembly of massive stars and discs, *MNRAS*, **373**, 1091 (2006)

Banerjee, R., Vázquez-Semadeni, E., Hennebelle, P., Klessen, R. S., Clump morphology and evolution in MHD simulations of molecular cloud formation, *MNRAS*, **398**, 1082 (2009)

Baraffe, I., Heger, A., Woosley, S. E., On the Stability of Very Massive Primordial Stars, *ApJ*, **550**, 890 (2001)

Bastian, N., Covery, K. R., Meyer, M. R., A Universal Stellar Initial Mass Function? A Critical Look at Variations, *ARA&A*, **48**, 339 (2010)

Bate, M. R., Bonnell, I. A., The origin of the initial mass function and its dependence on the mean Jeans mass in molecular clouds, *MNRAS*, **356**, 1201 (2005)

Bate, M. R., Bonnell, I. A., Bromm, V., The formation of a star cluster: predicting the properties of stars and brown dwarfs, *MNRAS*, **339**, 577 (2003)

Battersby C., Bally J., Jackson J. M., Ginsburg A., Shirley Y. L., Schlingman W., Glenn J., An Infrared Through Radio Study of the Properties and Evolution of IRDC Clumps, *ApJ*, **721**, 222 (2010)

Baumgardt, H., Klessen, R. S., The role of stellar collisions for the formation of massive stars, *MNRAS*, **413**, 1810 (2011)

Beaumont, C. N., Williams, J. P., Molecular Rings Around Interstellar Bubbles and the Thickness of Star-Forming Clouds, *ApJ*, **709**, 791 (2010)

Beaumont, C. N., Offner, S. S. R., Shetty, R., Glover, S. C. O., Goodman, A. A., Quantifying observational projection effects using molecular cloud simulations, *ApJ*, **777**, 173 (2013)

Bec, J., Khanin, K., Burgers turbulence, *Phys. Reports*, **447**, 1 (2007)

Beltrán, M. T., Cesaroni, R., Codella, C., Testi, L., Furuya, R. S., Olmi, L., Infall of gas as the formation mechanism of stars up to 20 times more massive than the Sun, *Nature*, **443**, 427 (2006)
Benz, W., Smoothed Particle Hydrodynamics: A Review, in: The Numerical Modelling of nonlinear Stellar Pulsations, edited by J. R. Buchler, p. 269, Kluwer Academic Publishers, The Netherlands (1990)
Bergin, E. A., Alves, J., Huard, T., Lada, C. J., N_2O Depletion in a Cold Dark Cloud, *ApJ*, **570**, L101 (2002)
Bergin, E. A., Hartmann, L. W., Raymond, J. C., Ballesteros-Paredes, J., Molecular Cloud Formation behind Shock Waves, *ApJ*, **612**, 921 (2004)
Bergin, E. A., Melnick, G. J., Stauffer, J. R., Ashby, M. L. N., Chin, G., Erickson, N. R., et al., Implications of Submillimeter Wave Astronomy Satellite Observations for Interstellar Chemistry and Star Formation, *ApJ*, **539**, L129 (2000)
Bergin, E. A., Tafalla, M., Cold Dark Clouds: The Initial Conditions for Star Formation, *ARA&A*, **45**, 339 (2007)
Bertram, E., Federrath, C., Banerjee, R., Klessen, R. S., Statistical analysis of the mass-to-flux ratio in turbulent cores: effects of magnetic field reversals and dynamo amplification, *MNRAS*, **420**, 3163 (2012)
Bertram, E., Shetty, R., Glover, S. C. O., Klessen, R. S., Roman-Duval, J., Federrath, C., Principal component analysis of molecular clouds: can CO reveal the dynamics?, *MNRAS*, **440**, 465 (2014)
Bestenlehner J. M., Vink J. S., Gräfener G., Najarro F., Evans C. J., Bastian N., Bonanos A. Z., Bressert E., Crowther P. A., et al., The VLT-FLAMES Tarantula Survey. III. A very massive star in apparent isolation from the massive cluster R136, *A&A*, **530**, L14 (2011)
Beuther, H., Churchwell, E. B., McKee, C. F., Tan, J. C., The Formation of Massive Stars, in: Protostars and Planets V, edited by B. Reipurth, D. Jewitt, K. Keil, p. 165 (2007)
Beuther, H., Schilke, P., Sridharan, T. K., Menten, K. M., Walmsley, C. M., Wyrowski, F., Massive molecular outflows, *A&A*, **383**, 892 (2002)
Bigiel, F., Leroy, A., Walter, F., Brinks, E., de Blok, W. J. G., Madore, B., Thornley, M. D., The Star Formation Law on sub-Kiloparsec Scales, *AJ*, **136**, 2846 (2008)
Bigiel, F., Leroy, A., Walter, F., Brinks, E., de Blok, W. J. G., Kramer, C., Rix, H. W., Schruba, A., Schuster, K.-F., Usero, A., Wiesemeyer, H. W., A Constant Molecular Gas Depletion Time in Nearby Disk Galaxies, *ApJ*, **730**, L13 (2011)
Bird, S., Vogelsberger, M., Sijacki, D., Zaldarriaga, M., Springel, V., Hernquist, L., Moving-mesh cosmology: properties of neutral hydrogen in absorption, *MNRAS*, **429**, 3341 (2013)
Black, J. H., The physical state of primordial intergalactic clouds, *MNRAS*, **197**, 553 (1981)
Black, J. H., Energy Budgets of Diffuse Clouds. In: The First Symposium on the Infrared Cirrus and Diffuse Interstellar Clouds (ASP Conf. Series, Vol. 58), edited by Cutri, R. M., Latter, W. B., p. 355 (1994)
Black, J. H., Dalgarno, A., Models of interstellar clouds. I - The Zeta Ophiuchi cloud, *ApJS*, **34**, 405 (1977)
Blasi, P., Recent Results in Cosmic Ray Physics and Their Interpretation, *Brazilian J. Phys.*, **44**, 426 (2014)
Blitz, L., Fukui, Y., Kawamura, A., Leroy, A., Mizuno, N., Rosolowsky, E., Giant Molecular Clouds in Local Group Galaxies, in: Protostars and Planets V, edited by B. Reipurth, D. Jewitt, K. Keil, 81 (2007)
Blitz, L., Shu, F. H., The Origin and Lifetime of Giant Molecular Cloud Complexes, *ApJ*, **238**, 148 (1980)
Blitz, L., Spergel, D. N., Teuben, P. J., Hartmann, D., Burton, W. B., High-velocity clouds: Building blocks of the local group, *ApJ*, **514**, 818 (1999)
Bodenheimer, P., Angular Momentum Evolution of Young Stars and Disks, *ARA&A*, **33**, 199 (1995)
Bolatto, A. D., Leroy, A. K., Rosolowsky, E., Walter, F., Blitz, L., The Resolved Properties of Extragalactic Giant Molecular Clouds, *ApJ*, **868**, 948 (2008)
Boldyrev, S., Linde, T., Polyakov, A., Velocity and velocity-difference distributions in burgers turbulence, *PRL*, **93**, 184503 (2004)

Boldyrev, S. A., Burgers turbulence, intermittency, and nonuniversality, Physics of Plasmas, **5**, 1681 (1998)

Bonanos, A. Z., Stanek, K. Z., Udalski, A., Wyrzykowski, L., Żebruń, K., Kubiak, M., Szymański, M. K., Szewczyk, O., Pietrzyński, G., Soszyński, I., WR 20a Is an Eclipsing Binary: Accurate Determination of Parameters for an Extremely Massive Wolf-Rayet System, *ApJ*, **611**, L33 (2004)

Bond, J. R., Cole, S., Efstathiou, G., Kaiser, N., Excursion set mass functions for hierarchical Gaussian fluctuations, *ApJ*, **379**, 440 (1991)

Bondi, H., On spherically symmetrical accretion, *MNRAS*, **112**, 195 (1952)

Bonnell, I. A., Bate, M. R., Accretion in stellar clusters and the collisional formation of massive stars, *MNRAS*, **336**, 659 (2002)

Bonnell, I. A., Bate, M. R., Star formation through gravitational collapse and competitive accretion, *MNRAS*, **370**, 488 (2006)

Bonnell, I. A., Bate, M. R., Clarke, C. J., Pringle, J. E., Competitive accretion in embedded stellar clusters, *MNRAS*, **323**, 785 (2001a)

Bonnell, I. A., Clarke, C. J., Bate, M. R., Pringle, J. E., Accretion in stellar clusters and the initial mass function, *MNRAS*, **324**, 573 (2001b)

Bonnell, I. A., Bate, M. R., Zinnecker, H., On the formation of massive stars, *MNRAS*, **298**, 93 (1998)

Bonnell, I. A., Clark, P., Bate, M. R., Gravitational fragmentation and the formation of brown dwarfs in stellar clusters, *MNRAS*, **389**, 1556 (2008)

Bonnell, I. A., Clarke, C. J., Bate, M. R., The Jeans mass and the origin of the knee in the IMF, *MNRAS*, **368**, 1296 (2006)

Bonnell, I. A., Vine, S. G., Bate, M. R., Massive star formation: nurture, not nature, *MNRAS*, **349**, 735 (2004)

Bonnor, W. B., Boyle's Law and gravitational instability, *MNRAS*, **116**, 351 (1956)

Bontemps, S., et al., ISOCAM observations of the rho Ophiuchi cloud: Luminosity and mass functions of the pre-main sequence embedded cluster, *A&A*, **372**, 173 (2001)

Bontemps, S., Andre, P., Terebey, S., Cabrit, S., Evolution of outflow activity around low-mass embedded young stellar objects, *A&A*, **311**, 858 (1996)

Borissova J., Georgiev L., Hanson M. M., Clarke J. R. A., Kurtev R., Ivanov V. D., Penaloza F., Hillier D. J., Zsargó J., et al., Obscured clusters. IV. The most massive stars in [DBS2003] 179, *A&A*, **546**, A110 (2012)

Bourke, T. L., Myers, P. C., Robinson, G., Hyland, A. R., New OH Zeeman Measurements of Magnetic Field Strengths in Molecular Clouds, *ApJ*, **554**, 916 (2001)

Brandl, B., Brandner, W., Eisenhauer, F., Moffat, A. F. J., Palla, F., Zinnecker, H., Low-mass stars in the massive H_{II} region NGC 3603 Deep NIR imaging with ANTU/ISAAC, *A&A*, **352**, L69 (1999)

Braun, R., Thilker, D. A., The WSRT wide-field H_I survey. II. local group features, *A&A*, **417**, 421 (2004)

Bromm, V., Yoshida, N., Hernquist, L., McKee, C. F., The formation of the first stars and galaxies, *Nature*, **459**, 49 (2009)

Bronstein, I. N., Semendjajew, K. A., Taschenbuch der Mathematik, Verlag Harri Deutsch, Thun & Frankfurt a. Main (1987)

Brunt, C. M., Large-scale turbulence in molecular clouds, *ApJ*, **583**, 280 (2003)

Brunt, C. M., Federrath, C., Price, D. J., A method for reconstructing the PDF of a 3D turbulent density field from 2D observations, *MNRAS*, **405**, L56 (2010)

Brunt, C. M., Heyer, M. H., Mac Low, M., Turbulent driving scales in molecular clouds, *A&A*, **504**, 883 (2009)

Burgers, J. M., Mathematical examples illustrating relations occurring in the theory of turbulent fluid motion., Verhandelingen der Koninklijke Nederlandse Akademie van Wetenschappen, Afdeeling Natuurkunde, **17**, 1 (1939)

Burke, J. R., Hollenbach, D. J., The gas-grain interaction in the interstellar medium - Thermal accommodation and trapping, *ApJ*, **265**, 223 (1983)

Burkert, A., Hartmann, L., Collapse and Fragmentation in Finite Sheets, *ApJ*, **616**, 288 (2004)
Burkert, A., Hartmann, L., The Dependence of Star Formation Efficiency on Gas Surface Density, *ApJ*, **773**, 48 (2013)
Burrows, A., Hubbard, W. B., Lunine, J. I., Liebert, J., The theory of brown dwarfs and extrasolar giant planets, Reviews of Modern Physics, 73, 719 (2001)
Burton, M. G., Hollenbach, D. J., Tielens, A. G. G. M., Line emission from clumpy photodissociation regions, *ApJ*, **365**, 620 (1990)
Caballero, J. A., Dynamical parallax of σ Ori AB: mass, distance and age, *MNRAS*, **383**, 750 (2008)
Caldú-Primo, A., Schruba, A., Walter, F., Leroy, A., Sandstrom, K., de Blok, W. J. G., Ianjamasimanana, R., Mogotsi, K. M., A high-dispersion molecular gas component in nearby galaxies, *AJ*, **146**, 150 (2013)
Camenzind, M., Magnetized Disk-Winds and the Origin of Bipolar Outflows., in: Reviews in Modern Astronomy, edited by G. Klare, p. 234 (1990)
Cardelli, J. A., Meyer, D. M., Jura, M., Savage, B. D., The abundance of interstellar carbon, *ApJ*, **467**, 334 (1996)
Carpenter, J. M., Meyer, M. R., Dougados, C., Strom, S. E., Hillenbrand, L. A., Properties of the Monoceros R2 Stellar Cluster, *AJ*, **114**, 198 (1997)
Carroll, J. J., Frank, A., Blackman, E. G., Isotropically driven versus outflow driven turbulence: Observational consequences for molecular clouds, *ApJ*, **722**, 145 (2010)
Caselli, P., Walmsley, C. M., Terzieva, R., Herbst, E., The Ionization Fraction in Dense Cloud Cores, *ApJ*, **499**, 234 (1998)
Casoli, F., Combes, F., Can giant molecular clouds form in spiral arms?, *A&A*, **110**, 287 (1982)
Casuso, E., Beckman, J. E., The K-dwarf problem and the time-dependence of gaseous accretion to the galactic disc, *A&A*, **419**, 181 (2004)
Cen, R., A hydrodynamic approach to cosmology - Methodology, *ApJS*, **78**, 341 (1992)
Cen, R., Fang, T., Where Are the Baryons? III. Nonequilibrium Effects and Observables, *ApJ*, **650**, 573 (2006)
Cernicharo, J., The Physical Conditions of Low Mass Star Forming Regions, in: NATO ASIC Proc. 342: The Physics of Star Formation and Early Stellar Evolution, p. 287 (1991)
Ceverino, D., Dekel, A., Bournaud, F., High-redshift clumpy disks and bulges in cosmological simulations, *MNRAS*, **404**, 2151 (2010)
Chabrier, G., Galactic Stellar and Substellar Initial Mass Function, *PASP*, **115**, 763 (2003a)
Chabrier G., The Galactic Disk Mass Function: Reconciliation of the Hubble Space Telescope and Nearby Determinations, *ApJ*, **586**, L133 (2003b)
Chabrier G., The Initial Mass Function: from Salpeter 1955 to 2005, in The Initial Mass Function 50 Years Later, eds. E. Corbelli and F. Palla, *Astrophys. Space Sc. Lib.*, **327**, 41 (2005)
Chakraborty, N., Fields, B. D., Inverse-Compton Contribution to the Star-forming Extragalactic Gamma-Ray Background, *ApJ*, **773**, 104 (2013)
Chabrier, G., Hennebelle, P., Star Formation: Statistical Measure of the Correlation between the Prestellar Core Mass Function and the Stellar Initial Mass Function, *ApJ*, **725**, L79 (2010)
Chandrasekhar S., The Gravitational Instability of an Infinite Homogeneous Turbulent Medium, *Proc. R. Soc. London A*, **210**, 26 (1951)
Chiappini, C., Renda, A., Matteucci, F., Evolution of deuterium, ^{3}He in the galaxy, *A&A*, **395**, 789 (2002)
Chini, R., Hoffmeister, V., Kimeswenger, S., Nielbock, M., Nürnberger, D., Schmidtobreick, L., Sterzik, M., The formation of a massive protostar through the disk accretion of gas, *Nature*, **429**, 155 (2004)
Chini, R., Hoffmeister, V. H., Nielbock, M., Scheyda, C. M., Steinacker, J., Siebenmorgen, R., Nürnberger, D., A Remnant Disk around a Young Massive Star, *ApJ*, **645**, L61 (2006)
Chira R.-A., Smith R. J., Klessen R. S., Stutz A. M., Shetty R., Line Profiles of Cores within Clusters. III. What is the most reliable tracer of core collapse in dense clusters? *MNRAS*, **444**, 874 (2014)

Cho, W., Kim, J., Enhanced core formation rate in a turbulent cloud by self-gravity, *MNRAS*, **410**, L8 (2011)
Cho J., Lazarian A., Compressible magnetohydrodynamic turbulence: mode coupling, scaling relations, anisotropy, viscosity-damped regime and astrophysical implications, *MNRAS*, **345**, 325 (2003)
Chokshi, A., Tielens, A. G. G. M., Werner, M. W., Castelaz, M. W., C_{II} 158 micron and O i 63 micron observations of NGC 7023 - A model for its photodissociation region, *ApJ*, **334**, 803 (1988)
Christensen, C., Quinn, T., Governato, F., Stilp, A., Shen, S., Wadsley, J., Implementing molecular hydrogen in hydrodynamic simulations of galaxy formation, *MNRAS*, **425**, 3058 (2012)
Clark, P. C., Bonnell, I. A., Clumpy shocks and the clump mass function, *MNRAS*, **368**, 1787 (2006)
Clark, P. C., Glover, S. C. O., On column density thresholds and the star formation rate, *MNRAS*, **444**, 2396 (2014)
Clark, P. C., Glover, S. C. O., Klessen, R. S., TreeCol: a novel approach to estimating column densities in astrophysical simulations, *MNRAS*, **420**, 745 (2012a)
Clark, P. C., Glover, S. C. O., Klessen, R. S., Bonnell, I. A., How long does it take to form a molecular cloud? *MNRAS*, **424**, 2599 (2012b)
Clark, P. C., Glover, S. C. O., Ragan, S. E., Shetty, R., Klessen, R.S., On the Temperature Structure of the Galactic Center Cloud G0.253+0.016, *ApJ*, **768**, L34 (2013)
Clark, P. C., Glover, S. C. O., Smith, R. J., Greif, T. H., Klessen, R. S., Bromm, V., The Formation and Fragmentation of Disks Around Primordial Protostars, *Science*, **331**, 1040 (2011)
Clark, P. C., Klessen, R. S., Bonnell, I. A., Clump lifetimes and the initial mass function, *MNRAS*, **379**, 57 (2007)
Clarke, C. J., Bromm, V., The characteristic stellar mass as a function of redshift, *MNRAS*, **343**, 1224 (2003)
Collins, D. C., Padoan, P., Norman, M. L., Xu, H., Mass and Magnetic Distributions in Self-gravitating Super-Alfvénic Turbulence with Adaptive Mesh Refinement, *ApJ*, **731**, 59 (2011)
Combes, F., Distribution of CO in the Milky Way, *ARA&A*, **29**, 195 (1991)
Commerçon B., Hennebelle P., Henning T., Collapse of Massive Magnetized Dense Cores Using Radiation Magnetohydrodynamics: Early Fragmentation Inhibition, *ApJ*, **742**, L9 (2011)
Congiu, E., Matar, E., Kristensen, L. E., Dulieu, F., Lemaire, J. L., Laboratory evidence for the non-detection of excited nascent H_2 in dark clouds, *MNRAS*, **397**, L96 (2009)
Cooke, R. J., Pettini, M., Jorgenson, R. A., Murphy, M. T., Steidel, C. C., Precision Measures of the Primordial Abundance of Deuterium, *ApJ*, **781**, 31 (2014)
Cowie, L. L., Songaila, A., High-resolution optical and ultraviolet absorption-line studies of interstellar gas, *ARA&A*, **24**, 499 (1986)
Crowther, P. A., Schurr, O., Mirschi, R., Yusof, N., Parker, R. J., Goodwin, S. P., Kassim, H. A., The emergence of super-canonical stars in R136-type starburst clusters, *MNRAS*, **408**, 731 (2010)
Crutcher, R., Heiles, C., Troland, T., Observations of Interstellar Magnetic Fields, in: Turbulence and Magnetic Fields in Astrophysics, vol. 614 of Lecture Notes in Physics, Berlin Springer Verlag, edited by E. Falgarone, T. Passot, p. 155 (2003)
Crutcher, R. M., Magnetic Fields in Molecular Clouds: Observations Confront Theory, *ApJ*, **520**, 706 (1999)
Crutcher, R. M., The Role of Magnetic Fields in Star Formation, in: http://www.mpia-hd.mpg.de/homes/stein/EPoS/Onlinematerial/crutcher.pdf.gz, edited by J. Steinacker, A. Bacmann (2010)
Crutcher, R. M., Hakobian, N., Troland, T. H., Testing Magnetic Star Formation Theory, *ApJ*, **692**, 844 (2009)
Crutcher, R. M., Hakobian, N., Troland, T. H., Self-consistent analysis of OH Zeeman observations, *MNRAS*, **402**, L64 (2010a)
Crutcher, R. M., Wandelt, B., Heiles, C., Falgarone, E., Troland, T. H., Magnetic Fields in Interstellar Clouds from Zeeman Observations: Inference of Total Field Strengths by Bayesian Analysis, *ApJ*, **725**, 466 (2010b)

Crutcher, R. M., Troland, T. H., OH Zeeman Measurement of the Magnetic Field in the L1544 Core, *ApJ*, **537**, L139 (2000)

Crutcher, R. M., Troland, T. H., Lazareff, B., Paubert, G., Kazès, I., Detection of the CN Zeeman Effect in Molecular Clouds, *ApJ*, **514**, L121 (1999)

Cunningham, A. J., Frank, A., Carroll, J., Blackman, E. G., Quillen, A. C., Protostellar Outflow Evolution in Turbulent Environments, *ApJ*, **692**, 816 (2008)

Dale, J. E., Bonnell, I., Ionizing feedback from massive stars in massive clusters: fake bubbles and untriggered star formation, *MNRAS*, **414**, 321 (2011)

Dale, J. E., Bonnell, I. A., Clarke, C. J., Bate, M. R., Photoionizing feedback in star cluster formation, *MNRAS*, **358**, 291 (2005)

Dalgarno, A., Black, J. H., Molecule formation in the interstellar gas, Rep. Prog. Phys., **39**, 573 (1976)

Dalgarno, A., Yan, M., Liu, W., Electron Energy Deposition in a Gas Mixture of Atomic and Molecular Hydrogen and Helium, *ApJS*, **125**, 237 (1999)

Dame, T. M., Hartmann, D., Thaddeus, P., The Milky Way in Molecular Clouds: A New Complete CO Survey, *ApJ*, **547**, 792 (2001)

Davies, B., Lumsden, S. L., Hoare, M. G., Oudmaijer, R. D., de Wit, W.-J., The circumstellar disc, envelope and bipolar outflow of the massive young stellar object W33A, *MNRAS*, **402**, 1504 (2010)

Dawson, J. R., McClure-Griffiths, N. M., Wong, T., Dickey, J. M., Hughes, A., Fukui, Y., Kawamura, A., Supergiant Shells and Molecular Cloud Formation in the Large Magellanic Cloud, *ApJ*, **763**, 56 (2013)

de Avillez, M. A., Breitschwerdt, D., Volume filling factors of the ISM phases in star forming galaxies. I. The role of the disk-halo interaction, *A&A*, **425**, 899 (2004)

de Avillez, M. A., Breitschwerdt, D., Global dynamical evolution of the ISM in star forming galaxies. I. High resolution 3D simulations: Effect of the magnetic field, *A&A*, **436**, 585 (2005)

de Avillez, M. A., Breitschwerdt, D., The Generation and Dissipation of Interstellar Turbulence: Results from Large-Scale High-Resolution Simulations, *ApJ*, **665**, L35 (2007)

de Avillez, M. A., Breitschwerdt, D., The Diagnostic O vi Absorption Line in Diffuse Plasmas: Comparison of Non-equilibrium Ionization Structure Simulations to FUSE Data, *ApJ*, **761**, L19 (2012)

de Avillez, M. A., Mac Low, M., Mixing Timescales in a Supernova-driven Interstellar Medium, *ApJ*, **581**, 1047 (2002)

Deharveng, L., Peña, M, Caplan, J., Costero, R., Oxygen and helium abundances in Galactic H_{II} regions - II. Abundance gradients, *MNRAS*, **311**, 329 (2000)

Deharveng, L., Schuller, F., Anderson, L. D., Zavagno, A., Wyrowski, F., Menten, K. M., Bronfman, L., Testi, L., Walmsley, C. M., Wienen, M., A gallery of bubbles. The nature of the bubbles observed by Spitzer and what ATLASGAL tells us about the surrounding neutral material, *A&A*, **523**, 6 (2010)

Désert, F.-X., Macías-Pérez, J.-F., Mayet, F., Giardino, G., Renault, C., Aumont, J., et al., Submillimetre point sources from the Archeops experiment: very cold clumps in the Galactic plane, *A&A*, **481**, 411 (2008)

Dekel, A., Birnboim, Y., Engel, G., Freundlich, J., Goerdt, T., Mumcuoglu, M., Neistein, E., Pichon, C., Teyssier, R., Zinger, E., Cold streams in early massive hot haloes as the main mode of galaxy formation, *Nature*, **457**, 451 (2009)

De Pree C. G., Peters T., Mac Low M.-M., Wilner D. J., Goss W. M., Galván-Madrid R., Keto E. R., Klessen R. S., Monsrud A., et al., Flickering of 1.3 cm Sources in Sgr B2: Toward a Solution to the Ultracompact H_{II} Region Lifetime Problem, *ApJ*, **781**, L36 (2014)

de Wit, W. J., Testi, L., Palla, F., Vanzi, L., Zinnecker, H., The origin of massive O-type field stars. I. A search for clusters, *A&A*, **425**, 937 (2004)

Dib, S., Burkert, A. On the Origin of the H_I Holes in the Interstellar Medium of Dwarf Irregular Galaxies, *ApJ*, **630**, 238 (2005)

Dickey, J. M., Lockman, F. J., H_I in the galaxy, *ARA&A*, **28**, 215 (1990)

Di Francesco, J., Evans, II, N. J., Caselli, P., Myers, P. C., Shirley, Y., Aikawa, Y., Tafalla, M., An Observational Perspective of Low-Mass Dense Cores I: Internal Physical and Chemical Properties, in: Protostars and Planets V, edited by B. Reipurth, D. Jewitt, K. Keil, p. 17 (2007)
Dobbs, C. L., Bonnell, I. A., Simulations of spiral galaxies with an active potential: molecular cloud formation and gas dynamics, *MNRAS*, **385**, 1893 (2008)
Dobbs, C. L., GMC formation by agglomeration and self gravity, *MNRAS*, **391**, 844 (2008)
Dobbs, C. L., Bonnell, I. A., Clark, P. C., Centrally condensed turbulent cores: massive stars or fragmentation?, *MNRAS*, **360**, 2 (2005)
Dobbs, C. L., Glover, S. C. O., Clark, P. C., Klessen, R. S., The ISM in spiral galaxies: can cooling in spiral shocks produce molecular clouds? *MNRAS*, **389**, 1097 (2008)
Dobbs, C. L., Krumholz, M. R., Ballesteros-Paredes, J., Bolatto, A. D., Fukui, Y., Heyer, M., Mac Low, M.-M., Ostriker, E. C., Vázquez-Semadeni, E., Formation of Molecular Clouds and Global Conditions for Star Formation. In: Protostars and Planets VI, edited by Beuther, H., Klessen, R. S., Dullemond, C. P., Henning, Th., University of Arizona Press, **3**, (2014)
Dobbs, C. L., Pringle, J. E., Burkert, A., Giant molecular clouds: what are they made from, and how do they get there?, *MNRAS*, **425**, 2157 (2012)
Dopcke, G., Glover, S. C. O., Clark, P. C., Klessen, R. S., On the Initial Mass Function of Low-metallicity Stars: The Importance of Dust Cooling, *ApJ*, **766**, 103 (2013)
Doran E. I., Crowther P. A., de Koter A., Evans C. J., McEvoy C., Walborn N. R., Bastian N., Bestenlehner J. M., Gräfener G., et al., The VLT-FLAMES Tarantula Survey. XI. A census of the hot luminous stars and their feedback in 30 Doradus, *A&A*, **558**, A134 (2013)
Draine, B. T., Photoelectric heating of interstellar gas, *ApJS*, **36**, 595 (1978)
Draine, B. T., Physics of the Interstellar and Intergalactic Medium, Princeton University Press (2011)
Draine, B. T., Bertoldi, F., Structure of Stationary Photodissociation Fronts, *ApJ*, **468**, 269 (1996)
Draine, B. T., Li, A., Infrared Emission from Interstellar Dust. IV. The Silicate-Graphite-PAH Model in the Post-Spitzer Era, *ApJ*, **657**, 810 (2007)
Draine, B. T., Lee, H. M., Optical properties of interstellar graphite and silicate grains, *ApJ*, **285**, 89 (1984)
Draine, B. T., Sutin, B., Collisional charging of interstellar grains, *ApJ*, **320**, 803 (1987)
Dupac, X., Bernard, J.-P., Boudet, N., Giard, M., Lamarre, J.-M., Mény, C., et al., Inverse temperature dependence of the dust submillimeter spectral index, *A&A*, **404**, L11 (2003)
Dziourkevitch, N., Elstner, D., Rüdiger, G., Interstellar turbulence driven by the magnetorotational instability, *A&A*, **423**, L29 (2004)
Ebert, R., Über die Verdichtung von H_I-Gebieten. Mit 5 Textabbildungen, *Z. Astrophys.*, **37**, 217 (1955)
Ekström, S., Georgy, C., Eggenberger, P., Meynet, G., Mowlavi, N., Wyttenbach, A., Granada, A., Decressin, T., Hirschi, R., Frischknecht, U., Charbonnel, C., Maeder, A., Grids of stellar models with rotation. I. Models from 0.8 to 120 $M_\odot$ at solar metallicity (Z = 0.014), *A&A*, **537**, 146 (2012)
Elmegreen, B. G., Elmegreen, D. M., H_I superclouds in the inner Galaxy, *ApJ*, **320**, 182 (1987)
Elmegreen, B. G., The H to H_2 transition in galaxies - Totally molecular galaxies, *ApJ*, **411**, 170 (1993)
Elmegreen, B. G., Falgarone E., A Fractal Origin for the Mass Spectrum of Interstellar Clouds, *ApJ*, **471**, 816 (1996)
Elmegreen, B. G., Intercloud Structure in a Turbulent Fractal Interstellar Medium, *ApJ*, **477**, 196 (1997a)
Elmegreen, B. G., The Initial Stellar Mass Function from Random Sampling in a Turbulent Fractal Cloud, *ApJ*, **486**, 944 (1997b)
Elmegreen, B. G., The Stellar Initial Mass Function from Random Sampling in Hierarchical Clouds. II. Statistical Fluctuations and a Mass Dependence for Starbirth Positions and Times, *ApJ*, **515**, 323 (1999)

Elmegreen, B. G., Two stellar mass functions combined into one by the random sampling model of the initial mass function, *MNRAS*, **311**, L5 (2000a)
Elmegreen, B. G., Star Formation in a Crossing Time, *ApJ*, **530**, 277 (2000b)
Elmegreen, B. G., A Fractal Origin for the Mass Spectrum of Interstellar Clouds. II. Cloud Models and Power-Law Slopes, *ApJ*, **564**, 773 (2002a)
Elmegreen, B. G., Star Formation from Galaxies to Globules, *ApJ*, **577**, 206 (2002b)
Elmegreen, B. G., On the Rapid Collapse and Evolution of Molecular Clouds, *ApJ*, **668**, 1064 (2007)
Elmegreen, B. G., Gravitational Instabilities in Two-component Galaxy Disks with Gas Dissipation, *ApJ*, **737**, 10 (2011)
Elmegreen, B. G., Burkert, A., Accretion-Driven Turbulence and the Transition to Global Instability in Young Galaxy Disks, *ApJ*, **712**, 294 (2010)
Elmegreen, B. G., Klessen, R. S., Wilson, C. D., On the Constancy of the Characteristic Mass of Young Stars, *ApJ*, **681**, 365 (2008)
Elmegreen, B. G., Lada, C. J., Sequential formation of subgroups in OB associations, *ApJ*, **214**, 725 (1977)
Elmegreen, B. G., Scalo, J., Interstellar Turbulence I: Observations and Processes, *ARA&A*, **42**, 211 (2004)
Enoch, M. L., Evans, N. J., Sargent, A. I., Glenn, J., Rosolowsky, E., Myers, P., The Mass Distribution and Lifetime of Prestellar Cores in Perseus, Serpens, and Ophiuchus, *ApJ*, **684**, 1240 (2008)
Evans, N. J., Physical Conditions in Regions of Star Formation, *ARA&A*, **37**, 311 (1999)
Evans, N. J., Heiderman, A., Vutisalchavakul, N., Star Formation Relations in Nearby Molecular Clouds, *ApJ*, **782**, 114 (2014)
Falgarone, E., Pineau des Forêts, G., Roueff, E., Chemical signatures of the intermittency of turbulence in low density interstellar clouds, *A&A*, **300**, 870 (1995)
Falgarone, E., Pety, J., Hily-Blant, P., Intermittency of interstellar turbulence: extreme velocity-shears and CO emission on milliparsec scale, *A&A*, **507**, 355 (2009)
Fatuzzo, M., Adams, F. C., Enhancement of Ambpolar Diffusion Rates through Field Fluctuations, *ApJ*, **570**, 210 (2002)
Faucher-Giguère, C.-A., Quataert, E., Hopkins, P., Feedback-regulated star formation in molecular clouds and galactic discs, *MNRAS*, **433**, 1970 (2013)
Federrath, C., The origin of physical variations in the star formation law, *MNRAS*, **463**, 3167 (2013)
Federrath, C., Klessen, R. S., The Star Formation Rate of Turbulent Magnetized Clouds: Comparing Theory, Simulations, and Observations, *ApJ*, **761**, 156 (2012)
Federrath, C., Klessen, R. S., On the Star Formation Efficiency of Turbulent Magnetized Clouds, *ApJ*, **763**, 51 (2013)
Federrath, C., Klessen, R. S., Schmidt, W., The Density Probability Distribution in Compressible Isothermal Turbulence: Solenoidal versus Compressive Forcing, *ApJ*, **688**, L79 (2008)
Federrath, C., Roman-Duval, J., Klessen, R. S., Schmidt, W., Mac Low, M.-M., Comparing the statistics of interstellar turbulence in simulations and observations. Solenoidal versus compressive turbulence forcing, *A&A*, **512**, A81 (2010)
Federrath, C., Schrön, M., Banerjee, R., Klessen, R. S., Modeling Jet and Outflow Feedback during Star Cluster Formation, *ApJ*, **790**, 128 (2014)
Fendt, C., Camenzind, M., On collimated stellar jet magnetospheres. II. dynamical structure of collimating wind flows, *A&A*, **313**, 591 (1996)
Feng, Y., Krumholz, M. R., Early turbulent mixing as the origin of chemical homogeneity in open star clusters, *Nature*, **513**, 523 (2014)
Ferreira, J., Magnetically-driven jets from Keplerian accretion discs, *A&A*, **319**, 340 (1997)
Ferriére, K. M., The interstellar environment of our galaxy, *Rev. Mod. Phys.*, **73**, 1031 (2001)
Ferrière, K., Gillard, W., Jean, P., Spatial distribution of interstellar gas in the innermost 3 kpc of our galaxy, *A&A*, **467**, 611 (2007)
Fich, M., Tremaine, S., The mass of the galaxy, *ARA&A*, **29**, 409 (1991)
Field, G. B., Thermal Instability, *ApJ*, **142**, 531 (1965)

Field, G. B., Blackman, E. G., Keto, E. R., A model of cloud fragmentation, *MNRAS*, **385**, 181 (2008)

Field, G. B., Goldsmith, D. W., Habing, H. J., Cosmic-Ray Heating of the Interstellar Gas, *ApJ*, **155**, L149 (1969)

Field, G. B., A Statistical Model of the Formation of Stars and Interstellar Clouds, *ApJ*, **142**, 568 (1965)

Figer, D. F., An upper limit to the masses of stars, *Nature*, **434**, 192 (2005)

Fixsen, D. J., Mather, J. C., The Spectral Results of the Far-Infrared Absolute Spectrophotometer Instrument on COBE, *ApJ*, **581**, 817 (2002)

Flower, D. R., The rotational excitation of CO by H_2, *J. Phys. B*, **34**, 2731 (2001)

Flower, D. R., Roueff, E., Rovibrational relaxation in collisions between H_2, *J. Phys. B*, **31**, 2935 (1998)

Flower, D. R., Roueff, E., Rovibrational relaxation in collisions between H_2 molecules: II. Influence of the rotational state of the perturber, *J. Phys. B*, **32**, 3399 (1999a)

Flower, D. R., Roueff, E., Rovibrational excitation of HD in collisions with atomic and molecular hydrogen, *MNRAS*, **309**, 833 (1999b)

Flower, D. R., Roueff, E., Zeippen, C. J., Rovibrational excitation of H_2 molecules by He atoms, *J. Phys. B*, **31**, 1105 (1998)

Flower, D. R., Le Bourlot, J., Pineau des Forêts, G., Cabrit, S., The contributions of J-type shocks to the H_2 emission from molecular outflow sources, *MNRAS*, **341**, 70 (2003)

Frank, A., Ray, T. P., Cabrit, S., Hartigan, P., Arce, H. G., Bacciotti, F., Bally, J., Benisty, M., Eislöffel, J., Güdel, M., Lebedev, S., Nisini, B., Raga, A., Jets and Outflows From Star to Cloud: Observations Confront Theory. In: Protostars and Planets VI, edited by Beuther, H., Klessen, R. S., Dullemond, C. P., Henning, Th., University of Arizona Press, **451**, (2014)

Frerking, M. A., Keene, J., Blake, G. A., Phillips, T. G., The abundances of atomic carbon and carbon monoxide compared with visual extinction in the Ophiuchus molecular cloud complex, *ApJ*, **344**, 311 (1989)

Frisch, U., Turbulence, Cambridge University Press (1996)

Fukui, Y., Kawamura, A., Wong, T., Murai, M., Iritani, H., Mizuno, N., Mizuno, Y., Onishi, T., Hughes, A., Ott, J., Muller, E., Staveley-Smith, L., Kim, S., Molecular and Atomic Gas in the Large Magellanic Cloud. II. Three-dimensional Correlation Between CO and H_I, *ApJ*, 705, 144 (2009)

Furuya, K., Aikawa, Y., Tomida, K., Matsumoto, T., Saigo, K., Tomisaka, K., Hersant, F., Wakelam, V., Chemistry in the First Hydrostatic Core Stage by Adopting Three-dimensional Radiation Hydrodynamic Simulations, *ApJ*, **758**, 86 (2012)

Gaensler, B. M., Madsen, G. J., Chatterjee, S., Mao, S. A., The Vertical Structure of Warm Ionised Gas in the Milky Way, *Pub. Astron. Soc. Aust.*, **25**, 184 (2008)

Gaisser, T. K., The Cosmic-ray Spectrum: from the knee to the ankle, *J. Phys. Conf. Ser.*, **47**, 15 (2006)

Galametz, M., Madden, S. C., Galliano, F., Hony, S., Bendo, G. J., Sauvage, M., Probing the dust properties of galaxies up to submillimetre wavelengths. II. Dust-to-gas mass ratio trends with metallicity and the submm excess in dwarf galaxies, *A&A*, **532**, 56 (2011)

Galli, D., Palla, F., Deuterium chemistry in the primordial gas, *Plan. Space Sci.*, **50**, 1197 (2002)

Galván-Madrid, R., Peters, T., Keto, E. R., Low, M.-M. M., Banerjee, R., Klessen, R. S., Time variability in simulated ultracompact and hypercompact H_{II} regions, *MNRAS*, **416**, 1033 (2011)

Gammie, C. F., Ostriker, E. C., Can nonlinear hydromagnetic waves support a self-gravitating cloud? *ApJ*, **466**, 814 (1996)

Genzel R., Physical conditions and heating/cooling processes in high mass star formation regions, NATO ASIC Proc. 342: The Physics of Star Formation and Early Stellar Evolution, eds. C. J. Lada and N. D. Kylafis, 155 (1991)

Georgy, C., Ekström, S., Meynet, G., Massey, P., Levesque, E. M., Hirschi, R., Eggenberger, P., Maeder, A., Grids of stellar models with rotation. II. WR populations and supernovae/GRB progenitors at Z = 0.014, *A&A*, **542**, 29 (2012)

Girichidis, P., Federrath, C., Allison, R., Banerjee, R., Klessen, R. S., Importance of the initial conditions for star formation - III. Statistical properties of embedded protostellar clusters, *MNRAS*, **420**, 3264 (2012a)

Girichidis, P., Federrath, C., Banerjee, R.,Klessen, R. S., Importance of the initial conditions for star formation - II. Fragmentation-induced starvation and accretion shielding, MNRAS, **420**, 613 (2012b)

Girichidis, P., Federrath, C., Banerjee, R., Klessen, R. S., Importance of the initial conditions for star formation - I. Cloud evolution and morphology, *MNRAS*, **413**, 2741 (2011)

Girichidis P., Konstandin L., Whitworth A. P., Klessen R. S., On the Evolution of the Density Probability Density Function in Strongly Self-gravitating Systems, *ApJ*, **781**, 91 (2014)

Glassgold, A. E., Langer, W. D., Heating of Molecular-Hydrogen Clouds by Cosmic Rays and X-Rays, *ApJ*, **186**, 859 (1973)

Glassgold, A. E., Langer, W. D., The C^+–CO transition in interstellar clouds, *ApJ*, **197**, 347 (1975)

Glassgold, A. E., Galli, D., Padovani, M., Cosmic-ray and X-ray heating of interstellar clouds and protoplanetary disks, *ApJ*, **756**, 157 (2012)

Glover, S. C. O., Comparing Gas-Phase and Grain-catalyzed H_2 Formation, *ApJ*, **584**, 331 (2003)

Glover, S. C. O., Abel, T., Uncertainties in H_2 and HD chemistry and cooling and their role in early structure formation, *MNRAS*, **388**, 1627 (2008)

Glover, S. C. O., Clark, P. C., Is molecular gas necessary for star formation? *MNRAS*, **421**, 9 (2012a)

Glover, S. C. O., Clark, P. C., Approximations for modelling CO chemistry in giant molecular clouds: a comparison of approaches, *MNRAS*, **421**, 116 (2012b)

Glover, S. C. O., Clark, P. C., Star formation in metal-poor gas clouds, *MNRAS*, **426**, 377 (2012c)

Glover, S. C. O., Clark, P. C., Molecular cooling in the diffuse interstellar medium, *MNRAS*, **437**, 9 (2014)

Glover, S. C. O., Clark, P. C., Micic, M., Molina, F. Z., Modelling [CI] emission from turbulent molecular clouds, *MNRAS*, **448**, 1607 (2015)

Glover, S. C. O., Federrath, C., Low, M.-M. M., Klessen, R. S., Modelling CO formation in the turbulent interstellar medium, *MNRAS*, **404**, 2 (2010)

Glover, S. C. O., Mac Low, M.-M., Simulating the Formation of Molecular Clouds. II. Rapid Formation from Turbulent Initial Conditions, *ApJ*, **659**, 1317 (2007b)

Glover, S. C. O., Mac Low, M.-M., Simulating the Formation of Molecular Clouds. I. Slow Formation by Gravitational Collapse from Static Initial Conditions, *ApJS*, **169**, 239 (2007a)

Glover, S. C. O., Jappsen, A.-K., Star Formation at Very Low Metallicity. I. Chemistry and Cooling at Low Densities, *ApJ*, **666**, 1 (2007)

Gnat, O., Ferland, G. J., Ion-by-ion Cooling Efficiencies, *ApJS*, **199**, 20 (2012)

Gnat, O., Sternberg, A., Time-dependent Ionization in Radiatively Cooling Gas, *ApJS*, **168**, 213 (2007)

Gnedin, N. Y., Hollon, N., Cooling and Heating Functions of Photoionized Gas, *ApJ*, **202**, 13 (2012)

Gnedin, N. Y., Tassis, K., Kravtsov, A. V., Modeling Molecular Hydrogen and Star Formation in Cosmological Simulations, *ApJ*, **697**, 55 (2009)

Godard, B., Falgarone, E., Pineau des Forêts, G., Models of turbulent dissipation regions in the diffuse interstellar medium, *A&A*, **495**, 847 (2009)

Goldbaum, N. J., Krumholz, M. R., Matzner, C. D., McKee, C. F., The Global Evolution of Giant Molecular Clouds. II. The Role of Accretion, *ApJ*, **738**, 101 (2011)

Goldreich, P., Kwan, J., Molecular Clouds, *ApJ*, **189**, 441 (1974)

Goldsmith, P. F., Temperatures and Densities in Interstellar Molecular Clouds, in: Molecular Clouds in the Milky Way and External Galaxies, edited by R. Chiao, p. 1, Springer, New York (1988)

Goldsmith, P. F., Molecular Depletion and Thermal Balance in Dark Cloud Cores, *ApJ*, **557**, 736 (2001)

Goldsmith, P. F., Langer, W. D., Molecular cooling and thermal balance of dense interstellar clouds, *ApJ*, **222**, 881 (1978)

Gómez, G. C., Cox, D. P., Three-dimensional magnetohydrodynamic modeling of the gaseous structure of the galaxy: Setup and initial results, *ApJ*, **580**, 235 (2002)

Gomez, M., Jones, B. F., Hartmann, L., Kenyon, S. J., Stauffer, J. R., Hewett, R., Reid, I. N., On the Ages of Pre-Main-Sequence Stars in Taurus, *AJ*, **104**, 762 (1992)
Gondhalekar, P. M., Phillips, A. P., Wilson, R., Observations of the interstellar ultraviolet radiation field from the S2/68 sky-survey telescope, *A&A*, **85**, 272 (1980)
Goodman, A. A., Barranco, J. A., Wilner, D. J., Heyer, M. H., Coherence in Dense Cores. II. The Transition to Coherence, *ApJ*, **504**, 223 (1998)
Goodwin, S. P., Kroupa, P., Limits on the primordial stellar multiplicity, *A&A*, **439**, 565 (2005)
Goodwin, S. P., Whitworth, A. P., Ward-Thompson, D., Simulating star formation in molecular cloud cores. I. The influence of low levels of turbulence on fragmentation and multiplicity, *A&A*, **414**, 633 (2004a)
Goodwin, S. P., Whitworth, A. P., Ward-Thompson, D., Simulating star formation in molecular cores. II. The effects of different levels of turbulence, *A&A*, **423**, 169 (2004b)
Goodwin, S. P., Whitworth A. P., Ward-Thompson D., Star formation in molecular cores. III. The effect of the turbulent power spectrum, *A&A*, **452**, 487 (2006)
Goto, M., Stecklum, B., Linz, H., Feldt, M., Henning, T., Pascucci, I., Usuda, T., High-Resolution Infrared Imaging of Herschel 36 SE: A Showcase for the Influence of Massive Stars in Cluster Environments, *ApJ*, **649**, 299 (2006)
Gould, R. J., Salpeter, E. E., The Interstellar Abundance of the Hydrogen Molecule. I. Basic Processes, *ApJ*, **138**, 393 (1963)
Gredel, R., Lepp, S., Dalgarno, A., The C/CO ratio in dense interstellar clouds, *ApJ*, **323**, L137 (1987)
Gredel, R., Lepp, S., Dalgarno, A., Herbst, E., Cosmic-ray-induced photodissociation and photoionization rates of interstellar molecules, *ApJ*, **347**, 289 (1989)
Greene, T. P., Meyer, M. R., An Infrared Spectroscopic Survey of the ρ-Ophiuchi Young Stellar Cluster: Masses and Ages from the H-R Diagram, *ApJ*, **450**, 233 (1995)
Greif, T. H., Bromm, V., Clark, P. C., Glover, S. C. O., Smith, R. J., Klessen, R. S., Yoshida, N., Springel, V., Formation and evolution of primordial protostellar systems, *MNRAS*, **424**, 399 (2012)
Greif, T. H., Springel, V., White, S. D. M., Glover, S. C. O., Clark, P. C., Smith, R. J., Klessen, R. S., Bromm, V., Simulations on a Moving Mesh: The Clustered Formation of Population III Protostars, *ApJ*, **737**, 75 (2011)
Guszejnov, D., Hopkins, P. F., Mapping the core mass function to the initial mass function, *MNRAS*, **450**, 4137 (2015)
Gutermuth, R. A., Pipher, J. L., Megeath, S. T., Myers, P. C., Allen, L. E., Allen, T. S., A Correlation between Surface Densities of Young Stellar Objects and Gas in Eight Nearby Molecular Clouds, *ApJ*, **739**, 84 (2011)
Habing, H. J., The interstellar radiation density between 912 Å and 2400 Å, *Bull. Astron. Inst. Netherlands*, **19**, 421 (1968)
Haffner, L. M., Dettmar, R.-J., Beckman, J. E., Wood, K., Slavin, J. D., Giammanco, C., et al., The warm ionized medium in spiral galaxies, *Rev. Mod. Phys.*, **81**, 969 (2009)
Hansen, C. J., Kawaler, S. D., Stellar Interiors: Physical principles, Structure, and Evolution., Springer Verlag, New York (1994)
Hartmann, L., Flows, Fragmentation, and Star Formation. I. Low-Mass Stars in Taurus, *ApJ*, **578**, 914 (2002)
Hartmann, L., Ballesteros-Paredes, J., Bergin, E. A., Rapid Formation of Molecular Clouds and Stars in the Solar Neighborhood, *ApJ*, **562**, 852 (2001)
Hatchell, J., Richer, J. S., Fuller, G. A., Qualtrough, C. J., Ladd, E. F., Chandler, C. J., Star formation in Perseus. Clusters, filaments and the conditions for star formation, *A&A*, **440**, 151 (2005)
Hawley, J. F., Gammie, C. F., Balbus, S. A., Local three-dimensional magnetohydrodynamic simulations of accretion disks, *ApJ*, **440**, 742 (1995)
Heiderman, A., Evans, N. J., Allen, L. E., Huard, T., Heyer, M., The Star Formation Rate and Gas Surface Density Relation in the Milky Way: Implications for Extragalactic Studies, *ApJ*, **723**, 1019 (2010)

Heiles, C., Troland, T. H., The Millennium Arecibo 21 Centimeter Absorption-Line Survey. II. Properties of the Warm and Cold Neutral Media, *ApJ*, **586**, 1067 (2003)
Heiles, C., Troland, T. H., The Millennium Arecibo 21 Centimeter Absorption-Line Survey. IV. Statistics of Magnetic Field, Column Density, and Turbulence, *ApJ*, **624**, 773 (2005)
Heiner, J. S., Vázquez-Semadeni, E., Applying a one-dimensional PDR model to the Taurus molecular cloud and its atomic envelope, *MNRAS*, **429**, 3584 (2013)
Heitsch, F., Burkert, A., Hartmann, L. W., Slyz, A. D., Devriendt, J. E. G., Formation of Structure in Molecular Clouds: A Case Study, *ApJ*, **633**, L113 (2005)
Heitsch, F., Hartmann, L., Rapid Molecular Cloud and Star Formation: Mechanisms and Movies, *ApJ*, **689**, 290 (2008)
Hartwig, T., Clark, P. C., Glover, S. C. O., Klessen, R. S., Sasaki, M., A new approach to determine optically thick H_2 cooling and its effect on primordial star formation, *ApJ*, **799**, 144 (2015)
Heitsch, F., Hartmann, L. W., Slyz, A. D., Devriendt, J. E. G., Burkert, A., Cooling, Gravity, and Geometry: Flow-driven Massive Core Formation, *ApJ*, **674**, 316 (2008)
Heitsch, F., Mac Low, M.-M., Klessen, R. S., Gravitational Collapse in Turbulent Molecular Clouds. II. Magnetohydrodynamical Turbulence, *ApJ*, **547**, 280 (2001a)
Heitsch, F., Zweibel, E. G., Mac Low, M.-M., Li, P., Norman, M. L., Magnetic Field Diagnostics Based on Far-Infrared Polarimetry: Tests Using Numerical Simulations, ApJ, **561**, 800 (2001b)
Heitsch, F., Putman, M. E., The fate of high-velocity clouds: Warm or cold cosmic rain? *ApJ*, **698**, 1485 (2009)
Heitsch, F., Slyz, A. D., Devriendt, J. E. G., Burkert, A., Cloud dispersal in turbulent flows, *MNRAS*, **373**, 1379 (2006)
Heitsch, F., Zweibel, E. G., Slyz, A. D., Devriendt, J. E. G., Turbulent Ambipolar Diffusion: Numerical Studies in Two Dimensions, *ApJ*, **603**, 165 (2004)
Hennebelle, P., Audit, E., On the structure of the turbulent interstellar atomic hydrogen. I. Physical characteristics. Influence and nature of turbulence in a thermally bistable flow, *A&A*, **465**, 431 (2007)
Hennebelle, P., Chabrier, G., Analytical Theory for the Initial Mass Function: CO Clumps and Prestellar Cores, *ApJ*, **684**, 395 (2008)
Hennebelle, P., Chabrier, G., Analytical Theory for the Initial Mass Function. II. Properties of the Flow, *ApJ*, **702**, 1428 (2009)
Hennebelle, P., Chabrier, G., Analytical Star Formation Rate from Gravoturbulent Fragmentation, *ApJ*, **743**, L29 (2011)
Hennebelle P., Chabrier G., Analytical Theory for the Initial Mass Function. III. Time Dependence and Star Formation Rate, *ApJ*, **770**, 150 (2013)
Hennebelle, P., Ciardi, A., Disk formation during collapse of magnetized protostellar cores, *A&A*, **506**, L29 (2009)
Hennebelle, P., Commerçon, B., Joos, M., Klessen, R. S., Krumholz, M., Tan, J. C., Teyssier, R., Collapse, outflows and fragmentation of massive, turbulent and magnetized prestellar barotropic cores, *A&A*, **528**, A72 (2011)
Hennebelle, P., Falgarone, E., Turbulent molecular clouds, *A&A Rev.*, **20**, 55 (2012)
Hennebelle, P., Fromang, S., Magnetic processes in a collapsing dense core. I. Accretion and ejection, *A&A*, **477**, 9 (2008)
Hennebelle, P., Iffrig, O., Simulations of magnetized multiphase galactic disk regulated by supernovae explosions, *A&A*, **570**, A81 (2014)
Hennebelle, P., Pérault, M., Dynamical condensation in a thermally bistable flow. Application to interstellar cirrus, *A&A*, **351**, 309 (1999)
Hennebelle, P., Pérault, M., Dynamical condensation in a magnetized and thermally bistable flow. Application to interstellar cirrus, *A&A*, **359**, 1124 (2000)
Hennebelle, P., Teyssier, R., Magnetic processes in a collapsing dense core. II. Fragmentation. Is there a fragmentation crisis?, *A&A*, **477**, 25 (2008)
Henning, Th., Influence of molecular outflows from young stellar objects on molecular clouds, *AN*, **310**, 363 (1989)

Henry, R. C., Anderson, R. C., Fastie, W. G., Far-ultraviolet studies. VII - The spectrum and latitude dependence of the local interstellar radiation field, *ApJ*, **239**, 859 (1980)

Herrera-Camus, R., Fisher, D. B., Bolatto, A. D., Leroy, A. K., Walter, F., Gordon, K. D., et al., Dust-to-gas Ratio in the Extremely Metal-poor Galaxy I Zw 18, *ApJ*, **752**, 112 (2012)

Heyer, M., Krawczyk, C., Duval, J., Jackson, J. M., Re-examining Larson's scaling relationships in galactic molecular clouds, *ApJ*, **699**, 1092 (2009)

Heyer, M. H., Brunt, C. M., The Universality of Turbulence in Galactic Molecular Clouds, *ApJ*, **615**, L45 (2004)

Heyer, M. H., Brunt, C. M., Trans-Alfvénic motions in the Taurus molecular cloud, *MNRAS*, **420**, 1562 (2012)

Heyer, M. H., Brunt, C., Snell, R. L., Howe, J. E., Schloerb, F. P., Carpenter, J. M., The Five College Radio Astronomy Observatory CO Survey of the Outer Galaxy, *ApJS*, **115**, 241 (1998)

Hillenbrand, L. A., On the Stellar Population and Star-Forming History of the Orion Nebula Cluster, *AJ*, **113**, 1733 (1997)

Hillenbrand, L. A., Hartmann, L. W., A Preliminary Study of the Orion Nebula Cluster Structure and Dynamics, *ApJ*, **492**, 540 (1998)

Hirota, T., Bushimata, T., Choi, Y. K., Honma, M., Imai, H., Iwadate, K., Jike, T., Kameno, S., Kameya, O., Kamohara, R., Kan-Ya, Y., Kawaguchi, N., Kijima, M., Kim, M. K., Kobayashi, H., Kuji, S., Kurayama, T., Manabe, S., Maruyama, K., Matsui, M., Matsumoto, N., Miyaji, T., Distance to Orion KL Measured with VERA, *PASJ*, **59**, 897 (2007)

Ho, P. T. P., Haschick, A. D., Formation of OB clusters: VLA observations, *ApJ*, **248**, 622 (1981)

Hocuk, S., Spaans, M., The impact of X-rays on molecular cloud fragmentation and the inital mass function, *A&A*, **522**, A24 (2010)

Hollenbach, D., Kaufman, M. J., Bergin, E. A., Melnick, G. J., Water, O_2, and Ice in Molecular Clouds, *ApJ*, **690**, 1497 (2009)

Hollenbach, D., McKee, C. F., Molecule formation and infrared emission in fast interstellar shocks. I Physical processes, *ApJS*, **41**, 555 (1979)

Hollenbach, D., McKee, C. F., Molecule formation and infrared emission in fast interstellar shocks. III - Results for J shocks in molecular clouds, *ApJ*, **342**, 306 (1989)

Holman K., Walch S. K., Goodwin S. P., Whitworth A. P., Mapping the core mass function on to the stellar initial mass function: multiplicity matters, *MNRAS*, **432**, 3534 (2013)

Hopkins P. F., An excursion-set model for the structure of giant molecular clouds and the interstellar medium, *MNRAS*, **423**, 2016 (2012a)

Hopkins P. F., The stellar initial mass function, core mass function and the last-crossing distribution, *MNRAS*, **423**, 2037 (2012b)

Hopkins P. F., Variations in the stellar CMF and IMF: from bottom to top, *MNRAS*, **433**, 170 (2013a)

Hopkins P. F., A general theory of turbulent fragmentation, *MNRAS*, **430**, 1653 (2013b)

Hopkins, A. M., McClure-Griffiths, N. M., Gaensler, B. M., Linked evolution of gas and star formation in galaxies over cosmic history, *ApJ*, **682**, L13 (2008)

Hosokawa T., Omukai K., Evolution of Massive Protostars with High Accretion Rates, *ApJ*, **691**, 823 (2009)

Hou, L. G., Gao, X. Y., A statistical study of gaseous environment of Spitzer interstellar bubbles, *MNRAS*, **438**, 426 (2014)

Hoyle, F., Ellis, G. R. A., On the Existence of an Ionized Layer about the Galactic Plane, *Aust. J. Phys.*, **16**, 1 (1963)

Hoyle, F., Lyttleton, R. A., The evolution of the stars, *Proceedings of the Cambridge Philosophical Society*, **35**, 592 (1939)

Hughes, A., Meidt, S. E., Colombo, D., Schinnerer, E., Pety, J., Leroy, A. K., Dobbs, C. L., García-Burillo, S., Thompson, T. A., Dumas, G., Schuster, K. F., Kramer, C., A Comparative Study of Giant Molecular Clouds in M51, M33, and the Large Magellanic Cloud, *ApJ*, **779**, 46 (2013)

Hughes A., Meidt S. E., Schinnerer E., Colombo, D., Schinnerer, E., Pety, P., Leroy, A. K., Dobbs, C. L., Garcia-Burillo, S., Thompson, T. A., Dumas, G., Schuster, K. L, Kramer, C., Probability

Distribution Functions of ^{12}0)) Brightness and Integrated Intensity in M51: The PAWS View, *ApJ*, **779**, 44 (2013)
Hunter, D. A., Shaya, E. J., Scowen, P., Hester, J. J., Groth, E. J., Lynds, R., O'Neil, Jr., E. J., Gas near the center of 30 Doradus as revealed by Hubble Space Telescope images, *ApJ*, **444**, 758 (1995)
Imara, N., Blitz, L., Angular Momentum in Giant Molecular Clouds. I. The Milky Way, *ApJ*, **732**, 78 (2011)
Indriolo, N., McCall, B. J., Investigating the cosmic-ray ionization rate in the Galactic diffuse interstellar medium through observations of H_3^+, *ApJ*, **745**, 91 (2012)
Inutsuka, S., The Mass Function of Molecular Cloud Cores, *ApJ*, **559**, L149 (2001)
Jappsen, A.-K., Klessen, R. S., Larson, R. B., Li, Y., Mac Low, M.-M., The stellar mass spectrum from non-isothermal gravoturbulent fragmentation, *A&A*, **435**, 611 (2005)
Jaquet, R., Staemmler, V., Smith, M. D., Flower, D. R., Excitation of the fine-structure transitions of O(^{3}), *J. Phys. B*, **25**, 285 (1992)
Jeans J. H., The Stability of a Spherical Nebula, *Phil. Trans. A.*, **199**, 1 (1902)
Jedamzik K., The Cloud-in-Cloud Problem in the Press-Schechter Formalism of Hierarchical Structure Formation, *ApJ*, **448**, 1 (1995)
Jenkins, E. B., A Unified Representation of Gas-Phase Element Depletions in the Interstellar Medium, *ApJ*, **700**, 1299 (2009)
Jenkins, E. B., The Fractional Ionization of the Warm Neutral Interstellar Medium, *ApJ*, **764**, 25 (2013)
Jensen, M. J., et al., Dissociative recombination of H_3, *ApJ*, **543**, 764 (2000)
Jiang, Z., Tamura, M., Hoare, M. G., Yao, Y., Ishii, M., Fang, M., Yang, J., Disks around Massive Young Stellar Objects: Are They Common?, *ApJ*, **673**, L175 (2008)
Jijina, J., Myers, P. C., Adams, F. C., Dense Cores Mapped in Ammonia: A Database, *ApJS*, **125**, 161 (1999)
Johnstone, D., Di Francesco, J., Kirk, H., An Extinction Threshold for Protostellar Cores in Ophiuchus, *ApJ*, **611**, L45 (2004)
Johnstone, D., Fich, M., Mitchell, G. F., Moriarty-Schieven, G., Large Area Mapping at 850 Microns. III. Analysis of the Clump Distribution in the Orion B Molecular Cloud, *ApJ*, **559**, 307 (2001)
Johnstone, D., Matthews, H., Mitchell, G. F., Large Area Mapping at 850 μm. IV. Analysis of the Clump Distribution in the Orion B South Molecular Cloud, *ApJ*, **639**, 259 (2006)
Johnstone, D., Wilson, C. D., Moriarty-Schieven, G., Joncas, G., Smith, G., Gregersen, E., Fich, M., Large-Area Mapping at 850 Microns. II. Analysis of the Clump Distribution in the ρ Ophiuchi Molecular Cloud, *ApJ*, **545**, 327 (2000)
Joung, M. K. R., Mac Low, M.-M., Turbulent Structure of a Stratified Supernova-driven Interstellar Medium, *ApJ*, **653**, 1266 (2006)
Joung, M. R., Mac Low, M.-M., Bryan, G. L., Dependence of interstellar turbulent pressure on the supernova rate, *ApJ*, **704**, 137 (2009)
Jura, M., Interstellar clouds containing optically thin H_2, *ApJ*, **197**, 575 (1975)
Kafatos, M., Time-Dependent Radiative Cooling of a Hot Low-Density Cosmic Gas, *ApJ*, **182**, 433 (1973)
Kahn, F. D., Cocoons around early-type stars, *A&A*, **37**, 149 (1974)
Kainulainen J., Beuther H., Banerjee R., Federrath C., Henning T., Probing the evolution of molecular cloud structure. II. From chaos to confinement, *A&A*, **530**, A64 (2011)
Kainulainen J., Federrath C., Henning T., Connection between dense gas mass fraction, turbulence driving, and star formation efficiency of molecular clouds, *A&A*, **553**, L8 (2013)
Kalberla, P. M. W., Dark Matter in the Milky Way. I. The Isothermal Disk Approximation, *ApJ*, **588**, 805 (2003)
Kalberla, P. M. W., Kerp, J., The H_I Distribution of the Milky Way, *ARA&A*, **47**, 27 (2009)
Kandori, R., Nakajima, Y., Tamura, M., Tatematsu, K., Aikawa, Y., Naoi, T., Sugitani, K., Nakaya, H., Nagayama, T., Nagata, T., Kurita, M., Kato, D., Nagashima, C., Sato, S., Near-Infrared Imaging Survey of Bok Globules: Density Structure, *AJ*, **130**, 2166 (2005)

Kawamura, A., Mizuno, Y., Minamidani, T., Filipovic, M. D., Staveley-Smith, L., Kim, S., Mizuno, N., Onishi, T., Mizuno, A., Fukui, Y., The Second Survey of the Molecular Clouds in the Large Magellanic Cloud by NANTEN. II. Star Formation, *ApJS*, **184**, 1 (2009)

Kennicutt, R. C., Evans, N. J., Star formation in the milky way and nearby galaxies, *ARA&A*, **50**, 531 (2012)

Keto, E., On the Evolution of Ultracompact H_{II} Regions, *ApJ*, **580**, 980 (2002)

Keto, E., The Formation of Massive Stars by Accretion through Trapped Hypercompact H_{II} Regions, *ApJ*, **599**, 1196 (2003)

Keto, E., The Formation of Massive Stars: Accretion, Disks, and the Development of Hypercompact H_{II} Regions, *ApJ*, **666**, 976 (2007)

Keto, E., Caselli, P., The Different Structures of the Two Classes of Starless Cores, *ApJ*, **683**, 238 (2008)

Keto, E., Caselli, P., Dynamics and depletion in thermally supercritical starless cores, *MNRAS*, **402**, 1625 (2010)

Keto, E., Field, G., Dark Cloud Cores and Gravitational Decoupling from Turbulent Flows, *ApJ*, **635**, 1151 (2005)

Keto, E., Wood, K., Observations on the Formation of Massive Stars by Accretion, *ApJ*, **637**, 850 (2006)

Kevlahan, N., Pudritz, R. E., Shock-generated vorticity in the interstellar medium and the origin of the stellar initial mass function, *ApJ*, **702**, 39 (2009)

Kim, J., Hong, S. S., Ryu, D., Jones, T. W., Three-dimensional Evolution of the Parker Instability under a Uniform Gravity, *ApJ*, **506**, L139 (1998)

Kim, J., Ryu, D., Jones, T. W., Three-dimensional Simulations of the Parker Instability in a Uniformly Rotating Disk, *ApJ*, **557**, 464 (2001)

Kim, W.-T., Ostriker, E. C., Stone, J. M., Three-dimensional Simulations of Parker, Magneto-Jeans, and Swing Instabilities in Shearing Galactic Gas Disks, *ApJ*, **581**, 1080 (2002)

Kippenhahn, R., Weigert, A., Weiss, A., Stellar Structure and Evolution, Springer-Verlag, Berlin (2012)

Kirk, H., Johnstone, D., Tafalla, M., Dynamics of Dense Cores in the Perseus Molecular Cloud, *ApJ*, **668**, 1042 (2007)

Klessen, R. S., One-Point Probability Distribution Functions of Supersonic Turbulent Flows in Self-gravitating Media, *ApJ*, **535**, 869 (2000)

Klessen, R. S., The Formation of Stellar Clusters: Time-Varying Protostellar Accretion Rates, *ApJ*, **550**, L77 (2001a)

Klessen, R. S., The Formation of Stellar Clusters: Mass Spectra from Turbulent Molecular Cloud Fragmentation, *ApJ*, **556**, 837 (2001b)

Klessen, R. S., Star Formation in Molecular Clouds, in: EAS Publications Series, vol. 51 of EAS Publications Series, edited by C. Charbonnel, T. Montmerle, p. 133 (2011)

Klessen, R. S., Ballesteros-Paredes, J., Vázquez-Semadeni, E., Durán-Rojas, C., Quiescent and Coherent Cores from Gravoturbulent Fragmentation, *ApJ*, **620**, 786 (2005)

Klessen R. S., Burkert A., Bate M. R., Fragmentation of Molecular Clouds: The Initial Phase of a Stellar Cluster, *ApJ*, **501**, L205 (1998)

Klessen, R. S., Burkert, A., The Formation of Stellar Clusters: Gaussian Cloud Conditions. I., *ApJS*, **128**, 287 (2000)

Klessen, R. S., Burkert, A., The Formation of Stellar Clusters: Gaussian Cloud Conditions. II., *ApJ*, **549**, 386 (2001)

Klessen, R. S., Heitsch, F., Mac Low, M., Gravitational Collapse in Turbulent Molecular Clouds. I. Gasdynamical Turbulence, *ApJ*, **535**, 887 (2000)

Klessen, R. S., Hennebelle, P., Accretion-driven turbulence as universal process: galaxies, molecular clouds, and protostellar disks, *A&A*, **520**, A17 (2010)

Klessen, R. S., Lin, D. N., Diffusion in supersonic turbulent compressible flows, *PRE*, **67**, 046311 (2003)

Kolmogorov, A. N., Dokl. Akad. Nauk SSSR, **30**, 301 (1941)

Konstandin, L., Federrath, C., Klessen, R. S., Schmidt, W., A New Density Variance - Mach Number Relation for Subsonic and Supersonic Isothermal Turbulence, *J. Fluid Mech.*, **692**, 183 (2012)

Könyves, V, et al., The Aquila prestellar core population revealed by Herschel, *A&A*, **518**, L106 (2010)

Koyama, H., Inutsuka, S., An Origin of Supersonic Motions in Interstellar Clouds, *ApJ*, **564**, L97 (2002)

Kramer, C., Cubick, M., Röllig, M., Sun, K., Yonekura, Y., Aravena, M., et al., Clumpy photon-dominated regions in Carina. I. [C_I] fields, *A&A*, **477**, 547 (2008)

Krasnopolsky, R., Li, Z.-Y., Blandford, R., Magnetocentrifugal launching of jets from accretion disks. I. Cold axisymmetric flows, *ApJ*, **526**, 631 (1999)

Kritsuk, A. G., Norman, M. L., Thermal Instability-induced Interstellar Turbulence, *ApJ*, **569**, L127 (2002a)

Kritsuk, A. G., Norman, M. L., Interstellar Phase Transitions Stimulated by Time-dependent Heating, *ApJ*, **580**, L51 (2002b)

Kritsuk, A. G., Norman, M. L., Wagner, R., On the Density Distribution in Star-forming Interstellar Clouds, *ApJ*, **727**, L20 (2011)

Kroupa P., On the variation of the initial mass function, *MNRAS*, **322**, 231 (2001)

Kroupa, P., The Initial Mass Function of Stars: Evidence for Uniformity in Variable Systems, *Science*, **295**, 82 (2002)

Kroupa, P., The Fundamental Building Blocks of Galaxies, in: The Three-Dimensional Universe with Gaia (ESA SP 576), edited by C. Turon, K. S. O'Flaherty, M. A. C. Perryman, ESA Publications (2002)

Kroupa, P., Tout, C. A., Gilmore, G., The effects of unresolved binary stars on the determination of the stellar mass function, *MNRAS*, **251**, 293 (1991)

Kroupa, P., Tout, C. A., Gilmore, G., The low-luminosity stellar mass function, *MNRAS*, **244**, 73 (1990)

Kroupa, P., Weidner, C., Pflamm-Altenburg, J., Thies, I., Dabringhausen, J., Marks, M., Maschberger, T., The Stellar and Sub-Stellar Initial Mass Function of Simple and Composite Populations, in: Planets, Stars and Stellar Systems. Volume 5: Galactic Structure and Stellar Populations, edited by T. D. Oswald, G. Gilmore, Springer Science+Business Media (2013)

Krumholz, M. R., Radiation Feedback and Fragmentation in Massive Protostellar Cores, *ApJ*, **641**, L45 (2006)

Krumholz, M. R., Star Formation in Atomic Gas, *ApJ*, **759**, 9 (2012)

Krumholz, M. R., The big problems in star formation: the star formation rate, stellar clustering, and the initial mass function, *Phys. Reports*, in press; arXiv:1402.0867 (2014)

Krumholz, M. R., Bate, M. R., Arce, H. G., Dale, J. E., Gutermuth, R., Klein, R. I., Li, Z.-Y., Nakamura, F., Zhang, Q., Star cluster formation and feedback. In: Protostars and Planets VI, edited by Beuther, H., Klessen, R. S., Dullemond, C. P., Henning, Th., University of Arizona Press, **243**, (2014)

Krumholz, M. R., Klein, R. I., McKee, C. F., Radiation-Hydrodynamic Simulations of Collapse and Fragmentation in Massive Protostellar Cores, *ApJ*, **656**, 959 (2007)

Krumholz, M., Burkert, A., On the Dynamics and Evolution of Gravitational Instability-dominated Disks, *ApJ*, **724**, 895 (2010)

Krumholz, M. R., Klein, R. I., McKee, C. F., Offner, S. S. R., Cunningham, A. J., The Formation of Massive Star Systems by Accretion, *Science*, **323**, 5915, 754 (2009)

Krumholz, M. R., Leroy, A. K., McKee, C. F., Which Phase of the Interstellar Medium Correlates with the Star Formation Rate? *ApJ*, **731**, 25 (2011)

Krumholz, M. R., Matzner, C. D., McKee, C. F., The global evolution of giant molecular clouds. I. Model formulation and quasi-equilibrium behavior, *ApJ*, **653**, 361 (2006)

Krumholz, M. R., McKee, C. F., A General Theory of Turbulence-regulated Star Formation, from Spirals to Ultraluminous Infrared Galaxies, *ApJ*, **630**, 250 (2005)

Krumholz, M. R., McKee, C. F., A Minimum Column Density of 1 g cm^{-2} for Massive Star Formation, *Nature*, **451**, 1082 (2008)

Krumholz, M. R., McKee, C. F., Klein, R. I., Bondi-Hoyle Accretion in a Turbulent Medium, *ApJ*, **638**, 369 (2006)

Krumholz, M. R., McKee, C. F., Tumlinson, J., The Atomic-to-Molecular Transition in Galaxies. I. An Analytic Approximation for Photodissociation Fronts in Finite Clouds, *ApJ*, **689**, 865 (2008)

Krumholz, M. R., McKee, C. F., Tumlinson, J., The Atomic-to-Molecular Transition in Galaxies. II: H_I Column Densities, *ApJ*, **693**, 216 (2009)

Krumholz, M. R., Stone, J. M., Gardiner, T. A., Magnetohydrodynamic Evolution of H_{II} Regions in Molecular Clouds: Simulation Methodology, Tests, and Uniform Media, *ApJ*, **671**, 518 (2007b)

Krumholz, M. R., Tan, J. C., Slow Star Formation in Dense Gas: Evidence and Implications, *ApJ*, **654**, 304 (2007)

Kuhlen, M., Krumholz, M. R., Madau, P., Smith, B. D., Wise, J., Dwarf Galaxy Formation with H_2-regulated Star Formation, *ApJ*, **749**, 36 (2012)

Kuiper, R., Klahr, H., Beuther, H., Henning, T., Circumventing the Radiation Pressure Barrier in the Formation of Massive Stars via Disk Accretion, *ApJ*, **722**, 1556 (2010)

Kuiper, R., Klahr, H., Beuther, H., Henning, T., Three-dimensional simulation of massive star formation in the disk accretion scenario, *ApJ*, **732**, 20 (2011)

Kurtz, S., Churchwell, E., Wood, D. O. S., Ultracompact H_{II} regions. II. New high-resolution radio images, *ApJS*, **91**, 659 (1994)

Kwan, J., The mass spectrum of interstellar clouds, *ApJ*, **229**, 567 (1979)

Kwan, J., Valdes, F., Spiral gravitational potentials and the mass growth of molecular clouds, *ApJ*, **271**, 604 (1983)

Lada, C. J., Stellar Multiplicity and the Initial Mass Function: Most Stars Are Single, *ApJ*, **640**, L63 (2006)

Lada, C. L., Alves, J. F., Lombardi, M., Lada, E. A., Near-Infrared Extinction and the Structure and Nature of Molecular Clouds, in: Protostars and Planets V, edited by B. Reipurth, D. Jewitt, K. Keil (2006)

Lada, C. J., Bergin, E. A., Alves, J. F., Huard, T. L., The Dynamical State of Barnard 68: A Thermally Supported, Pulsating Dark Cloud, *ApJ*, **586**, 286 (2003)

Lada, C. J., Lada, E. A., Embedded Clusters in Molecular Clouds, *ARA&A*, **41**, 57 (2003)

Lada, C. J., Lombardi, M., Alves, J. F., On the Star Formation Rates in Molecular Clouds, *ApJ*, **724**, 687 (2010)

Lada, C. J., Lombardi, M., Roman-Zuniga, C., Forbrich, J., Alves, J. F., Schmidt's Conjecture and Star Formation in Molecular Clouds, *ApJ*, **778**, 133 (2013)

Lada, C. J., Muench, A. A., Rathborne, J., Alves, J. F., Lombardi, M., The Nature of the Dense Core Population in the Pipe Nebula: Thermal Cores Under Pressure, *ApJ*, **672**, 410 (2008)

Lamers, H. J. G. L. M., Cassinelli, J. P., Introduction to Stellar Winds, Cambridge University Press (1999)

Landau, L. D., Lifshitz, E. M., Fluid Mechanics, Pergamon Press, Oxford (1959)

Langer, W., The carbon monoxide abundance in interstellar clouds, *ApJ*, **206**, 699 (1976)

Langer, W. D., Velusamy, T., Li, D., Goldsmith, P. F., Star Forming Conditions of Quiescent Pre-Stellar Cores in Orion, in: Protostars and Planets V, edited by B. Reipurth, D. Jewitt, K. Keil, p. 8179 (2005)

Larson, R. B., Turbulence and star formation in molecular clouds, *MNRAS*, **194**, 809 (1981)

Larson, R. B., Cloud fragmentation and stellar masses, *MNRAS*, **214**, 379 (1985)

Larson, R. B., Thermal physics, cloud geometry and the stellar initial mass function, *MNRAS*, **359**, 211 (2005)

Larson, R. B., Starrfield, S., On the Formation of Massive Stars and the Upper Limit of Stellar Masses, *A&A*, **13**, 190 (1971)

Launhardt, R., Stutz, A. M., Schmiedeke, A., Henning, Th., Krause, O., Balog, Z., Beuther, H., Birkmann, S., Hennemann, M., Kainulainen, J., Khanzadyan, T., Linz, H., Lippok, N., Nielbock, M., Pitann, J., Ragan, S., Risacher, C., Schmalzl, M., Shirley, Y. L., Stecklum, B., Steinacker, J., Tackenberg, J., The Earliest Phases of Star Formation (EPoS): a Herschel key project. The thermal structure of low-mass molecular cloud cores, *A&A*, **551**, 98 (2013)

Le Bourlot, J., Le Petit, F., Pinto, C., Roueff, E., Roy, F., Surface chemistry in the interstellar medium. I. H_2 formation by Langmuir-Hinshelwood and Eley-Rideal mechanisms, *A&A*, **541**, A76 (2012)

Lee, C. W., Myers, P. C., Tafalla, M., A Survey of Infall Motions toward Starless Cores. I. CS (2–1) and N_2 (1–0) Observations, *ApJ*, **526**, 788 (1999)

Lee, M.-Y., Stanimirović, S., Douglas, K. A., Knee, L. B. G., Di Francesco, J., Gibson, S. J., et al., A High-resolution Study of the H_I Transition across the Perseus Molecular Cloud, *ApJ*, **748**, 75 (2012)

Leger, A., Jura, M., Omont, A., Desorption from interstellar grains, *A&A*, **144**, 147 (1985)

Leroy, A. K., Bigiel, F., de Blok, W. J. G., Boissier, S., Bolatto, A., Brinks, E., Madore, B., Munoz-Mateos, J.-C., Murphy, E., Sandstrom, K., Schruba, A., Walter, F., Estimating the star formation rate at 1 kpc scales in nearby galaxies, *AJ*, **144**, 3 (2012)

Leroy, A., Walter, F., Bigiel, F., Brinks, E., de Blok, W. J. G., Madore, B., Star Formation in THINGS, the H_I Nearby Galaxy Survey, *ApJ*, **136**, 2782 (2008)

Leroy, A. K., Walter, F., Brinks, E., Bigiel, F., de Blok, W. J. G., Madore, B., Thornley, M. D., The star formation efficiency in nearby galaxies: Measuring where gas forms stars effectively, *AJ*, **136**, 2782 (2008)

Leroy, A. K., Walter, F., Sandstrom, K., Schruba, A., Munoz-Mateos, J.-C., Bigiel, F., Bolatto, A., Brinks, E., de Blok, W. J. G., Meidt, S., Rix, H.-W., Rosolowsky, E., Schinnerer, E., Schuster, K.-F., Usero, A., Molecular gas and star formation in nearby disk galaxies, *AJ*, **146**, 19 (2013)

Lesieur, M., Turbulence in Fluids, Kluwer Academic Publishers, Dordrecht (1997)

Leung, C. M., Radiation transport in dense interstellar dust clouds. I - Grain temperature, *ApJ*, **199**, 340 (1975)

Levrier, F., Le Petit, F., Hennebelle, P., Lesaffre, P., Gerin, M., Falgarone, E., UV-driven chemistry in simulations of the interstellar medium. I. Post-processed chemistry with the Meudon PDR code, *A&A*, **544**, A22 (2012)

Li, Z.-Y., Banerjee, R., Pudritz, R. E., Jørgensen, J. K., Shang, H., Krasnopolsky, R., Maury, A., The Earliest Stages of Star and Planet Formation: Core Collapse, and the Formation of Disks and Outflows. In: Protostars and Planets VI, edited by Beuther, H., Klessen, R. S., Dullemond, C. P., Henning, Th., University of Arizona Press, **149**, (2014)

Li, Y., Klessen, R. S., Mac Low, M.-M., The Formation of Stellar Clusters in Turbulent Molecular Clouds: Effects of the Equation of State, *ApJ*, **592**, 975 (2003)

Li, Y., Mac Low, M.-M., Klessen, R. S., Star Formation in Isolated Disk Galaxies. I. Models and Characteristics of Nonlinear Gravitational Collapse, *ApJ*, **626**, 823 (2005)

Li, Y., Mac Low, M.-M., Klessen, R. S., Star Formation in Isolated Disk Galaxies. I. Schmidt Laws and Efficiency of Gravitational Collapse, *ApJ*, **639**, 879 (2006)

Li, P. S., Myers, A., McKee, C. F., Ambipolar Diffusion Heating in Turbulent Systems, *ApJ*, **760**, 33 (2012)

Li, Z.-Y., Nakamura, F., Magnetically Regulated Star Formation in Turbulent Clouds, *ApJ*, **609**, L83 (2004)

Li, Z.-Y., Nakamura, F., Cluster formation in protostellar outflow-driven turbulence, *ApJ*, **640**, L187 (2006)

Lin, C. C., Shu, F. H., On the Spiral Structure of Disk Galaxies, *ApJ*, **140**, 646 (1964)

Lin, C. C., Yuan, C., Shu, F. H., On the Spiral Structure of Disk Galaxies. III. Comparison with Observations, *ApJ*, **155**, 721 (1969)

Linsky, J. L., Atomic deuterium/hydrogen in the galaxy, *Space Science Reviews*, **106**, 49 (2003)

Linsky, J. L., Draine, B. T., Moos, H. W., Jenkins, E. B., Wood, B. E., Oliveira, C., et al., What Is the Total Deuterium Abundance in the Local Galactic Disk? *ApJ*, **647**, 1106 (2006)

Linz, H., Stecklum, B., Henning, T., Hofner, P., Brandl, B., The G9.62+0.19-F hot molecular core. The infrared view on very young massive stars, *A&A*, **429**, 903 (2005)

Little, L. T., Gibb, A. G., Heaton, B. D., Ellison, B. N., Claude, S. M. X., The C_I/CO Ratio in the Molecular Cloud G:34.3+0.2, *MNRAS*, **271**, 649 (1994)

Lodders, K., Solar System Abundances and Condensation Temperatures of the Elements, *ApJ*, **591**, 1220 (2003)

Longmore, S. N., Kruijssen, J. M. D., Bastian, N., Bally, J., Rathborne, J., Testi, L., Stolte, A., Dale, J., Bressert, E., Alves, J., The formation and early evolution of young massive clusters. In: Protostars and Planets VI, edited by Beuther, H., Klessen, R. S., Dullemond, C. P., Henning, Th., University of Arizona Press, **291**, (2014)

Lubowich, D. A., Pasachoff, J. M., Balonek, T. J., Millar, T. J., Tremonti, C., Roberts, H., Galloway, R. P., Deuterium in the galactic centre as a result of recent infall of low-metallicity gas, *Nature*, **405**, 1025 (2000)

Lunttila, T., Padoan, P., Juvela, M., Nordlund, Å., The super-alfvénic model of molecular clouds: Predictions for mass-to-flux and turbulent-to-magnetic energy ratios, *ApJ*, **702**, L37 (2009)

Lynden-Bell, D., Kalnajs, A. J., On the generating mechanism of spiral structure, *MNRAS*, **157**, 1 (1972)

Maciel, W. J., Costa, R. D. D., Metallicity gradients in the Milky Way, in Chemical Abundances in the Universe (IAU Symposium 265), p. 317 (2010)

Mac Low, M., The Energy Dissipation Rate of Supersonic, Magnetohydrodynamic Turbulence in Molecular Clouds, *ApJ*, **524**, 169 (1999)

Mac Low, M.-M., Glover, S. C. O., The Abundance of Molecular Hydrogen and Its Correlation with Midplane Pressure in Galaxies: Non-equilibrium, Turbulent, Chemical Models, *ApJ*, **746**, 135 (2012)

Mac Low, M., Klessen, R. S., Control of star formation by supersonic turbulence, *Rev. Mod. Phys.*, **76**, 125 (2004)

Mac Low, M.-M., Klessen, R. S., Burkert, A., Smith, M. D., Kinetic Energy Decay Rates of Supersonic and Super-Alfvénic Turbulence in Star-Forming Clouds, *PRL*, **80**, 2754 (1998)

Machida, M. N., Binary Formation in Star-forming Clouds with Various Metallicities, *ApJ*, **682**, L1 (2008)

Maio, U., Dolag, K., Ciardi, B., Tornatore, L., Metal and molecule cooling in simulations of structure formation, *MNRAS*, **379**, 963 (2007)

Maeder, A., Meynet, G., Rotating massive stars: From first stars to gamma ray bursts, *Rev. Mod. Phys.*, **84**, 25 (2012)

Maret, S., Bergin, E. A., Lada, C. J., Using Chemistry to Unveil the Kinematics of Starless Cores: Complex Radial Motions in Barnard 68, *ApJ*, **670**, L25 (2007)

Marks, M., Kroupa, P., Inverse dynamical population synthesis. Constraining the initial conditions of young stellar clusters by studying their binary populations, *A&A*, **543**, A8 (2012)

Marsh K. A., Griffin M. J., Palmeirim P., et al., Properties of starless and prestellar cores in Taurus revealed by Herschel: SPIRE/PACS imaging, *MNRAS*, **439**, 3683 (2014)

Martos, M. A., Cox, D. P., Magnetohydrodynamic modeling of a galactic spiral arm as a combination shock and hydraulic jump, *ApJ*, **509**, 703 (1998)

Maschberger T., On the function describing the stellar initial mass function, *MNRAS*, **429**, 1725 (2013a)

Maschberger, T., On the mass function of stars growing in a flocculent medium, *MNRAS*, **436**, 1381 (2013b)

Massey, P., Massive Stars in the Local Group: Implications for Stellar Evolution and Star Formation, *ARA&A*, **41**, 15 (2003)

Mathis, J. S., Mezger, P. G., Panagia, N., Interstellar radiation field and dust temperatures in the diffuse interstellar matter and in giant molecular clouds, *A&A*, **128**, 212 (1983)

Mathis, J. S., Rumpl, W., Nordsieck, K. H., The size distribution of interstellar grains, *ApJ*, **217**, 425 (1977)

Matzner, C. D., On the Role of Massive Stars in the Support and Destruction of Giant Molecular Clouds, *ApJ*, **566**, 302 (2002)

Matzner, C. D., McKee, C. F., Efficiencies of Low-Mass Star and Star Cluster Formation, *ApJ*, **545**, 364 (2000)

McCall, B. J., et al., Observations of H_3^+ in the Diffuse Interstellar Medium, *ApJ*, **567**, 391 (2002)

McCall, B. J., Huneycutt, A. J., Saykally, R. J., Djuric, N., Dunn, G. H., Semaniak, J., et al., Dissociative recombination of rotationally cold H_3^+, *Phys. Rev. A*, **70**, 052716 (2004)

McCaughrean, M., The Trapezium Cluster: A Laboratory for Star Formation, in: From Darkness to Light: Origin and Evolution of Young Stellar Clusters, vol. 243 of Astronomical Society of the Pacific Conference Series, edited by T. Montmerle, P. André, p. 449 (2001)

McElroy, D., Walsh, C., Markwick, A. J., Cordiner, M. A., Smith, K., Millar, T. J., The UMIST database for astrochemistry 2012, *A&A*, **550**, 36 (2013)

McKee, C. F., Photoionization-regulated star formation and the structure of molecular clouds, *ApJ*, **345**, 782 (1989)

McKee, C. F., Krumholz, M. R., The Atomic-to-Molecular Transition in Galaxies. III. A New Method for Determining the Molecular Content of Primordial and Dusty Clouds, *ApJ*, **709**, 308 (2010)

McKee, C. F., Ostriker, E. C., Theory of Star Formation, *ARA&A*, **45**, 565 (2007)

McKee, C. F., Ostriker, J. P., A theory of the interstellar medium - Three components regulated by supernova explosions in an inhomogeneous substrate, *ApJ*, **218**, 148 (1977)

McKee, C. F., Williams, J. P., The Luminosity Function of OB Associations in the Galaxy, *ApJ*, **476**, 144 (1997)

McKee C. F., Zweibel E. G., On the virial theorem for turbulent molecular clouds, *ApJ*, **399**, 551 (1992)

Mellon, R. R., Li, Z.-Y., Magnetic Braking and Protostellar Disk Formation: Ambipolar Diffusion, *ApJ*, **698**, 922 (2009)

Menten, K. M., Reid, M. J., Forbrich, J., Brunthaler, A., The distance to the Orion Nebula, *A&A*, **474**, 515 (2007)

Mestel L., Spitzer L., Star formation in magnetic dust clouds, *MNRAS*, **116**, 503 (1956)

Meyerdierks, H., Heithausen, A., Reif, K., The North Celestial Pole Loop, *A&A*, **245**, 247 (1991)

Meynet, G., Physics of rotation in stellar models, The Rotation of Sun and Stars, *Lecture Notes in Physics*, **765**, 139 (2009)

Micic, M., Glover, S. C. O., Federrath, C., Klessen, R. S., Modelling H_2 formation in the turbulent interstellar medium: solenoidal versus compressive turbulent forcing, *MNRAS*, **421**, 2531 (2012)

Micic, M., Glover, S. C. O., Banerjee, R., Klessen, R. S., Cloud formation in colliding flows: influence of the choice of cooling function, *MNRAS*, **432**, 626 (2013)

Mierkiewicz, E. J., Reynolds, R. J., Roesler, F. L., Harlander, J. M., Jaehnig, K. P., Detection of Diffuse Interstellar [O_{II}] Emission from the Milky Way Using Spatial Heterodyne Spectroscopy, *ApJ*, **650**, L63 (2006)

Miller, G. E., Scalo, J., The Initial Mass Function and the Stellar Birthrate in the Solar Neighborhood, *ApJS*, **41**, 513 (1979)

Moffat, A. F. J., Corcoran, M. F., Stevens, I. R., Skalkowski, G., Marchenko, S. V., Mücke, A., Ptak, A., Koribalski, B. S., Brenneman, L., Mushotzky, R., Pittard, J. M., Pollock, A. M. T., Brandner, W., Galactic Starburst NGC 3603 from X-Rays to Radio, *ApJ*, **573**, 191 (2002)

Molina, F. Z., Glover, S. C. O., Federrath, C., Klessen, R. S., The density variance-Mach number relation in supersonic turbulence - I. Isothermal, magnetized gas, *MNRAS*, **423**, 2680 (2012)

Molinari, S., et al., The Milky Way as a Star Formation Engine. In: Protostars and Planets VI, edited by Beuther, H., Klessen, R. S., Dullemond, C. P., Henning, Th., University of Arizona Press, **125**, (2014)

Morris, M., Serabyn, E., The Galactic Center Environment, *ARA&A*, **34**, 645 (1996)

Motte, F., Andre, P., Neri, R., The initial conditions of star formation in the rho Ophiuchi main cloud: wide-field millimeter continuum mapping, *A&A*, **336**, 150 (1998)

Motte, F., Bontemps, S., Schneider, N., Schilke, P., Menten, K. M., Massive Infrared-Quiet Dense Cores: Unveiling the Initial Conditions of High-Mass Star Formation, in: Massive Star Formation: Observations Confront Theory, vol. 387 of Astronomical Society of the Pacific Conference Series, edited by H. Beuther, H. Linz, T. Henning, p. 22 (2008)

Motte, F., Nguyen Luong, Q., Schneider, N., Heitsch, F., Glover, S., Carlhoff, P., Hill, T., Bontemps, S., Schilke, P., Louvet, F., Hennemann, M., Didelon, P., Beuther, H., The formation of the W43

complex: constraining its atomic-to-molecular transition and searching for colliding clouds, *A&A*, **571**, A35 (2014)
Mouschovias T. C., Static Equilibria of the Interstellar Gas in the Presence of Magnetic and Gravitational Fields: Large-Scale Condensations, *ApJ*, **192**, 37 (1974)
Mouschovias, T. C., Nonhomologous contraction and equilibria of self-gravitating, magnetic interstellar clouds embedded in an intercloud medium: Star formation. II - Results, *ApJ*, **207**, 141 (1976)
Mouschovias, T. C., Ambipolar diffusion in interstellar clouds - A new solution, *ApJ*, **228**, 475 (1979)
Mouschovias, T. C., Cosmic Magnetism and the Basic Physics of the Early Stages of Star Formation, in: NATO ASIC Proc. 342: The Physics of Star Formation and Early Stellar Evolution, edited by C. J. Lada, N. D. Kylafis, p. 61 (1991a)
Mouschovias, T. C., Single-Stage Fragmentation and a Modern Theory of Star Formation, in: NATO ASIC Proc. 342: The Physics of Star Formation and Early Stellar Evolution, edited by C. J. Lada, N. D. Kylafis, p. 449 (1991b)
Mouschovias, T. Ch., Kunz, M. W., Christie, D. A., Formation of interstellar clouds: Parker instability with phase transitions, *MNRAS*, **397**, 14 (2009)
Mouschovias, T. C., Paleologou, E. V., Ambipolar diffusion in interstellar clouds - Time-dependent solutions in one spatial dimension, *ApJ*, **246**, 48 (1981)
Mouschovias T. C., Shu, F. H., Woodward, P. R., On the Formation of Interstellar Cloud Complexes, OB Associations and Giant H_{II} Regions, *A&A*, **33**, 73 (1974)
Mouschovias T. C., Spitzer L., Jr., Note on the collapse of magnetic interstellar clouds, *ApJ*, **210**, 326 (1976)
Mundt, R., Buehrke, T., Solf, J., Ray, T. P., Raga, A. C., Optical jets and outflows in the HL Tauri region, *A&A*, **232**, 37 (1990)
Mundt, R., Ray, T. P., Raga, A. C., Collimation of stellar jets - constraints from the observed spatial structure - part two - observational results, *A&A*, **252**, 740 (1991)
Myers, P. C., Dense Cores in Dark Clouds. III - Subsonic Turbulence, *ApJ*, **270**, 105 (1983)
Naab, T., Ostriker, J. P., A simple model for the evolution of disc galaxies: the milky way, *MNRAS*, **366**, 899 (2006)
Nakamura, F., Li, Z.-Y., Protostellar Turbulence Driven by Collimated Outflows, *ApJ*, **662**, 395 (2007)
Nakamura, F., Li, Z.-Y., Magnetically regulated star formation in three dimensions: The case of the Taurus molecular cloud complex, *ApJ*, **687**, 354 (2008)
Nelson, R. P., Langer, W. D., The Dynamics of Low-Mass Molecular Clouds in External Radiation Fields, *ApJ*, **482**, 796 (1997)
Nelson, R. P., Langer, W. D., On the Stability and Evolution of Isolated BOK Globules, *ApJ*, **524**, 923 (1999)
Neufeld, D. A., Kaufman, M. J., Radiative Cooling of Warm Molecular Gas, *ApJ*, **418**, 263 (1993)
Neufeld, D. A., Lepp, S., Melnick, G. J., Thermal Balance in Dense Molecular Clouds: Radiative Cooling Rates and Emission-Line Luminosities, *ApJS*, 100, 132 (1995)
Nisini, B., Benedettini, M., Codella, C., Giannini, T., Liseau, R., Neufeld, D., et al., Water cooling of shocks in protostellar outflows. Herschel-PACS map of L1157, *A&A*, 518, L120 (2010)
Norman, C. A., Ferrara, A., The turbulent interstellar medium: Generalizing to a scale-dependent phase continuum, *ApJ*, **467**, 280 (1996)
Norman, C., Silk, J., Clumpy molecular clouds - A dynamic model self-consistently regulated by T Tauri star formation, *ApJ*, **238**, 158 (1980)
Ntormousi, E., Burkert, A., Fierlinger, K., Heitsch, F., Formation of Cold Filamentary Structure from Wind-blown Superbubbles, *ApJ*, **731**, 13 (2011)
Nugis, T., Lamers, H. J. G. L. M., Mass-loss rates of wolf-rayet stars as a function of stellar parameters, *A&A*, **360**, 227 (2000)
Nutter, D., Ward-Thompson, D., A SCUBA survey of Orion - the low-mass end of the core mass function, *MNRAS*, **374**, 1413 (2007)

Oey, M. S., Clarke, C. J., Statistical Confirmation of a Stellar Upper Mass Limit, *ApJ*, **620**, L43 (2005)

Offner, S. S. R., Bisbas, T. G., Viti, S., Bell, T. A., Modeling the atomic-to-molecular transition and chemical distributions of turbulent star-forming clouds, *ApJ*, **770**, 49 (2013)

Offner, S. S. R., Clark, P. C., Hennebelle, P., Bastian, N., Bate, M. R., Hopkins, P., Moraux, E., Whitworth, A., The Origin and Universality of the Stellar Initial Mass Function. In: Protostars and Planets VI, edited by Beuther, H., Klessen, R. S., Dullemond, C. P., Henning, Th., University of Arizona Press, **53**, (2014)

Offner, S. S. R., Klein, R. I., McKee, C. F., Driven and Decaying Turbulence Simulations of Low-Mass Star Formation: From Clumps to Cores to Protostars, *ApJ*, **686**, 1174 (2008)

Øksendal, B., Stochastic Differential Equations, 5th edition, Springer Verlag, Heidelberg, New York (2000)

Onishi, T., Mizuno, A., Kawamura, A., Ogawa, H., Fukui, Y., A $C^{18}O$ Survey of Dense Cloud Cores in Taurus: Star Formation, *ApJ*, **502**, 296 (1998)

Oort, J. H., Outline of a theory on the origin and acceleration of interstellar clouds and O associations, *Bull. Astron. Inst. Netherlands*, **12**, 177 (1954)

Oppenheimer, B. D., Schaye, J., Non-equilibirum ionization and cooling of metal-enriched gas in the presence of a photoionization background, *MNRAS*, **434**, 1043 (2013)

Ossenkopf, V., The Sobolev approximation in molecular clouds, *New. Astron.* **2**, 365 (1997)

Ossenkopf, V., Molecular line emission from turbulent clouds, *A&A*, **391**, 295 (2002)

Ossenkopf, V., Henning, Th., Dust opacities for protostellar cores, *A&A*, **291**, 943 (1994)

Ossenkopf, V., Mac Low, M.-M., Turbulent velocity structure in molecular clouds, *A&A*, **390**, 307 (2002)

Osterbrock, D. E., Astrophysics of gaseous nebulae and active galactic nuclei, University Science Books (1989)

Ostriker, E. C., Stone, J. M., Gammie, C. F., Density, Velocity, and Magnetic Field Structure in Turbulent Molecular Cloud Models, *ApJ*, **546**, 980 (2001)

Ostriker, J. P., Tinsley, B. M., Is deuterium of cosmological or of galactic origin?, *ApJ*, **201**, L51 (1975)

Ouyed, R., Clarke, D. A., Pudritz, R. E., Three-dimensional Simulations of Jets from Keplerian Disks: Self-regulatory Stability, *ApJ*, **582**, 292 (2003)

Ouyed, R., Pudritz, R. E., Numerical Simulations of Astrophysical Jets from Keplerian Disks. I. Stationary Models, *ApJ*, **482**, 712 (1997)

Padoan, P., Federrath, C., Chabrier, G., Evans, N. J., Johnstone, D., Jørgensen, J. K., McKee, C. F., Nordlund, Å., The star formation rate of molecular clouds. In: Protostars and Planets VI, edited by Beuther, H., Klessen, R. S., Dullemond, C. P., Henning, Th., University of Arizona Press, **77**, (2014)

Padoan, P., Juvela, M., Goodman, A. A., Nordlund, Å., The Turbulent Shock Origin of Proto-Stellar Cores, *ApJ*, **553**, 227 (2001)

Padoan, P., Nordlund, A., Jones, B. J. T., The universality of the stellar initial mass function, *MNRAS*, **288**, 145 (1997)

Padoan, P., Nordlund, Å., A Super-Alfvénic Model of Dark Clouds, *ApJ*, **526**, 279 (1999)

Padoan, P., Nordlund, Å., The Stellar Initial Mass Function from Turbulent Fragmentation, *ApJ*, **576**, 870 (2002)

Padoan, P., Nordlund, Å., The Star Formation Rate of Supersonic Magnetohydrodynamic Turbulence, *ApJ*, **730**, 40 (2011)

Padoan, P., Nordlund, Å., Kritsuk, A. G., Norman, M. L., Li, P. S., Two Regimes of Turbulent Fragmentation and the Stellar Initial Mass Function from Primordial to Present-Day Star Formation, *ApJ*, **661**, 972 (2007)

Padoan, P., Scalo, J., Confinement-driven Spatial Variations in the Cosmic-Ray Flux, *ApJ*, **624**, L97 (2005)

Padoan, P., Zweibel, E., Nordlund, Å., Ambipolar Drift Heating in Turbulent Molecular Clouds, *ApJ*, **540**, 332 (2000)

Padovani, M., Galli, D., Glassgold, A. E., Cosmic-ray ionization of molecular clouds, *A&A*, **501**, 619 (2009)

Palla, F., Stahler, S. W., Star Formation in the Orion Nebula Cluster, *ApJ*, **525**, 772 (1999)

Pan, L., Padoan, P., The Temperature of Interstellar Clouds from Turbulent Heating, *ApJ*, **692**, 594 (2009)

Papadopoulos, P. P., A Cosmic-ray-dominated Interstellar Medium in Ultra Luminous Infrared Galaxies: New Initial Conditions for Star Formation, *ApJ*, **720**, 226 (2010)

Parker, E. N., The Dynamical State of the Interstellar Gas and Field, *ApJ*, **145**, 811 (1966)

Parravano, A., Hollenbach, D. J., McKee, C. F., Time Dependence of the Ultraviolet Radiation Field in the Local Interstellar Medium, *ApJ*, **584**, 797 (2003)

Passot, T., Vázquez-Semadeni, E., Density probability distribution in one-dimensional polytropic gas dynamics, *PRE*, **58**, 4501 (1998)

Pauldrach, A. W. A., Puls, J., Radiation-driven winds of hot stars. VIII - The bistable wind of the luminous blue variable P Cygni (B1 Ia/+/), *A&A*, **237**, 409 (1990)

Pavlyuchenkov, Y., Wiebe, D., Launhardt, R., Henning, T., CB 17: Inferring the Dynamical History of a Prestellar Core with Chemodynamical Models, *ApJ*, **645**, 1212 (2006)

Peek, J. E. G., Hitting the bull's-eye: The radial profile of accretion and star formation in the Milky Way, *ApJ*, **698**, 1429 (2009)

Persi, P., Marenzi, A. R., Olofsson, G., Kaas, A. A., Nordh, L., Huldtgren, M., Abergel, A., André, P., Bontemps, S., Boulanger, F., Burggdorf, M., Casali, M. M., Cesarsky, C. J., Copet, E., Davies, J., Falgarone, E., Montmerle, T., Perault, M., Prusti, T., Puget, J. L., Sibille, F., ISOCAM observations of the Chamaeleon I dark cloud, *A&A*, **357**, 219 (2000)

Peters, T., Banerjee, R., Klessen, R. S., Mac Low, M., The Interplay of Magnetic Fields, Fragmentation, and Ionization Feedback in High-mass Star Formation, *ApJ*, **729**, 72 (2011)

Peters, T., Banerjee, R., Klessen, R. S., Mac Low, M.-M., Galván-Madrid, R., Keto, E. R., H_{II} regions: Witnesses to massive star formation, *ApJ*, **711**, 1017 (2010a)

Peters, T., Klessen, R. S., Mac Low, M., Banerjee, R., Limiting Accretion onto Massive Stars by Fragmentation-induced Starvation, *ApJ*, **725**, 134 (2010b)

Peters, T., Mac Low, M.-M., Banerjee, R., Klessen, R. S., Dullemond, C. P., Understanding spatial and spectral morphologies of ultracompact H_{II} regions, *ApJ*, **719**, 831 (2010c)

Pety J., Schinnerer E., Leroy A. K., Hughes A., Meidt S. E., Colombo D., Dumas G., García-Burillo S., Schuster K. F., Kramer C., Dobbs C. L., Thompson T. A., The Plateau de Bure + 30 m Arcsecond Whirlpool Survey Reveals a Thick Disk of Diffuse Molecular Gas in the M51 Galaxy, *ApJ*, **779**, 43 (2013)

Pflamm-Altenburg J., Kroupa P., A highly abnormal massive star mass function in the Orion Nebula cluster and the dynamical decay of trapezium systems, *MNRAS*, **373**, 295 (2006)

Phillips, J. P., Rotation in molecular clouds, *A&AS*, **134**, 241 (1999)

Pineda, J. L., Langer, W. D., Velusamy, T., Goldsmith, P. F., A Herschel [C_{II}] Galactic plane survey. I. The global distribution of ISM gas components, *A&A*, **554**, 103 (2013)

Piontek, R. A., Ostriker, E. C., Thermal and Magnetorotational Instability in the Interstellar Medium: Two-dimensional Numerical Simulations, *ApJ*, **601**, 905 (2004)

Piontek, R. A., Ostriker, E. C., Saturated-state turbulence and structure from thermal and magnetorotational instability in the ISM: Three-dimensional numerical simulations, *ApJ*, **629**, 849 (2005)

Planck Collaboration; Abergel, A., Ade, P. A. R., Aghanim, N., Alina, D., Alves, M. I. R., Aniano, G., et al., Planck intermediate results. XVII. Emission of dust in the diffuse interstellar medium from the far-infrared to microwave frequencies, *A&A*, **566**, A55 (2014)

Pon, A., Johnstone, D., Kaufman, M. J., Molecular Tracers of Turbulent Shocks in Giant Molecular Clouds, *ApJ*, **748**, 25 (2012)

Pope, S. B., Turbulent Flows, Cambridge University Press (2000)

Poppel, W., The Gould Belt System and the Local Interstellar Medium, Fundamentals of Cosmic Physics, **18**, 1 (1997)

Portegies Zwart, S. F., McMillan, S. L. W., Gieles, M., Young Massive Star Clusters, *ARA&A*, **48**, 431 (2010)

Prasad, S. S., Tarafdar, S. P., UV radiation field inside dense clouds - Its possible existence and chemical implications, *ApJ*, **267**, 603 (1983)

Preibisch, T., Zinnecker, H., The History of Low-Mass Star Formation in the Upper Scorpius OB Association, *AJ*, **117**, 2381 (1999)

Press, W. H., Schechter, P., Formation of Galaxies and Clusters of Galaxies by Self-Similar Gravitational Condensation, *ApJ*, **187**, 425 (1974)

Price, D. J., Bate, M. R., The effect of magnetic fields on the formation of circumstellar discs around young stars, *Astrophys. Space Sci.*, **311**, 75 (2007a)

Price, D. J., Bate, M. R., The impact of magnetic fields on single and binary star formation, *MNRAS*, **377**, 77 (2007b)

Price, D. J., Bate, M. R., The effect of magnetic fields on star cluster formation, *MNRAS*, **385**, 1820 (2008)

Pringle, J. E., Allen, R. J., Lubow, S. H., The Formation of Molecular Clouds, *MNRAS*, **327**, 663 (2001)

Prochaska, J. X., Wolfe, A. M., On the (non)evolution of H_I gas in galaxies over cosmic time, *ApJ*, **696**, 1543 (2009)

Prodanović, T., Steigman, G., Fields, B. D., The deuterium abundance in the local interstellar medium, *MNRAS*, **406**, 1108 (2010)

Pudritz, R. E., Ouyed, R., Fendt, C., Brandenburg, A., Disk Winds, Jets, and Outflows: Theoretical and Computational Foundations, in: Protostars and Planets V, edited by B. Reipurth, D. Jewitt, K. Keil, p. 277 (2007)

Puls, J., Kudritzki, R.-P., Herrero, A., Pauldrach, A. W. A., Haser, S. M., Lennon, D. J., Gabler, R., Voels, S. A., Vilchez, J. M., Wachter, S., Feldmeier, A., O-star mass-loss and wind momentum rates in the galaxy and the magellanic clouds observations and theoretical predictions., *A&A*, **305**, 171 (1996)

Putman, M. E., Potential condensed fuel for the Milky Way, *ApJ*, **645**, 1164 (2006)

Rafikov, R. R., The local axisymmetric instability criterion in a thin, rotating, multicomponent disc, *MNRAS*, **323**, 445 (2001)

Ragan S., Henning T., Krause O., et al., The Earliest Phases of Star Formation (EPoS): a Herschel key program. The precursors to high-mass stars and clusters, *A&A*, **547**, A49 (2012)

Ragan S. E., Henning T., Beuther H., APEX/SABOCA observations of small-scale structure of infrared-dark clouds. I. Early evolutionary stages of star-forming cores, *A&A*, **559**, A79 (2013)

Rauw, G., Crowther, P. A., De Becker, M., Gosset, E., Nazé, Y., Sana, H., van der Hucht, K. A., Vreux, J.-M., Williams, P. M., The spectrum of the very massive binary system WR 20a (WN6ha + WN6ha): Fundamental parameters and wind interactions, *A&A*, **432**, 985 (2005)

Rees, M. J., Opacity-limited hierarchical fragmentation and the masses of protostars, *MNRAS*, **176**, 483 (1976)

Reipurth B., Clarke C. J., Boss A. P., Goodwin S. P., Rodriguez L. F., Stassun K. G., Tokovinin A., Zinnecker H., Multiplicity in Early Stellar Evolution. In: Protostars and Planets VI, edited by Beuther, H., Klessen, R. S., Dullemond, C. P., Henning, Th., University of Arizona Press, **267**, (2014)

Reynolds, R. J., The column density and scale height of free electrons in the galactic disk, *ApJ*, **339**, L29 (1989)

Reynolds, R. J., Scherb, F., Roesler, F. L., Observations of Diffuse Galactic Hα] emission, *ApJ*, **185**, 869 (1973)

Richardson, L. F., The Supply of Energy from and to Atmospheric Eddies, *Royal Society of London Proceedings Series A*, **97**, 354 (1920)

Richings, A. J., Schaye, J., Oppenheimer, B. D., Non-equilibrium chemistry and cooling in the diffuse interstellar medium I: Optically thin regime, *MNRAS*, **440**, 3349 (2014)

Rix, H.-W., Bovy, J., The Milky Way's Stellar Disk. Mapping and Modeling the Galactic Disk, *A&A Rev.*, **21**, 61 (2013)

Roberts, W. W., Large-Scale Shock Formation in Spiral Galaxies and its Implications on Star Formation, *ApJ*, **158**, 123 (1969)
Rolleston, W. R. J., Smartt, S. J., Dufton, P. L., Ryans, R. S. I., The Galactic metallicity gradient, *A&A*, **363**, 537 (2000)
Roman-Duval, J., Federrath, C., Brunt, C., Heyer, M., Jackson, J., Klessen, R. S., The turbulence spectrum of molecular clouds in the galactic ring survey: A density-dependent principal component analysis calibration, *ApJ*, **740**, 120 (2011)
Roser, J. E., Swords, S., Vidali, G., Manicò, G., Pirronello, V., Measurement of the kinetic energy of hydrogen molecules from amorphous water ice, *ApJ*, **596**, L55 (2003)
Rosolowsky, E. W., Pineda, J. E., Foster, J. B., Borkin, M. A., Kauffmann, J., Caselli, P., Myers, P. C., Goodman, A. A., An Ammonia Spectral Atlas of Dense Cores in Perseus, *ApJS*, **175**, 509 (2008)
Roueff, E., Zeippen, C. J., Rotational excitation of HD molecules by He atoms, *A&A*, **343**, 1005 (1999)
Roy, N., Kanekar, N., Chengalur, J. N., The temperature of the diffuse H_I 21 cm absorption spectra, *MNRAS*, **436**, 2366 (2013)
Rudolph, A. L., Fich, M., Bell, G. R., Norsen, T., Simpson, J. P., Haas, M. R., Erickson, E. F., Abundance Gradients in the Galaxy, *ApJS*, **162**, 346 (2006)
Ruffert, M., Arnett, D., Three-Dimensional Hydrodynamic Bondi-Hoyle Accretion. 2: Homogeneous Medium at mach 3 with γ =5/3, *ApJ*, **427**, 351 (1994)
Rybicki, G. B., Lightman, A. P., Radiative Processes in Astrophysics, Wiley-VCH (1986)
Salpeter, E. E., The Luminosity Function and Stellar Evolution., *ApJ*, **121**, 161 (1955)
Sandstrom, K. M., Peek, J. E. G., Bower, G. C., Bolatto, A. D., Plambeck, R. L., A Parallactic Distance of 389^{+24}_{-21} Parsecs to the Orion Nebula Cluster from Very Long Baseline Array Observations, *ApJ*, **667**, 1161 (2007)
Savage, B. D., Bohlin, R. C., Drake, J. F., Budich, W., A survey of interstellar molecular hydrogen, *ApJ*, **216**, 291 (1977)
Savage, B. D., Sembach, K. R., Interstellar Abundances from Absorption-Line Observations with the Hubble Space Telescope, *ARA&A*, **34**, 279 (1996)
Scalo, J., The Stellar Initial Mass Function, *Fund. Cosm. Phys.*, **11**, 1 (1986)
Scalo, J., Elmegreen, B. G., Interstellar Turbulence II: Implications and Effects, *ARA&A*, **42**, 275 (2004)
Schekochihin, A. A., Cowley, S. C., Maron, J. L., McWilliams, J. C., Critical magnetic prandtl number for small-scale dynamo, *PRL*, **92**, 54502 (2004a)
Schekochihin, A. A., Cowley, S. C., Taylor, S. F., Maron, J. L., McWilliams, J. C., Simulations of the small-scale turbulent dynamo, *ApJ*, **612**, 276 (2004b)
Schilke, P., Keene, J., Le Bourlot, J., Pineau des Forêts, G., Roueff, E., Atomic carbon in a dark cloud: TMC-1, *A&A*, **294**, L17 (1995)
Schmeja, S., Klessen, R. S., Protostellar mass accretion rates from gravoturbulent fragmentation, *A&A*, **419**, 405 (2004)
Schmidt, W., Federrath, C., Hupp, M., Kern, S., Niemeyer, J. C., Numerical simulations of compressively driven interstellar turbulence. I. Isothermal gas, *A&A*, **494**, 127 (2009)
Schneider, R., Omukai, K., Metals, dust and the cosmic microwave background: fragmentation of high-redshift star-forming clouds, *MNRAS*, **402**, 429 (2010)
Schneider N., Csengeri T., Hennemann M., et al., Cluster-formation in the Rosette molecular cloud at the junctions of filaments, *A&A*, **540**, L11 (2012)
Schneider, N., et al., What Determines the Density Structure of Molecular Clouds? A Case Study of Orion B with Herschel, *ApJ*, **766**, L17 (2013)
Schneider N., Ossenkopf V., Csengeri T., et al., Understanding star formation in molecular clouds I. A universal probability distribution of column densities?, *A&A*, **575**, A79 (2015)
Schober, J., Schleicher, D., Bovino, S., Klessen, R. S., Small-scale dynamo at low magnetic Prandtl numbers, *PRE*, **86**, 66412 (2012a)

Schober, J., Schleicher, D., Federrath, C., Klessen, R., Banerjee, R., Magnetic field amplification by small-scale dynamo action: Dependence on turbulence models and Reynolds and Prandtl numbers, *PRE*, **85**, 26303 (2012b)

Schöier, F. L., van der Tak, F. F. S., van Dishoeck, E. F., Black, J. H., An atomic and molecular database for analysis of submillimetre line observations, *A&A*, **432**, 369 (2005)

Schroder, K., Staemmler, V., Smith, M. D., Flower, D. R., Jaquet, R., Excitation of the fine-structure transitions of C in collisions with ortho- and para-H_2, *J. Phys. B*, **24**, 2487 (1991)

Schruba, A., Leroy, A. K., Walter, F., Bigiel, F., Brins, E., de Blok, W. J. G., Dumas, G., Kramer, C., Rosolowsky, E., Sandstrom, K., Schuster, K., Usero, A., Weiss, A., Wiesemeyer, H., A Molecular Star Formation Law in the Atomic-gas-dominated Regime in Nearby Galaxies, *AJ*, **142**, 37 (2011)

Seifried, D., Schmidt, W., Niemeyer, J. C., Forced turbulence in thermally bistable gas: a parameter study, *A&A*, **526**, A14 (2011)

Seifried D., Banerjee R., Klessen R. S., Duffin D., Pudritz R. E., Magnetic fields during the early stages of massive star formation - I. Accretion and disc evolution, *MNRAS*, **417**, 1054 (2011)

Seifried D., Banerjee R., Pudritz R. E., Klessen R. S., Disc formation in turbulent massive cores: circumventing the magnetic braking catastrophe, *MNRAS*, **423**, L40 (2012a)

Seifried D., Pudritz R. E., Banerjee R., Duffin D., Klessen R. S., Magnetic fields during the early stages of massive star formation - II. A generalized outflow criterion, *MNRAS*, **422**, 347 (2012b)

Seifried D., Banerjee R., Pudritz R. E., Klessen R. S., Turbulence-induced disc formation in strongly magnetized cloud cores, *MNRAS*, **432**, 3320 (2013)

Sellwood, J. A., Balbus, S. A., Differential rotation and turbulence in extended H_I disks, *ApJ*, **511**, 660 (1999)

Sembach, K. R., Howk, J. C., Ryans, R. S. I., Keenan, F. P., Modeling the Warm Ionized Interstellar Medium and Its Impact on Elemental Abundance Studies, *ApJ*, **528**, 310 (2000)

Selman, F. J., Melnick, J., The Scale-Free Character of the Cluster Mass Function and the Universality of the Stellar Initial Mass Function, *ApJ*, **689**, 816 (2008)

Sheffer, Y., Rogers, M., Federman, S. R., Abel, N. P., Gredel, R., Lambert, D. L., Shaw, G., Ultraviolet Survey of CO and H_2 in Diffuse Molecular Clouds: The Reflection of Two Photochemistry Regimes in Abundance Relationships, *ApJ*, **687**, 1075 (2008)

Shetty, R., Beaumont, C. N., Burton, M. G., Kelly, B. C., Klessen, R. S., The linewidth-size relationship in the dense interstellar medium of the central molecular zone, *MNRAS*, **425**, 720 (2012)

Shetty R., Clark P. C., Klessen R. S., Interpreting the sub-linear Kennicutt-Schmidt relationship: the case for diffuse molecular gas, *MNRAS*, **442**, 2208 (2014)

Shetty, R., Kauffmann, J., Schnee, S., Goodman, A. A., The Effect of Noise on the Dust Temperature-Spectral Index Correlation, *ApJ*, **696**, 676 (2009a)

Shetty, R., Kauffmann, J., Schnee, S., Goodman, A. A., Ercolano, B., The Effect of Line-of-Sight Temperature Variation and Noise on Dust Continuum Observations, *ApJ*, **696**, 2234 (2009b)

Shetty, R., Kelly, B. C., Bigiel, F., Evidence for a non-universal Kennicutt-Schmidt relationship using hierarchical Bayesian linear regression, *MNRAS*, **430**, 288 (2013)

Shetty, R., Kelly, B. C., Rahman, N., Bigiel, F., Bolatto, A. D., Clark, P. C., Klessen, R. S., Konstandin, L. K., Indications of a sub-linear and non-universal Kennicutt-Schmidt relationship, *MNRAS*, **437**, L61 (2014)

Shu, F., Najita, J., Ostriker, E., Wilkin, F., Ruden, S., Lizano, S., Magnetocentrifugally driven flows from young stars and disks. 1: A generalized model, *ApJ*, **429**, 781 (1994)

Shu, F. H., Adams, F. C., Lizano, S., Star formation in molecular clouds - Observation and theory, *ARA&A*, **25**, 23 (1987)

Shu, F. H., Najita, J. R., Shang, H., Li, Z.-Y., X-Winds Theory and Observations, Protostars and Planets IV, edited by V. Mannings, A. P. Boss, S. S. Russell, p. 789 (2000)

Simon, R., Schneider, N., Stutzki, J., Güsten, R., Graf, U. U., Hartogh, P., Guan, X., Staguhn, J. G., Benford, D. J., SOFIA observations of S106: dynamics of the warm gas, *A&A*, **542**, L12 (2012)

Smith, M. C., Ruchti, G. R., Helmi, A., Wyse, R. F. G., Fulbright, J. P., Freeman, K. C., Navarro, J. F., Seabroke, G. M., Steinmetz, M., Williams, M., Bienaymé, O., Binney, J., Bland-Hawthorn, J., Dehnen, W., Gibson, B. K., Gilmore, G., Grebel, E. K., Munari, U., Parker, Q. A., Scholz, R.-D.,

Siebert, A., Watson, F. G., Zwitter, T., The RAVE survey: constraining the local galactic escape speed, *MNRAS*, **379**, 755 (2007)

Smith, B. D., Turk, M. J., Sigurdsson, S., O'Shea, B. W., Norman, M. L., Three Modes of Metal-Enriched Star Formation in the Early Universe, *ApJ*, **691**, 441 (2009)

Smith, R. J., Clark, P. C., Bonnell, I. A., The structure of molecular clouds and the universality of the clump mass function, *MNRAS*, **391**, 1091 (2008)

Smith R. J., Glover S. C. O., Clark P. C., Klessen R. S., Springel V., CO-dark gas and molecular filaments in Milky Way-type galaxies, *MNRAS*, **441**, 1628 (2014)

Smith R. J., Shetty R., Beuther H., Klessen R. S., Bonnell I. A., Line Profiles of Cores within Clusters. II. Signatures of Dynamical Collapse during High-mass Star Formation, *ApJ*, **771**, 24 (2013)

Smith R. J., Shetty R., Stutz A. M., Klessen R. S., Line Profiles of Cores within Clusters. I. The Anatomy of a Filament, *ApJ*, **750**, 64 (2012)

Snell, R. L., Howe, J. E., Ashby, M. L. N., Bergin, E. B., Chin, G., Erickson, N. R., et al., Submillimeter Wave Astronomy Satellite Observations of Extended Water Emission in Orion, *ApJ*, **539**, L93 (2000)

Snow, T. P., McCall, B. J., Diffuse Atomic and Molecular Clouds, *ARA&A*, **44**, 367 (2006)

Snowden, S. L., Egger, R., Freyberg, M. J., McCammon, D., Plucinsky, P. P., Sanders, W. T., Schmitt, J. H. M. M., Trümper, J., Voges, W., ROSAT Survey Diffuse X-Ray Background Maps. II, *ApJ*, **485**, 125 (1997)

Sobolev, V. V., The Diffusion of Lα Radiation in Nebulae and Stellar Envelopes, *Sov. Astron.*, **1**, 678 (1957)

Sofia, U. J., Interstellar Abundances and Depletions, in: Astrophysics of Dust (ASP Conf. Series, Vol. 309), edited by Witt, A. N., Clayton, G. C., Draine, B. T., p. 393 (2004)

Solomon, P. M., Rivolo, A. R., Barrett, J., Yahil, A., Mass, luminosity, and line width relations of Galactic molecular clouds, *ApJ*, **319**, 730 (1987)

Spitzer, L., Physical Processes in the Interstellar Medium, Wiley-Interscience, New York (1978)

Springel, V., Smoothed particle hydrodynamics in astrophysics, *ARA&A*, **48**, 391 (2010)

Stacy A., Bromm V., Constraining the statistics of Population III binaries, *MNRAS*, **433**, 1094 (2013)

Stamatellos, D., Whitworth, A. P., Ward-Thompson, D., The dust temperatures of the pre-stellar cores in the ρ Oph main cloud and in other star-forming regions: consequences for the core mass function, *MNRAS*, **379**, 1390 (2007)

Stanke, T., McCaughrean, M. J., Zinnecker, H., An unbiased H_2 survey for protostellar jets in Orion A. II. The infrared survey data, *A&A*, **392**, 239 (2002)

Stecher, T. P., Williams, D. A., Photodestruction of Hydrogen Molecules in H$_{\rm I}$ Regions, *ApJ*, **149**, L29 (1967)

Stephens, T. L., Dalgarno, A., Kinetic Energy in the Spontaneous Radiative Dissociation of Molecular Hydrogen, *ApJ*, **186**, 165 (1973)

Sternberg, A., Dalgarno, A., Chemistry in Dense Photon-dominated Regions, *ApJS*, **99**, 565 (1995)

Stone, J. M., Norman, M. L., ZEUS-2D: A radiation magnetohydrodynamics code for astrophysical flows in two space dimensions. I - The hydrodynamic algorithms and tests, *ApJS*, **80**, 753 (1992a)

Stone, J. M., Norman, M. L., ZEUS-2D: A Radiation Magnetohydrodynamics Code for Astrophysical Flows in Two Space Dimensions. II. The Magnetohydrodynamic Algorithms and Tests, *ApJS*, **80**, 791 (1992b)

Stone, J. M., Ostriker, E. C., Gammie, C. F., Dissipation in Compressible Magnetohydrodynamic Turbulence, *ApJ*, **508**, L99 (1998)

Stutzki, J., Stacey, G. J., Genzel, R., Harris, A. I., Jaffe, D. T., Lugten, J. B., Submillimeter and far-infrared line observations of M17 SW - A clumpy molecular cloud penetrated by ultraviolet radiation, *ApJ*, **332**, 379 (1988)

Sugitani, K., Nakamura, F., Tamura, M., Watanabe, M., Kandori, R., Nishiyama, S., Kusakabe, N., Hashimoto, J., Nagata, T., Sato, S., Near-infrared Imaging Polarimetry of the Serpens Cloud Core: Magnetic Field Structure, Outflows, and Inflows in a Cluster Forming Clump, *ApJ*, **716**, 299 (2010)

Sutherland, R. S., Dopita, M. A., Cooling functions for low-density astrophysical plasmas, *ApJS*, **88**, 253 (1993)

Szűcs L., Glover S. C. O., Klessen R. S., The ^{12}CO ratio in turbulent molecular clouds, *MNRAS*, **445**, 4055 (2014)

Tafalla, M., Myers, P. C., Caselli, P., Walmsley, C. M., Comito, C., Systematic Molecular Differentiation in Starless Cores, *ApJ*, **569**, 815 (2002)

Tafalla, M., Santiago-García, J., Myers, P. C., Caselli, P., Walmsley, C. M., Crapsi, A., On the internal structure of starless cores. II. A molecular survey of L1498 and L1517B, *A&A*, **455**, 577 (2006)

Tamburro, D., Rix, H.-W., Leroy, A. K., Low, M.-M. M., Walter, F., Kennicutt, R. C., Brinks, E., de Blok, W. J. G., What is driving the H_I velocity dispersion? *AJ*, **137**, 4424 (2009)

Tasker, E. J., Tan, J. C., Star Formation in Disk Galaxies. I. Formation and Evolution of Giant Molecular Clouds via Gravitational Instability and Cloud Collisions, *ApJ*, **700**, 358 (2009)

Testi, L., Palla, F., Prusti, T., Natta, A., Maltagliati, S., A search for clustering around Herbig Ae/Be stars, *A&A*, **320**, 159 (1997)

Testi, L., Sargent, A. I., Star Formation in Clusters: A Survey of Compact Millimeter-Wave Sources in the Serpens Core, *ApJ*, **508**, L91 (1998)

Tegmark, M., Silk, J., Rees, M. J., Blanchard, A., Abel, T., Palla, F., How Small Were the First Cosmological Objects? *ApJ*, **474**, 1 (1997)

Thompson, R., Nagamine, K., Jaacks, J., Choi, J.-H., Molecular Hydrogen Regulated Star Formation in Cosmological Smoothed Particle Hydrodynamics Simulations, *ApJ*, **780**, 145 (2014)

Thornton, K., Gaudlitz, M., Janka, H.-T., Steinmetz, M., Energy input and mass redistribution by supernovae in the interstellar medium, *ApJ*, **500**, 95 (1998)

Tielens, A. G. G. M., The Physics and Chemistry of the Interstellar Medium, Cambridge University Press (2010)

Tielens, A. G. G. M., Hollenbach, D., Photodissociation regions. I. Basic model, *ApJ*, **291**, 722 (1985)

Tomisaka, K., Coagulation of interstellar clouds in spiral gravitational potential and formation of giant molecular clouds, *PASJ*, **36**, 457 (1984)

Toomre, A., On the gravitational stability of a disk of stars, *ApJ*, **139**, 1217 (1964)

Tóth, L. V., Haas, M., Lemke, D., Mattila, K., Onishi, T., Very cold cores in the Taurus Molecular Ring as seen by ISO, *A&A*, **420**, 533 (2004)

Townsley, L. K., Broos, P. S., Feigelson, E. D., Garmire, G. P., Getman, K. V., A Chandra ACIS Study of 30 Doradus. II. X-Ray Point Sources in the Massive Star Cluster R136 and Beyond, *AJ*, **131**, 2164 (2006)

Troland, T. H., Crutcher, R. M., Goodman, A. A., Heiles, C., Kazes, I., Myers, P. C., The Magnetic Fields in the Ophiuchus and Taurus Molecular Clouds, *ApJ*, **471**, 302 (1996)

van der Tak, F. F. S., van Dishoeck, E. F., Limits on the cosmic-ray ionization rate toward massive young stars, *A&A*, **358**, L79 (2000)

van der Werf, P., H_2 Emission as a Diagnostic of Physical Processes in Starforming Galaxies, in: Molecular Hydrogen in Space, edited by F. Combes, G. Pineau Des Forets, p. 307 (2000)

van Dishoeck, E. F., Photodissociation processes of astrophysical molecules, in *Astrochemistry*, IAU Symposium, vol. 120 (D. Reidel, Dordrecht), p. 51 (1987)

van Dishoeck, E. F., Black, J. H., The photodissociation and chemistry of interstellar CO, *ApJ*, **334**, 771 (1988)

van Weeren, R. J., Brinch, C., Hogerheijde, M. R., Modeling the chemical evolution of a collapsing prestellar core in two spatial dimensions, *A&A*, **497**, 773 (2009)

van Zadelhoff, G. J., Dullemond, C. P., van der Tak, F. F. S., Yates, J. A., Doty, S. D., Ossenkopf, V., et al., Numerical methods for non-LTE line radiative transfer: Performance and convergence characteristics, *A&A*, **395**, 373 (2002)

van Zee, L., Bryant, J., Neutral gas distribution and kinematics of the nearly face-on spiral galaxy NGC 1232, *AJ*, **118**, 2172 (1999)

Vázquez-Semadeni, E., Hierarchical Structure in Nearly Pressureless Flows as a Consequence of Self-similar Statistics, *ApJ*, **423**, 681 (1994)

Vázquez-Semadeni, E., Ballesteros-Paredes, J., Klessen, R. S., A Holistic Scenario of Turbulent Molecular Cloud Evolution and Control of the Star Formation Efficiency: First Tests, *ApJ*, **585**, L131 (2003)

Vázquez-Semadeni, E., Gazol, A., Scalo, J., Is thermal instability significant in turbulent galactic gas? *ApJ*, **540**, 271 (2000)

Vázquez-Semadeni, E., Gómez, G. C., Jappsen, A.-K., Ballesteros-Paredes, J., Klessen, R. S., High- and Low-Mass Star-Forming Regions from Hierarchical Gravitational Fragmentation. High Local Star Formation Rates with Low Global Efficiencies, *ApJ*, **707**, 1023 (2009)

Vázquez-Semadeni, E., Ryu, D., Passot, T., González, R. F., Gazol, A., Molecular Cloud Evolution. I. Molecular Cloud and Thin Cold Neutral Medium Sheet Formation, *ApJ*, **643**, 245 (2006)

Veltchev, T. V., Klessen, R. S., Clark, P. C., Stellar and substellar initial mass function: a model that implements gravoturbulent fragmentation and accretion, *MNRAS*, **411**, 301 (2011)

Verma, M. K., Intermittency exponents and energy spectrum of the burgers and kpz equations with correlated noise, *Physica A*, **277**, 359 (2000)

Verschuur, G. L., Zeeman Effect Observations of H_I Emission Profiles. I. Magnetic Field Limits for Three Regions Based on Observations Corrected for Polarized Beam Structure, *ApJ*, **451**, 624 (1995a)

Verschuur, G. L., Zeeman Effect Observations of H_I Emission Profiles. II. Results of an Attempt to Confirm Previous Claims of Field Detections, *ApJ*, **451**, 645 (1995b)

Vink, J. S., de Koter, A., Lamers, H. J. G. L. M., New theoretical mass-loss rates of O and B stars, *A&A*, **362**, 295 (2000)

Vink, J. S., de Koter, A., Lamers, H. J. G. L. M., Mass-loss predictions for O and B stars as a function of metallicity, *A&A*, **369**, 574 (2001)

Vink J. S., Heger A., Krumholz M. R., et al., Very Massive Stars in the Local Universe, Hightlights of Astronomy, **16**, 51 (2015)

Visser, R., van Dishoeck, E. F., Black, J. H., The photodissociation and chemistry of CO isotopologues: applications to interstellar clouds and circumstellar disks, *A&A*, **503**, 323 (2009)

von Weizsäcker C. F., The Evolution of Galaxies and Stars, *ApJ*, **114**, 165 (1951)

Wada, K., Instabilities of spiral shocks. II. A quasi-steady state in the multiphase inhomogeneous ISM, *ApJ*, **675**, 188 (2008)

Wada, K., Meurer, G., Norman, C. A., Gravity-driven turbulence in galactic disks, *ApJ*, **577**, 197 (2002)

Wakker, B. P., Howk, J. C., Savage, B. D., van Woerden, H., Tufte, S. L., Schwarz, U. J., Benjamin, R., Reynolds, R. J., Peletier, R. F., Kalberla, P. M. W., Accretion of low-metallicity gas by the milky way, *Nature*, **402**, 388 (1999)

Walborn, N. R., Blades, J. C., Spectral Classification of the 30 Doradus Stellar Populations, *ApJS*, **112**, 457 (1997)

Walch, S., Whitworth, A. P., Bisbas, T. G., Wünsch, R., Hubber, D. A., Clumps and triggered star formation in ionized molecular clouds, *MNRAS*, **435**, 917 (2013)

Walch, S. K., Whitworth, A. P., Bisbas, T., Wünsch, R., Hubber, D., Dispersal of molecular clouds by ionizing radiation, *MNRAS*, **427**, 625 (2012)

Walmsley, C. M., Ungerechts, H., Ammonia as a molecular cloud thermometer, *A&A*, **122**, 164 (1983)

Walter, F., Brinks, E., de Blok, W. J. G., Bigiel, F., Kennicutt, R. C., Jr., Thornley, M. D., Leroy, A. K., THINGS: The H_I Nearby Galaxy Survey, *AJ*, **136**, 2563 (2008)

Wang, P., Li, Z.-Y., Abel, T., Nakamura, F., Outflow feedback regulated massive star formation in parsec-scale cluster-forming clumps, *ApJ*, **709**, 27 (2010)

Wannier, P. G., Lichten, S. M., Morris, M., Warm H_I halos around molecular clouds, *ApJ*, **268**, 727 (1983)

Ward-Thompson, D., André, P., Crutcher, R., Johnstone, D., Onishi, T., Wilson, C., An Observational Perspective of Low-Mass Dense Cores II: Evolution Toward the Initial Mass Function, in: Protostars and Planets V, edited by B. Reipurth, D. Jewitt, K. Keil, p. 33 (2007)

Ward-Thompson, D., André, P., Kirk, J. M., The initial conditions of isolated star formation - V. ISOPHOT imaging and the temperature and energy balance of pre-stellar cores, *MNRAS*, **329**, 257 (2002)

Ward-Thompson, D., Motte, F., Andre, P., The initial conditions of isolated star formation - III. Millimetre continuum mapping of pre-stellar cores, *MNRAS*, **305**, 143 (1999)

Warin, S., Benayoun, J. J., Viala, Y. P., Photodissociation and rotational excitation of interstellar CO, *A&A*, **308**, 535 (1996)

Weidner, C., Kroupa, P., Evidence for a fundamental stellar upper mass limit from clustered star formation, *MNRAS*, **348**, 187 (2004)

Weidner, C., Kroupa, P., The maximum stellar mass, star-cluster formation and composite stellar populations, *MNRAS*, **365**, 1333 (2006)

Weidner, C., Kroupa, P., Bonnell, I. A. D., The relation between the most-massive star and its parental star cluster mass, *MNRAS*, **401**, 275 (2010)

Weingartner, J. C., Draine, B. T., Dust Grain-Size Distributions and Extinction in the Milky Way, Large Magellanic Cloud, and Small Magellanic Cloud *ApJ*, **548**, 296 (2001a)

Weingartner, J. C., Draine, B. T., Photoelectric Emission from Interstellar Dust: Grain Charging and Gas Heating, *ApJS*, **134**, 263 (2001b)

Welty, D. E., Hobbs, L. M., Lauroesch, J. T., Morton, D. C., Spitzer, L., York, D. G., The Diffuse Interstellar Clouds toward 23 Orionis, *ApJS*, **124**, 465 (1999)

Wernli, M., Valiron, P., Faure, A., Wiesenfeld, L., Jankowski, P., Szalewicz, K., Improved low-temperature rate constants for rotational excitation of CO by H_2, *A&A*, **446**, 367 (2006)

Wiersma, R. P. C., Schaye, J., Smith, B. D., The effect of photoionization on the cooling rates of enriched, astrophysical plasmas, *MNRAS*, **393**, 99 (2009)

Wiesenfeld, L., Goldsmith, P. F., C^+ in the interstellar medium: collisional excitation by H_2 revisited, *ApJ*, **780**, 183 (2014)

Wilden, B. S., Jones, B. F., Lin, D. N. C., Soderblom, D. R., Evolution of the Lithium Abundance of Solar-Type Stars. X. Does Accretion Affect the Lithium Dispersion in the Pleiades? *AJ*, **124**, 2799 (2002)

Williams, J. P., Bergin, E. A., Caselli, P., Myers, P. C., Plume, R., The Ionization Fraction in Dense Molecular Gas. I. Low-Mass Cores, *ApJ*, **503**, 689 (1998)

Williams, J. P., Blitz, L., McKee, C. F., The Structure and Evolution of Molecular Clouds: from Clumps to Cores to the IMF, in Protostars and Planets IV, edited by V. Mannings, A. P. Boss, S. S. Russell, p. 97 (2000)

Wilson, B. A., Dame, T. M., Masheder, M. R. W., Thaddeus, P., A uniform co survey of the molecular clouds in orion and monoceros, *A&A*, **430**, 523 (2005)

Wolfire, M. G., Hollenbach, D., McKee, C. F., The Dark Molecular Gas, *ApJ*, **716**, 1191 (2010)

Wolfire, M. G., Hollenbach, D., McKee, C. F., Tielens, A. G. G. M., Bakes, E. L. O., The neutral atomic phases of the interstellar medium, *ApJ*, **443**, 152 (1995)

Wolfire, M. G., McKee, C. F., Hollenbach, D., Tielens, A. G. G. M., Neutral Atomic Phases of the Interstellar Medium in the Galaxy, *ApJ*, **587**, 278 (2003)

Wong, T., Blitz, L., The Relationship between Gas Content and Star Formation in Molecule-rich Spiral Galaxies, *ApJ*, **569**, 157 (2002)

Wood, D. O. S., Churchwell, E., The morphologies and physical properties of ultracompact H_{II} regions, *ApJS*, **69,** 831 (1989)

Xue, X. X., Rix, H. W., Zhao, G., Fiorentin, P. R., Naab, T., Steinmetz, M., van den Bosch, F. C., Beers, T. C., Lee, Y. S., Bell, E. F., Rockosi, C., Yanny, B., Newberg, H., Wilhelm, R., Kang, X., Smith, M. C., Schneider, D. P., The Milky Way's circular velocity curve to 60 kpc and an estimate of the dark matter halo mass from the kinematics of 2400 SDSS blue horizontal-branch stars, *ApJ*, **684**, 1143 (2008)

Yeh, S. C. C., Matzner, C. D., Ionization parameter as a diagnostic of radiation and wind pressures in H_{II} regions and starburst galaxies, *ApJ*, **757**, 108 (2012)
Yorke, H. W., Sonnhalter, C., On the Formation of Massive Stars, *ApJ*, **569**, 846 (2002)
Zhukovska, S., Gail, H.-P., Trieloff, M., Evolution of interstellar dust and stardust in the solar neighbourhood, *A&A*, **479**, 453 (2008)
Zinnecker, H., Star Formation from Hierarchical Cloud Fragmentation - A Statistical Theory of the Log-Normal Initial Mass Function, *MNRAS*, **210**, 43 (1984)
Zinnecker, H., Yorke, H. W., Toward Understanding Massive Star Formation, *ARA&A*, **45**, 481 (2007)
Zuckerman, B., Evans, N. J., Models of massive molecular clouds, *ApJ*, **192**, L149 (1974)
Zweibel, E. G., Ambipolar drift in a turbulent medium, *ApJ*, **567**, 962 (2002)
Zweibel, E. G., Josafatsson, K., Hydromagnetic wave dissipation in molecular clouds, *ApJ*, **270**, 511 (1983)

High Performance Computing and Numerical Modelling

Volker Springel

1 Preamble

Numerical methods play an ever more important role in astrophysics. This can be easily demonstrated through a cursory comparison of a random sample of paper abstracts from today and 20 years ago, which shows that a growing fraction of studies in astronomy is based, at least in part, on numerical work. This is especially true in theoretical works, but of course, even in purely observational projects, data analysis without massive use of computational methods has become unthinkable. For example, cosmological inferences of large CMB experiments routinely use very large Monte-Carlo simulations as part of their Baysian parameter estimation.

The key utility of computer simulations comes from their ability to solve complex systems of equations that are either intractable with analytic techniques or only amenable to highly approximative treatments. Thanks to the rapid increase of the performance of computers, the technical limitations faced when attacking the equations numerically (in terms of calculational time, memory use, numerical resolution, etc.) become progressively smaller. But it is important to realize that they will always stay with us at some level. Computer simulations are therefore best viewed as a powerful complement to analytic reasoning, and as the method of choice to model systems that feature enormous physical complexity—such as star formation in evolving galaxies, the topic of this *43rd Saas Fee Advanced Course*.

The organizers asked me to lecture about *High performance computing and numerical modelling* in this winter school, which took place March 11–16, 2013, in Villars-sur-Ollon, Switzerland. As my co-lecturers Ralf Klessen und Nick Gnedin should focus on the physical processes in the interstellar medium and on galactic scales, my task was defined as covering the basics of numerically treating gravity and hydrodynamics, and on making some remarks on the use of high performance

V. Springel (✉)
ZAH, ARI, Heidelberg University, Heidelberg, Germany
e-mail: volker.springel@h-its.org

Y. Revaz et al. (eds.), *Star Formation in Galaxy Evolution: Connecting Numerical Models to Reality*, Saas-Fee Advanced Course 43,
DOI 10.1007/978-3-662-47890-5_3

computing techniques in general. In a nutshell, my lectures hence intend to cover the basic numerical methods necessary to simulate evolving galaxies. This is still a vast field, and I necessarily had to make a selection of a subset of the relevant material. I have tried to strike a compromise between what I considered most useful for the majority of students and what I could cover in the available time.

In particular, my lectures concentrate on techniques to compute gravitational dynamics of collisionless fluids composed of dark matter and stars in galaxies. I also spend a fair amount of time explaining basic concepts of various solvers for Eulerian gas dynamics. Due to lack of time, I am not discussing collisional N-body dynamics as applicable to star cluster, and I omit a detailed discussion of different schemes to implement adaptive mesh refinement.

The written notes presented here quite closely follow the lectures as held in Villars-sur-Ollon, apart from being expanded somewhat in detail where this seemed adequate. I note that the shear breadth of the material made it impossible to include detailed mathematical discussions and proofs of all the methods. The discussion is therefore often at an introductory level, but hopefully still useful as a general overview for students working on numerical models of galaxy evolution and star formation. Interested readers are referred to some of the references for a more detailed and mathematically sound exposition of the numerical techniques.

2 Collisionless N-Body Dynamics

According to the ΛCDM paradigm, the matter density of our Universe is dominated by *dark matter*, which is thought to be composed of a yet unidentified, non-baryonic elementary particle (e.g. Bertone et al. 2005). A full description of the dark mass in a galaxy would hence be based on following the trajectories of each dark matter particle—resulting in a gigantic N-body model. This is clearly impossible due to the large number of particles involved. Similarly, describing all the stars in a galaxy as point masses would require of order 10^{11} bodies. This may come within reach in a few years, but at present it is still essentially infeasible. In this section we discuss why we can nevertheless describe both of these galactic components as discrete N-body systems, but composed of far fewer particles than there are in reality.

2.1 *The Hierarchy of Particle Distribution Functions*

The state of an N-particle ensemble at time t can be specified by the *exact* particle distribution function (Hockney and Eastwood 1988), in the form

$$F(\mathbf{r}, \mathbf{v}, t) = \sum_{i=1}^{N} \delta(\mathbf{r} - \mathbf{r}_i(t)) \cdot \delta(\mathbf{v} - \mathbf{v}_i(t)), \tag{1}$$

where $\mathbf{r}_i$ and $\mathbf{v}_i$ denote the position and velocity of particle i, respectively. This effectively gives the number density of particles at point $(\mathbf{r}, \mathbf{v})$ at time t. Let now

$$p(\mathbf{r}_1, \mathbf{r}_2, \ldots, \mathbf{r}_N, \mathbf{v}_1, \mathbf{v}_2, \ldots \mathbf{v}_N)\, \mathrm{d}\mathbf{r}_1\, \mathrm{d}\mathbf{r}_2 \cdots \mathrm{d}\mathbf{r}_N\, \mathrm{d}\mathbf{v}_1\, \mathrm{d}\mathbf{v}_2 \cdots \mathrm{d}\mathbf{v}_N, \tag{2}$$

be the probability that the system is in the given state at time t. Then a reduced statistical description is obtained by *ensemble averaging*:

$$f_1(\mathbf{r}, \mathbf{v}, t) = \langle F(\mathbf{r}, \mathbf{v}, t) \rangle = \int F \cdot p \cdot \mathrm{d}\mathbf{r}_1\, \mathrm{d}\mathbf{r}_2 \cdots \mathrm{d}\mathbf{r}_N\, \mathrm{d}\mathbf{v}_1\, \mathrm{d}\mathbf{v}_2 \cdots \mathrm{d}\mathbf{v}_N. \tag{3}$$

We can integrate out one of the Dirac delta-functions in F to obtain

$$f_1(\mathbf{r}, \mathbf{v}, t) = N \int p(\mathbf{r}, \mathbf{r}_2, \ldots, \mathbf{r}_N, \mathbf{v}, \mathbf{v}_2, \ldots \mathbf{v}_N)\, \mathrm{d}\mathbf{r}_2 \cdots \mathrm{d}\mathbf{r}_N\, \mathrm{d}\mathbf{v}_2 \cdots \mathrm{d}\mathbf{v}_N. \tag{4}$$

Note that as all particles are equivalent we can permute the arguments in p where $\mathbf{r}$ and $\mathbf{v}$ appear. $f_1(\mathbf{r}, \mathbf{v}, t)\, \mathrm{d}\mathbf{r}\, \mathrm{d}\mathbf{v}$ now gives the *mean number* of particles in a phase-space volume $\mathrm{d}\mathbf{r}\, \mathrm{d}\mathbf{v}$ around $(\mathbf{r}, \mathbf{v})$.

Similarly, the ensemble-averaged two-particle distribution ("the mean product of the numbers of particles at $(\mathbf{r}, \mathbf{v})$ and $(\mathbf{r}', \mathbf{v}')$") is given by

$$\begin{aligned} f_2(\mathbf{r}, \mathbf{v}, \mathbf{r}', \mathbf{v}', t) &= \big\langle F(\mathbf{r}, \mathbf{v}, t) F(\mathbf{r}', \mathbf{v}', t) \big\rangle \qquad (5) \\ &= N(N-1) \int p(\mathbf{r}, \mathbf{r}', \mathbf{r}_3, \ldots, \mathbf{r}_N, \mathbf{v}, \mathbf{v}', \mathbf{v}_3, \ldots \mathbf{v}_N)\, \mathrm{d}\mathbf{r}_3 \cdots \mathrm{d}\mathbf{r}_N\, \mathrm{d}\mathbf{v}_3 \cdots \mathrm{d}\mathbf{v}_N. \end{aligned}$$

Likewise one may define $f_3, f_4, \ldots$ and so on. This yields the so-called BBGKY (Bogoliubov-Born-Green-Kirkwood-Yvon) chain (e.g. Kirkwood 1946), see also Hockney and Eastwood (1988) for a detailed discussion.

Uncorrelated (collisionless) systems The simplest closure for the BBGKY hierarchy is to assume that particles are *uncorrelated*, i.e., that we have

$$f_2(\mathbf{r}, \mathbf{v}, \mathbf{r}', \mathbf{v}', t) = f_1(\mathbf{r}, \mathbf{v}, t)\, f_1(\mathbf{r}', \mathbf{v}', t). \tag{6}$$

Physically, this means that a particle at $(\mathbf{r}, \mathbf{v})$ is completely unaffected by one at $(\mathbf{r}', \mathbf{v}')$. Systems in which this is approximately the case include stars in a galaxy, dark matter particles in the universe, or electrons in a plasma. We will later consider in more detail under which conditions a system is collisionless.

Let's now go back to the probability density $p(\mathbf{w})$ which depends on the N-particle phase-space state $\mathbf{w} = (\mathbf{r}_1, \mathbf{r}_2, \ldots, \mathbf{r}_N, \mathbf{v}_1, \mathbf{v}_2, \ldots \mathbf{v}_N)$. The conservation of probability in phase-space means that it fulfills a continuity equation

$$\frac{\partial p}{\partial t} + \nabla_{\mathbf{w}} \cdot (p\, \dot{\mathbf{w}}) = 0. \tag{7}$$

We can cast this into

$$\frac{\partial p}{\partial t} + \sum_i \left(p \frac{\partial \dot{\mathbf{r}}_i}{\partial \mathbf{r}_i} + \frac{\partial p}{\partial \mathbf{r}_i} \dot{\mathbf{r}}_i + p \frac{\partial \dot{\mathbf{v}}_i}{\partial \mathbf{v}_i} + \frac{\partial p}{\partial \mathbf{v}_i} \dot{\mathbf{v}}_i \right) = 0. \tag{8}$$

Because only conservative gravitational fields are involved, the system is described by classical mechanics as a so-called Hamiltonian system. Recalling the equations of motion $\dot{\mathbf{r}} = \frac{\partial H}{\partial \mathbf{p}}$ and $\dot{\mathbf{p}} = -\frac{\partial H}{\partial \mathbf{r}}$ of Hamiltonian dynamics (Goldstein 1950), we can differentiate them to get $\frac{\partial \dot{\mathbf{r}}}{\partial \mathbf{r}} = \frac{\partial^2 H}{\partial \mathbf{r} \partial \mathbf{p}}$, and $\frac{\partial \dot{\mathbf{p}}}{\partial \mathbf{p}} = -\frac{\partial^2 H}{\partial \mathbf{r} \partial \mathbf{p}}$. Hence it follows $\frac{\partial \dot{\mathbf{r}}}{\partial \mathbf{r}} = -\frac{\partial \dot{\mathbf{v}}}{\partial \mathbf{v}}$. Using this we get

$$\frac{\partial p}{\partial t} + \sum_i \left(\mathbf{v}_i \frac{\partial p}{\partial \mathbf{r}_i} + \mathbf{a}_i \frac{\partial p}{\partial \mathbf{v}_i} \right) = 0, \tag{9}$$

where $\mathbf{a}_i = \dot{\mathbf{v}}_i = \mathbf{F}_i / m_i$ is the particle acceleration and m_i is the particle mass. This is *Liouville's theorem.*

Now, in the collisionless/uncorrelated limit, this directly carries over to the one-point distribution function $f = f_1$ if we integrate out all particle coordinates except for one as in Eq. (4), yielding the *Vlasov equation*, also known as collisionless Boltzmann equation:

$$\frac{\partial f}{\partial t} + \mathbf{v} \frac{\partial f}{\partial \mathbf{r}} + \mathbf{a} \frac{\partial f}{\partial \mathbf{v}} = 0. \tag{10}$$

The close relation to Liouville's equation means that also here the phase space-density stays constant along characteristics of the system (i.e., along orbits of individual particles).

What about the acceleration? In the limit of a collisionless system, the acceleration $\mathbf{a}$ in the above equation cannot be due to another single particle, as this would imply local correlations. However, *collective effects*, for example from the gravitational field produced by the whole system are still allowed.

For example, the source field of self-gravity (i.e., the mass density) can be described as

$$\rho(\mathbf{r}, t) = m \int f(\mathbf{r}, \mathbf{v}, t) \, \mathrm{d}\mathbf{v}. \tag{11}$$

This then produces a gravitational field through Poisson's equation,

$$\nabla^2 \Phi(\mathbf{r}, t) = 4\pi G \rho(\mathbf{r}, t), \tag{12}$$

which gives the accelerations as

$$\mathbf{a} = -\frac{\partial \Phi}{\partial \mathbf{r}}. \tag{13}$$

One can also combine these equations to yield the Poisson-Vlasov system, given by

$$\frac{\partial f}{\partial t} + \mathbf{v}\frac{\partial f}{\partial \mathbf{r}} - \frac{\partial \Phi}{\partial \mathbf{r}}\frac{\partial f}{\partial \mathbf{v}} = 0, \tag{14}$$

$$\nabla^2 \Phi = 4\pi G m \int f(\mathbf{r}, \mathbf{v}, t)\, \mathrm{d}\mathbf{v}. \tag{15}$$

This holds in an analogous way also for a plasma where the mass density is replaced by a charge density.

It is interesting to note that in this description the particles have basically completely vanished and have been replaced with a continuum fluid description. Later, for the purpose of solving the equations, we will have to reintroduce particles as a means of discretizing the equations—but these are then not the real physical particles any more, rather they are fiducial macro particles that sample the phase-space in a Monte-Carlo fashion.

2.2 The Relaxation Time—When Is a System Collisionless?

Consider a system of size R containing N particles. The time for one crossing of a particle through the system is of order

$$t_{\mathrm{cross}} = \frac{R}{v}, \tag{16}$$

where v is the typical particle velocity (Binney and Tremaine 1987, 2008). For a self-gravitating system of that size we expect

$$v^2 \simeq \frac{GNm}{R} = \frac{GM}{R}, \tag{17}$$

where $M = N\, m$ is the total mass.

We now want to estimate the rate at which a particle experiences weak deflections by other particles, which is the process that violates perfect collisionless behavior and which induces relaxation. We calculate the deflection in the impulse approximation where the particle's orbit is taken as a straight path, as sketched in Fig. 1.

To get the deflection, we compute the transverse momentum acquired by the particle as it flies by the perturber (assumed to be stationary for simplicity):

$$\Delta p = m\Delta v = \int F_{\perp} \mathrm{d}t = \int \frac{Gm^2}{x^2 + b^2} \frac{b}{\sqrt{x^2 + b^2}} \frac{\mathrm{d}x}{v} = \frac{2Gm^2}{bv}. \tag{18}$$

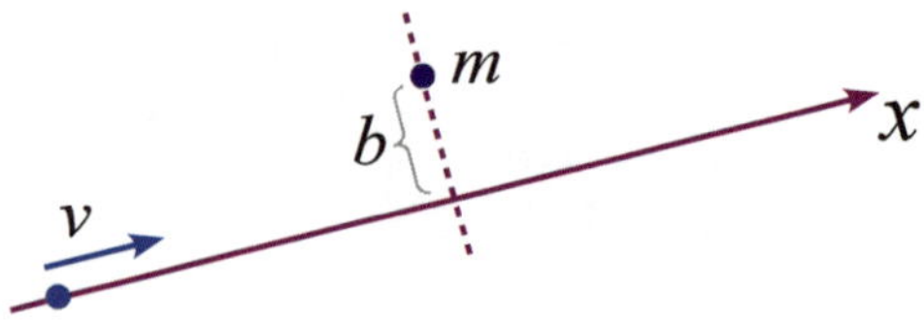

Fig. 1 Sketch of a two-body encounter, in which a particle passes another particle (assumed to be at rest) with impact parameter b and velocity v

How many encounters do we expect in one crossing? For impact parameters between $[b, b + \mathrm{d}b]$ we have

$$\mathrm{d}n = N \frac{2\pi b \, \mathrm{d}b}{\pi R^2} \tag{19}$$

targets. The velocity perturbations from each encounter have random orientations, so they add up in quadrature. Per crossing we hence have for the quadratic velocity perturbation:

$$(\Delta v)^2 = \int \left(\frac{2Gm}{bv} \right)^2 \mathrm{d}n = 8N \left(\frac{Gm}{Rv} \right)^2 \ln \Lambda, \tag{20}$$

where

$$\ln \Lambda = \ln \frac{b_{\rm max}}{b_{\rm min}} \tag{21}$$

is the so-called Coulomb logarithm, and $b_{\rm max}$ and $b_{\rm min}$ are the adopted integration limits. We can now define the relaxation time as

$$t_{\rm relax} \equiv \frac{v^2}{(\Delta v)^2 / t_{\rm cross}}, \tag{22}$$

i.e., after this time the individual perturbations have reached ~100 % of the typical squared velocity, and one can certainly not neglect the interactions any more. With our result for $(\Delta v)^2$, and using Eq. (17) this now becomes

$$t_{\rm relax} = \frac{N}{8 \ln \Lambda} t_{\rm cross}. \tag{23}$$

But we still have to clarify what we can sensibly use for $b_{\rm min}$ and $b_{\rm max}$ in the Coulomb logarithm. For $b_{\rm max}$, we can set the size of the system, i.e., $b_{\rm max} \simeq R$. For $b_{\rm min}$, we can use as a lower limit the b where very strong deflections ensue, which is given by

$$\frac{2Gm}{b_{\rm min} v} \simeq v, \tag{24}$$

i.e where the transverse velocity perturbation becomes as large as the velocity itself (see Eq. 18). This then yields $b_{\rm min} = 2R/N$. We hence get for the Coulomb logarithm

$\ln \Lambda \simeq \ln(N/2)$. But a factor of 2 in the logarithm might as well be neglected in this coarse estimate, so that we obtain $\ln \Lambda \sim N$. We hence arrive at the final result (Chandrasekhar 1943):

$$t_{\rm relax} = \frac{N}{8 \ln N} t_{\rm cross}. \tag{25}$$

A system can be viewed as collisionless if $t_{\rm relax} \gg t_{\rm age}$, where $t_{\rm age}$ is the time of interest. We note that $t_{\rm cross}$ depends only on the size and mass of the system, but *not* on the particle number N or the individual masses of the N-body particles. We therefore clearly see that the primary requirement to obtain a collisionless system is to use a sufficiently large N.

Examples:

- globular star clusters have $N \sim 10^5$, $t_{\rm cross} \sim \frac{3\,\rm pc}{6\,\rm km/sec} \simeq 0.5\,\rm Myr$. This implies that such systems are strongly affected by collisions over the age of the Universe, $t_{\rm age} = \frac{1}{H_0} \sim 10\,\rm Gyr$, where H_0 is the Hubble constant.
- stars in a typical galaxy: Here we have $N \sim 10^{11}$ and $t_{\rm cross} \sim \frac{1}{100\,H_0}$. This means that these large stellar systems are collisionless over the age of the Universe to extremely good approximation.
- dark matter in a galaxy: Here we have $N \sim 10^{77}$ if the dark matter is composed of a $\sim$ 100 GeV weakly interacting massive particle (WIMP). In addition, the crossing time is longer than for the stars, $t_{\rm cross} \sim \frac{1}{10\,H_0}$, due to the larger size of the 'halo' relative to the embedded stellar system. Clearly, dark matter represents the crème de la crème of collisionless systems.

2.3 N-Body Models and Gravitational Softening

We now reintroduce particles in order to discretize the collisionless fluid described by the Poisson-Vlasov system. We use however *far fewer* particles than in real physical systems, and we correspondingly give them a higher mass. These are hence fiducial macro-particles. Their equations of motion in the case of gravity take on the form:

$$\ddot{\mathbf{x}}_i = -\nabla_i \Phi(\mathbf{r}_i), \tag{26}$$

$$\Phi(\mathbf{r}) = -G \sum_{j=1}^{N} \frac{m_j}{[(\mathbf{r} - \mathbf{r}_j)^2 + \epsilon^2]^{1/2}}. \tag{27}$$

A few comments are in order here:

- Provided we can ensure $t_{\rm relax} \gg t_{\rm sim}$, where $t_{\rm sim}$ is the simulated time-space, the numerical model keeps behaving as a collisionless system over $t_{\rm sim}$ despite a smaller N than in the real physical system. In this limit, the collective gravitational potential is sufficiently smooth.

- Note that the mass of a macro-particle used to discretize the collision system drops out from its equation of motion (because there is no self-force). Provided there are enough particles to describe the gravitational potential accurately, the orbits of the macro-particles will be just as valid as the orbits of the real physical particles.
- The N-body model gives only one (quite noisy) realization of the one-point function. It does not give the ensemble average directly (this would require multiple simulations).
- The equations of motion contain a **softening length** ϵ. The purpose of the force softening is to avoid large angle scatterings and the numerical expense that would be needed to integrate the orbits with sufficient accuracy in singular potentials. Also, we would like to prevent the possibility of the formation of bound particle pairs—they would obviously be highly correlated and hence strongly violate collisionless behavior. We don't get bound pairs if

$$\left\langle v^2 \right\rangle \gg \frac{Gm}{\epsilon}, \tag{28}$$

  which can be viewed as a necessary (but not in general sufficient) condition on reasonable softening settings (Power et al. 2003). The adoption of a softening length also implies the introduction of a smallest resolved length-scale. The specific softening choice one makes ultimately represents a compromise between spatial resolution, discreteness noise in the orbits and the gravitational potential, computational cost, and the relaxation effects that adversely influence results.

2.4 N-Body Equations in Cosmology

In cosmological simulations, it is customary to use comoving coordinates $\mathbf{x}$ instead of physical coordinates $\mathbf{r}$. The two are related by

$$\mathbf{r} = a(t)\,\mathbf{x}, \tag{29}$$

where $a = 1/(1+z)$ is the cosmological scale factor. Its evolution is governed by the Hubble rate

$$\frac{\dot{a}}{a} = H(a), \tag{30}$$

which in turn is given by $H(a) = [\Omega_0 a^{-3} + (1-\Omega_0-\Omega_\Lambda)a^{-2} + \Omega_\Lambda]^{1/2}$ in standard Friedmann-Lemaitre models (e.g. Peacock 1999; Mo et al. 2010).

In an (infinite) expanding space, modelled through period replication of a box of size L, one can then show (e.g. Springel et al. 2001) that the Newtonian equations of motion in comoving coordinates can be written as

$$\frac{\mathrm{d}}{\mathrm{d}t}(a^2\dot{\mathbf{x}}) = -\frac{1}{a}\nabla_i \phi(\mathbf{x}_i), \tag{31}$$

$$\nabla^2\phi(\mathbf{x}) = 4\pi G \sum_i m_i \left[-\frac{1}{L^3} + \sum_{\mathbf{n}} \delta(\mathbf{x} - \mathbf{x}_i - \mathbf{n}L) \right], \tag{32}$$

where the sum over i extends over N particles in the box, and ϕ is the *peculiar gravitational potential*. It corresponds to the Newtonian potential of density deviations around a constant mean background density. Note that the sum over all particles for calculating the potential extends also over all of their period images, with $\mathbf{n} = (n_1, n_2, n_3)$ being a vector of integer triples. The term $-1/L^3$ is simply needed to ensure that the mean density sourcing the Poisson equation vanishes, otherwise there would be no solution for an infinite space.

2.5 Calculating the Dynamics of an N-Body System

Once we have discretized a collisionless fluid in terms of an N-body system, two questions come up:

1. How do we integrate the equations of motion in time?
2. How do we compute the right hand side of the equations of motion, i.e., the gravitational forces?

For the first point, we can use an integration scheme for ordinary differential equations, preferably a symplectic one since we are dealing with a Hamiltonian system. We shall briefly discuss elementary aspects of these time integration methods in the following section.

The second point seems also straightforward at first, as the accelerations (forces) can be readily calculated through *direct summation*. In the isolated case this reads as

$$\ddot{\mathbf{r}}_i = -G \sum_{j=1}^{N} \frac{m_j}{[(\mathbf{r}_i - \mathbf{r}_j)^2 + \epsilon^2]^{3/2}} (\mathbf{r}_i - \mathbf{r}_j). \tag{33}$$

For a periodic space, the force kernel is slightly different but in principle the same summation applies (Hernquist et al. 1991). This calculation is *exact*, but for each of the N equations we have to calculate a sum with N partial forces, yielding a computational cost of order $\mathcal{O}(N^2)$. This quickly becomes prohibitive for large N, and causes a conflict with our urgent need to have a large N!

Perhaps a simple example is in order to show how bad the N^2 scaling really is in practice. Suppose you can do $N = 10^6$ in a month of computer time, which is close to the maximum that one may want to do in practice. A particle number of $N = 10^{10}$ would then already take of order 10 million years.

We hence need faster, *approximative* force calculation schemes. We shall discuss a number of different possibilities for this in Sect. 4, namely:

- Particle-mesh (PM) algorithms
- Fourier-transform based solvers of Poisson's equations
- Iterative solvers for Poisson's equation (multigrid-methods)
- Hierarchical multipole methods ("tree-algorithms")
- So-called TreePM methods

Various combinations of these approaches may also be used, and sometimes they are also applied together with direct summation on small scales. The latter may also be accelerated with special-purpose hardware (e.g. the GRAPE board; Makino et al. 2003), or with graphics processing units (GPUs) that are used as fast number-crushers.

3 Time Integration Techniques

We discuss in the following some basic methods for the integration of *ordinary differential equations* (ODEs). These are relations between an unknown scalar or vector-values function $\mathbf{y}(t)$ and its derivatives with respect to an independent variable, t in this case (the following discussion associates the independent variable with 'time', but this could of course be also any other quantity). Such equations hence formally take the form

$$\frac{\mathrm{d}\mathbf{y}}{\mathrm{d}t} = \mathbf{f}(\mathbf{y}, t), \tag{34}$$

and we seek the solution $\mathbf{y}(t)$, subject to boundary conditions.

Many simple dynamical problems can be written in this form, including ones that involve second or higher derivatives. This is done through a procedure called *reduction to 1st order*. One does this by adding the higher derivatives, or combinations of them, as further rows to the vector $\mathbf{y}$.

For example, consider a simple pendulum of length l with the equation of motion

$$\ddot{q} = -\frac{g}{l}\sin(q), \tag{35}$$

where q is the angle with respect to the vertical. Now define $p \equiv \dot{q}$, yielding a state vector

$$\mathbf{y} \equiv \begin{pmatrix} q \\ p \end{pmatrix}, \tag{36}$$

and a first order ODE of the form:

$$\frac{\mathrm{d}\mathbf{y}}{\mathrm{d}t} = \mathbf{f}(\mathbf{y}) = \begin{pmatrix} p \\ -\frac{g}{l}\sin(q) \end{pmatrix}. \tag{37}$$

A numerical approximation to the solution of an ODE is a set of values $\{y_0, y_1, y_2, \ldots\}$ at discrete times $\{t_0, t_1, t_2, \ldots\}$, obtained for certain boundary conditions.

The most common boundary condition for ODEs is the *initial value problem* (IVP), where the state of **y** is known at the beginning of the integration interval. It is however also possible to have mixed boundary conditions where **y** is partially known at both ends of the integration interval.

There are many different methods for obtaining a discrete solution of an ODE system (e.g. Press et al. 1992). We shall here discuss some of the most basic ones, restricting ourselves to the IVP problem, for simplicity, as this is the one naturally appearing in cosmological simulations.

3.1 Explicit and Implicit Euler Methods

Explicit Euler This solution method, sometimes also called "forward Euler", uses the iteration

$$y_{n+1} = y_n + f(y_n)\Delta t, \tag{38}$$

where y can also be a vector. Δt is the integration step.

- This approach is the simplest of all.
- The method is called *explicit* because y_{n+1} is computed with a right-hand-side that only depends on quantities that are already known.
- The stability of the method can be a sensitive function of the step size, and will in general only be obtained for a sufficiently small step size.
- It is recommended to refrain from using this scheme in practice, since there are other methods that offer higher accuracy at the same or lower computational cost. The reason is that the Euler method is only *first order accurate*. To see this, note that the truncation error in a single step is of order $\mathcal{O}_s(\Delta t^2)$, which follows simply from a Taylor expansion. To integrate over a time interval T, we need however $N_s = T/\Delta t$ steps, producing a total error that scales as $N_s \mathcal{O}_s(\Delta t^2) = \mathcal{O}_T(\Delta t)$.
- The method is also not time-symmetric, which makes it prone to accumulation of secular integration errors.

We remark in passing that for a method to reach a global error that scales as $\mathcal{O}_T(\Delta t^n)$ (which is then called an "nth order accurate" scheme), a local truncation error of one order higher is required, i.e., $\mathcal{O}_s(\Delta t^{n+1})$.

Implicit Euler In a so-called "backwards Euler" scheme, one uses

$$y_{n+1} = y_n + f(y_{n+1})\Delta t, \tag{39}$$

which seemingly represents only a tiny change compared to the explicit scheme.

- This approach has excellent stability properties, and for some problems, it is in fact essentially always stable even for extremely large timestep. Note however that the accuracy will usually nevertheless become very bad when using such large steps.

- This stability property makes implicit Euler sometimes useful for *stiff equations*, where the derivatives (suddenly) can become very large.
- The implicit equation for y_{n+1} that needs to be solved here corresponds in many practical applications to a non-linear equation that can be complicated to solve for y_{n+1}. Often, the root of the equation has to be found numerically, for example through an iterative technique.
- The method is still first order accurate, and also lacks time-symmetry, just like the explicit Euler scheme.

Implicit midpoint rule If we use

$$y_{n+1} = y_n + f\left(\frac{y_n + y_{n+1}}{2}\right)\Delta t, \tag{40}$$

we obtain the implicit midpoint rule, which can be viewed as a symmetrized variant of explicit and implicit Euler. This is *second order accurate*, but still implicit, so difficult to use in practice. Interestingly, it is also time-symmetric, i.e., one can formally integrate backwards and recover exactly the same steps (modulo floating point round-off errors) as in a forward integration.

3.2 Runge-Kutta Methods

The Runge-Kutta schemes form a whole class of versatile integration methods (e.g. Atkinson 1978; Stoer and Bulirsch 2002). Let's derive one of the simplest Runge-Kutta schemes.

1. We start from the exact solution,

$$y_{n+1} = y_n + \int_{t_n}^{t_{n+1}} f\,(y(t))\,\mathrm{d}t. \tag{41}$$

2. Next, we approximate the integral with the (implicit) trapezoidal rule:

$$y_{n+1} = y_n + \frac{f(y_n) + f(y_{n+1})}{2}\Delta t. \tag{42}$$

3. Runge (1895) proposed to predict the unknown y_{n+1} on the right hand side by an Euler step, yielding a *2nd order accurate Runge-Kutta scheme*, sometimes also called predictor-corrector scheme:

$$k_1 = f(y_n, t_n), \tag{43}$$

$$k_2 = f(y_n + k_1\Delta t, t_{n+1}), \tag{44}$$

$$y_{n+1} = \frac{k_1 + k_2}{2}\Delta t. \tag{45}$$

Here the step done with the derivate of Eq. (43) is called the 'predictor' and the one done with Eq. (44) is the corrector step.

Higher order Runge-Kutta schemes A variety of further Runge-Kutta schemes of different order can be defined. Perhaps the most commonly used is the classical 4th-order Runge-Kutta scheme:

$$k_1 = f(y_n, t_n), \tag{46}$$

$$k_2 = f\left(y_n + k_1\frac{\Delta t}{2}, t_n + \frac{\Delta t}{2}\right), \tag{47}$$

$$k_3 = f\left(y_n + k_2\frac{\Delta t}{2}, t_n + \frac{\Delta t}{2}\right), \tag{48}$$

$$k_4 = f\left(y_n + k_3\Delta t, t_n + \Delta t\right). \tag{49}$$

These four function evaluations per step are then combined in a weighted fashion to carry out the actual update step:

$$y_{n+1} = y_n + \left(\frac{k_1}{6} + \frac{k_2}{3} + \frac{k_3}{3} + \frac{k_4}{6}\right)\Delta t + \mathcal{O}(\Delta t^5). \tag{50}$$

We note that the use of higher order schemes also entails more function evaluations per step, i.e., the individual steps become more complicated and expensive. Because of this, higher order schemes are not always better; they usually are up to some point, but sometimes even a simple second-order accurate scheme can be the best choice for certain problems.

3.3 The Leapfrog

Suppose we have a second order differential equation of the type

$$\ddot{x} = f(x). \tag{51}$$

This could of course be brought into standard form, $\dot{\mathbf{y}} = \tilde{\mathbf{f}}(\mathbf{y})$, by defining something like $\mathbf{y} = (x, \dot{x})$ and $\tilde{\mathbf{f}} = (\dot{x}, f(x))$, followed by applying a Runge-Kutta scheme as introduced above.

However, there is also another approach in this case, which turns out to be particularly simple and interesting. Let's define $v \equiv \dot{x}$. Then the so-called Leapfrog integration scheme is the mapping $(x_n, v_n) \rightarrow (x_{n+1}, v_{n+1})$ defined as:

$$v_{n+\frac{1}{2}} = v_n + f(x_n)\frac{\Delta t}{2}, \tag{52}$$

$$x_{n+1} = x_n + v_{n+\frac{1}{2}}\,\Delta t, \tag{53}$$

$$v_{n+1} = v_{n+\frac{1}{2}} + f(x_{n+1})\frac{\Delta t}{2}. \tag{54}$$

- This scheme is 2nd-order accurate (proof through Taylor expansion).
- It requires only 1 evaluation of the right hand side per step (note that $f(x_{n+1})$ can be reused in the next step.
- The method is time-symmetric, i.e., one can integrate backwards in time and arrives at the initial state again, modulo numerical round-off errors.
- The scheme can be written in a number of alternative ways, for example by combining the two half-steps of two subsequent steps. One then gets:

$$x_{n+1} = x_n + v_{n+\frac{1}{2}}\,\Delta t, \tag{55}$$

$$v_{n+\frac{3}{2}} = v_{n+\frac{1}{2}} + f(x_{n+1})\,\Delta t. \tag{56}$$

  One here sees the time-centered nature of the formulation very clearly, and the interleaved advances of position and velocity give it the name leapfrog.

The performance of the leapfrog in certain problems is found to be surprisingly good, better than that of other schemes such as Runge-Kutta which have formally the same or even a better error order. This is illustrated in Fig. 2 for the Kepler problem, i.e., the integration of the motion of a small point mass in the gravitational field of a large mass. We see that the long-term evolution is entirely different. Unlike the RK schemes, the leapfrog does not build up a large energy error. So why is the leapfrog behaving here so much better than other 2nd order or even 4th order schemes?

3.4 Symplectic Integrators

The reason for these beneficial properties lies in the fact that the leapfrog is a so-called symplectic method. These are structure-preserving integration methods (e.g. Saha and Tremaine 1992; Hairer et al. 2002) that observe important special properties of Hamiltonian systems: Such systems have first conserved integrals (such as the energy), they also exhibit phase-space conservation as described by the Liouville theorem, and more generally, they preserve Poincare's integral invariants.

Symplectic transformations

- A linear map $F : \mathbb{R}^{2d} \rightarrow \mathbb{R}^{2d}$ is called symplectic if $\omega(F\xi, F\eta) = \omega(\xi, \eta)$ for all vectors $\xi, \eta \in \mathbb{R}^{2d}$, where ω gives the area of the parallelogram spanned by the two vectors.
- A differentiable map $g : U \rightarrow \mathbb{R}^{2d}$ with $U \in \mathbb{R}$ is called symplectic if its Jacobian matrix is everywhere symplectic, i.e., $\omega(g'\xi, g'\eta) = \omega(\xi, \eta)$.
- *Poincare's theorem* states that the time evolution generated by a Hamiltonian in phase-space is a symplectic transformation.

The above suggests that there is a close connection between exact solutions of Hamiltonians and symplectic transformations. Also, two consecutive symplectic transformations are again symplectic.

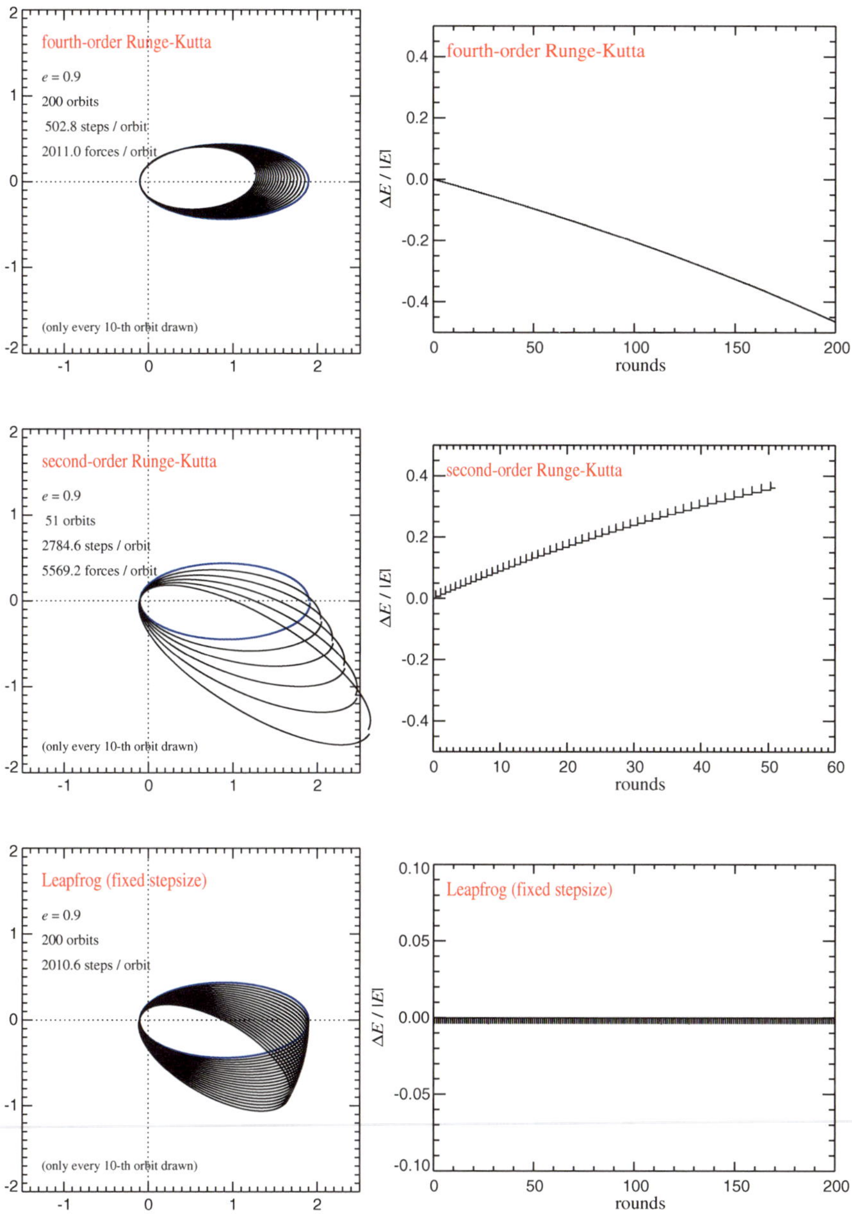

Fig. 2 Kepler problem integrated with different integration schemes (Springel, 2005). The panels on *top* are for a 4th-order Runge Kutta scheme, the *middle* for a 2nd order Runge-Kutta, and the *bottom* for a 2nd-order leapfrog. The leapfrog does not show a secular drift of the total energy, and is hence much more suitable for long-term integration of this Hamiltonian system

Separable Hamiltonians Dynamical problems that are described by Hamiltonians of the form

$$H(p,q) = \frac{p^2}{2m} + U(q) \tag{57}$$

are quite common. These systems have separable Hamiltonians that can be written as

$$H(p,q) = H_{\rm kin}(p) + H_{\rm pot}(q). \tag{58}$$

Now we will allude to the general idea of *operator splitting* (Strang 1968). Let's try to solve the two parts of the Hamiltonian individually:

1. For the part $H = H_{\rm kin} = \frac{p^2}{2m}$, the equations of motion are

$$\dot{q} = \frac{\partial H}{\partial p} = \frac{p}{m}, \tag{59}$$

$$\dot{p} = -\frac{\partial H}{\partial q} = 0. \tag{60}$$

   These equations are straightforwardly solved and give

$$q_{n+1} = q_n + \frac{p_n}{m}\Delta t, \tag{61}$$

$$p_{n+1} = p_n. \tag{62}$$

   Note that this solution is exact for the given Hamiltonian, for arbitrarily long time intervals Δt. Given that it is a solution of a Hamiltonian, the solution constitutes a symplectic mapping.
2. The potential part, $H = H_{\rm pot} = U(q)$, leads to the equations

$$\dot{q} = \frac{\partial H}{\partial p} = 0, \tag{63}$$

$$\dot{p} = -\frac{\partial H}{\partial q} = -\frac{\partial U}{\partial q}. \tag{64}$$

   This is solved by

$$q_{n+1} = q_n, \tag{65}$$

$$p_{n+1} = p_n - \frac{\partial U}{\partial q}\Delta t. \tag{66}$$

   Again, this is an exact solution independent of the size of Δt, and therefore a symplectic transformation.

Let's now introduce an operator $\varphi_{\Delta t}(H)$ that describes the mapping of phase-space under a Hamiltonian H that is evolved over a time interval Δt. Then it is easy to see that the leapfrog is given by

$$\varphi_{\Delta t}(H) = \varphi_{\frac{\Delta t}{2}}(H_{\text{pot}}) \circ \varphi_{\Delta t}(H_{\text{kin}}) \circ \varphi_{\frac{\Delta t}{2}}(H_{\text{pot}}) \tag{67}$$

for a separable Hamiltonian $H = H_{\text{kin}} + H_{\text{pot}}$.

- Since each individual step of the leapfrog is symplectic, the concatenation of Eq. (67) is also symplectic.
- In fact, the leapfrog generates the exact solution of a modified Hamiltonian H_{leap}, where $H_{\text{leap}} = H + H_{\text{err}}$. The difference lies in the 'error Hamiltonian' H_{err}, which is given by

$$H_{\text{err}} \propto \frac{\Delta t^2}{12} \left\{ \{H_{\text{kin}}, H_{\text{pot}}\}, H_{\text{kin}} + \frac{1}{2} H_{\text{pot}} \right\} + \mathcal{O}(\Delta t^3), \tag{68}$$

  where the curly brackets are Poisson brackets (Goldstein 1950). This can be demonstrated by expanding

$$\mathrm{e}^{(H+H_{\text{err}})\Delta t} = \mathrm{e}^{H_{\text{pot}}\frac{\Delta t}{2}} \, \mathrm{e}^{H_{\text{kin}}\Delta t} \, \mathrm{e}^{H_{\text{pot}}\frac{\Delta t}{2}} \tag{69}$$

  with the help of the Baker-Campbell-Hausdorff formula (Campbell 1897; Saha and Tremaine 1992).
- The above property explains the superior long-term stability of the integration of conservative systems with the leapfrog. Because it respects phase-space conservation, secular trends are largely absent, and the long-term energy error stays bounded and reasonably small.

4 Gravitational Force Calculation

As mentioned earlier, calculating the gravitational forces exactly for a large number of bodies becomes computational prohibitive very quickly. Fortunately, in the case of collisionless systems, this is also not necessary, because comparatively large force errors can be tolerated. All they do is to shorten the relaxation time slightly by an insignificant amount (Hernquist et al. 1993). In this section, we discuss a number of the most commonly employed approximate force calculation schemes, beginning with the so-called particle mesh techniques (White et al. 1983; Klypin and Shandarin 1983) that were originally pioneered in plasma physics (Hockney and Eastwood 1988).

4.1 Particle Mesh Technique

An important approach to accelerate the force calculation for an N-body system lies in the use of an auxiliary mesh. Conceptually, this so-called particle-mesh (PM) technique involves four steps:

1. Construction of a density field ρ on a suitable mesh.
2. Computation of the potential on the mesh by solving the Poisson equation.
3. Calculation of the force field from the potential.
4. Calculation of the forces at the original particle positions.

We shall now discuss these four steps in turn. An excellent coverage of the material in this section is given by Hockney and Eastwood (1988).

4.1.1 Mass Assignment

We want to put N particles with mass m_i and coordinates $\mathbf{r}_i$ $(i = 1, 2, \ldots, N)$ onto a mesh with uniform spacing $h = L/N_{\rm g}$ (Fig. 3). For simplicity, we will assume a cubical calculational domain with extension L and a number of $N_{\rm g}$ grid cells per dimension. Let $\{\mathbf{r}_\mathbf{p}\}$ denote the set of discrete cell-centers, with $\mathbf{p} = (p_x, p_y, p_z)$ being a suitable integer index $(0 \le p_{x,y,z} < N_g)$. Note that one may equally well identify the $\{\mathbf{r}_\mathbf{p}\}$ with the lower left corner of a mesh cell, if this is more practical.

We associate a shape function $S(\mathbf{x})$ with each particle, normalized according to

$$\int S(\mathbf{x})\,\mathrm{d}\mathbf{x} = 1. \tag{70}$$

To each mesh-cell, we then assign the fraction $W_\mathbf{p}(\mathbf{x}_i)$ of particle i's mass that falls into the cell indexed by $\mathbf{p}$. This is given by the overlap of the mesh cell with the shape function, namely:

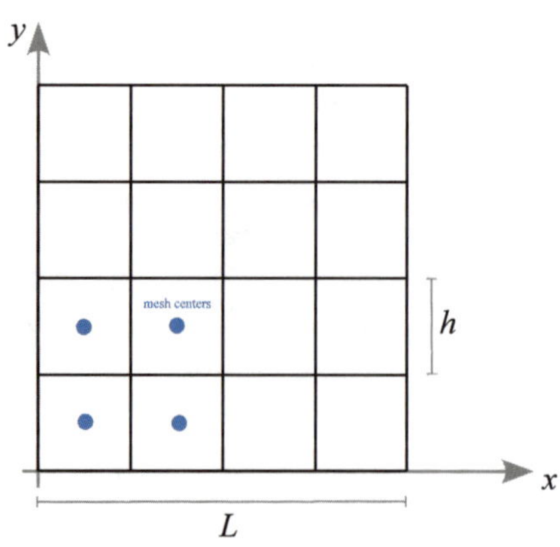

Fig. 3 Sketch of the mesh geometry used in typical particle-mesh techniques with Cartesian grids

$$W_{\mathbf{p}}(\mathbf{x}_i) = \int_{\mathbf{x_p}-\frac{h}{2}}^{\mathbf{x_p}+\frac{h}{2}} S(\mathbf{x} - \mathbf{x}_i)\, \mathrm{d}\mathbf{x}. \quad (71)$$

The integration extends here over the cubical cell **p**. By introducing the top-hat function

$$\Pi(\mathbf{x}) = \begin{cases} 1 & \text{for } |\mathbf{x}| \le \frac{1}{2}, \\ 0 & \text{otherwise,} \end{cases} \quad (72)$$

we can extend the integration boundaries to all space and write instead:

$$W_{\mathbf{p}}(\mathbf{x}_i) = \int \Pi\left(\frac{\mathbf{x} - \mathbf{x_p}}{h}\right) S(\mathbf{x} - \mathbf{x}_i)\, \mathrm{d}\mathbf{x}. \quad (73)$$

Note that this also shows that the assignment function W is a convolution of Π with S. The full density in grid cell **p** is then given

$$\rho_{\mathbf{p}} = \frac{1}{h^3} \sum_{i=1}^{N} m_i W_{\mathbf{p}}(\mathbf{x}_i). \quad (74)$$

These general formula evidently depend on the specific choice one makes for the shape function $S(\mathbf{x})$. Below, we discuss a few of the most commonly employed low-order assignment schemes.

4.1.2 Nearest Grid Point (NGP) Assignment

The simplest possible choice for S is a Dirac δ-function. One then gets:

$$W_{\mathbf{p}}(\mathbf{x}_i) = \int \Pi\left(\frac{\mathbf{x} - \mathbf{x_p}}{h}\right) \delta(\mathbf{x} - \mathbf{x}_i)\, \mathrm{d}\mathbf{x} = \Pi\left(\frac{\mathbf{x}_i - \mathbf{x_p}}{h}\right). \quad (75)$$

In other words, this means that $W_{\mathbf{p}}$ is either 1 (if the coordinate of particle i lies inside the cell), or otherwise it is zero. Consequently, the mass of particle i is fully assigned to exactly one cell—the nearest grid point, as sketched in Fig. 4.

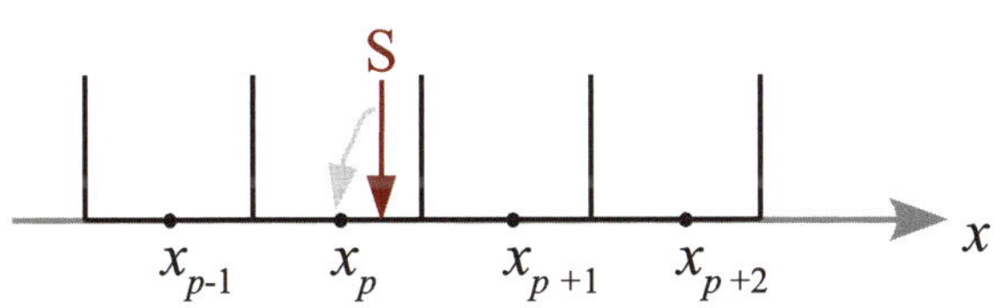

Fig. 4 Sketch of the nearest grid point (NGP) assignment scheme. This simple binning scheme simply assigns the mass of a particle completely to the one mesh cell in which it falls

4.1.3 Clouds-in-cell (CIC) Assignment

Here one adopts as shape function

$$S(\mathbf{x}) = \frac{1}{h^3}\Pi\left(\frac{\mathbf{x}}{h}\right), \tag{76}$$

which is the same cubical 'cloud' shape as that of individual mesh cells. The assignment function is

$$W_{\mathbf{p}}(\mathbf{x}_i) = \int \Pi\left(\frac{\mathbf{x}-\mathbf{x}_{\mathbf{p}}}{h}\right)\frac{1}{h^3}\Pi\left(\frac{\mathbf{x}-\mathbf{x}_i}{h}\right)\,\mathrm{d}\mathbf{x}, \tag{77}$$

which only has a non-zero (and then constant) integrand if the cubes centered on $\mathbf{x}_i$ and $\mathbf{x}_{\mathbf{p}}$ overlap. How can this overlap be calculated? The 1D sketch of Fig. 5 can help to make this clear.

Recall that for one of the dimensions we have $x_p = (p_x + 1/2)h$, for $p \in \{0, 1, 2, \ldots, N-1\}$. For a given particle coordinate x_i we may first calculate a 'floating point index' by inverting this relation, yielding $p_f = x_i/h - 1/2$. The index of the left cell of the two cells with some overlap is then given by $p = \lfloor p_f \rfloor$, where the brackets denote the integer floor, i.e., the largest integer not larger than p_f. We may then further define $p^\star \equiv p_f - p$, which is a number between 0 and 1. From the sketch, we see that the length of the overlap of the particle's cloud with the cell p is $h - hp^\star$, hence the assignment function at cell p takes on the value $W_p = 1 - p^\star$ for this location of the particle, whereas the assignment function for the neighboring cell $p+1$ will take on the value $W_{p+1} = p^\star$.

These considerations readily generalize to 2D and 3D. For example, in 2D (as sketched in Fig. 6), we first assign to the y_i-coordinate of point i a 'floating point index' $q_f = y_i/h - 1/2$. We can then use this to compute a cell index as the integer floor $q = \lfloor q_f \rfloor$, and a fractional contribution $q^\star = q_f - q$. Finally, we obtain the following weights for the assignment of a particle's mass to the four cells its 'cloud' touches in 2D (as sketched):

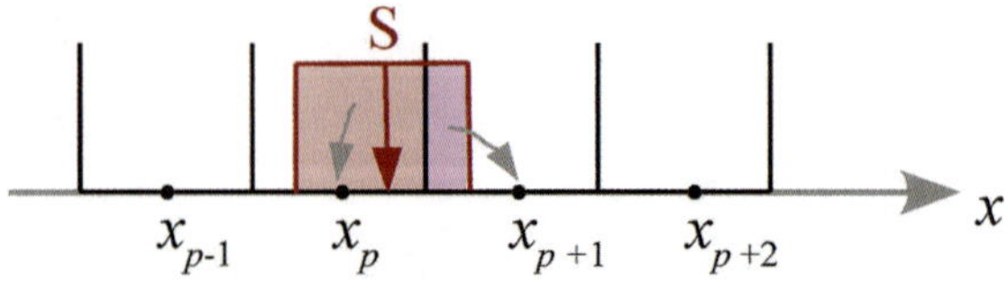

Fig. 5 Sketch of the clouds-in-cell (CIC) assignment scheme. The fraction of mass assigned to a given cell is given by the fraction of the cubical cloud shape of the particle that overlaps with the cell

Fig. 6 Sketch of CIC assignment of a particle to a two-dimensional mesh

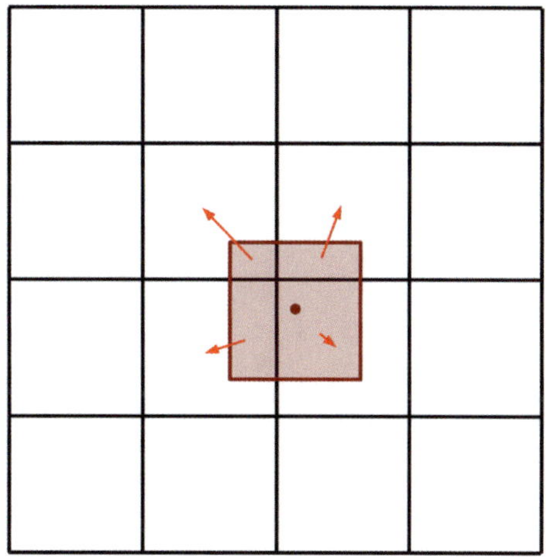

$$\begin{aligned} W_{p,q} &= (1-p^\star)(1-q^\star) && (78)\\ W_{p+1,q} &= p^\star(1-q^\star) && (79)\\ W_{p,q+1} &= (1-p^\star)q^\star && (80)\\ W_{p+1,q+1} &= p^\star q^\star && (81)\end{aligned}$$

In the corresponding 3D case, each particle contributes to the weight functions of 8 cells, or in other words, it is spread over 8 cells.

4.1.4 Triangular Shaped Clouds (TSC) Assignment

One can construct a systematic sequence of ever higher-order shape functions by adding more convolutions with the top-hat kernel. For example, the next higher order (in 3D) is given by

$$W_{\mathbf{p}}(\mathbf{x}_i) = \int \Pi\left(\frac{\mathbf{x}-\mathbf{x}_{\mathbf{p}}}{h}\right) \frac{1}{h^3}\Pi\left(\frac{\mathbf{x}_i-\mathbf{x}-\mathbf{x}'}{h}\right) \frac{1}{h^3}\Pi\left(\frac{\mathbf{x}'}{h}\right) \mathrm{d}\mathbf{x}\,\mathrm{d}\mathbf{x}' \quad (82)$$

$$= \frac{1}{h^6}\int \Pi\left(\frac{\mathbf{x}-\mathbf{x}_{\mathbf{p}}}{h}\right) \Pi\left(\frac{\mathbf{x}_i-\mathbf{x}}{h}\right) \Pi\left(\frac{\mathbf{x}'-\mathbf{x}}{h}\right) \mathrm{d}\mathbf{x}\,\mathrm{d}\mathbf{x}'. \quad (83)$$

This still has a simple geometric interpretation. If one pictures the kernel shape as a triangle with total base length $2h$, then the fraction assigned to a certain cell is given by the area of overlap of this triangle with the cell of interest (see Fig. 7). The triangle will now in general touch 3 cells per dimension, making an evaluation correspondingly more expensive. In 3D, 27 cells are touched for every particle.

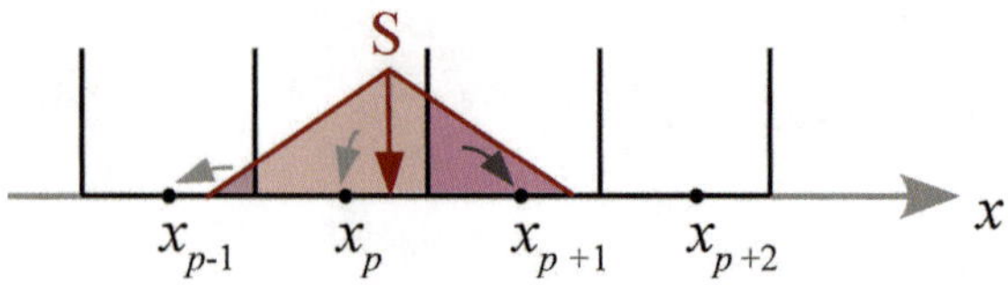

Fig. 7 Sketch of triangular-shaped-clouds (TSC) assignment. Here a particle is spread to three cells in one dimension

Table 1 Commonly used shape functions

Name	Cloud shape $S(x)$	# of cells used	Assignment function shape
NGP	$\delta(x)$	1^d	Π
CIC	$\frac{1}{h^d}\Pi\left(\frac{\mathbf{x}}{h}\right)$	2^d	$\Pi \star \Pi$
TSC	$\frac{1}{h^d}\Pi\left(\frac{\mathbf{x}}{h}\right) \star \frac{1}{h^d}\Pi\left(\frac{\mathbf{x}}{h}\right)$	3^d	$\Pi \star \Pi \star \Pi$

What's the advantage of using TSC over CIC, if any? Or should one stick with the computationally cheap NGP? The assignment schemes differ in the smoothness and differentiability of the reconstructed density field. In particular, for NGP, the assigned density and hence the resulting force jump discontinuously when a particle crosses a cell boundary. The resulting force law will then at best be piece-wise constant.

In contrast, the CIC scheme produces a force that is piece-wise linear and continuous, but its first derivative jumps. Here the information where a particle is inside a certain cell is not completely lost, unlike in NGP.

Finally, TSC is yet smoother, and also the first derivative of the force is continuous. See Table 1 for a brief summary of these assignment schemes. Which of these schemes is the preferred choice is ultimately problem-dependent. In most cases, CIC and TSC are quite good options, providing sufficient accuracy with still reasonably small (and hence computationally efficient) assignment kernels. The latter get invariably more extended for higher-order assignment schemes, which not only is computationally ever more costly but also invokes additional communication overheads in parallelization schemes.

4.1.5 Solving for the Gravitational Potential

Once the density field is obtained, we would like to solve Poisson's equation

$$\nabla^2 \Phi = 4\pi G \rho, \tag{84}$$

and obtain the gravitational potential discretized on the same mesh. There are primarily two methods that are in widespread use for this.

First, there are Fourier-transform based methods which exploit the fact that the potential can be viewed as a convolution of a Green's function with the density field. In

Fourier-space, one can then use the convolution theorem and cast the computationally expensive convolution into a cheap algebraic multiplication. Due to the importance of this approach, we will discuss it extensively in Sect. 4.2.

Second, there are also iterative solvers for Poisson's equation which yield a solution directly in real-space. Simple versions of such iteration schemes use Jacobi or Gauss-Seidel iteration, more complicated ones employ a sophisticated multi-grid approach to speed up convergence. We shall discuss these methods in Sect. 4.3.

4.1.6 Calculation of the Forces

Let's assume for the moment that we already obtained the gravitational potential Φ on the mesh, with one of the methods mentioned above. We would then like to get the acceleration field from

$$\mathbf{a} = -\nabla\Phi. \tag{85}$$

One can achieve this by calculating a numerical derivative of the potential by *finite differencing*. For example, the simplest estimate of the force in the x-direction would be

$$a_x^{(i,j,k)} = -\frac{\Phi^{(i+1,j,k)} - \Phi^{(i-1,j,k)}}{2h}, \tag{86}$$

where $\mathbf{p} = (i, j, k)$ is a cell index. The truncation error of this expression is $\mathcal{O}(h^2)$, hence the estimate of the derivative is second-order accurate.

Alternatively, one can use larger *stencils* to obtain a more accurate finite difference approximation of the derivative, at greater computational cost. For example, the 4-point expression

$$a_x^{(i,j,k)} = -\frac{1}{2h}\left\{\frac{4}{3}\left[\Phi^{(i+1,j,k)} - \Phi^{(i-1,j,k)}\right] - \frac{1}{6}\left[\Phi^{(i+2,j,k)} - \Phi^{(i-2,j,k)}\right]\right\} \tag{87}$$

can be used, which has a truncation error of $\mathcal{O}(h^4)$, as verified through simple Taylor expansions.

For the y- and z-dimensions, corresponding formulae, where j or k are varied and the other cell coordinates are held fixed, can be used. Whether a second- or fourth-order discretization formula should be used depends again on the question which compromise between accuracy and speed is best for a given problem. In many collisionless simulation set-ups, the residual truncation error of the second-order finite difference approximation of the force will be negligible compared to other errors inherent in the simulation methodology, hence the second-order formula would then be expected to be sufficient. But this cannot be generalized to all situations and simulation setups; if in doubt, it is best to explicitly test for this source of error.

4.1.7 Interpolating from the Mesh to the Particles

Once we have the force field on a mesh, we are not yet fully done. We actually desire the forces at the particle coordinates of the N-body system, not at the coordinates of the mesh cells of our auxiliary computational grid. We are hence left with the problem of interpolating the forces from the mesh to the particle coordinates.

Recall that we defined the density field in terms of mass assignment functions, of the form

$$\rho_{\mathbf{p}} = \frac{1}{h^3}\sum_i W_{\mathbf{p}}(\mathbf{x}_i) = \frac{1}{h^3}\sum_i W(\mathbf{x}_i - \mathbf{x}_{\mathbf{p}}). \tag{88}$$

Here we introduced in the last expression an alternative notation for the weight assignment function.

Assume that we have computed the acceleration field on the grid, $\{\mathbf{a}_{\mathbf{p}}\}$. It turns out to be very important to *use the same* assignment kernel as used in the density construction also for the force interpolation, i.e., the force at coordinate $\mathbf{x}$ for a mass m needs to be computed as

$$\mathbf{F}(\mathbf{x}) = m\sum_{\mathbf{p}} \mathbf{a}_{\mathbf{p}} W(\mathbf{x} - \mathbf{x}_{\mathbf{p}}), \tag{89}$$

where W denotes the assignment function used for computing the density field on the mesh. This requirement results from the desire to have a vanishing *self-force*, as well as pairwise antisymmetric forces between every particle pair. The self-force is the force that a particle would feel if just it alone would be present in the system. If numerically this force would evaluate to a non-zero value, the particle would accelerate all by itself, violating momentum conservation. Likewise, for two particles, we require that the forces they mutually exert on each other are equal in magnitude and opposite in direction, such that momentum conservation is manifest.

We now show that using the same kernels for the mass assignment and force interpolation protects against these numerical artefacts (Hockney and Eastwood 1988). We start by noting that the acceleration field at a mesh point $\mathbf{p}$ depends linearly on the mass at another mesh point $\mathbf{p}'$, which is a manifestation of the superposition principle (this can, for example, also be seen when Fourier techniques are used to solve the Poisson equation). We can hence express the field as

$$\mathbf{a}_{\mathbf{p}} = \sum_{\mathbf{p}'} \mathbf{d}(\mathbf{p}, \mathbf{p}')\, h^3 \rho_{\mathbf{p}'}, \tag{90}$$

with a Green's function $\mathbf{d}(\mathbf{p}, \mathbf{p}')$. This vector-valued Green's function for the force is antisymmetric, i.e., it changes sign when the two points in the arguments are swapped. Note that $h^3\rho_{\mathbf{p}'}$ is simply the mass contained in mesh cell $\mathbf{p}'$.

We can now calculate the self-force resulting from the density assignment and interpolation steps:

$$\mathbf{F}_{\rm self}(\mathbf{x}_i) = m_i \mathbf{a}_i(\mathbf{x}_i) = m_i \sum_{\mathbf{p}} W(\mathbf{x}_i - \mathbf{x}_{\mathbf{p}})\mathbf{a}_{\mathbf{p}} \tag{91}$$

$$= m_i \sum_{\mathbf{p}} W(\mathbf{x}_i - \mathbf{x}_{\mathbf{p}}) \sum_{\mathbf{p}'} \mathbf{d}(\mathbf{p}, \mathbf{p}') h^3 \rho_{\mathbf{p}'} \tag{92}$$

$$= m_i \sum_{\mathbf{p}} W(\mathbf{x}_i - \mathbf{x}_{\mathbf{p}}) \sum_{\mathbf{p}'} \mathbf{d}(\mathbf{p}, \mathbf{p}') m_i W(\mathbf{x}_i - \mathbf{x}_{\mathbf{p}'}) \tag{93}$$

$$= m_i^2 \sum_{\mathbf{p},\mathbf{p}'} \mathbf{d}(\mathbf{p}, \mathbf{p}') W(\mathbf{x}_i - \mathbf{x}_{\mathbf{p}}) W(\mathbf{x}_i - \mathbf{x}_{\mathbf{p}'}) \tag{94}$$

$$= 0. \tag{95}$$

Here we have started out with the interpolation from the mesh-based acceleration field, and then inserted the expansion of the latter as convolution over the density field of the mesh. Finally, we put in the density contribution created by the particle i at a mesh cell $\mathbf{p}'$. We then see that the double sum vanishes because of the antisymmetry of $\mathbf{d}$ and the symmetry of the kernel product under exchange of $\mathbf{p}$ and $\mathbf{p}'$. Note that this however only works because the kernels used for force interpolation and density assignment are indeed equal—it would have not worked out if they would be different, which brings us back to the point emphasized above.

Now let's turn to the force antisymmetry. The force exerted on a particle 1 of mass m_1 at location $\mathbf{x}_1$ due to a particle 2 of mass m_2 at location $\mathbf{x}_2$ is given by

$$\mathbf{F}_{12} = m_1 \mathbf{a}(\mathbf{x}_1) = m_1 \sum_{\mathbf{p}} W(\mathbf{x}_1 - \mathbf{x}_{\mathbf{p}})\mathbf{a}_{\mathbf{p}} \tag{96}$$

$$= m_1 \sum_{\mathbf{p}} W(\mathbf{x}_1 - \mathbf{x}_{\mathbf{p}}) \sum_{\mathbf{p}'} \mathbf{d}(\mathbf{p}, \mathbf{p}') h^3 \rho_{\mathbf{p}'} \tag{97}$$

$$= m_1 \sum_{\mathbf{p}} W(\mathbf{x}_1 - \mathbf{x}_{\mathbf{p}}) \sum_{\mathbf{p}'} \mathbf{d}(\mathbf{p}, \mathbf{p}') m_2 W(\mathbf{x}_2 - \mathbf{x}_{\mathbf{p}'}) \tag{98}$$

$$= m_1 m_2 \sum_{\mathbf{p},\mathbf{p}'} \mathbf{d}(\mathbf{p}, \mathbf{p}') W(\mathbf{x}_1 - \mathbf{x}_{\mathbf{p}}) W(\mathbf{x}_2 - \mathbf{x}_{\mathbf{p}'}). \tag{99}$$

Likewise, we obtain for the force experienced by particle 2 due to particle 1:

$$\mathbf{F}_{21} = m_1 m_2 \sum_{\mathbf{p}',\mathbf{p}} \mathbf{d}(\mathbf{p}, \mathbf{p}') W(\mathbf{x}_2 - \mathbf{x}_{\mathbf{p}}) W(\mathbf{x}_1 - \mathbf{x}_{\mathbf{p}'}). \tag{100}$$

We may swap the summation indices through relabeling and exploiting the antisymmetry of $\mathbf{d}$, obtaining:

$$\mathbf{F}_{21} = -m_1 m_2 \sum_{\mathbf{p}',\mathbf{p}} \mathbf{d}(\mathbf{p}, \mathbf{p}') W(\mathbf{x}_1 - \mathbf{x}_{\mathbf{p}}) W(\mathbf{x}_2 - \mathbf{x}_{\mathbf{p}'}). \tag{101}$$

Hence we have $\mathbf{F}_{12} + \mathbf{F}_{21} = 0$, independent on where the points are located on the mesh.

4.2 Fourier Techniques

Fourier transforms provide a powerful tool for solving certain partial differential equations. In this subsection we shall consider the particularly important example of using them to solve Poisson's equation, but we note that the basic technique can be used in similar form also for other systems of equations.

4.2.1 Convolution Problems

Suppose we want to solve Poisson's equation,

$$\nabla^2 \Phi = 4\pi G \rho, \tag{102}$$

for a given density distribution ρ. Actually, we can readily write down a solution for a non-periodic space, since we know the Newtonian potential of a point mass, and the equation is linear. The potential is simply a linear superposition of contributions from individual mass elements, which in the continuum can be written as the integration:

$$\Phi(\mathbf{x}) = -\int G \frac{\rho(\mathbf{x}')\,\mathrm{d}\mathbf{x}'}{|\mathbf{x} - \mathbf{x}'|}. \tag{103}$$

This is recognized to be a convolution integral of the form

$$\Phi(\mathbf{x}) = \int g(\mathbf{x} - \mathbf{x}')\,\rho(\mathbf{x}')\,\mathrm{d}\mathbf{x}', \tag{104}$$

where

$$g(\mathbf{x}) = -\frac{G}{|\mathbf{x}|} \tag{105}$$

is the *Green's function* of Newtonian gravity. The convolution may also be formally written as:

$$\Phi = g \star \rho. \tag{106}$$

We now recall the *convolution theorem*, which says that the Fourier transform of the convolution of two functions is equal to the product of the individual Fourier transforms of the two functions, i.e.,

$$\mathcal{F}(f \star g) = \mathcal{F}(f) \cdot \mathcal{F}(g), \tag{107}$$

where $\mathcal{F}$ denotes the Fourier transform and f and g are the two functions. A convolution in real space can hence be transformed to a much simpler, point-by-point multiplication in Fourier space.

There are many problems where this can be exploited to arrive at efficient calculational schemes, for example in solving Poisson's equation for a given density field. Here the central idea is to compute the potential through

$$\Phi = \mathcal{F}^{-1}\left[\mathcal{F}(g)\cdot\mathcal{F}(\rho)\right], \tag{108}$$

i.e., in Fourier space, with $\hat{\Phi}(\mathbf{k}) \equiv \mathcal{F}(\Phi)$, we have the simple equation

$$\hat{\Phi}(\mathbf{k}) = \hat{g}(\mathbf{k})\cdot\hat{\rho}(\mathbf{k}). \tag{109}$$

4.2.2 The Continuous Fourier Transform

But how do we solve this in practice? Let's first assume that we have *periodic boundary conditions* with a box of size L in each dimension. The continuous $\rho(\mathbf{x})$ can in this case be written as a Fourier series of the form

$$\rho(\mathbf{x}) = \sum_{\mathbf{k}} \rho_{\mathbf{k}}\,\mathrm{e}^{i\mathbf{k}\mathbf{x}}, \tag{110}$$

where the sum over the $\mathbf{k}$-vectors extends over a discrete spectrum of wave vectors, with

$$\mathbf{k} \in \frac{2\pi}{L}\begin{pmatrix} n_1 \\ n_2 \\ n_3 \end{pmatrix}, \tag{111}$$

where n_1, n_2, n_3 are from the set of positive and negative integer numbers. The allowed modes in $\mathbf{k}$ hence form an infinitely extended Cartesian grid with spacing $2\pi/L$. Because of the periodicity condition, only these waves 'fit' into the box. For a real field such as ρ, there is also a reality constraint of the form $\rho_{\mathbf{k}} = \rho^{\star}_{-\mathbf{k}}$, hence the modes are not all independent. The Fourier coefficients can be calculated as

$$\rho_{\mathbf{k}} = \frac{1}{L^3}\int_V \rho(\mathbf{x})\,\mathrm{e}^{-i\mathbf{k}\mathbf{x}}\mathrm{d}\mathbf{x}, \tag{112}$$

where the integration is over one instance of the periodic box.

More generally, the periodic Fourier series features the following orthogonality and closure relationships:

$$\frac{1}{L^3}\int \mathrm{d}\mathbf{x}\;\mathrm{e}^{i(\mathbf{k}-\mathbf{k}')\mathbf{x}} = \delta_{\mathbf{k},\mathbf{k}'}, \tag{113}$$

$$\frac{1}{L^3}\sum_{\mathbf{k}} \mathrm{e}^{i\mathbf{k}\mathbf{x}} = \delta(\mathbf{x}), \tag{114}$$

where the first relation gives a Kronecker delta, the second a Dirac δ-function.

Let's now look at the Poisson equation again and replace the potential and the density field with their corresponding Fourier series:

$$\nabla^2\left(\sum_{\mathbf{k}} \Phi_{\mathbf{k}}\,\mathrm{e}^{i\mathbf{k}\mathbf{x}}\right) = 4\pi G\left(\sum_{\mathbf{k}} \rho_{\mathbf{k}}\,\mathrm{e}^{i\mathbf{k}\mathbf{x}}\right). \tag{115}$$

We see that we can easily carry out the spatial derivate on the left hand side, yielding:

$$\sum_{\mathbf{k}}\left(-\mathbf{k}^2\Phi_{\mathbf{k}}\right)\mathrm{e}^{i\mathbf{k}\mathbf{x}} = 4\pi G\sum_{\mathbf{k}} \rho_{\mathbf{k}}\,\mathrm{e}^{i\mathbf{k}\mathbf{x}}. \tag{116}$$

The equality must hold for each of the Fourier modes separately, hence we infer

$$\Phi_{\mathbf{k}} = -\frac{4\pi G}{\mathbf{k}^2}\rho_{\mathbf{k}}. \tag{117}$$

Comparing with Eq. (109), this means we have identified the Green's function of the Poisson equation in a periodic space as

$$g_{\mathbf{k}} = -\frac{4\pi G}{\mathbf{k}^2}. \tag{118}$$

4.2.3 The Discrete Fourier Transform (DFT)

The above considerations were still for a continuous density field. On a computer, we will usually only have a discretized version of the field $\rho(\mathbf{x})$, defined at a set of points. Assuming we have N equally spaced points per dimension, the $\mathbf{x}$ positions may only take on the discrete positions

$$\mathbf{x}_{\mathbf{p}} = \frac{L}{N}\begin{pmatrix} p_1 \\ p_2 \\ p_3 \end{pmatrix} \quad \text{where } p_1, p_2, p_3 \in \{0, 1, \ldots, N-1\}. \tag{119}$$

With the replacement $\mathrm{d}^3\mathbf{x} \to (L/N)^3$, we can cast the Fourier integral (112) into a discrete sum:

$$\rho_{\mathbf{k}} = \frac{1}{N^3}\sum_{\mathbf{p}} \rho_{\mathbf{p}}\,\mathrm{e}^{-i\mathbf{k}\mathbf{x}_{\mathbf{p}}}. \tag{120}$$

Because of the periodicity and the finite number of density values that is summed over, it turns out that this also restricts the number of $\mathbf{k}$ values that give different answers—shifting $\mathbf{k}$ in any of the dimensions by N times the fundamental mode $2\pi/L$ gives again the same result. We may then for example select as primary set of $\mathbf{k}$-modes the values

$$\mathbf{k}_\mathbf{l} = \frac{2\pi}{L}\begin{pmatrix} l_1 \\ l_2 \\ l_3 \end{pmatrix} \quad \text{where } l_1, l_2, l_3 \in \{0, 1, \ldots, N-1\}, \tag{121}$$

and the construction of ρ through the Fourier series becomes a finite sum over these N^3 modes. We have now arrived at the *discrete Fourier transform* (DFT), which can equally well be written as:

$$\hat{\rho}_\mathbf{l} = \frac{1}{N^3}\sum_\mathbf{p} \rho_\mathbf{p}\, \mathrm{e}^{-i\frac{2\pi}{N}\mathbf{l}\mathbf{p}}, \tag{122}$$

$$\rho_\mathbf{p} = \sum_\mathbf{l} \hat{\rho}_\mathbf{l}\, \mathrm{e}^{i\frac{2\pi}{N}\mathbf{l}\mathbf{p}}. \tag{123}$$

Here are some notes about different aspects of the Fourier pair defined by these relations:

- The two transformations are an invertible linear mapping of a set of N^3 (or N in 1D) complex values $\rho_\mathbf{p}$ to N^3 complex values $\hat{\rho}_\mathbf{l}$, and vice versa.
- To label the frequency values, $\mathbf{k} = (2\pi/L)\cdot\mathbf{l}$, one often conventionally uses the set $l \in \{-N/2, \ldots, -1, 0, 1, \ldots, \frac{N}{2}-1\}$ instead of $l \in \{0, 1, \ldots, N-1\}$, which is always possible because shifting l by multiples of N does not change anything as this yields only a 2π phase factor. With this convention, the occurrence of both negative and positive frequencies is made more explicit, and they are arranged quasi-symmetrically in a box in $\mathbf{k}$-space centered on $\mathbf{k} = (0, 0, 0)$. The box extends out to

$$k_{\rm max} = \frac{N}{2}\frac{2\pi}{L}, \tag{124}$$

  which is the so-called Nyquist frequency (e.g. Diniz et al. 2002). Adding waves beyond the Nyquist frequency in a reconstruction of ρ on a given grid would add redundant information that could not be unambiguously recovered from the discretized density field. (Instead, the power in these waves would be erroneously mapped to lower frequencies—this is called *aliasing*, see also the so-called *sampling theorem.*)
- Parseval's theorem relates the quadratic norms of the transform pair, namely

$$\sum_\mathbf{p} |\rho_\mathbf{p}|^2 = N^3 \sum_\mathbf{l} |\hat{\rho}_\mathbf{l}|^2. \tag{125}$$

- The $1/N^3$ normalization factor could equally well be placed in front of the Fourier series instead of the Fourier transform, or one may split it symmetrically and introduce a factor $1/\sqrt{N^3}$ in front of both. This is just a matter of convention, and all of these alternative conventions are sometimes used.

- In fact, many computer libraries for the DFT will omit the factor N completely and leave it up to the user to introduce it where needed. Commonly, the DFT library functions define as forward transform of a set of N complex numbers x_j, with $j \in \{0, \ldots, N-1\}$, the set of N complex numbers:

$$y_k = \sum_{j=0}^{N-1} x_j \, \mathrm{e}^{-i \frac{2\pi}{N} j \cdot k}. \tag{126}$$

The backwards transform is then defined as

$$y_k = \sum_{j=0}^{N-1} x_j \, \mathrm{e}^{i \frac{2\pi}{N} j \cdot k}. \tag{127}$$

This form of writing the Fourier transform is now nicely symmetric, with the *only difference* between forward and backward transforms being the sign in the exponential function. However, in this case we have that $\mathcal{F}^{-1}(\mathcal{F}(\mathbf{x})) = N\mathbf{x}$, i.e., to get back to the original input vector $\mathbf{x}$ one must eventually divide by N. Note that the multi-dimensional transforms are simply Cartesian products of one-dimensional transforms, i.e., those are obtained as straightforward generalizations of the one-dimensional definition.

- Computing the DFT of N numbers has in principal a computational cost of order $\mathcal{O}(N^2)$. This is because for each of the N numbers one has to calculate N terms and sum them up. Fortunately, in 1965, the *Fast Fourier Transform* (FFT) algorithm (Cooley and Tukey 1965) has been discovered (interestingly, Gauss had already known something similar; Gauss 1866). This method for calculating the DFT subdivides the problem recursively into smaller and smaller blocks. It turns out that this divide and conquer strategy can reduce the computational cost to $\mathcal{O}(N \log N)$, which is a very significant difference. The result of the FFT algorithm is mathematically identical to the DFT. But actually, in practice, the FFT is even better than a direct computation of the DFT, because as an aside the FFT algorithm also reduces the numerical floating point round-off error that would otherwise be incurred. It is ultimately only because of the existence of the FFT algorithm that Fourier methods are so widely used in numerical calculations and applicable to even very large problem sizes.

4.2.4 Storage Conventions for the DFT

Most numerical libraries for computing the FFT store both the original field and its Fourier transform as simple arrays indexed by $k \in \{0, \ldots, N-1\}$. The negative frequencies will then be stored in the upper half of the array, consistent with what

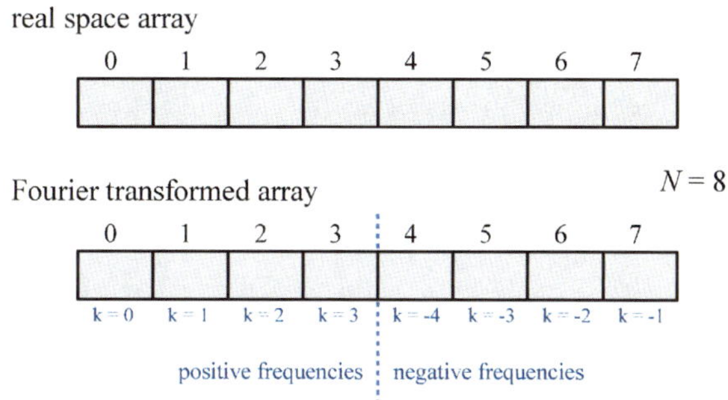

Fig. 8 Commonly employed storage convention for DFTs. The positive frequencies are stored in the *lower half* of the array, the negative ones in the *upper half*

one obtains by subtracting N from the linear index. The example shown in Fig. 8 for $N = 8$ in 1D may help to make this clear.

Correspondingly, in 2D, the grid of real-space values is mapped to a grid of k-space values of the same dimensions. Again, negative frequencies seem to be stored 'backwards', with the smallest negative frequency having the largest linear index, and the most negative frequency appearing as first value past the middle of the mesh. But note that this is consistent with the translational invariance in k-space with respect to shifts of the indices by multiples of N.

Finally, when we have a real real-space field (such as the physical density), the discrete Fourier transform fulfills a reality constraint of the form $\hat{\rho}_{\mathbf{k}} = \hat{\rho}^{\star}_{-\mathbf{k}}$. This implies a set of relations between the complex values that make up the Fourier transform of ρ, reducing the number of values that can be chosen arbitrarily. What does this imply in the discrete case? Consider the sketch shown in Fig. 9, in which regions of like color are related to each other by the reality constraint. Note that $k_x = N/2$ indices are aliased to themselves under complex conjugation, i.e., negating this gives $k_x = -N/2$, but since N can be added, this mode really maps again to $k_x = N/2$. Nevertheless, for the yellow regions there are always different partner cells when one considers the corresponding $-\mathbf{k}$ cell. Only for the orange cells this

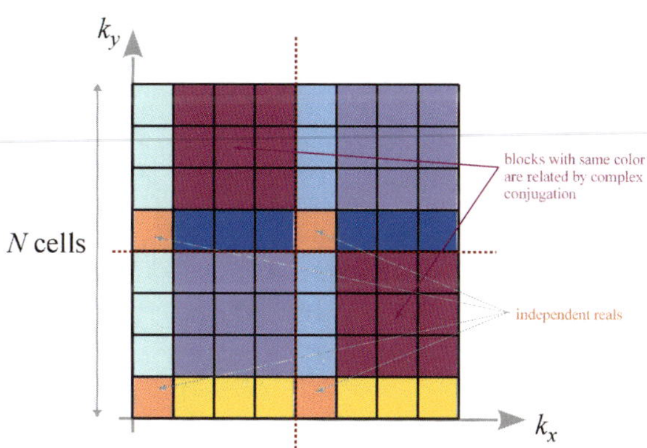

Fig. 9 Sketch illustrating the implications of the reality constraint for the FFT of a field of reals in 2D. Different pairs of cells are related to each other as complex conjugate numbers (labeled as *colored blocks*), and some are aliased to themselves (*orange*) so that they end up being real

is not the case; those are mapped to themselves and are hence real due to the reality constraint.

If we now count how many independent numbers we have in the Fourier transformed grid of a 2D real field, we find

$$2\left(\frac{N}{2}-1\right)^2 \times 2 + 4\left(\frac{N}{2}-1\right) \times 2 + 4 \times 1. \tag{128}$$

The first term accounts for the two square-shaped regions that have different mirrored regions. Those contain $(\frac{N}{2}-1)^2$ complex numbers, each with two independent real and imaginary values. Then there are 4 different sections of rows and columns that are related to each other by mirroring in k-space. Those contain $(\frac{N}{2}-1)$ complex numbers each. Finally, there are 4 independent cells that are real and hence account for one independent value each. Reassuringly, the sum of Eq. (128) works out to N^2, which is the result we expect: the number of independent values in Fourier space must be exactly equal to the N^2 real values we started out with, otherwise we would not expect a strictly reversible transformation.

4.2.5 Non-periodic Problems with 'Zero Padding'

Can we use the FFT/DFT techniques discussed above also to calculate non-periodic force fields? At first, this may seem impossible since the DFT is intrinsically periodic. However, through the so-called zero-padding trick one can circumvent this limitation.

Let's discuss the procedure based on a 2D example (it works also in 1D or 3D, of course):

1. We need to arrange our mesh such that the source distribution lives only in one quarter of the mesh, the rest of the density field needs to be zeroed out. Schematically we hence have the situation depicted in Fig. 10.

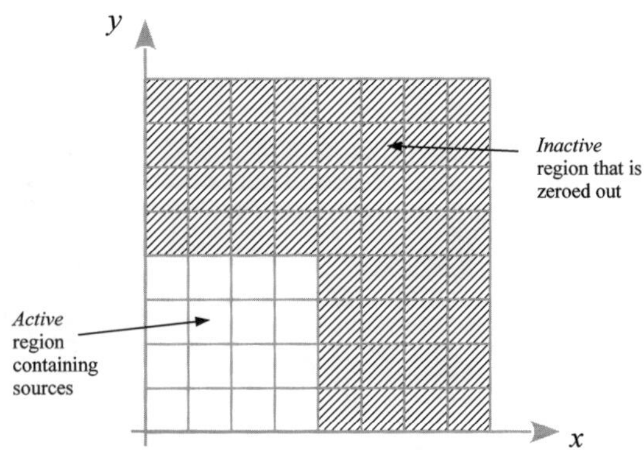

Fig. 10 Sketch of zero padding used to treat non-periodic problems with the discrete Fourier transform

2. We now set up our desired real-space Green's function, i.e., this is the response of a mass at the origin. The Green's function for the whole mesh is set-up as $g_{N-i,j} = g_{i,N-j} = g_{N-i,N-j} = g_{i,j}$ where $0 \leq i,\, j \leq N/2$. This is equivalent to defining g everywhere on the mesh, and using as relevant distance the distance to the *nearest periodic image* of the origin. Note that by replicating g with the condition of periodicity, the tessellated mesh then effectively yields a Green's function that is nicely symmetric around the origin.
3. We now want to carry out the real-space convolution

$$\phi = g \star \rho \tag{129}$$

by using the definition of the discrete, periodic convolution

$$\Phi_{\mathbf{p}} = \sum_{\mathbf{n}} g_{\mathbf{p}-\mathbf{n}}\, \rho_{\mathbf{n}}, \tag{130}$$

where both g and ρ are treated as periodic fields for which adding multiples of N to the indices does not change anything. We see that this sum indeed yields the correct result for the non-periodic potential in the quarter of the mesh that contains our source distribution. This is because the Green's function 'sees' only one copy of the source distribution in this sector; the zero-padded region is big enough to prevent any cross-talk from the (existing) periodic images of the source distribution. This is different in the other three quadrants of the mesh. Here we obtain incorrect potential values that are basically useless and need to be discarded.
4. Given that Eq. (130) yields the correct result in the region of the mesh covered by the sources, we may now just as well use periodic FFTs in the usual way to carry out this convolution quickly! A downside of this procedure is that it features an enlarged cost in terms of CPU and memory usage. Because we have to effectively double the mesh-size compared to the corresponding periodic problem, the cost goes up by a factor of 4 in 2D, and by a factor of 8 in 3D.
5. We note that James (1977) proposed an ingenious trick based that allows a more efficient treatment of isolated source distributions. Through suitably determined correction masses on the boundaries, the memory and CPU cost can be reduced compared to the zero-padding approach described above.

4.3 Multigrid Techniques

Let's return once more to the problem of solving Poisson's equation,

$$\nabla^2 \Phi = 4\pi G \rho, \tag{131}$$

and consider first the one-dimensional problem, i.e.,

$$\frac{\partial^2 \Phi}{\partial x^2} = 4\pi G\rho(x). \tag{132}$$

The spatial derivative on the left hand-side can be approximated as

$$\left(\frac{\partial^2 \Phi}{\partial x^2}\right)_i \simeq \frac{\Phi_{i+1} - 2\Phi_i + \Phi_{i-1}}{h^2}, \tag{133}$$

where we have assumed that Φ is discretized with N points on a regular mesh with spacing h, and i is the cell index. This means that we have the equations

$$\frac{\Phi_{i+1} - 2\Phi_i + \Phi_{i-1}}{h^2} = 4\pi G\rho_i. \tag{134}$$

There are N of these equations, for the N unknowns Φ_i, with $i \in \{0, 1, \ldots, N-1\}$. This means we should in principle be able to solve this algebraically! In other words, the system of equations can be rewritten as a standard linear set of equations, in the form

$$\mathbf{A}\mathbf{x} = \mathbf{b}, \tag{135}$$

with a vector of unknowns, $\mathbf{x} = (\Phi_i)$, and a right hand side $\mathbf{b} = \frac{4\pi G}{h^2}\boldsymbol{\rho}$. In the 1D case, the matrix $\mathbf{A}$ (assuming periodic boundary conditions) is explicitly given as

$$\mathbf{A} = \begin{pmatrix} -2 & 1 & & & & 1 \\ 1 & -2 & 1 & & & \\ & 1 & -2 & 1 & & \\ & & \cdots & & & \\ & & & 1 & -2 & 1 \\ 1 & & & & 1 & -2 \end{pmatrix}. \tag{136}$$

Solving equation (135) directly constitutes a matrix inversion that can in principle be carried out by LU-decomposition or Gauss elimination with pivoting (e.g. Press et al. 1992). However, the computational cost of these procedures is of order $\mathcal{O}(N^3)$, meaning that it becomes extremely costly with growing N, and rather sooner than later infeasible, already for problems of small to moderate size.

4.3.1 Jacobi Iteration

However, if we are satisfied with an approximate solution, then we can turn to iterative solvers that are much faster. Suppose we decompose the matrix $\mathbf{A}$ as

$$\mathbf{A} = \mathbf{D} - (\mathbf{L} + \mathbf{U}), \tag{137}$$

where $\mathbf{D}$ is the diagonal part, $\mathbf{L}$ is the (negative) lower diagonal part and $\mathbf{U}$ is the upper diagonal part. Then we have

$$[\mathbf{D} - (\mathbf{L} + \mathbf{U})]\,\mathbf{x} = \mathbf{b}, \tag{138}$$

and from this

$$\mathbf{x} = \mathbf{D}^{-1}\mathbf{b} + \mathbf{D}^{-1}(\mathbf{L} + \mathbf{U})\mathbf{x}. \tag{139}$$

We can use this to define an iterative sequence of vectors $\mathbf{x}^n$:

$$\mathbf{x}^{(n+1)} = \mathbf{D}^{-1}\mathbf{b} + \mathbf{D}^{-1}(\mathbf{L} + \mathbf{U})\mathbf{x}^{(n)}. \tag{140}$$

This is called Jacobi iteration (e.g. Saad 2003). Note that $\mathbf{D}^{-1}$ is trivially obtained because $\mathbf{D}$ is diagonal. I.e., here $(\mathbf{D}^{-1})_{ii} = 1/\mathbf{A}_{ii}$.

The scheme converges if and only if the so-called convergence matrix

$$\mathbf{M} = \mathbf{D}^{-1}(\mathbf{L} + \mathbf{U}) \tag{141}$$

has only eigenvalues that are less than 1, or in other words, that the spectral radius $\rho_s(\mathbf{M})$ fullfils

$$\rho_s(\mathbf{M}) \equiv \max_i |\lambda_i| < 1. \tag{142}$$

We can easily derive this condition by considering the error vector of the iteration. At step n it is defined as

$$\mathbf{e}^{(n)} \equiv \mathbf{x}_{\text{exact}} - \mathbf{x}^{(n)}, \tag{143}$$

where $\mathbf{x}_{\text{exact}}$ is the exact solution. We can use this to write the error at step $n+1$ of the iteration as

$$\mathbf{e}^{(n+1)} = \mathbf{x}_{\text{exact}} - \mathbf{x}^{(n+1)} = \mathbf{x}_{\text{exact}} - \mathbf{D}^{-1}\mathbf{b} - \mathbf{D}^{-1}(\mathbf{L}+\mathbf{U})\mathbf{x}^{(n)} = \mathbf{M}\mathbf{x}_{\text{exact}} - \mathbf{M}\mathbf{x}^{(n)} = \mathbf{M}\mathbf{e}^{(n)} \tag{144}$$

Hence we find

$$\mathbf{e}^{(n)} = \mathbf{M}^n\mathbf{e}^{(0)}. \tag{145}$$

This implies $|\mathbf{e}^{(n)}| \leq [\rho_s(\mathbf{M})]^n |\mathbf{e}^{(0)}|$, and hence convergence if the spectral radius is smaller than 1.

For completeness, we state the Jacobi iteration rule for the Poisson equation in 3D when a simple 2-point stencil is used in each dimension for estimating the corresponding derivatives:

$$\begin{aligned} \Phi_{i,j,k}^{(n+1)} &= \frac{1}{6}\Big(\Phi_{i+1,j,k} + \Phi_{i-1,j,k} + \Phi_{i,j+1,k} + \Phi_{i,j-1,k} + \Phi_{i,j,k+1} + \Phi_{i,j,k-1} \\ &\quad - 4\pi G h^2 \rho_{i,j,k}\Big). \end{aligned} \tag{146}$$

4.3.2 Gauss-Seidel Iteration

The central idea of Gauss-Seidel iteration is to use the updated values as soon as they become available for computing further updated values. We can formalize this as follows. Adopting the same decomposition of $\mathbf{A}$ as before, we can write

$$(\mathbf{D}-\mathbf{L})\mathbf{x} = \mathbf{U}\mathbf{x} + \mathbf{b}, \tag{147}$$

from which we obtain

$$\mathbf{x} = (\mathbf{D}-\mathbf{L})^{-1}\mathbf{U}\mathbf{x} + (\mathbf{D}-\mathbf{L})^{-1}\mathbf{b}, \tag{148}$$

suggesting the iteration rule

$$\mathbf{x}^{(n+1)} = (\mathbf{D}-\mathbf{L})^{-1}\mathbf{U}\mathbf{x}^{(n)} + (\mathbf{D}-\mathbf{L})^{-1}\mathbf{b}. \tag{149}$$

This seems at first problematic, because we can't easily compute $(\mathbf{D}-\mathbf{L})^{-1}$. But we can modify the last equation as follows:

$$\mathbf{D}\mathbf{x}^{(n+1)} = \mathbf{U}\mathbf{x}^{(n)} + \mathbf{L}\mathbf{x}^{(n+1)} + \mathbf{b}. \tag{150}$$

From which we get the alternative form:

$$\mathbf{x}^{(n+1)} = \mathbf{D}^{-1}\mathbf{U}\mathbf{x}^{(n)} + \mathbf{D}^{-1}\mathbf{L}\mathbf{x}^{(n+1)} + \mathbf{D}^{-1}\mathbf{b}. \tag{151}$$

Again, this may seem of little help because it looks like $\mathbf{x}^{(n+1)}$ would only be implicitly given. However, if we start computing the new elements in the first row $i = 1$ of this matrix equation, we see that no values of $\mathbf{x}^{(n+1)}$ are actually needed, because $\mathbf{L}$ has only elements below the diagonal. For the same reason, if we then proceed with the second row $i = 2$, then with $i = 3$, etc., only elements of $\mathbf{x}^{(n+1)}$ from rows above the current one are needed. So we can calculate things in this order without problem and make use of the already updated values. It turns out that this speeds up the convergence quite a bit, with one Gauss-Seidel step often being close to two Jacobi steps.

4.3.3 Red Black Ordering

A problematic point about Gauss-Seidel is that the equations have to be solved in a specific sequential order, meaning that this part cannot be parallelized. Also, the result will in general depend on which element is selected to be the first. To overcome this problem, one can sometimes use so-called red-black ordering, which effectively is a compromise between Jacobi and Gauss-Seidel.

Certain update rules, such as that for the Poisson equation, allow a decomposition of the cells into disjoint sets whose update rules depend only on cells from other

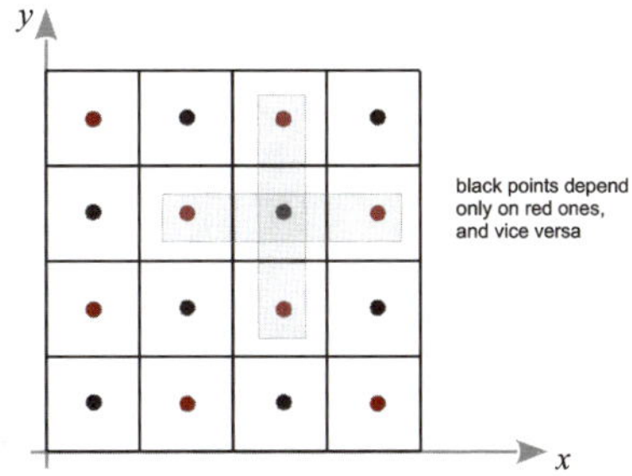

Fig. 11 *Red-black* ordering in which two interleaved chessboard-like patterns are formed that can be independently processed with immediate updating

sets, as shown in Fig. 11. For example, for the Poisson equation, this is the case for a chess-board like pattern of 'red' and 'black' cells.

One can then first update all the black points (which rely only on the red points), followed by an update of all the red points (which rely only on the black ones). In the second of these two half-steps, one can then use the updated values from the first half-step, making it intuitively clear why such a scheme can almost double the convergence rate relative to Jacobi.

4.3.4 The Multigrid Technique

Iterative solvers like Jacobi or Gauss-Seidel often converge quite slowly, in fact, the convergence seems to "stall" after a few steps and proceeds only anemically. One also observes that high-frequency errors in the solution are damped out quickly by the iterations, but long-wavelength errors die out much more slowly. Intuitively this is not unexpected: In every iteration, only neighboring points communicate, so the information "travels" only by one cell (or more generally, one stencil length) per iteration. And for convergence, it has to propagate back and forth over the whole domain a few times.

Idea By going to a coarser mesh, we may be able to compute an improved initial guess which may help to speed up the convergence on the fine grid (Brandt 1977). Note that on the coarser mesh, the relaxation will be computationally cheaper (since there are only $1/8$ as many points in 3D, or $1/4$ in 2D), and the convergence rate should be faster, too, because the perturbation is there less smooth and effectively on a smaller scale relative to the coarser grid.

So schematically, we, for example, might imagine an iteration scheme where we first iterate the problem $\mathbf{Ax} = \mathbf{b}$ on a mesh with cells $4h$, i.e., for times coarser than the fine mesh. Once we have a solution there, we continue to iterate it on a mesh coarsened with cell size $2h$, and only finally we iterate to solution on the fine mesh with cell size h.

A couple of questions immediately come up when we want to work out the details of this basic idea:

1. How do we get from a coarse solution to a guess on a finer grid?
2. How should we solve $\mathbf{Ax} = \mathbf{b}$ on the coarsened mesh?
3. What if there is still an error left with long wavelength on the fine grid?

In order to make things work, we clearly need mappings from a finer grid to a coarser one, and vice versa. This is the most important issue to solve.

4.3.5 Prolongation and Restriction Operations

Coarse-to-fine This transition is an interpolation step, or in the language of multigrid methods (Briggs et al. 2000), it is called *prolongation*. Let $\mathbf{x}^{(h)}$ be a vector defined on a mesh $\Omega^{(h)}$ with N cells and spacing h, covering our computational domain. Similarly, let $\mathbf{x}^{(2h)}$ be a vector living on a coarser mesh $\Omega^{(2h)}$ with twice the spacing and half as many points per dimension. We now define a linear interpolation operator $\mathbf{I}_{2h}^{h}$ that maps points from the coarser to the fine mesh, as follows:

$$\mathbf{I}_{2h}^{h}\,\mathbf{x}^{(2h)} = \mathbf{x}^{(h)}. \tag{152}$$

A simple realization of this operator in 2D would be the following:

$$\mathbf{I}_{2h}^{h}: \quad \begin{aligned} x_{2i}^{(h)} &= x_i^{(2h)} \\ x_{2i+1}^{(h)} &= \tfrac{1}{2}(x_i^{(2h)} + x_{i+1}^{(2h)}) \end{aligned} \qquad \text{for } 0 \le i < \tfrac{N}{2}. \tag{153}$$

Here, every second point is simply injected from the coarse to the fine mesh, and the intermediate points are linearly interpolated from the neighboring points, which here boils down to a simple arithmetic average.

Fine-to-coarse The converse mapping represents a smoothing operation, or a *restriction* in multigrid-language. We can define the restriction operator as

$$\mathbf{I}_{h}^{2h}\,\mathbf{x}^{(h)} = \mathbf{x}^{(2h)}, \tag{154}$$

which hence takes a vector defined on the fine grid $\Omega^{(h)}$ to one that lives on the coarse grid $\Omega^{(2h)}$. Again, let's give a simple realization example in 2D:

$$\mathbf{I}_{h}^{2h}: \; x_i^{(2h)} = \frac{x_{2i-1}^{(h)} + 2x_{2i}^{(h)} + x_{2i+1}^{(h)}}{4} \quad \text{for } 0 \le i < \frac{N}{2}. \tag{155}$$

Evidently, this is a smoothing operation with a simple 3-point stencil.

One usually chooses these two operators such that the transpose of one is proportional to the other, i.e., they are related as follows:

$$\mathbf{I}_{h}^{2h} = c\,[\mathbf{I}_{2h}^{h}]^{\mathrm{T}}, \tag{156}$$

where c is a real number.

In a shorter notation, the above prolongation operator can be written as

$$\text{1D-prolongation,} \quad \mathbf{I}_{2h}^{h} : \quad \left] \frac{1}{2} \;\; 1 \;\; \frac{1}{2} \right[, \tag{157}$$

which means that every coarse point is added with these weights to three points of the fine grid. The fine-grid points accessed with weight $1/2$ will get contributions from two coarse grid points. Similarly, the restriction operator can be written with the short-hand notation

$$\text{1D-restriction,} \quad \mathbf{I}_{h}^{2h} : \quad \left[\frac{1}{4} \;\; \frac{1}{2} \;\; \frac{1}{4} \right] . \tag{158}$$

This expresses that every coarse grid point is a weighted sum of three fine grid points.

For reference, we also state the corresponding low-order prolongation and restriction operators in 2D:

$$\text{2D-prolongation,} \quad \mathbf{I}_{2h}^{h} : \quad \left] \begin{array}{ccc} \frac{1}{4} & \frac{1}{2} & \frac{1}{4} \\ \frac{1}{2} & 1 & \frac{1}{2} \\ \frac{1}{4} & \frac{1}{2} & \frac{1}{4} \end{array} \right[\tag{159}$$

$$\text{2D-restriction,} \quad \mathbf{I}_{h}^{2h} : \quad \left[\begin{array}{ccc} \frac{1}{16} & \frac{1}{8} & \frac{1}{16} \\ \frac{1}{8} & \frac{1}{4} & \frac{1}{8} \\ \frac{1}{16} & \frac{1}{8} & \frac{1}{16} \end{array} \right] \tag{160}$$

Corresponding extensions to 3D can be readily derived.

4.3.6 The Multigrid V-Cycle

An important role in the multigrid approach plays the error vector, defined as

$$\mathbf{e} \equiv \mathbf{x}_{\text{exact}} - \tilde{\mathbf{x}}, \tag{161}$$

where $\mathbf{x}_{\text{exact}}$ is the exact solution, and $\tilde{\mathbf{x}}$ the (current) approximate solution. Another important concept is the *residual*, defined as

$$\mathbf{r} \equiv \mathbf{b} - \mathbf{A}\tilde{\mathbf{x}}. \tag{162}$$

Note that the pair of error and residual are solutions of the original linear system, i.e., we have

$$\mathbf{A}\mathbf{e} = \mathbf{r}. \tag{163}$$

Coarse-grid correction scheme We now define a function that is supposed to return an improved solution $\tilde{\mathbf{x}}'^{(h)}$ for the problem $\mathbf{A}^{(h)}\mathbf{x}^{(h)} = \mathbf{b}^{(h)}$ on grid level h, based on some starting guess $\tilde{\mathbf{x}}^{(h)}$ and a right hand side $\mathbf{b}^{(h)}$. This so-called *coarse grid correction*,

$$\tilde{\mathbf{x}}'^{(h)} = \mathrm{CG}(\tilde{\mathbf{x}}^{(h)}, \mathbf{b}^{(h)}), \tag{164}$$

proceeds along the following steps:

1. Carry out a relaxation step on h (for example by using one Gauss-Seidel or one Jacobi iteration).
2. Compute the residual: $\mathbf{r}^{(h)} = \mathbf{b}^{(h)} - \mathbf{A}^{(h)}\tilde{\mathbf{x}}^{(h)}$.
3. Restrict the residual to a coarser mesh: $\mathbf{r}^{(2h)} = \mathbf{I}_h^{2h}\,\mathbf{r}^{(h)}$.
4. Solve $\mathbf{A}^{(2h)}\mathbf{e}^{(2h)} = \mathbf{r}^{(2h)}$ on the coarsened mesh, with $\tilde{\mathbf{e}}^{(2h)} = 0$ as initial guess.
5. Prolong the obtained error $\mathbf{e}^{(2h)}$ to the finer mesh, $\mathbf{e}^{(h)} = \mathbf{I}_{2h}^{h}\,\mathbf{e}^{(2h)}$, and use it to correct the current solution on the fine grid: $\tilde{\mathbf{x}}'^{(h)} = \tilde{\mathbf{x}}^{(h)} + \mathbf{e}^{(h)}$.
6. Carry out a further relaxation step on the fine mesh h.

How do we carry out step 4 in this scheme? We can use recursion! Because what we have to do in step 4 is exactly the job description of the function CG(., .). However, we also need a stopping condition for the recursion, which is simply a prescription that tells us under which conditions we should skip steps 2–5 in the above scheme. We can do this by simply saying that further coarsening of the problem should stop once we have reached a minimum number of cells N. At this point we either just do the relaxation steps, or we solve the remaining problem exactly.

V-Cycle When the coarse grid correction scheme is recursively called, we arrive at the schematic diagram shown in Fig. 12 for how the iteration progresses, which is called a V-cycle. It turns out that the V-cycle rather dramatically speeds up the convergence rate of simple iterative solvers for linear systems of equations. It is easy

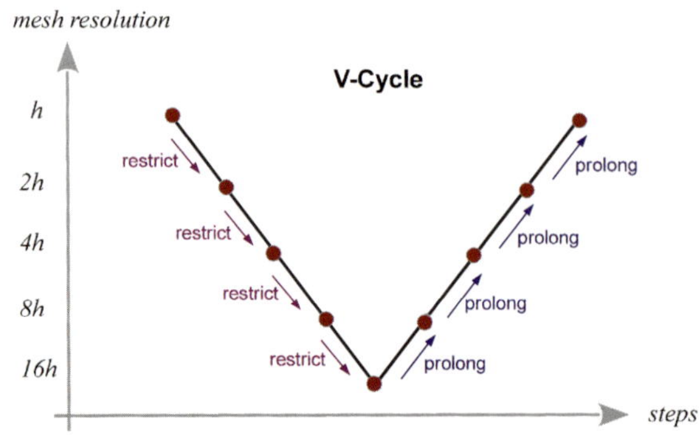

Fig. 12 The typical V-cycle of a multigrid iteration scheme. The current solution on a fine mesh is recursively restricted to coarser meshes. Coarse-grid corrections are then prolonged back up to the finer meshes, interleaved with one Gauss-Seidel or Jacobi iteration at the corresponding mesh level

to show that the computational cost of one V-cycle is of order $\mathcal{O}(N_{\rm grid})$, where $N_{\rm grid}$ is the number of grid cells on the fine mesh. A convergence to truncation error (i.e., machine precision) requires several V-cycles and involves a computational cost of order $\mathcal{O}(N_{\rm grid} \log N_{\rm grid})$. For the Poisson equation, this is the same cost scaling as one gets with FFT-based methods. In practice, good implementations of the two schemes should roughly be equally fast. In cosmology, a multigrid solver is for example used by the MLAPM (Knebe et al. 2001) and RAMSES codes (Teyssier 2002). An interesting advantage of multigrid is that it requires less data communication when parallelized on distributed memory machines.

One problem we haven't addressed yet is how one finds the operator $\mathbf{A}^{(2h)}$ required on the coarse mesh. The two most commonly used options for this are:

- Direct coarse grid approximation: Here one simply uses the same discrete equations on the coarse grid as on the fine grid, just scaled by the grid resolution h as needed. In this case, the stencil of the matrix does not change.
- Galerkin coarse grid approximation: Here one defines the coarse operator as

$$\mathbf{A}^{(2h)} = \mathbf{I}_h^{2h}\, \mathbf{A}^{(h)}\, \mathbf{I}_{2h}^{h}, \tag{165}$$

  which is formally the most consistent way of defining $\mathbf{A}^{(2h)}$, and in this sense optimal. However, computing the matrix in this way can be a bit cumbersome, and it may involve a growing size of the stencil, which then leads to an enlarged computational cost.

4.3.7 The Full Multigrid Method

The V-cycle scheme discussed thus far still relies on an initial guess for the solution, and if this guess is bad, one has to do more V-cycles to reach satisfactory convergence. This raises the question on how one may get a good guess. If one is dealing with the task of repeatedly having to solve the same problem over and over again with only small changes from solution to solution (as will often be the case in dynamical simulation problems) one may be able to simply use the solution from the previous timestep as a guess. In all other cases, one can allude to the following idea: Let's get a good guess by solving the problem on a coarser grid first, and then interpolate the coarse solution to the fine grid as a starting guess.

But at the coarser grid, one is then again confronted with the task to solve the problem without a starting guess. Well, we can simply recursively apply the idea again, and delegate the finding of a good guess to a yet coarser grid, etc. This then yields the *full multigrid cycle*, as depicted in Fig. 13. It involves the following steps:

1. Initialize the right hand side on all grid levels, $\mathbf{b}^{(h)}, \mathbf{b}^{(2h)}, \mathbf{b}^{(4h)}, \ldots, \mathbf{b}^{(H)}$, down to some coarsest level H.
2. Solve the problem (exactly) on the coarsest level H.
3. Given a solution on level i with spacing $2h$, map it to the next level $i+1$ with spacing h and obtain the initial guess $\tilde{\mathbf{x}}^{(h)} = I_{2h}^{h}\, \mathbf{x}^{(2h)}$.

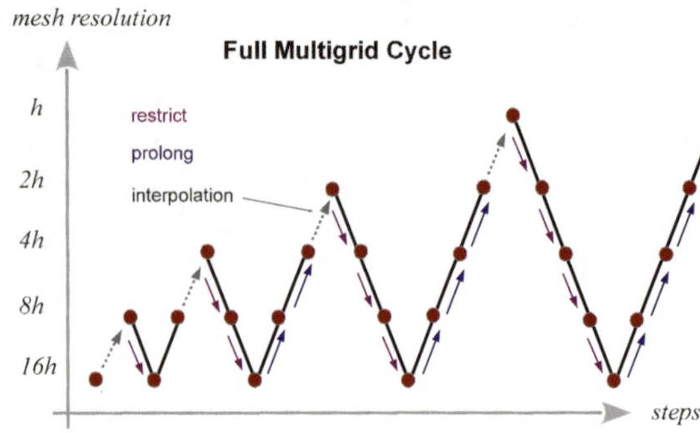

Fig. 13 The full multigrid cycle in which also the problem of finding an adequate starting guess is addressed

4. Use this starting guess to solve the problem on the level $i + 1$ with one V-cycle.
5. Repeat Step 3 until the finest level is reached.

The computational cost of such a full multigrid cycle is still of order the number of mesh cells, as before.

4.4 Hierarchical Multipole Methods ("tree Codes")

Another approach for a real-space evaluation of the gravitational field are so-called tree codes (Barnes and Hut 1986). In cosmology, they are for example used in the PKD-GRAV/GASOLINE (Wadsley et al. 2004) and GADGET (Springel et al. 2001; Springel 2005) codes.

4.4.1 Multipole Expansion

The central idea is here to use the multipole expansion of a distant group of particle to describe its gravity, instead of summing up the forces from all individual particles.

The potential of the group is given by

$$\Phi(\mathbf{r}) = -G \sum_i \frac{m_i}{|\mathbf{r} - \mathbf{x}_i|}, \tag{166}$$

which we can re-write as

$$\Phi(\mathbf{r}) = -G \sum_i \frac{m_i}{|\mathbf{r} - \mathbf{s} + \mathbf{s} - \mathbf{x}_i|}. \tag{167}$$

Now we expand the denominator assuming $|\mathbf{x}_i - \mathbf{s}| \ll |\mathbf{r} - \mathbf{s}|$, which will be the case provided the *opening angle* θ under which the group is seen is sufficiently small, as sketched in Fig. 14. We can then use the Taylor expansion

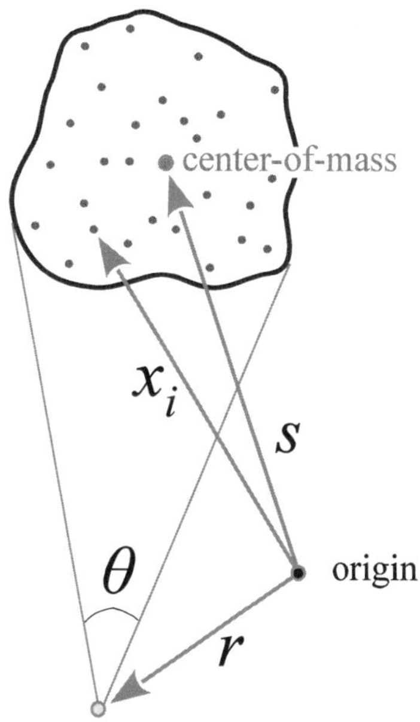

Fig. 14 Multipole expansion for a group of distant particles. Provided the reference point **r** is sufficiently far away, the particles are seen under a small opening angle θ, and the field created by the particle group can be approximated by the monopole term at its center of mass, augmented with higher order multipole corrections if desired

$$\frac{1}{|\mathbf{y}+\mathbf{s}-\mathbf{x}_i|} = \frac{1}{|\mathbf{y}|} - \frac{\mathbf{y}\cdot(\mathbf{s}-\mathbf{x}_i)}{|\mathbf{y}|^3} + \frac{1}{2}\frac{\mathbf{y}^T\left[3(\mathbf{s}-\mathbf{x}_i)(\mathbf{s}-\mathbf{x}_i)^T - (\mathbf{s}-\mathbf{x}_i)^2\right]\mathbf{y}}{|\mathbf{y}|^5} + \cdots, \tag{168}$$

where we introduced $\mathbf{y} \equiv \mathbf{r} - \mathbf{s}$ as a short-cut. The first term on the right hand side gives rise to the monopole moment, the second to the dipole moment, and the third to the quadrupole moment. If desired, one can continue the expansion to ever higher order terms.

These multipole moments then become properties of the group of particles:

$$\text{monopole:} \quad M = \sum_i m_i, \tag{169}$$

$$\text{quadrupole:} \quad Q_{ij} = \sum_k m_k \left[3(\mathbf{s}-\mathbf{x}_k)_i(\mathbf{s}-\mathbf{x}_k)_j - \delta_{ij}(\mathbf{s}-\mathbf{x}_k)^2\right]. \tag{170}$$

The dipole vanishes, because we carried out the expansion relative to the center-of-mass, defined as

$$\mathbf{s} = \frac{1}{M}\sum_i m_i \mathbf{x}_i. \tag{171}$$

If we restrict ourselves to terms of up to quadrupole order, we hence arrive at the expansion

$$\Phi(\mathbf{r}) = -G\left(\frac{M}{|\mathbf{y}|} + \frac{1}{2}\frac{\mathbf{y}^T\mathbf{Q}\mathbf{y}}{|\mathbf{y}|^5}\right), \qquad \mathbf{y} = \mathbf{r} - \mathbf{s}, \tag{172}$$

from which also the force can be readily obtained through differentiation. Recall that we expect the expansion to be accurate if

$$\theta \simeq \frac{\langle |\mathbf{x}_i - \mathbf{s}| \rangle}{|\mathbf{y}|} \simeq \frac{l}{y} \ll 1, \tag{173}$$

where l is the radius of the group.

4.4.2 Hierarchical Grouping

Tree algorithms are based on a hierarchical grouping of the particles, and for each group, one then pre-computes the multipole moments for later use in approximations of the force due to distant groups. Usually, the hierarchy of groups is organized with the help of a tree-like data structure, hence the name "tree algorithms".

There are different strategies for defining the groups. In the popular Barnes and Hut (1986) oct-tree, one starts out with a cube that contains all the particles. This cube is then subdivided into 8 sub-cubes of half the size in each spatial dimension. One continues with this refinement recursively until each subnode contains only a single particle. Empty nodes (sub-cubes) need not be stored. Figure 15 shows a schematic sketch how this can look like in two dimensions.

- We note that the oct-tree is not the only possible grouping strategy. Sometimes kd-trees (Stadel 2001), or binary trees where subdivisions are done along alternating spatial axes are used.
- An important property of such hierarchical, tree-based groupings is that they are geometrically highly flexible and adjust to any clustering state the particles may have. They are hence automatically adaptive.
- Also, there is no significant slow-down when severe clustering starts.
- The simplest way to construct the hierarchical grouping is to sequentially insert particles into the tree, and then to compute the multipole moments recursively.

4.4.3 Tree Walk

The force calculation with the tree then proceeds by *walking the tree*. Starting at the root node, one checks for every node whether the opening angle under which it is seen is smaller than a prescribed tolerance angle θ_c. If this is the case, the multipole expansion of the node can be accepted, and the corresponding partial force is evaluated and added to an accumulation of the total force. The tree walk along this branch of the tree can then be stopped. Otherwise, one must open the tree node and consider all its sub-nodes in turn.

The resulting force is *approximate* by construction, but the overall size of the error can be conveniently controlled by the tolerance opening angle θ_c (see also Salmon and Warren 1994). If one makes this smaller, more nodes will have to be opened. This

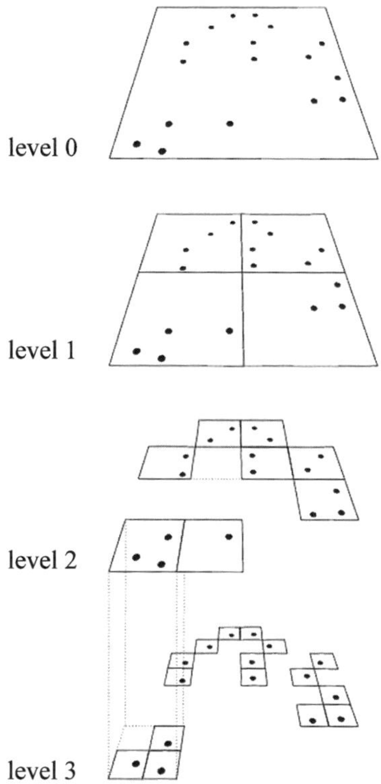

Fig. 15 Organization of the Barnes and Hut (1986) tree in two dimensions (quad tree). All particles are enclosed in a *square-shaped box*. This is then hierarchically subdivided until each particle finds itself in a node on its own. Empty cells do not need to be stored

will make the residual force errors smaller, but at the price of a higher computational cost. In the limit of $\theta_c \to 0$ one gets back to the expensive direct summation force.

An interesting variant of this approach to walk the tree is obtained by not only expanding the potential on the source side into a multipole expansion, but also around the target coordinate. This can yield a substantial additional acceleration and results in so-called fast multipole methods (FFM). The FALCON code of Dehnen (2000, 2002) employs this approach. A further advantage of the FFM formulation is that force anti-symmetry is manifest, so that momentum conservation to machine precision can be achieved. Unfortunately, the speed advantages of FFM compared to ordinary tree codes are significantly alleviated once individual time-step schemes are considered. Also, FFM is more difficult to parallelize efficiently on distributed memory machines.

4.4.4 Cost of the Tree-Based Force Computations

How do we expect the total cost of the tree algorithm to scale with particle number N? For simplicity, let's consider a sphere of size R containing N particles that are approximately homogeneously distributed. The mean particle spacing of these particles will then be

$$d = \left[\frac{(4\pi/3)R^3}{N} \right]^{1/3}. \tag{174}$$

We now want to estimate the number of nodes that we need for calculating the force on a central particle in the middle of the sphere. We can identify the computational cost with the number of interaction terms that are needed. Since the used nodes must tessellate the sphere, their number can be estimated as

$$N_{\rm nodes} = \int_d^R \frac{4\pi r^2 \, {\rm d}r}{l^3(r)}, \tag{175}$$

where $l(r)$ is the expected node size at distance r, and d is the characteristic distance of the nearest particle. Since we expect the nodes to be close to their maximum allowed size, we can set $l \simeq \theta_c r$. We then obtain

$$N_{\rm nodes} = \frac{4\pi}{\theta_c^3} \ln \frac{R}{d} \propto \frac{\ln N}{\theta_c^3}. \tag{176}$$

The total computational cost for a calculation of the forces for all particles is therefore expected to scale as $\mathcal{O}(N \ln N)$. This is a very significant improvement compared with the N^2-scaling of direct summation.

We may also try to estimate the expected typical force errors. If we keep only monopoles, the error in the force per unit mass from one node should roughly be of order the truncation error, i.e., about

$$\Delta F_{\rm node} \sim \frac{G M_{\rm node}}{r^2} \theta^2. \tag{177}$$

The errors from multipole nodes will add up in quadrature, hence

$$(\Delta F_{\rm tot})^2 \sim N_{\rm node} (\Delta F_{\rm node})^2 = N_{\rm node} \left(\frac{G M_{\rm node}}{r^2} \theta^2 \right)^2 \propto \frac{\theta^4}{N_{\rm node}} \propto \theta^7. \tag{178}$$

The force error for a monopoles-only scheme therefore scales as $(\Delta F_{\rm tot}) \propto \theta^{3.5}$, roughly inversely as the invested computational cost. A much more detailed analysis of the performance characteristics of tree codes can be found, for example, in Hernquist (1987).

4.5 TreePM Schemes

While the high adaptivity of tree algorithms is particularly ideal for strongly clustered particle distributions and when a high spatial force accuracy is desired, the mesh-based approaches are usually faster when only a coarsely resolved gravitational field on large scales is required. In particular, the particle-mesh (PM) approach based on Fourier techniques is probably the fastest method to calculate the gravitational field on a homogenous mesh. The obvious limitation of this method is however that the force resolution cannot be better than the size of one mesh cell, and the latter cannot be easily made small enough to resolve all the scales of interest in cosmological simulations.

One interesting idea is to try to combine both approaches into a unified scheme, where the gravitational field on large scales is calculated with a PM algorithm, while the short-range forces are delivered by a hierarchical tree method. Such TreePM schemes have first been proposed by Xu (1995) and Bagla (2002), and a version similar to that of Bagla (2002) is implemented in the GADGET2 code (Springel 2005).

In order to achieve a clean separation of scales, one can consider the potential in Fourier space. The individual modes $\Phi_{\mathbf{k}}$ can be decomposed into a long-range and a short-range part, as follows:

$$\Phi_{\mathbf{k}} = \Phi_{\mathbf{k}}^{\mathrm{long}} + \Phi_{\mathbf{k}}^{\mathrm{short}}, \tag{179}$$

where

$$\Phi_{\mathbf{k}}^{\mathrm{long}} = \Phi_{\mathbf{k}} \exp(-\mathbf{k}^2 r_s^2), \tag{180}$$

and

$$\Phi_{\mathbf{k}}^{\mathrm{short}} = \Phi_{\mathbf{k}}[1 - \exp(-\mathbf{k}^2 r_s^2)], \tag{181}$$

with r_s describing the spatial scale of the force-split. Due to the exponential cut-off of the Fourier-spectrum of the long-range force, a PM grid of finite size can be used to fully resolve this force component (this is achieved once the cell size is a few times smaller than r_s). Compared to the ordinary PM-scheme, the only change is that the Greens function in Fourier-space gets an additional exponential smoothing factor. Thanks to this force-shaping factor, inaccuracies such as force anisotropies from the mesh geometry can be made arbitrarily small, so that the long-range force in the transition region between the force components is accurately computed by the PM scheme.

To calculate the short-range force, one transforms Eq. (181) back to real space. Assuming a single point mass m somewhere in a periodic box of size L, this becomes for $r_s \ll L$:

$$\Phi^{\mathrm{short}}(\mathbf{x}) = -G\frac{m}{r}\mathrm{erfc}\left(\frac{r}{2r_s}\right), \tag{182}$$

where $r = \min(|\mathbf{x} - \mathbf{r} - \mathbf{n}L|)$ is defined as the smallest distance of any of the periodic images ($\mathbf{n}$ is an arbitrary integer triplet) of the point mass at $\mathbf{r}$ relative to the point $\mathbf{x}$. Now, this is recognized as the ordinary Newtonian potential, modified with a truncation factor that rapidly turns off the force at a finite distance of order r_s. In fact, the force drops to about 1 % of its Newtonian value for $r \simeq 4.5r_s$, and quickly becomes completely negligible at still larger separations.

The potential (182) can still be treated with a hierarchical tree algorithm, except for the simplification that any tree node more distant than a finite cut-off range (of order $\sim 5r_s$) can be immediately discarded in the tree walk. This can yield a significant speed-up relative to a plain tree code, because the tree-walk can now be restricted to a small region around the target particle as opposed to having to be carried out for the full volume. Also, periodic boundary conditions do not have to be included explicitly through Ewald summation (Hernquist et al. 1991) any more, rather they are absorbed in the periodic PM force. Another advantage is that for close to homogeneous particle distributions, the PM method used for long-range forces delivers a precise force quickly, whereas a pure tree code struggles in this regime to reach the required force accuracy, simply because here large forces in all directions, which almost completely compensate in the end, need to be evaluated with high relative accuracy, otherwise they do not cancel out properly. Finally, the hybrid TreePM scheme also offers the possibility to split the time integration into a less frequent evaluation of the long-range force, and a more frequent evaluation of the short-range tree force, because the former is associated with longer dynamical time scales than the latter. This can be exploited to realize additional efficiency gains, and can in principle even be done in a symplectic fashion (Saha and Tremaine 1992; Springel 2005).

5 Basic Gas Dynamics

Gravity is the dominant driver behind cosmic structure formation (e.g. Mo et al. 2010), but at small scales hydrodynamics in the baryonic components becomes very important, too. In this section we very briefly review the basic equations and some prominent phenomena related to gas dynamics in order to make the discussion of the numerical fluid solvers used in galaxy evolution more accessible. For a detailed introduction to hydrodynamics, the reader is referred to the standard textbooks on this subject (e.g. Landau and Lifshitz 1959; Shu 1992).

5.1 Euler and Navier-Stokes Equations

The gas flows in astrophysics are often of extremely low density, making internal friction in the gas extremely small. In the limit of assuming internal friction to be completely absent, we arrive at the so-called ideal gas dynamics as described by the Euler equations. Most calculations in cosmology and galaxy formation are carried out under this assumption. However, in certain regimes, viscosity may still become important (for example in the very hot plasma of rich galaxy clusters), hence we shall also briefly discuss the hydrodynamical equations in the presence of physical viscosity, the Navier-Stokes equations, which in a sense describe *real* fluids as opposed to ideal ones. Phenomena such as fluid instabilities or turbulence are also best understood if one does not neglect viscosity completely.

5.1.1 Euler Equations

If internal friction in a gas flow can be neglected, the dynamics of the fluid is governed by the Euler equations:

$$\frac{\partial \rho}{\partial t} + \nabla(\rho \mathbf{v}) = 0, \tag{183}$$

$$\frac{\partial}{\partial t}(\rho \mathbf{v}) + \nabla(\rho \mathbf{v}\mathbf{v}^T + P) = 0, \tag{184}$$

$$\frac{\partial}{\partial t}(\rho e) + \nabla[(\rho e + P)\mathbf{v}] = 0, \tag{185}$$

where $e = u + \mathbf{v}^2/2$ is the total energy per unit mass, and u is the thermal energy per unit mass. Each of these equations is a continuity law, one for the mass, one for the momentum, and one for the total energy. The equations hence form a set of hyperbolic conservation laws. In the form given above, they are not yet complete, however. One still needs a further expression that gives the pressure in terms of the other thermodynamic variables. For an ideal gas, the pressure law is

$$P = (\gamma - 1)\rho u, \tag{186}$$

where $\gamma = c_p/c_v$ is the ratio of specific heats. For a monoatomic gas, we have $\gamma = 5/3$.

5.1.2 Navier-Stokes Equations

Real fluids have internal stresses, due to *viscosity*. The effect of viscosity is to dissipate relative motions of the fluid into heat. The Navier-Stokes equations are then given by

$$\frac{\partial \rho}{\partial t} + \nabla(\rho \mathbf{v}) = 0, \tag{187}$$

$$\frac{\partial}{\partial t}(\rho \mathbf{v}) + \nabla(\rho \mathbf{v}\mathbf{v}^T + P) = \nabla \mathbf{\Pi}, \tag{188}$$

$$\frac{\partial}{\partial t}(\rho e) + \nabla[(\rho e + P)\mathbf{v}] = \nabla(\mathbf{\Pi}\mathbf{v}). \tag{189}$$

Here $\mathbf{\Pi}$ is the so-called viscous stress tensor, which is a material property. For $\mathbf{\Pi} = 0$, the Euler equations are recovered. To first order, the viscous stress tensor must be a linear function of the velocity derivatives (Landau and Lifshitz 1959). The most general tensor of rank-2 of this type can be written as

$$\mathbf{\Pi} = \eta \left[\nabla \mathbf{v} + (\nabla \mathbf{v})^T - \frac{2}{3}(\nabla \cdot \mathbf{v})\mathbf{1} \right] + \xi(\nabla \cdot \mathbf{v})\mathbf{1}, \tag{190}$$

where **1** is the unit matrix. Here η scales the traceless part of the tensor and describes the shear viscosity. ξ gives the strength of the diagonal part, and is the so-called bulk viscosity. Note that η and ξ can in principle be functions of local fluid properties, such as ρ, T, etc.

Incompressible fluids In the following we shall assume constant viscosity coefficients. Also, we specialize to incompressible fluids with $\nabla \cdot \mathbf{v} = 0$, which is a particularly important case in practice. Let's see how the Navier-Stokes equations simplify in this case. Obviously, ξ is then unimportant and we only need to deal with shear viscosity. Now, let us consider one of the components of the viscous shear force described by Eq. (188):

$$\begin{aligned}\frac{1}{\eta}(\nabla\,\mathbf{\Pi})_x &= \frac{\partial}{\partial x}\left(2\frac{\partial v_x}{\partial x}\right) + \frac{\partial}{\partial y}\left(\frac{\partial v_x}{\partial y} + \frac{\partial v_y}{\partial x}\right) + \frac{\partial}{\partial z}\left(\frac{\partial v_x}{\partial z} + \frac{\partial v_z}{\partial x}\right) \\ &= \left(\frac{\partial^2}{\partial x^2} + \frac{\partial^2}{\partial y^2} + \frac{\partial^2}{\partial z^2}\right) v_x = \nabla^2 v_x, \end{aligned} \tag{191}$$

where we made use of the $\nabla \cdot \mathbf{v} = 0$ constraint. If we furthermore introduce the *kinematic viscosity* ν as

$$\nu \equiv \frac{\eta}{\rho}, \tag{192}$$

we can write the equivalent of Eq. (188) in the compact form

$$\frac{\mathrm{D}\mathbf{v}}{\mathrm{D}t} = -\frac{\nabla P}{\rho} + \nu \nabla^2 \mathbf{v}, \tag{193}$$

where the derivative on the left-hand side is the Lagrangian derivative,

$$\frac{\mathrm{D}}{\mathrm{D}t} = \frac{\partial}{\partial t} + \mathbf{v} \cdot \nabla. \tag{194}$$

We hence see that the motion of individual fluid elements responds to pressure gradients and to viscous forces. The form (193) of the equation is also often simply referred to as the Navier-Stokes equation.

5.1.3 Scaling Properties of Viscous Flows

Consider the Navier-Stokes equations for some flow problem that is characterized by some characteristic length L_0, velocity V_0, and density scale ρ_0. We can then define dimensionless fluid variables of the form

$$\hat{\mathbf{v}} = \frac{\mathbf{v}}{V_0}, \qquad \hat{\mathbf{x}} = \frac{\mathbf{x}}{L_0}, \qquad \hat{P} = \frac{P}{\rho_0 V_0^2}. \tag{195}$$

Similarly, we define a dimensionless time, a dimensionless density, and a dimensionless Nabla operator:

$$\hat{t} = \frac{t}{L_0/V_0}, \qquad \hat{\rho} = \frac{\rho}{\rho_0}, \qquad \hat{\nabla} = L_0 \nabla. \tag{196}$$

Inserting these definitions into the Navier-Stokes equation (193), we obtain the dimensionless equation

$$\frac{D\hat{\mathbf{v}}}{D\hat{t}} = -\frac{\hat{\nabla}\hat{P}}{\hat{\rho}} + \frac{\nu}{L_0 V_0}\hat{\nabla}^2\hat{\mathbf{v}}. \tag{197}$$

Interestingly, this equation involves one number,

$$\mathrm{Re} \equiv \frac{L_0 V_0}{\nu}, \tag{198}$$

which characterizes the flow and determines the structure of the possible solutions of the equation. This is the so-called Reynolds number. Problems which have similar Reynolds number are expected to exhibit very similar fluid behavior. One then has *Reynolds-number similarity*. In contrast, the Euler equations ($\mathrm{Re} \to \infty$) exhibit always scale similarity because they are invariant under scale transformations.

One intuitive interpretation one can give the Reynolds number is that it measures the importance of inertia relative to viscous forces. Hence:

$$\mathrm{Re} \approx \frac{\text{inertial forces}}{\text{viscous forces}} \approx \frac{D\mathbf{v}/Dt}{\nu\nabla^2\mathbf{v}} \approx \frac{V_0/(L_0/V_0)}{\nu V_0/L_0^2} = \frac{L_0 V_0}{\nu}. \tag{199}$$

If we have $\mathrm{Re} \sim 1$, we are completely dominated by viscosity. On the other hand, for $\mathrm{Re} \to \infty$ viscosity becomes unimportant and we approach an ideal gas.

5.2 Shocks

An important feature of hydrodynamical flows is that they can develop shock waves in which the density, velocity, temperature and specific jump by finite amounts (e.g. Toro 1997). In the case of the Euler equations, such shocks are true mathematical discontinuities. Interestingly, shocks can occur even from perfectly smooth initial conditions, which is a typical feature of hyperbolic partial differential equations. In fact, acoustic waves with sufficiently large amplitude will suffer from wave-steeping (because the slightly hotter wave crests travel faster than the colder troughs), leading eventually to shocks. Of larger practical importance in astrophysics are however the shocks that occur when flows collide supersonically; here kinetic energy is irreversibly transferred into thermal energy, a process that also manifests itself with an increase in entropy.

In the limit of vanishing viscosity (i.e., for the Euler equations), the differential form of the fluid equations breaks down at the discontinuity of a shock, but the integral form (the *weak formulation*) remains valid. In other words this means that the flux of mass, momentum and energy must remain continuous at a shock front. Assuming that the shock connects two piecewise constant states, this leads to the Rankine-Hugoniot jump conditions (Rankine 1870). If we select a frame of reference where the shock is stationary ($v_s = 0$) and denote the pre-shock state with (v_1, P_1, ρ_1), and the post-shock state as (v_2, P_2, ρ_2) (hence $v_1, v_2 > 0$), we have

$$\rho_1 v_2 = \rho_2 v_2, \tag{200}$$

$$\rho_1 v_1^2 + P_1 = \rho_2 v_2^2 + P_2, \tag{201}$$

$$(\rho_1 e_1 + P_1) v_1 = (\rho_2 e_2 + P_2) v_2. \tag{202}$$

For an ideal gas, the presence of a shock requires that the pre-shock gas streams supersonically into the discontinuity, i.e., $v_1 > c_1$, where $c_1^2 = \gamma P_1/\rho_1$ is the sound speed in the pre-shock phase. The Mach number

$$\mathcal{M} = \frac{v_1}{c_1} \tag{203}$$

measures the strength of the shock ($\mathcal{M} > 1$). The shock itself decelerates the fluid and compresses it, so that we have $v_2 < v_1$ and $\rho_2 > \rho_1$. It also heats it up, so that $T_2 > T_1$, and makes the postshock flow subsonic, with $v_2/c_2 < 1$. Manipulating Eqs. (200)–(202), we can express the relative jumps in the thermodynamic quantities (density, temperature, entropy, etc.) through the Mach number alone, for example:

$$\frac{\rho_2}{\rho_1} = \frac{(\gamma + 1)\mathcal{M}^2}{(\gamma - 1)\mathcal{M}^2 + 2}. \tag{204}$$

5.3 Fluid Instabilities

In many situations, gaseous flows can be subject to fluid instabilities in which small perturbations can rapidly grow, thereby tapping a source of free energy. An important example of this are Kelvin-Helmholtz and Rayleigh-Taylor instabilities, which we briefly discuss in this subsection.

Stability of a shear flow We consider a flow in the x-direction, which in the lower half-space $z < 0$ has velocity U_1 and density ρ_1, whereas in the upper half-space the gas streams with U_2 and has density ρ_2. In addition there can be a homogeneous gravitational field $\mathbf{g}$ pointing into the negative z-direction, as sketched in Fig. 16.

The stability of the flow can be analysed through perturbation theory. To this end, one can for example treat the flow as an incompressible potential flow, and carry out

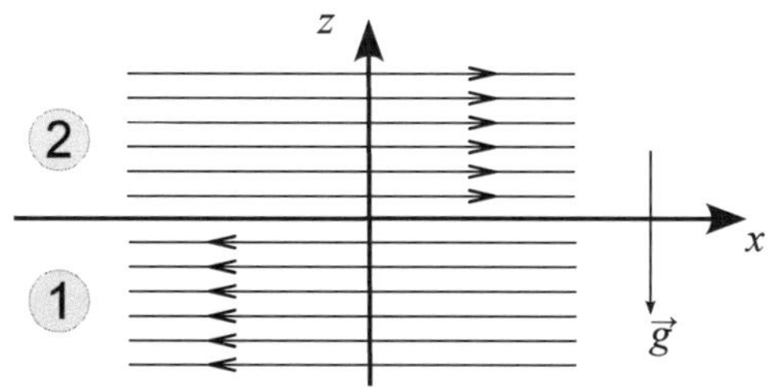

Fig. 16 Geometry of a generic shear flow

an Eigenmode analysis in Fourier space. With the help of Bernoulli's theorem one can then derive an equation for a function $\xi(x,t) = z$ that describes the z-location of the interface between the two phases of the fluid. Details of this calculation can for example be found in Pringle and King (2007). For a single perturbative Fourier mode

$$\xi = \hat{\xi}\exp[i(kx - \omega t)], \tag{205}$$

one then finds that non-trivial solutions with $\hat{\xi} \neq 0$ are possible for

$$\omega^2(\rho_1 + \rho 2) - 2\omega k(\rho_1 U_1 + \rho_2 U_2) + k^2(\rho_1 U_1^2 + \rho_2 U_2^2) + (\rho_2 - \rho 1)kg = 0, \tag{206}$$

which is the *dispersion relation*. Unstable, exponentially growing mode solutions appear if there are solutions for ω with positive imaginary part. Below, we examine the dispersion relation for a few special cases.

Rayleigh-Taylor instability Let us consider the case of a fluid at rest, $U_1 = U_2 = 0$. The dispersion relation simplifies to

$$\omega^2 = \frac{(\rho_1 - \rho_2)kg}{\rho_1 + \rho_2}. \tag{207}$$

We see that for $\rho_2 > \rho_1$, i.e., the denser fluid lies on top, unstable solutions with $\omega^2 < 0$ exist. This is the so-called Rayleigh-Taylor instability. It is in essence buoyancy driven and leads to the rise of lighter material underneath heavier fluid in a stratified atmosphere, as illustrated in the simulation shown in Fig. 17. The free energy that is tapped here is the potential energy in the gravitational field. Also notice that for an ideal gas, arbitrary small wavelengths are unstable, and those modes will grow fastest. If on the other hand we have $\rho_1 > \rho_2$, then the interface is stable and will only oscillate when perturbed.

Kelvin-Helmholtz instability If we set the gravitational field to zero, $g = 0$, we have the situation of a pure shear flow. In this case, the solutions of the dispersion relation are given by

$$\omega_{1/2} = \frac{k(\rho_1 U_1 + \rho_2 U_2)}{\rho_1 + \rho_2} \pm ik\frac{\sqrt{\rho_1\rho_2}}{\rho_1 + \rho_2}|U_1 - U_2|. \tag{208}$$

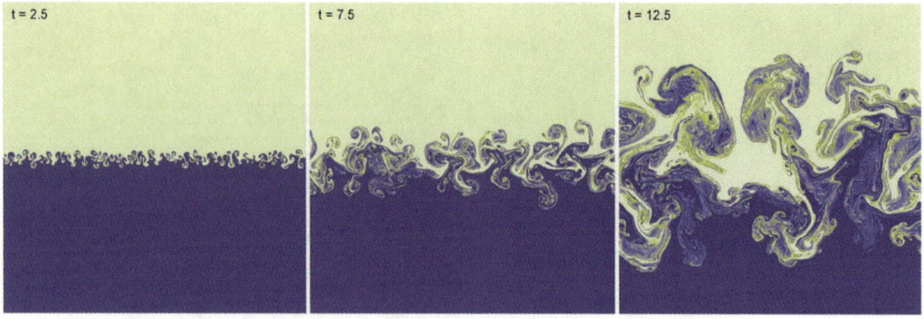

Fig. 17 A growing Rayleigh-Taylor instability in which a lighter fluid (*blue*) is covered by a heaver fluid (*yellow*)

Interestingly, in an ideal gas there is an imaginary growing mode component for every $|U_1 - U_2| > 0$! This means that a small wave-like perturbation at an interface will grow rapidly into large waves that take the form of characteristic Kelvin-Helmholtz "billows". In the non-linear regime reached during the subsequent evolution of this instability the waves are rolled up, leading to the creation of vortex like structures, as seen in Fig. 18. As the instability grows fastest for small scales (high k), the billows tend to get larger and larger with time.

Because the Kelvin-Helmholtz instability basically means that any sharp velocity gradient in a shear flow is unstable in a freely streaming fluid, this instability is particularly important for the creation of fluid turbulence. Under certain conditions, some modes can however be stabilized against the instability. This happens for example if we consider shearing with $U_1 \neq U_2$ in a gravitational field $g > 0$. Then the dispersion relation has the solutions

$$\omega = \frac{k(\rho_1 U_1 + \rho_2 U_2)}{\rho_1 + \rho_2} \pm \frac{\sqrt{-k^2 \rho_1 \rho_2 (U_1 - U_2)^2 - (\rho_1 + \rho_2)(\rho_2 - \rho_1) k g}}{\rho_1 + \rho_2}. \quad (209)$$

Stability is possible if two conditions are met. First, we need $\rho_1 > \rho_2$, i.e., the lighter fluid needs to be on top (otherwise we would have in any case a Rayleigh-Taylor instability). Second, the condition

Fig. 18 Characteristic Kelvin-Helmholtz billows arising in a shear flow

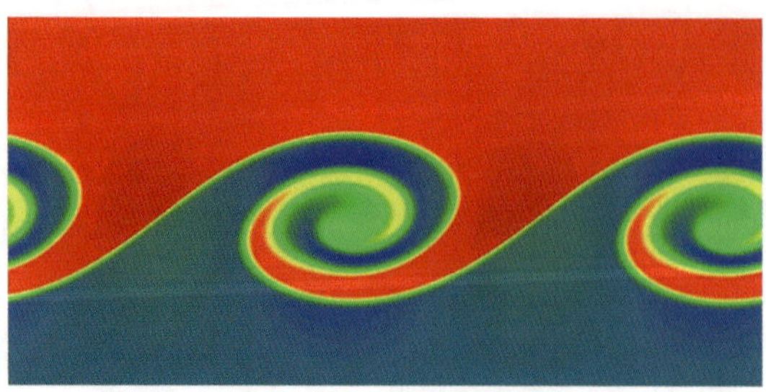

$$(U_1 - U_2)^2 < \frac{(\rho_1 + \rho_2)(\rho_1 - \rho_2)g}{k\rho_1\rho_2} \tag{210}$$

must be fulfilled. Compared to the ordinary Kelvin-Holmholtz instability without a gravitational field, we hence see that sufficiently small wavelengths are stabilized below a threshold wavelength. The larger the shear becomes, the further this threshold moves to small scales.

The Rayleigh-Taylor and Kelvin-Helmholtz instabilities are by no means the only fluid instabilities that can occur in an ideal gas (Pringle and King 2007). For example, there is also the Richtmyer-Meshov instability, which can occur when an interface is suddenly accelerated, for example due to the passage of a shock wave. In self-gravitating gases, there is the Jeans instability, which occurs when the internal gas pressure is not strong enough to prevent a positive density perturbation from growing and collapsing under its own gravitational attraction. This type of instability is particularly important in cosmic structure growth and star formation. If the gas dynamics is coupled to external sources of heat (e.g., through a radiation field), a number of further instabilities are possible. For example, a thermal instability (Field 1965) can occur when a radiative cooling function has a negative dependence on temperature. If the temperature drops somewhere a bit more through cooling than elsewhere, the cooling rate of this cooler patch will increase such that it is cooling even faster. In this way, cool clouds can drop out of the background gas.

5.4 Turbulence

Fluid flow which is unsteady, irregular, seemingly random, and chaotic is called *turbulent* (Pope 2000). Familiar examples of such situations include the smoke from a chimney, a waterfall, or the wind field behind a fast car or airplane. The characteristic feature of turbulence is that the fluid velocity varies significantly and irregularly both in position and time. As a result, turbulence is a statistical phenomenon and is best described with statistical techniques.

If the turbulent motions are subsonic, the flow can often be approximately treated as being incompressible, even for an equation of state that is not particularly stiff. Then only solenoidal motions that are divergence free can occur, or in other words, only shear flows are present. We have already seen that such flows are subject to fluid instabilities such as the Kelvin-Helmholtz instability, which can easily produce swirling motions on many different scales. Such vortex-like motions, also called *eddies*, are the conceptual building blocks of Kolmogorov's theory of incompressible turbulence (Kolmogorov 1941), which yields a surprisingly accurate description of the basic phenomenology of turbulence, even though many aspects of turbulence are still not fully understood.

5.4.1 Kolmogorov's Theory of Incompressible Turbulence

We consider a fully turbulent flow with characteristic velocity U_0 and length scale L_0. We assume that a quasi-stationary state for the turbulence is achieved by some kind of driving process on large scales, which in a time-averaged way injects an energy ϵ per unit mass. We shall also assume that the Reynolds number Re is large. We further imagine that the turbulent flow can be considered to be composed of eddies of different size l, with characteristic velocity $u(l)$, and associated timescale $\tau(l) = l/u(l)$.

For the largest eddies, $l \sim L_0$ and $u(l) \sim U_0$, hence viscosity is unimportant for them. But large eddies are unstable und break up, transferring their energy to somewhat smaller eddies. This continues to yet smaller scales, until

$$\mathrm{Re}(l) = \frac{l u(l)}{\nu} \tag{211}$$

reaches of order unity, where ν is the kinematic viscosity. For these eddies, viscosity will be very important so that their kinetic energy is dissipated away. We will see that this transfer of energy to smaller scales gives rise to the *energy cascade* of turbulence. But several important questions are still unanswered:

1. What is the actual size of the smallest eddies that dissipate the energy?
2. How do the velocities $u(l)$ of the eddies vary with l when the eddies become smaller?

Kolmogorov's hypotheses Kolmogorov conjectured a number of hypotheses that can answer these questions. In particular, he proposed:

- For high Reynolds number, the small-scale turbulent motions ($l \ll L_0$) become statistically isotropic. Any memory of large-scale boundary conditions and the original creation of the turbulence on large scales is lost.
- For high Reynolds number, the statistics of small-scale turbulent motions has a universal form and is only determined by ν and the energy injection rate per unit mass, ϵ.

From ν and ϵ, one can construct characteristic Kolmogorov length, velocity and timescales. Of particular importance is the *Kolmogorov length*:

$$\eta \equiv \left(\frac{\nu^3}{\epsilon}\right)^{1/4}. \tag{212}$$

Velocity and timescales are given by

$$u_\eta = (\epsilon\nu)^{1/4}, \qquad \tau_\eta = \left(\frac{\nu}{\epsilon}\right)^{1/2}. \tag{213}$$

We then see that the Reynolds number at the Kolmogorov scales is

$$\mathrm{Re}(\eta) = \frac{\eta u_\eta}{\nu} = 1, \tag{214}$$

showing that they describe the dissipation range. Kolmogorov has furthermore made a second similarity hypothesis, as follows:

- For high Reynolds number, there is a range of scales $L_0 \gg l \gg \eta$ over which the statistics of the motions on scale l take a universal form, and this form is *only* determined by ϵ, *independent* of ν.

In other words, this also means that viscous effects are unimportant over this range of scales, which is called the *inertial range*. Given an eddy size l in the inertial range, one can construct its characteristic velocity and timescale just from l and ϵ:

$$u(l) = (\epsilon l)^{1/3}, \qquad \tau(l) = \left(\frac{l^2}{\epsilon}\right)^{1/3}. \tag{215}$$

One further consequence of the existence of the inertial range is that here the energy transfer rate

$$T(l) \sim \frac{u^2(l)}{\tau(l)} \tag{216}$$

of eddies to smaller scales is expected to be scale-invariant. Indeed, putting in the expected characteristic scale dependence we get $T(l) \sim \epsilon$, i.e., $T(l)$ is equal to the energy injection rate. This also implies that we have

$$\epsilon \sim \frac{U_0^3}{L_0}. \tag{217}$$

With this result we can also work out what we expect for the ratio between the characteristic quantities of the largest and smallest scales:

$$\frac{\eta}{L_0} \sim \left(\frac{\nu^3}{\epsilon L_0^4}\right)^{1/4} = \left(\frac{\nu^3}{U_0^3 L_0^3}\right)^{1/4} = \mathrm{Re}^{-\frac{3}{4}}, \tag{218}$$

$$\frac{u_\eta}{U_0} \sim \left(\frac{\epsilon \nu}{U_0^4}\right)^{1/4} = \left(\frac{U_0^3 \nu}{L_0 U_0^4}\right)^{1/4} = \mathrm{Re}^{-\frac{1}{4}}, \tag{219}$$

$$\frac{\tau_\eta}{\tau} \sim \left(\frac{\nu U_0^2}{\epsilon L_0^2}\right)^{1/2} = \left(\frac{\nu U_0^2 L_0}{U_0^3 L_0^2}\right)^{1/2} = \mathrm{Re}^{-\frac{1}{2}}. \tag{220}$$

This shows that the Reynolds number directly sets the dynamic range of the inertial range.

5.4.2 Energy Spectrum of Kolmogorov Turbulence

Eddy motions on a length-scale l correspond to wavenumber $k = 2\pi/l$. The kinetic energy ΔE contained between two wave numbers k_1 and k_2 can be described by

$$\Delta E = \int_{k_1}^{k_2} E(k)\,\mathrm{d}k, \tag{221}$$

where $E(k)$ is the so-called energy spectrum. For the inertial range in Kolmogorov's theory, we know that $E(k)$ is a universal function that only depends on ϵ and k. Hence $E(k)$ must be of the form

$$E(k) = C\,\epsilon^a\,k^b, \tag{222}$$

where C is a dimensionless constant. Through dimensional analysis it is easy to see that one must have $a = 2/3$ and $b = -5/3$. We hence obtain the famous $-5/3$ slope of the Kolmogorov energy power spectrum:

$$E(k) = C\,\epsilon^{2/3}\,k^{-5/3}. \tag{223}$$

The constant C is universal in Kolmogorov's theory, but cannot be computed from first principles. Experiment and numerical simulations give $C \simeq 1.5$ (Pope 2000).

Actually, if we recall Kolmogorov's first similarity hypothesis, it makes the stronger claim that the statistics for all small scale motion is universal. This means that also the dissipation part of the turbulence must have a universal form. To include this in the description of the spectrum (Fig. 19), we can for example write

$$E(k) = C\,\epsilon^{2/3}\,k^{-5/3}\,f_\eta(k\eta), \tag{224}$$

where $f_\eta(k\eta)$ is a universal function with $f_\eta(x) = 1$ for $x \ll 1$, and with $f_\eta(x) \to 0$ for $x \to \infty$. This function has to be determined experimentally or numerically. A good fit to different results is given by

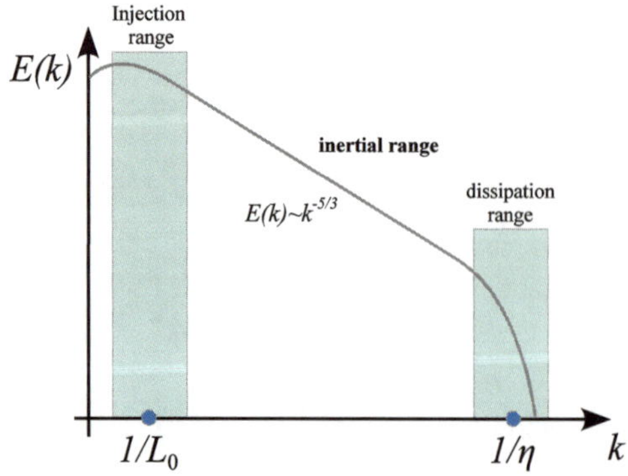

Fig. 19 Schematic energy spectrum of Kolmogorov turbulence

$$f_\eta(x) = \exp\left(-\beta[(x^4 + c^4)^{1/4} - c]\right), \tag{225}$$

with $\beta_0 \sim 5.2$ and $c \sim 0.4$ (Pope 2000).

6 Eulerian Hydrodynamics

Many physical theories are expressed as partial differential equations (PDEs), including some of the most fundamental laws of nature, such as fluid dynamics (Euler and Navier Stokes equations), electromagnetism (Maxwell's equations) or general relativity/gravity (Einstein's field equations). Broadly speaking, partial differential equations (PDE) are equations describing relations between partial derivatives of a dependent variable with respect to several independent variables. Unlike for ordinary differential equations (ODEs), there is no simple unified theory for PDEs. Rather, there are different types of PDEs which exhibit special features (Renardy and Rogers 2004).

The Euler equations, which will be the focus of this section, are so-called hyperbolic conservation laws. They are non-linear, because they contain non-linear terms in the unknown functions and/or its partial derivatives. We note that a full characterization of the different types of PDEs goes beyond the scope of these lecture notes.

6.1 Solution Schemes for PDEs

Unfortunately, for partial differential equations one cannot give a general solution method that works equally well for all types of problems. Rather, each type requires different approaches, and certain PDEs encountered in practice may even be best addressed with special custom techniques built by combining different elements from standard techniques. Important classes of solution schemes include the following:

- **Finite difference methods**: Here the differential operators are approximated through finite difference approximations, usually on a regular (cartesian) mesh, or some other kind of structured mesh (for example a polar grid). An example we already previously discussed is Poisson's equation treated with iterative (multigrid) methods.
- **Finite volume methods**: These may be seen as a subclass of finite difference methods. They are particularly useful for hyperbolic conservation laws. We shall discuss examples for this approach in applications to fluid dynamics later in this section.
- **Spectral methods**: Here the solution is represented by a linear combination of functions, allowing the PDE to be transformed to algebraic equations or ordinary differential equations. Often this is done by applying Fourier techniques. For

example, solving the Poisson equation with FFTs, as we discussed earlier, is a spectral method.

- **Method of lines**: This is a semi-discrete approach where all derivatives except for one are approximated with finite differences. The remaining derivative is then the only one left, so that the remaining problem forms a set of ordinary differential equations (ODEs). Very often, this approach is used in time-dependent problems. One here discretizes space in terms of a set of N points x_i, and for each of these points one obtains an ODE that describes the time evolution of the function at this point. The PDE is transformed in this way into a set of N coupled ODEs. For example, consider the heat diffusion equation in one dimension,

$$\frac{\partial u}{\partial t} + \lambda \frac{\partial^2 u}{\partial x^2} = 0. \tag{226}$$

If we discretize this into a set of points that are spaced h apart, we obtain N equations

$$\frac{\mathrm{d}u_i}{\mathrm{d}t} + \lambda \frac{u_{i+1} + u_{i-1} - 2u_i}{h^2} = 0. \tag{227}$$

These differential equations can now be integrated in time as an ODE system. Note however that this is not necessarily stable. Some problems may require upwinding, i.e., asymmetric forms for the finite difference estimates to recover stability.

- **Finite element methods**: Here the domain is subdivided into "cells" (elements) of fairly arbitrary shape. The solution is then represented in terms of simple, usually polynomial functions on the element, and then the PDE is transformed to an algebraic problem for the coefficients in front of these simple functions. This is hence similar in spirit to spectral methods, except that the expansion is done in terms of highly localized functions on an element by element basis, and is truncated already at low order.

In practice, many different variants of these basic methods exist, and sometimes also combinations of them are used.

6.2 Simple Advection

First-order equations of hyperbolic type are particularly useful for introducing the numerical difficulties that then also need to be addressed for more complicated non-linear conservation laws (e.g. Toro 1997; LeVeque 2002; Stone et al. 2008). The simplest equation of this type is the *advection equation* in one dimension. This is given by

$$\frac{\partial u}{\partial t} + v \cdot \frac{\partial u}{\partial x} = 0, \tag{228}$$

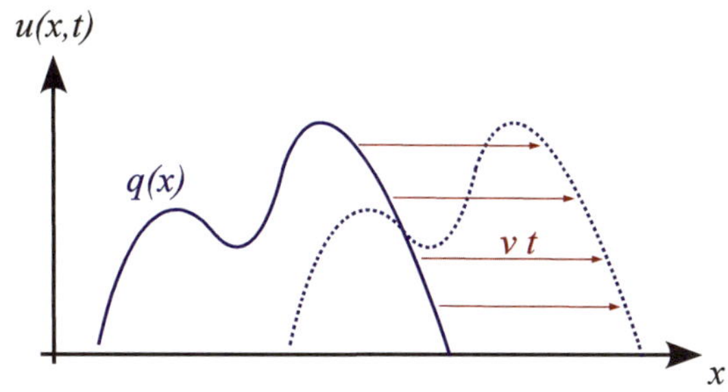

Fig. 20 Simple advection with constant velocity to the right

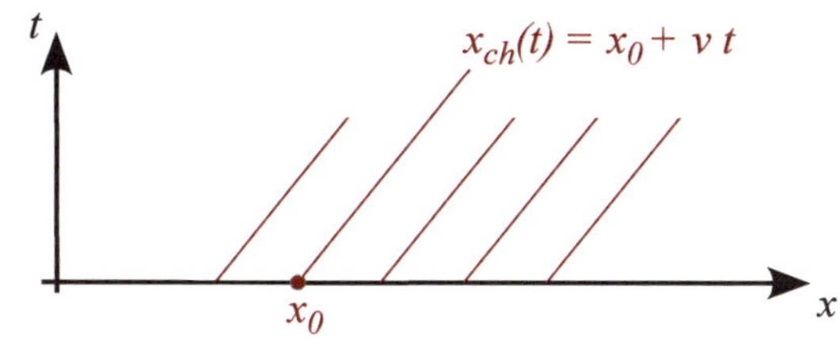

Fig. 21 A set of flow characteristics for advection to the right with constant velocity v

where $u = u(x, t)$ is a function of x and t, and v is a constant parameter. This equation is hyperbolic because the so-called coefficient matrix[1] is real and trivially diagonalizable.

If we are given any function $q(x)$, then

$$u(x, t) = q(x - vt) \tag{230}$$

is a solution of the PDE, as one can easily check. We can interpret $u(x, t = 0) = q(x)$ as initial condition, and the solution at a later time is then an exact copy of q, simply translated by $v\,t$ along the x-direction, as shown in Fig. 20.

Points that start at a certain coordinate x_0 are advected to a new location $x_{\rm ch}(t) = vt + x_0$. These so-called *characteristics* (see Fig. 21), which can be viewed as mediating the propagation of information in the system, are straight lines, all oriented in the downstream direction. Note that "downstream" refers to the direction in which the flow goes, whereas "upstream" is from where the flow comes.

Let's now assume we want to solve the advection problem numerically. (Strictly speaking this is of course superfluous as we have an analytic solution in this case, but we want to see how well a numerical technique would perform here.) We can

---

[1] A linear system of first-order PDEs can be written in the generic form

$$\frac{\partial u_i}{\partial t} + \sum_j A_{ij} \frac{\partial u_i}{\partial x_j} = 0, \tag{229}$$

where A_{ij} is the coefficient matrix.

approach this with a straightforward discretization of u on a special mesh, using for example the method of lines. This gives us:

$$\frac{\mathrm{d}u_i}{\mathrm{d}t} + v\frac{u_{i+1} - u_{i-1}}{2h} = 0. \tag{231}$$

If we go one step further and also discretize the time derivative with a simple Euler scheme, we get

$$u_i^{(n+1)} = u_i^{(n)} - v\frac{u_{i+1}^{(n)} - u_{i-1}^{(n)}}{2h}\Delta t. \tag{232}$$

This is a complete update formula which can be readily applied to a given initial state on the grid. The big surprise is that this turns out to be quite violently unstable! For example, if one applies this to the advection of a step function, one invariably obtains strong oscillatory errors in the downstream region of the step, quickly rendering the numerical solution into complete garbage. What is the reason for this fundamental failure?

- First note that all characteristics (signals) propagate downstream in this problem, or in other words, information strictly travels in the flow direction in this problem.
- But, the information to update u_i is derived both from the upstream (u_{i-1}) and the downstream (u_{i+1}) side.
- According to how the information flows, u_i should not really depend on the downstream side at all, which in some sense is causally disconnected. So let's try to get rid off this dependence by going to a one-sided approximation for the spatial derivative, of the form:

$$\frac{\mathrm{d}u_i}{\mathrm{d}t} + v\frac{u_i - u_{i-1}}{h} = 0. \tag{233}$$

  This is called *upwind differencing*. Interestingly, now the stability problems are completely gone!
- But there are still some caveats to observe: First of all, the discretization now depends on the sign of v. For negative v, one instead has to use

$$\frac{\mathrm{d}u_i}{\mathrm{d}t} + v\frac{u_{i+1} - u_i}{h} = 0. \tag{234}$$

  The other is that the solution is not advected in a perfectly faithful way, instead it is quite significantly smoothed out, through a process one calls *numerical diffusion*.

We can actually understand where this strong diffusion in the 1st-order upwind scheme comes from. To this end, let's rewrite the upwind finite difference approximation of the spatial derivative as

$$\frac{u_i - u_{i-1}}{h} = \frac{u_{i+1} - u_{i-1}}{2h} - \frac{u_{i+1} - 2u_i + u_{i-1}}{2h}. \tag{235}$$

Hence our stable upwind scheme can also be written as

$$\frac{\mathrm{d}u_i}{\mathrm{d}t} + v\frac{u_{i+1} - u_{i-1}}{2h} = \frac{vh}{2}\frac{u_{i+1} - 2u_i + u_{i-1}}{h^2}. \tag{236}$$

But recall from Eq. (133) that

$$\left(\frac{\partial^2 u}{\partial x^2}\right)_i \simeq \frac{u_{i+1} - 2u_i + u_{i-1}}{h^2}, \tag{237}$$

so if we define a diffusion constant $D = (vh)/2$, we are effectively solving the following problem,

$$\frac{\partial u}{\partial t} + v \cdot \frac{\partial u}{\partial x} = D\frac{\partial^2 u}{\partial x^2}, \tag{238}$$

and not the original advection problem. The diffusion term on the right hand side is here a byproduct of the numerical algorithm that we have used. We needed to add this numerical diffusion in order to obtain stability of the integration.

Note however that for better grid resolution, $h \to 0$, the diffusion becomes smaller, so in this limit one obtains an ever better solution. Also note that the diffusivity becomes larger for larger velocity v, so the faster one needs to advect, the stronger the numerical diffusion effects become.

Besides the upwinding requirement, integrating a hyperbolic conservation law with an explicit method in time also requires the use of a sufficiently small integration timestep, not only to get sufficiently good accuracy, but also for reasons of *stability*. In essence, there is a maximum timestep that may be used before the integration brakes down. How large can we make this timestep? Again, we can think about this in terms of information travel. If the timestep exceeds $\Delta t_{\max} = h/v$, then the updating of u_i would have to include information from u_{i-2}, but if we don't do this, the updating will likely become unstable.

This leads to the so-called *Courant-Friedrichs-Levy* (CFL) timestep condition (Courant et al. 1928), which for this problem takes the form

$$\Delta t \leq \frac{h}{v}. \tag{239}$$

This is a necessary but not sufficient condition for any explicit finite different approach of the hyperbolic advection equation. For other hyperbolic conservation laws, similar CFL-conditions apply.

Hyperbolic conservation laws We now consider a hyperbolic conservation law, such as the continuity equation for the mass density of a fluid:

$$\frac{\partial \rho}{\partial t} + \nabla \cdot (\rho \mathbf{v}) = 0. \tag{240}$$

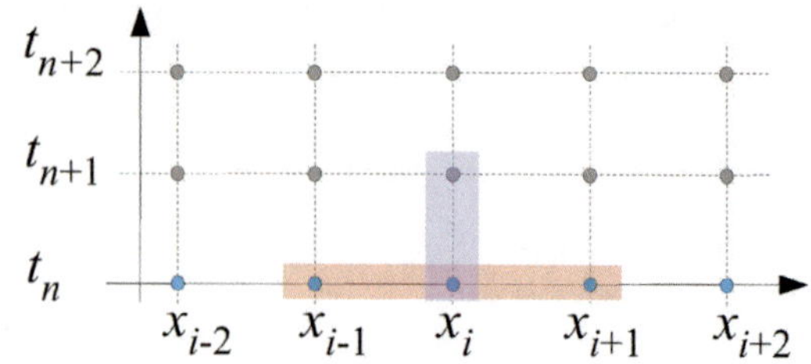

Fig. 22 A discretization scheme for the continuity equation in one spatial dimension. The *red* and *blue boxes* mark the stencils that are applied for calculating the spatial and time derivatives

We see that this is effectively the advection equation, but with a spatially variable velocity $\mathbf{v} = \mathbf{v}(\mathbf{x})$. Here $\mathbf{F} = \rho\mathbf{v}$ is the mass flux.

Let's study the problem in one spatial dimension, and consider a discretization both of the x- and t-axis (Fig. 22). This corresponds to

$$\frac{\rho_i^{(n+1)} - \rho_i^{(n)}}{\Delta t} + \frac{F_{i+1}^{(n)} - F_{i-1}^{(n)}}{2\Delta x} = 0, \tag{241}$$

leading to the update rule

$$\rho_i^{(n+1)} = \rho_i^{(n)} + \frac{\Delta t}{2\Delta x}\left(F_{i-1}^{(n)} - F_{i+1}^{(n)}\right). \tag{242}$$

This is again found to be highly unstable, for the same reasons as in the plain advection problem: we have not observed in 'which direction the wind blows', or in other words, we have ignored in which direction the local characteristics point. For example, if the mass flux is to the right, we know that the characteristics point also to the right. The upwind direction is therefore towards negative x, and by using only this information in making our spatial derivative one-sided, we should be able to resurrect stability.

Now, for the mass continuity equation identifying the local characteristics is quite easy, and in fact, their direction can simply be inferred from the sign of the mass flux. However, in more general situations for systems of non-linear PDEs, this is far less obvious. Here we need to use a so-called Riemann solvers to give us information about the local solution and the local characteristics (Toro 1997). This then also implicitly identifies the proper upwinding that is needed for stability.

6.3 Riemann Problem

The Riemann problem is an initial value problem for a hyperbolic system, consisting of two piece-wise constant states (two half-spaces) that meet at a plane at $t = 0$. The task is then to solve for the subsequent evolution at $t > 0$.

An important special case is the Riemann problem for the Euler equations (i.e., for ideal gas dynamics). Here the left and right states of the interface, can, for example, be uniquely specified by giving the three "primitive" variables density, pressure and velocity, viz.

$$U_L = \begin{pmatrix} \rho_L \\ P_L \\ \mathbf{v}_L \end{pmatrix}, \quad U_R = \begin{pmatrix} \rho_R \\ P_R \\ \mathbf{v}_R \end{pmatrix}. \tag{243}$$

Alternatively one can also specify density, momentum density, and energy density. For an ideal gas, this initial value problem can be solved analytically (Toro 1997), modulo an implicit equation which requires numerical root-finding, i.e., the solution cannot be written down explicitly. The solution always contains characteristics for three self-similar waves, as shown schematically in Fig. 23. Some notes on this:

- The middle wave is always present and is a contact wave that marks the boundary between the original fluid phases from the left and right sides.
- The contact wave is sandwiched between a shock or a rarefaction wave on either side (it is possible to have shocks on both sides, or rarefactions on both sides, or one of each). The rarefaction wave is not a single characteristic but rather a rarefaction fan with a beginning and an end.
- These waves propagate with constant speed. If the solution is known at some time $t > 0$, it can also be obtained at any other time through a suitable scaling transformation. An important corollary is that at $x = 0$, the fluid quantities $(\rho^\star, P^\star, \mathbf{v}^\star)$ are *constant in time* for $t > 0$.
- For $\mathbf{v}_L = \mathbf{v}_R = 0$, the Riemann problem simplifies and becomes the 'Sod shock tube' problem.

Let's consider an example how this wave structure looks in a real Riemann problem. We consider, for definiteness, a Riemann problem with $\rho_L = 1.0$, $P_L = 1.0$, $v_L = 0$, and $\rho_R = 0.25$, $P_R = 0.1795$, $v_R = 0$ (which is of Sod-shock type). The adiabatic exponent is taken to be $\gamma = 1.4$. We hence deal at $t = 0.0$ with the initial state displayed in Fig. 24. After time $t = 5.0$, the wave structure formed by a rarefaction to the left (location marked in green), a contact in the middle (blue) and a shock to the right (red) can be nicely seen in Fig. 25.

Some general properties of the waves appearing in the Riemann problem can be summarized as follows:

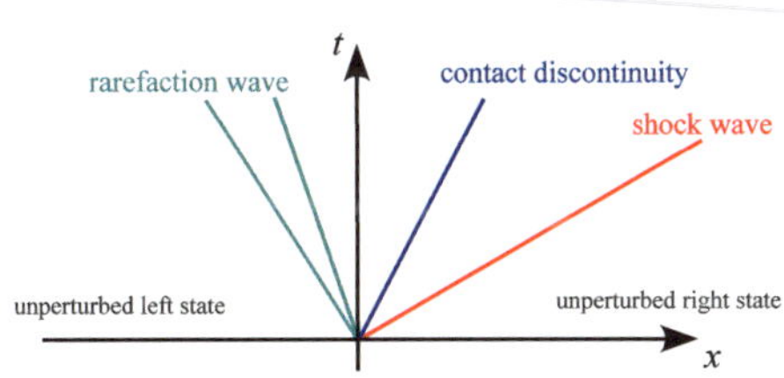

Fig. 23 Wave structure of the solution of the Riemann problem. The central contact wave separates the original fluid phases. On the *left* and the *right*, there is either a shock or a rarefaction wave

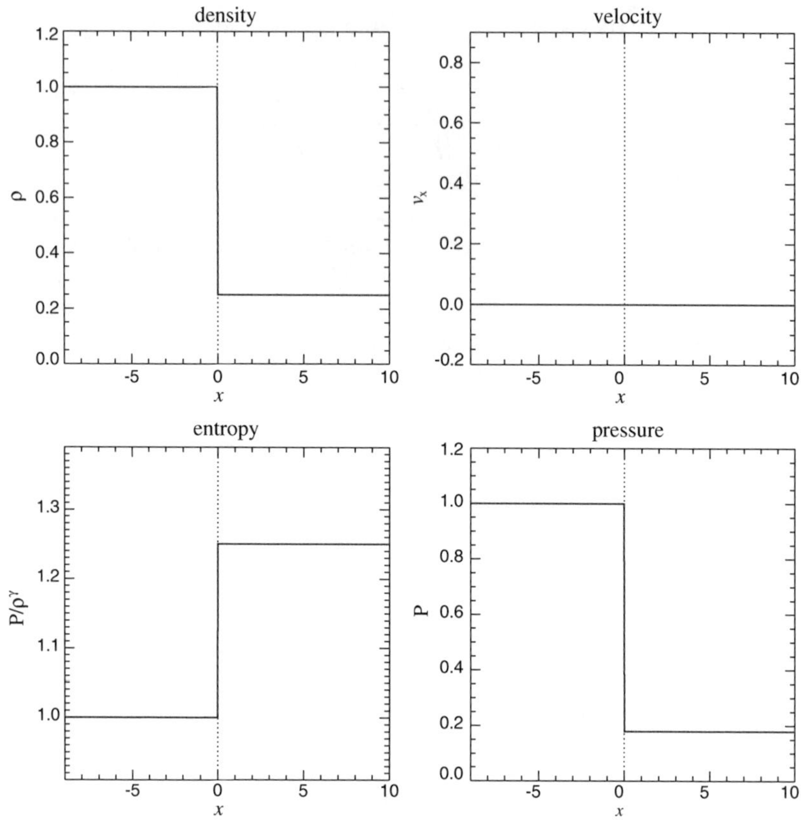

Fig. 24 Initial state of an example Riemann problem, composed of two phases in different states that are brought into contact at $x = 0$ at time $t = 0$. (Since $v_x = 0$, the initial conditions are actually an example of the special case of a Sod shock-tube problem)

- *Shock*: This is a sudden compression of the fluid, associated with an irreversible conversion of kinetic energy to heat, i.e., here entropy is produced. The density, normal velocity component, pressure, and entropy all change discontinuously at a shock.
- *Contact discontinuity*: This traces the original separating plane between the two fluid phases that have been brought into contact. Pressure as well as the normal velocity are constant across a contact, but density, entropy and temperature can jump.
- *Rarefaction wave*: This occurs when the gas (suddenly) expands. The rarefaction wave smoothly connects two states over a finite spatial region; there are no discontinuities in any of the fluid variables.

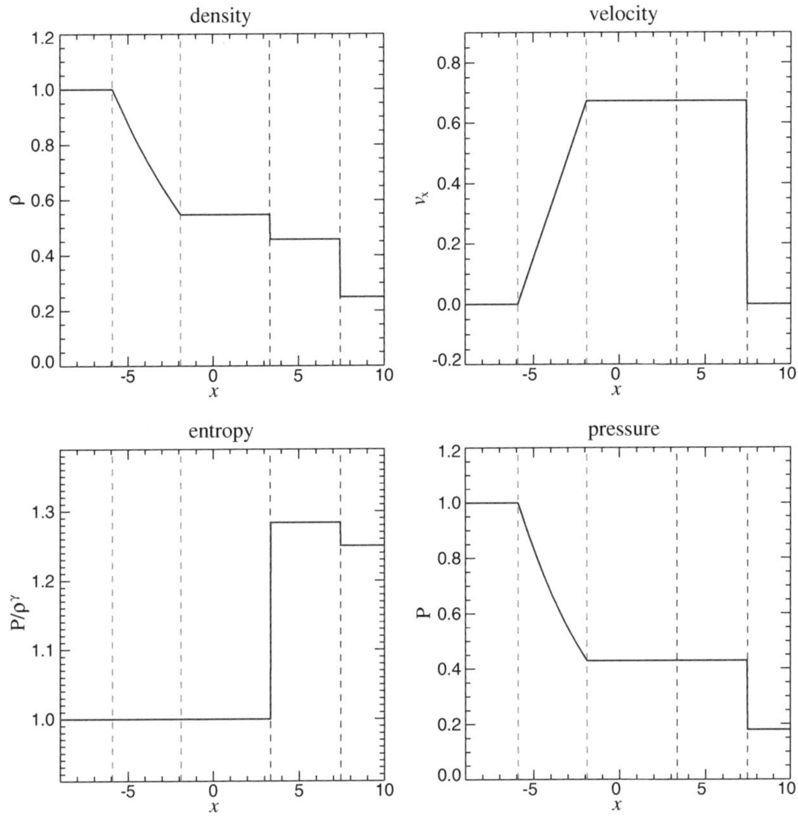

Fig. 25 Evolved state at $t = 5.0$ of the initial fluid state displayed in Fig. 24. The *blue dashed line* marks the position of the contact wave, the *green dashed lines* give the location of the rarefaction fan, and the *red dashed line* marks the shock

6.4 Finite Volume Discretization

Let's now take a look how Riemann solvers can be used in the finite volume discretization approach to the PDEs of fluid dynamics. Recall that we can write our hyperbolic conservation laws as

$$\frac{\partial \mathbf{U}}{\partial t} + \nabla \cdot \mathbf{F} = 0. \tag{244}$$

Here **U** is a state vector and **F** is the flux vector. For example, the Euler equations of Sect. 5.1.1 can be written in the form

$$\mathbf{U} = \begin{pmatrix} \rho \\ \rho \mathbf{v} \\ \rho e \end{pmatrix}, \qquad \mathbf{F} = \begin{pmatrix} \rho \mathbf{v} \\ \rho \mathbf{v}\mathbf{v}^T + P \\ (\rho e + P)\mathbf{v} \end{pmatrix}, \tag{245}$$

with the specific energy $e = u + \mathbf{v}^2/2$ and u being the thermal energy per unit mass. The ideal gas equation gives the pressure as $P = (\gamma - 1)\rho u$ and provides a closure for the system.

In a finite volume scheme, we describe the system through the averaged state over a set of finite cells. These cell averages are defined as

$$\mathbf{U}_i = \frac{1}{V_i} \int_{\text{cell } i} \mathbf{U}(\mathbf{x})\, \mathrm{d}V. \tag{246}$$

Let's now see how we could divise an update scheme for these cell-averaged quantities.

1. We start by integrating the conservation law over a cell, and over a finite interval in time:
$$\int_{x_{i-\frac{1}{2}}}^{x_{i+\frac{1}{2}}} \mathrm{d}x \int_{t_n}^{t_{n+1}} \mathrm{d}t \left(\frac{\partial \mathbf{U}}{\partial t} + \frac{\partial \mathbf{F}}{\partial x} \right) = 0. \tag{247}$$

2. This gives
$$\int_{x_{i-\frac{1}{2}}}^{x_{i+\frac{1}{2}}} \mathrm{d}x \left[\mathbf{U}(x, t_{n+1}) - \mathbf{U}(x, t_n) \right] + \int_{t_n}^{t_{n+1}} \mathrm{d}t \left[\mathbf{F}(x_{i+\frac{1}{2}}, t) - \mathbf{F}(x_{i-\frac{1}{2}}, t) \right] = 0. \tag{248}$$
In the first term, we recognize the definition of the cell average:
$$\mathbf{U}_i^{(n)} \equiv \frac{1}{\Delta x} \int_{x_{i-\frac{1}{2}}}^{x_{i+\frac{1}{2}}} \mathbf{U}(x, t_n) \mathrm{d}x. \tag{249}$$
Hence we have
$$\Delta x \left[\mathbf{U}_i^{(n+1)} - \mathbf{U}_i^{(n)} \right] + \int_{t_n}^{t_{n+1}} \mathrm{d}t \left[\mathbf{F}(x_{i+\frac{1}{2}}, t) - \mathbf{F}(x_{i-\frac{1}{2}}, t) \right] = 0. \tag{250}$$

3. Now, $\mathbf{F}(x_{i+\frac{1}{2}}, t)$ for $t > t_n$ is given by the solution of the Riemann problem with left state $\mathbf{U}_i^{(n)}$ and right state $\mathbf{U}_{i+1}^{(n)}$. At the interface, this solution is *independent* of time. We can hence write
$$\mathbf{F}(x_{i+\frac{1}{2}}, t) = \mathbf{F}^{\star}_{i+\frac{1}{2}}, \tag{251}$$

where $\mathbf{F}^{\star}_{i+\frac{1}{2}} = \mathbf{F}_{\text{Riemann}}(\mathbf{U}_i^{(n)}, \mathbf{U}_{i+1}^{(n)})$ is a short-hand notation for the corresponding Riemann solution sampled at the interface. Hence we now get

$$\Delta x \left[\mathbf{U}_i^{(n+1)} - \mathbf{U}_i^{(n)}\right] + \Delta t \left[\mathbf{F}^{\star}_{i+\frac{1}{2}} - \mathbf{F}^{\star}_{i-\frac{1}{2}}\right] = 0. \tag{252}$$

Or alternative, as an explicit update formula:

$$\mathbf{U}_i^{(n+1)} = \mathbf{U}_i^{(n)} + \frac{\Delta t}{\Delta x}\left[\mathbf{F}^{\star}_{i-\frac{1}{2}} - \mathbf{F}^{\star}_{i+\frac{1}{2}}\right]. \tag{253}$$

The first term in the square bracket gives the flux that flows from left into the cell, the second term is the flux out of the cell on its right side. The idea to use the Riemann solution in the updating step is due to Godunov, that's why such schemes are often called *Godunov schemes*.

It is worthwhile to note that we haven't really made any approximation in the above (yet). In particular, if we calculate $\mathbf{F}_{\text{Riemann}}$ analytically (and hence exactly), then the above seems to account for the correct fluxes for arbitrarily long times. So does this mean that we get a perfectly accurate result even for very large timesteps? This certainly sounds too good to be true, so there must be a catch somewhere.

Indeed, there is. First of all, we have assumed that the Riemann problems are independent of each other and each describe infinite half-spaces. This is not true once we consider finite volume cells, but it is still ok for a while as long t_{n+1} is close enough to t_n such that the waves emanating in one interface have not yet arrived at the next interface left or right. This then leads to a CFL-timestep criterion, were $\Delta t \leq \Delta x / c_{\max}$ and $c_{\max}$ is the maximum wavespeed.

Another point is more subtle and comes into play when we consider more than one timestep. We assumed that the $\mathbf{U}_i^{(n)}$ describe piece-wise constant states which can then be fed to the Riemann solver to give us the flux. However, even when this is true initially, we have just seen that after one timestep it will not be true anymore. By ignoring this in the subsequent timestep (which is done by performing an averaging step that washes out the cell substructure that developed as part of the evolution during the previous timestep) we make some error.

6.5 Godunov's Method and Riemann Solvers

It is useful to introduce another interpretation of common finite-volume discretizations of fluid dynamics, so-called Reconstruct-Evolve-Average (REA) schemes. We also use this here for a short summary of Godunov's important method, and the way Riemann solvers come into play in it.

An REA update scheme of a hydrodynamical system discretized on a mesh can be viewed as a sequence of three steps:

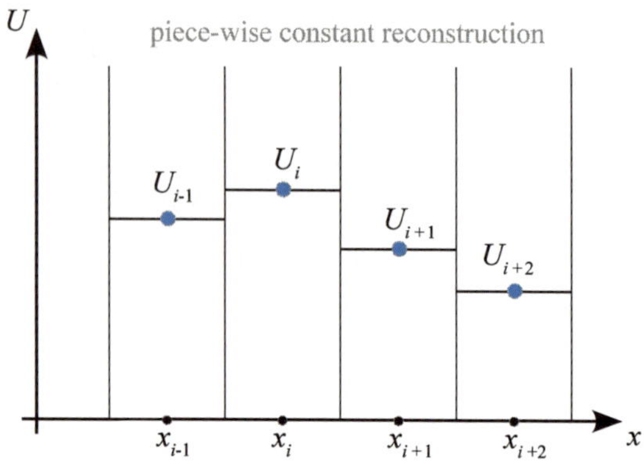

Fig. 26 Piece-wise constant states of a fluid forming the simplest possible reconstruction of its state based on a set of discrete values U_i known at spatial positions x_i

1. *Reconstruct*: Using the cell-averaged quantities (as shown in Fig. 26), this defines the run of these quantities everywhere in the cell. In the sketch, a piece-wise constant reconstruction is assumed, which is the simplest procedure one can use and leads to 1st order accuracy.
2. *Evolve*: The reconstructed state is then evolved forward in time by Δt. In Godunov's approach, this is done by treating each cell interface as a piece-wise constant initial value problem which is solved with the Riemann solver exactly or approximately. This solution is formally valid as long as the waves emanating from opposite sides of a cell do not yet start to interact. In practice, one therefore needs to limit the timestep Δt such that this does not happen.
3. *Average*: The wave structure resulting from the evolution over timestep Δt is spatially averaged in a conservative fashion to compute new states $\mathbf{U}^{n+1}$ for each cell. Fortunately, the averaging step does not need to be done explicitly; instead it can simply be carried out by accounting for the fluxes that enter or leave the control volume of the cell. Then the whole cycle repeats again.

What is needed for the *evolve* step is a prescription to either exactly or approximately solve the Riemann problem for a piece-wise linear left and right state that are brought into contact at time $t = t_n$. Formally, this can be written as

$$\mathbf{F}^{\star} = \mathbf{F}_{\text{Riemann}}(\mathbf{U}_L, \mathbf{U}_R). \tag{254}$$

In practice, a variety of approximate Riemann solvers $\mathbf{F}_{\text{Riemann}}$ are commonly used in the literature (Rusanov 1961; Harten et al. 1983; Toro 1997; Miyoshi and Kusano 2005). For the ideal gas and for isothermal gas, it is also possible to solve the Riemann problem exactly, but not in closed form (i.e., the solution involves an iterative root finding of a non-linear equation).

There are now two main issues left:

- How can this be extended to multiple spatial dimensions?
- How can it be extended such that a higher order integration accuracy both in space and time is reached?

We'll discuss these issues next.

6.6 Extensions to Multiple Dimensions

So far, we have considered *one-dimensional* hyperbolic conservation laws of the form

$$\partial_t \mathbf{U} + \partial_x \mathbf{F}(\mathbf{U}) = 0, \tag{255}$$

where ∂_t is a short-hand notation for $\partial_t = \frac{\partial}{\partial t}$, and similarly $\partial_x = \frac{\partial}{\partial x}$. For example, for isothermal gas with soundspeed c_s, the state vector $\mathbf{U}$ and flux vector $\mathbf{F}(\mathbf{U})$ are given as

$$\mathbf{U} = \begin{pmatrix} \rho \\ \rho u \end{pmatrix}, \quad \mathbf{F} = \begin{pmatrix} \rho u \\ \rho u^2 + \rho c_s^2 \end{pmatrix}, \tag{256}$$

where u is the velocity in the x-direction.

In three dimensions, the PDEs describing a fluid become considerably more involved. For example, the Euler equations for an ideal gas are given in explicit form as

$$\partial_t \begin{pmatrix} \rho \\ \rho u \\ \rho v \\ \rho w \\ \rho e \end{pmatrix} + \partial_x \begin{pmatrix} \rho u \\ \rho u^2 + P \\ \rho u v \\ \rho u w \\ \rho u(\rho e + P) \end{pmatrix} + \partial_y \begin{pmatrix} \rho v \\ \rho u v \\ \rho v^2 + P \\ \rho v w \\ \rho v(\rho e + P) \end{pmatrix} + \partial_z \begin{pmatrix} \rho w \\ \rho u w \\ \rho v w \\ \rho w^2 + P \\ \rho w(\rho e + P) \end{pmatrix} = 0, \tag{257}$$

where $e = e_{\text{therm}} + (u^2 + v^2 + w^2)/2$ is the total specific energy per unit mass, e_{therm} is the thermal energy per unit mass, and $P = (\gamma - 1)\rho\, e_{\text{therm}}$ is the pressure. These equations are often written in the following notation:

$$\partial_t \mathbf{U} + \partial_x \mathbf{F} + \partial_y \mathbf{G} + \partial_z \mathbf{H} = 0. \tag{258}$$

Here the functions $\mathbf{F}(\mathbf{U})$, $\mathbf{G}(\mathbf{U})$ and $\mathbf{H}(\mathbf{U})$ give the flux vectors in the x-, y- and z-direction, respectively.

6.6.1 Dimensional Splitting

Let us now consider the three dimensionally split problems derived from Eq. (258):

$$\partial_t \mathbf{U} + \partial_x \mathbf{F} = 0, \tag{259}$$

$$\partial_t \mathbf{U} + \partial_y \mathbf{G} = 0, \tag{260}$$

$$\partial_t \mathbf{U} + \partial_z \mathbf{H} = 0. \tag{261}$$

Note that the vectors appearing here have still the same dimensionality as in the full equations. They are *augmented* one-dimensional problems, i.e., the transverse variables still appear but spatial differentiation happens only in one direction. Because of this, these additional transverse variables do not make the 1D problem more difficult compared to the 'pure' 1D problem considered earlier, but the fluxes appearing in them still need to be included.

Now let us assume that he have a method to solve/advance each of these one-dimensional problems. We can for example express this formally through time-evolution operators $\mathcal{X}(\Delta t)$, $\mathcal{Y}(\Delta t)$, and $\mathcal{Z}(\Delta t)$, which advance the solution by a timestep Δt. Then the full time advance of the system can for example be approximated by

$$\mathbf{U}^{n+1} \simeq \mathcal{Z}(\Delta t)\mathcal{Y}(\Delta t)\mathcal{X}(\Delta t)\mathbf{U}^{n}. \tag{262}$$

This is one possible dimensionally split update scheme. In fact, this is the exact solution if Eqs. (259)–(260) represent the linear advection problem, but for more general non-linear equations it only provides a first order approximation. However, higher-order dimensionally split update schemes can also be easily constructed. For example, in two-dimensions,

$$\mathbf{U}^{n+1} = \frac{1}{2}[\mathcal{X}(\Delta t)\mathcal{Y}(\Delta t) + \mathcal{Y}(\Delta t)\mathcal{X}(\Delta t)]\mathbf{U}^{n} \tag{263}$$

and

$$\mathbf{U}^{n+1} = \mathcal{X}(\Delta t/2)\mathcal{Y}(\Delta t)\mathcal{X}(\Delta t/2)\mathbf{U}^{n} \tag{264}$$

are second-order accurate. Similarly, for three dimensions the scheme

$$\mathbf{U}^{n+1} = \mathcal{X}(\Delta t/2)\mathcal{Y}(\Delta t/2)\mathcal{Z}(\Delta t)\mathcal{Y}(\Delta t/2)\mathcal{X}(\Delta t/2)\mathbf{U}^{n} \tag{265}$$

is second-order accurate. As a general rule of thumb, the time evolution operators have to be applied alternatingly in reverse order to reach second-order accuracy. We see that the dimensionless splitting reduces the problem effectively to a sequence of one-dimensional solution operations which are applied to multi-dimensional domains. Note that each one-dimensional operator leads to an update of $\mathbf{U}$, and is a complete step for the corresponding augmented one-dimensional problem. Gradients, etc., that are needed for the next step then have to be recomputed before the next time-evolution operator is applied. In practical applications of mesh codes, these one-dimensional solves are often called *sweeps*.

6.6.2 Unsplit Schemes

In an unsplit approach, all flux updates of a cell are applied simultaneously to a cell, not sequentially. This is for example illustrated in 2D in the situations depicted in Fig. 27. The unsplit update of cell i, j in the Cartesian case is then given by

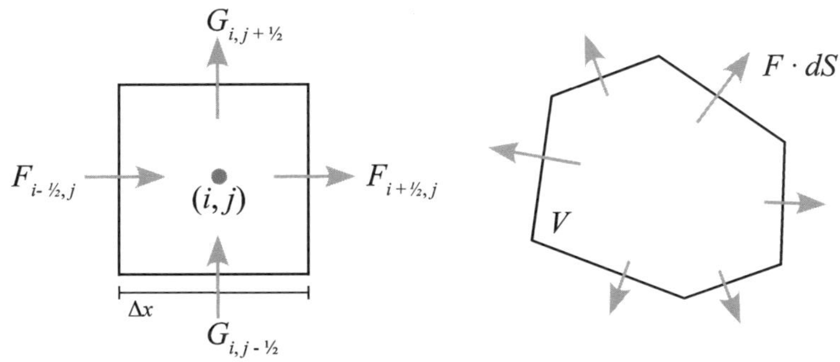

Fig. 27 Sketch of unsplit finite-volume update schemes. On the *left*, the case of a structured Cartesian grid is shown, the case on the *right* is for an unstructured grid

$$U_{i,j}^{n+1} = U_{i,j}^{n} + \frac{\Delta t}{\Delta x}\left(\mathbf{F}_{i-\frac{1}{2},j} - \mathbf{F}_{i+\frac{1}{2},j}\right) + \frac{\Delta t}{\Delta y}\left(\mathbf{G}_{i,j-\frac{1}{2}} - \mathbf{G}_{i,j+\frac{1}{2}}\right). \tag{266}$$

Unsplit approaches can also be used for irregular shaped cells like those appearing in unstructured meshes (see Fig. 27). For example, integrating over a cell of volume V and denoting with $\mathbf{U}$ the cell average, we can write the cell update with the divergence theorem as

$$\mathbf{U}^{n+1} = \mathbf{U}^{n} - \frac{\Delta t}{V}\int \mathbf{F}\cdot \mathbf{dS}, \tag{267}$$

where the integration is over the whole cell surface, with outwards pointing face area vectors $\mathbf{dS}$.

6.7 Extensions for High-Order Accuracy

We should first clarify what we mean with higher order schemes. Loosely speaking, this refers to the convergence rate of a scheme in smooth regions of a flow. For example, if we know the analytic solution $\rho(x)$ for some problem, and then obtain a numerical result ρ_i at a set of N points at locations x_i, we can ask what the typical error of the solution is. One possibility to quantify this would be through a L1 error norm, for example in the form

$$L1 = \frac{1}{N}\sum_i |\rho_i - \rho(x_i)|, \tag{268}$$

which can be interpreted as the average error per cell. If we now measure this error quantitatively for different resolutions of the applied discretization, we would like to find that L1 decreases with increasing N. In such a case our numerical scheme is converging, and provided we use sufficient numerical resources, we have a chance to get below any desired absolute error level. But the *rate of convergence* can be very different between different numerical schemes when applied to the same problem. If a method shows a $L1 \propto N^{-1}$ scaling, it is said to be first-order accurate; a doubling of the number of cells will then cut the error in half. A second-order method has $L1 \propto N^{-2}$, meaning that a doubling of the number of cells can actually reduce the error by a factor of 4. This much better convergence rate is of course highly desirable. It is also possible to construct schemes with still higher convergence rates, but they tend to quickly become much more complex and computationally involved, so that one eventually reaches a point of diminishing return, depending on the specific type of problem. But the extra effort one needs to make to go from first to second-order is often very small, sometimes trivially small, so that one basically should always strive to try at least this.

A first step in constructing a 2nd order extension of Godunov's method is to replace the piece-wise constant with a piece-wise linear reconstruction (Fig. 28). This requires that one first estimates gradients for each cell (usually by a simple finite difference formula). These are then slope-limited if needed such that the linear extrapolations of the cell states to the cell interfaces do not introduce new extrema. This slope-limiting procedure is quite important; it needs to be done to avoid that real fluid discontinuities introduce large spurious oscillations into the fluid.

Given slope limited gradients, for example $\nabla\rho$ for the density, one can then estimate the left and right states adjacent to an interface $x_{i+\frac{1}{2}}$ by spatial extrapolation from the centers of the cells left and right from the interface:

$$\rho^L_{i+\frac{1}{2}} = \rho_i + (\nabla\rho)_i \,\frac{\Delta x}{2}, \tag{269}$$

$$\rho^R_{i+\frac{1}{2}} = \rho_{i+1} - (\nabla\rho)_{i+1} \,\frac{\Delta x}{2}. \tag{270}$$

The next step would in principle be to use these states in the Riemann solver. In doing this we will ignore the fact that our reconstruction has now a gradient over the cell; instead we still pretend that the fluid state can be taken as piece-wise constant left and right of the interface as far as the Riemann solver is concerned. However, it turns out

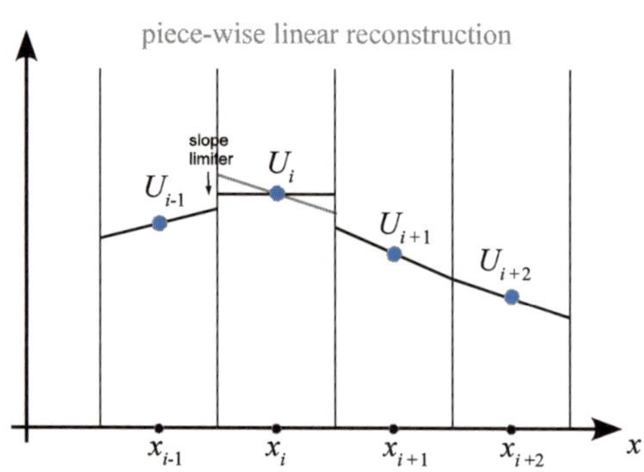

Fig. 28 Piece-wise linear reconstruction scheme applied to a fluid state represented through a regular mesh

that the spatial extrapolation needs to be augmented with a temporal extrapolation one half timestep into the future, such that the flux estimate is now effectively done in the middle of the timestep. This is necessary both to reach second-order accuracy in time and also for stability reasons. Hence we really need to use

$$\rho^L_{i+\frac{1}{2}} = \rho_i + (\nabla\rho)_i \frac{\Delta x}{2} + \left(\frac{\partial\rho}{\partial t}\right)_i \frac{\Delta t}{2}, \tag{271}$$

$$\rho^R_{i+\frac{1}{2}} = \rho_{i+1} - (\nabla\rho)_{i+1} \frac{\Delta x}{2} + \left(\frac{\partial\rho}{\partial t}\right)_{i+1} \frac{\Delta t}{2}, \tag{272}$$

for extrapolating to the interfaces. More generally, this has to be done for the whole state vector of the system, i.e.,

$$\mathbf{U}^L_{i+\frac{1}{2}} = \mathbf{U}_i + (\partial_x\mathbf{U})_i \frac{\Delta x}{2} + (\partial_t\mathbf{U})_i \frac{\Delta t}{2}, \tag{273}$$

$$\mathbf{U}^R_{i+\frac{1}{2}} = \mathbf{U}_{i+1} - (\partial_x\mathbf{U})_{i+1} \frac{\Delta x}{2} + (\partial_t\mathbf{U})_{i+1} \frac{\Delta t}{2}. \tag{274}$$

Note that here the quantity $(\partial_x\mathbf{U})_i$ is a (slope-limited) *estimate* of the gradient in cell i, based on finite-differences plus a slope limiting procedure. Similarly, we somehow need to estimate the time derivative encoded in $(\partial_x\mathbf{U})_i$. How can this be done? One way to do this is to exploit the Jacobian matrix of the Euler equations. We can write the Euler equations as

$$\partial_t\mathbf{U} = -\partial_x\mathbf{F}(\mathbf{U}) = -\frac{\partial\mathbf{F}}{\partial\mathbf{U}}\,\partial_x\mathbf{U} = -\mathbf{A}(\mathbf{U})\,\partial_x\mathbf{U}, \tag{275}$$

where $\mathbf{A}(\mathbf{U})$ is the Jacobian matrix. Using this, we can simply estimate the required time-derivative based on the spatial derivatives:

$$(\partial_t\mathbf{U})_i = -\mathbf{A}(\mathbf{U}_i)\,(\partial_x\mathbf{U})_i. \tag{276}$$

Hence the extrapolation can be done as

$$\mathbf{U}^L_{i+\frac{1}{2}} = \mathbf{U}_i + \left[\frac{\Delta x}{2} - \frac{\Delta t}{2}\mathbf{A}(\mathbf{U}_i)\right](\partial_x\mathbf{U})_i, \tag{277}$$

$$\mathbf{U}^R_{i+\frac{1}{2}} = \mathbf{U}_{i+1} + \left[-\frac{\Delta x}{2} - \frac{\Delta t}{2}\mathbf{A}(\mathbf{U}_{i+1})\right](\partial_x\mathbf{U})_{i+1}. \tag{278}$$

This procedure defines the so-called MUSCL-Hancock scheme (van Leer 1984; Toro 1997; van Leer 2006), which is a 2nd-order accurate extension of Godunov's method.

Higher-order extensions such as the piece-wise parabolic method (PPM) start out with a higher order polynomial reconstruction. In the case of PPM, parabolic shapes are assumed in each cell instead of piece-wise linear states. The reconstruction is still guaranteed to be conservative, i.e., the integral underneath the reconstruction recovers the total values of the conserved variables individually in each cell. So-called ENO and WENO schemes (e.g. Balsara et al. 2009) use yet higher-order polynomials to reconstruct the state in a conservative fashion. Here many more cells in the environment need to be considered (i.e., the so-called *stencil* of these methods is much larger) to robustly determine the coefficients of the reconstruction. This can for example involve a least-square fitting procedure (Ollivier-Gooch 1997).

7 Smoothed Particle Hydrodynamics

Smoothed Particle Hydrodynamics (SPH) is a technique for approximating the continuum dynamics of fluids through the use of particles, which may also be viewed as interpolation points (SPH; Lucy 1977; Gingold and Monaghan 1977; Monaghan 1992; Springel 2010b). The principal idea of SPH is to treat hydrodynamics in a completely mesh-free fashion, in terms of a set of sampling particles. Hydrodynamical equations of motion are then derived for these particles, yielding a quite simple and intuitive formulation of gas dynamics. Moreover, it turns out that the particle representation of SPH has excellent conservation properties. Energy, linear momentum, angular momentum, mass, and entropy (if no artificial viscosity operates) are all simultaneously conserved. In addition, there are no advection errors in SPH, and the scheme is fully Galilean-invariant, unlike alternative mesh-based Eulerian techniques. Due to its Lagrangian character, the local resolution of SPH follows the mass flow automatically, a property that is convenient in representing the large density contrasts often encountered in astrophysical problems.

7.1 Kernel Interpolation

At the heart of smoothed particle hydrodynamics lie so-called kernel interpolants. In particular, we use a kernel summation interpolant for estimating the density, which then determines the rest of the basic SPH equations through the variational formalism.

For any field $F(\mathbf{r})$, we may define a smoothed interpolated version, $F_s(\mathbf{r})$, through a convolution with a kernel $W(\mathbf{r}, h)$:

$$F_s(\mathbf{r}) = \int F(\mathbf{r}')\, W(\mathbf{r} - \mathbf{r}', h)\, \mathrm{d}\mathbf{r}'. \tag{279}$$

Here h describes the characteristic width of the kernel, which is normalized to unity and approximates a Dirac δ-function in the limit $h \to 0$. We further require that the kernel is symmetric and sufficiently smooth to make it at least differentiable twice. One possibility for W is a Gaussian. However, most current SPH implementations are based on kernels with a finite support. Usually a cubic spline is adopted with

$W(r, h) = w(\frac{r}{2h})$, and

$$w_{3D}(q) = \frac{8}{\pi} \begin{cases} 1 - 6q^2 + 6q^3, & 0 \le q \le \frac{1}{2}, \\ 2\,(1-q)^3, & \frac{1}{2} < q \le 1, \\ 0, & q > 1, \end{cases} \tag{280}$$

in three-dimensional normalization, but recent work also considered various alternative kernels (Read et al. 2010; Dehnen and Aly 2012). Through Taylor expansion, it is easy to see that the above kernel interpolant is second-order accurate for regularly distributed points due to the symmetry of the kernel (Fig. 29).

Suppose now we know the field at a set of points $\mathbf{r}_i$, i.e., $F_i = F(\mathbf{r}_i)$. The points have an associated mass m_i and density ρ_i, such that $V_i \sim m_i/\rho_i$ is their associated finite volume element. Provided the points sufficiently densely sample the kernel volume, we can approximate the integral in Eq. (279) with the sum

$$F_s(\mathbf{r}) \simeq \sum_j \frac{m_j}{\rho_j} F_j\, W(\mathbf{r} - \mathbf{r}_j, h). \tag{281}$$

This is effectively a Monte-Carlo integration, except that thanks to the comparatively regular distribution of points encountered in practice, the accuracy is better than for a random distribution of the sampling points. In particular, for points in one dimension with equal spacing d, one can show that for $h = d$ the sum of Eq. (281) provides a second order accurate approximation to the real underlying function. Unfortunately, for the irregular yet somewhat ordered particle configurations encountered in real applications, a formal error analysis is not straightforward. It is clear however, that at the very least one should have $h \ge d$, which translates to a minimum of $\sim$33 neighbors in 3D if a Cartesian point distribution is assumed.

Importantly, we see that the estimate for $F_s(\mathbf{r})$ is defined everywhere (not only at the underlying points), and is differentiable thanks to the differentiability of the kernel, albeit with a considerably higher interpolation error for the derivative. Moreover,

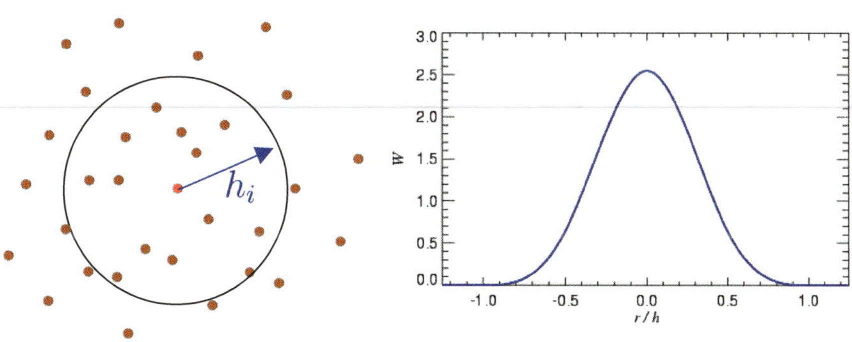

Fig. 29 Kernel interpolation with a B-spline kernel

if we set $F(\mathbf{r}) = \rho(\mathbf{r})$, we obtain

$$\rho_s(\mathbf{r}) \simeq \sum_j m_j W(\mathbf{r} - \mathbf{r}_j, h), \tag{282}$$

yielding a density estimate just based on the particle coordinates and their masses. In general, the smoothing length can be made variable in space, $h = h(\mathbf{r}, t)$, to account for variations in the sampling density. This adaptivity is one of the key advantages of SPH and is essentially always used in practice. There are two options to introduce the variability of h into Eq. (282). One is by adopting $W(\mathbf{r} - \mathbf{r}_j, h(\mathbf{r}))$ as kernel, which corresponds to the so-called 'scatter' approach (Hernquist and Katz 1989). It has the advantage that the volume integral of the smoothed field recovers the total mass, $\int \rho_s(\mathbf{r})\,\mathrm{d}\mathbf{r} = \sum_i m_i$. On the other hand, the so-called 'gather' approach, where we use $W(\mathbf{r}-\mathbf{r}_j, h(\mathbf{r}_i))$ as kernel in Eq. (282), requires only knowledge of the smoothing length $h_i = h(\mathbf{r}_i)$ for estimating the density of particle i, which leads to computationally convenient expressions when the variation of the smoothing length is consistently included in the SPH equations of motion. Since the density is only needed at the coordinates of the particles and the total mass is conserved anyway (since it is tied to the particles), it is not important that the volume integral of the gather form of $\rho_s(\mathbf{r})$ exactly equals the total mass.

In the following we drop the subscript s for indicating the smoothed field, and adopt as SPH estimate of the density of particle i the expression

$$\rho_i = \sum_{j=1}^{N} m_j\, W(\mathbf{r}_i - \mathbf{r}_j, h_i). \tag{283}$$

It is clear now why kernels with a finite support are preferred. They allow the summation to be restricted to the N_{ngb} neighbors that lie within the spherical region of radius $2h$ around the target point $\mathbf{r}_i$, corresponding to a computational cost of order $\mathcal{O}(N_{\mathrm{ngb}}\, N)$ for the full density estimate. Normally this number N_{ngb} of neighbors within the support of the kernel is approximately (or exactly) kept constant by choosing the h_i appropriately. N_{ngb} hence represents an important parameter of the SPH method and needs to be made large enough to provide sufficient sampling of the kernel volumes. Kernels like the Gaussian on the other hand would require a summation over all particles N for every target particle, resulting in a $\mathcal{O}(N^2)$ scaling of the computational cost.

If SPH was really a Monte-Carlo method, the accuracy expected from the interpolation errors of the density estimate would be rather problematic. But the errors are much smaller because the particles do not sample the fluid in a Poissonian fashion. Instead, their distances tend to equilibrate due to the pressure forces, which makes the interpolation errors much smaller (Price 2012). Yet, they remain a significant source of error in SPH and are ultimately the primary origin of the noise inherent in SPH results (Bauer and Springel 2012).

Even though we have based most of the above discussion on the density, the general kernel interpolation technique can also be applied to other fields, and to the construction of differential operators. For example, we may write down a smoothed velocity field and take its derivative to estimate the local velocity divergence, yielding:

$$(\nabla \cdot \mathbf{v})_i = \sum_j \frac{m_j}{\rho_j} \mathbf{v}_j \cdot \nabla_i W(\mathbf{r}_i - \mathbf{r}_j, h). \tag{284}$$

However, an alternative estimate can be obtained by considering the identity $\rho \nabla \cdot \mathbf{v} = \nabla(\rho \mathbf{v}) - \mathbf{v} \cdot \nabla \rho$, and computing kernel estimates for the two terms on the right hand side independently. Their difference then yields

$$(\nabla \cdot \mathbf{v})_i = \frac{1}{\rho_i} \sum_j m_j (\mathbf{v}_j - \mathbf{v}_i) \cdot \nabla_i W(\mathbf{r}_i - \mathbf{r}_j, h). \tag{285}$$

This pair-wise formulation turns out to be more accurate in practice. In particular, it has the advantage of always providing a vanishing velocity divergence if all particle velocities are equal.

7.2 SPH Equations of Motion

The Euler equations for inviscid gas dynamics in Lagrangian form are given by

$$\frac{\mathrm{d}\rho}{\mathrm{d}t} + \rho \nabla \cdot \mathbf{v} = 0, \tag{286}$$

$$\frac{\mathrm{d}\mathbf{v}}{\mathrm{d}t} + \frac{\nabla P}{\rho} = 0, \tag{287}$$

$$\frac{\mathrm{d}u}{\mathrm{d}t} + \frac{P}{\rho} \nabla \cdot \mathbf{v} = 0, \tag{288}$$

where $\mathrm{d}/\mathrm{d}t = \partial/\partial t + \mathbf{v} \cdot \nabla$ is the convective derivative. This system of partial differential equations expresses conservation of mass, momentum and energy. Eckart (1960) has shown that the Euler equations for an inviscid ideal gas follow from the Lagrangian

$$L = \int \rho \left(\frac{\mathbf{v}^2}{2} - u \right) \mathrm{d}V. \tag{289}$$

This opens up an interesting route for obtaining discretized equations of motion for gas dynamics. Instead of working with the continuum equations directly and trying to heuristically work out a set of accurate difference formulas, one can discretize the Lagrangian and then derive SPH equations of motion by applying the variational principals of classical mechanics (Springel and Hernquist 2002). Using a Lagrangian also immediately guarantees certain conservation laws and retains the geometric structure imposed by Hamiltonian dynamics on phase space.

We start by discretizing the Lagrangian in terms of fluid particles of mass m_i, yielding

$$L_{\mathrm{SPH}} = \sum_i \left(\frac{1}{2} m_i \mathbf{v}_i^2 - m_i u_i \right), \tag{290}$$

where it has been assumed that the thermal energy per unit mass of a particle can be expressed through an entropic function A_i of the particle, which simply labels its specific thermodynamic entropy. The pressure of the particles is

$$P_i = A_i \rho_i^\gamma = (\gamma - 1)\rho_i u_i, \tag{291}$$

where γ is the adiabatic index. Note that for isentropic flow (i.e., in the absence of shocks, and without mixing or thermal conduction) we expect the A_i to be constant. We hence define u_i, the thermal energy per unit mass, in terms of the density estimate as

$$u_i(\rho_i) = A_i \frac{\rho_i^{\gamma-1}}{\gamma - 1}. \tag{292}$$

This raises the question of how the smoothing lengths h_i needed for estimating ρ_i should be determined. As we discussed above, we would like to ensure adaptive kernel sizes, meaning that the number of points in the kernel should be approximately constant. In much of the older SPH literature, the number of neighbors was allowed to vary within some (small) range around a target number. Sometimes the smoothing length itself was evolved with a differential equation in time, exploiting the continuity relation and the expectation that ρh^3 should be approximately constant. In case the number of neighbors outside the kernel happened to fall outside the allowed range, h was suitably readjusted, at the price of some errors in energy conservation.

A better method is to require that the mass in the kernel volume should be constant, viz.

$$\rho_i h_i^3 = const \tag{293}$$

for three dimensions. Since $\rho_i = \rho_i(\mathbf{r}_1, \mathbf{r}_2, \ldots \mathbf{r}_N, h_i)$ is only a function of the particle coordinates and of h_i, this equation implicitly defines the function $h_i = h_i(\mathbf{r}_1, \mathbf{r}_2, \ldots \mathbf{r}_N)$ in terms of the particle coordinates.

We can then proceed to derive the equations of motion from

$$\frac{\mathrm{d}}{\mathrm{d}t}\frac{\partial L}{\partial \dot{\mathbf{r}}_i} - \frac{\partial L}{\partial \mathbf{r}_i} = 0. \tag{294}$$

This first gives

$$m_i \frac{\mathrm{d}\mathbf{v}_i}{\mathrm{d}t} = -\sum_{j=1}^{N} m_j \frac{P_j}{\rho_j^2} \frac{\partial \rho_j}{\partial \mathbf{r}_i}, \tag{295}$$

where the derivative $\partial \rho_j / \partial \mathbf{r}_i$ stands for the total variation of the density with respect to the coordinate $\mathbf{r}_i$, including any variation of h_j this may entail. We can hence write

$$\frac{\partial \rho_j}{\partial \mathbf{r}_i} = \nabla_i \rho_j + \frac{\partial \rho_j}{\partial h_j} \frac{\partial h_j}{\partial \mathbf{r}_i}, \tag{296}$$

where the smoothing length is kept constant in the first derivative on the right hand side (in our notation, the Nabla operator $\nabla_i = \partial / \partial \mathbf{r}_i$ means differentiation with respect to $\mathbf{r}_i$ holding the smoothing lengths constant). On the other hand, differentiation of $\rho_j h_j^3 = const$ with respect to $\mathbf{r}_i$ yields

$$\frac{\partial \rho_j}{\partial h_j} \frac{\partial h_j}{\partial \mathbf{r}_i} \left[1 + \frac{3\rho_j}{h_j} \left(\frac{\partial \rho_j}{\partial h_j} \right)^{-1} \right] = -\nabla_i \rho_j. \tag{297}$$

Combining Eqs. (296) and (297) we then find

$$\frac{\partial \rho_j}{\partial \mathbf{r}_i} = \left(1 + \frac{h_j}{3\rho_j} \frac{\partial \rho_j}{\partial h_j} \right)^{-1} \nabla_i \rho_j. \tag{298}$$

Using

$$\nabla_i \rho_j = m_i \nabla_i W_{ij}(h_j) + \delta_{ij} \sum_{k=1}^{N} m_k \nabla_i W_{ki}(h_i), \tag{299}$$

we finally obtain the equations of motion

$$\frac{\mathrm{d}\mathbf{v}_i}{\mathrm{d}t} = -\sum_{j=1}^{N} m_j \left[f_i \frac{P_i}{\rho_i^2} \nabla_i W_{ij}(h_i) + f_j \frac{P_j}{\rho_j^2} \nabla_i W_{ij}(h_j) \right], \tag{300}$$

where the f_i are defined by

$$f_i = \left[1 + \frac{h_i}{3\rho_i} \frac{\partial \rho_i}{\partial h_i} \right]^{-1}, \tag{301}$$

and the abbreviation $W_{ij}(h) = W(|\mathbf{r}_i - \mathbf{r}_j|, h)$ has been used. Note that the correction factors f_i can be easily calculated alongside the density estimate, all that is required is an additional summation to get $\partial \rho_i / \partial \mathbf{r}_i$ for each particle. This quantity is in fact also useful to get the correct smoothing radii by iteratively solving $\rho_i h_i^3 = const$ with a Newton-Raphson iteration (Springel and Hernquist 2002).

The equations of motion (300) for inviscid hydrodynamics are remarkably simple. In essence, we have transformed a complicated system of partial differential equations into a much simpler set of ordinary differential equations. Furthermore, we only have to solve the momentum equation explicitly. The mass conservation equation as well as the total energy equation (and hence the thermal energy equation) are already taken care of, because the particle masses and their specific entropies stay constant for reversible gas dynamics. However, later we will introduce an artificial viscosity that is needed to allow a treatment of shocks. This will introduce additional terms in the equation of motion and requires the time integration of one thermodynamic quantity per particle, which can either be chosen as entropy or thermal energy. Indeed, the above formulation can also be equivalently expressed in terms of thermal energy instead of entropy. This follows by taking the time derivative of Eq. (292), which first yields

$$\frac{\mathrm{d}u_i}{\mathrm{d}t} = \frac{P_i}{\rho_i^2} \sum_j \mathbf{v}_j \cdot \frac{\partial \rho_i}{\partial \mathbf{r}_j}. \tag{302}$$

Using Eqs. (298) and (299) then gives the evolution of the thermal energy as

$$\frac{\mathrm{d}u_i}{\mathrm{d}t} = f_i \frac{P_i}{\rho_i^2} \sum_j m_j (\mathbf{v}_i - \mathbf{v}_j) \cdot \nabla W_{ij}(h_i), \tag{303}$$

which needs to be integrated along the equation of motion if one wants to use the thermal energy as independent thermodynamic variable. There is no difference however to using the entropy; the two are completely equivalent in the variational formulation.

Note that the above formulation readily fulfills the conservation laws of energy, momentum and angular momentum. This can be shown based on the discretized form of the equations, but it is also manifest due to the symmetries of the Lagrangian that was used as a starting point. The absence of an explicit time dependence gives the energy conservation, the translational invariance implies momentum conservation, and the rotational invariance gives angular momentum conservation.

7.3 Artificial Viscosity

Even when starting from perfectly smooth initial conditions, the gas dynamics described by the Euler equations may readily produce true discontinuities in the form of shock waves and contact discontinuities. At such fronts the differential form of the Euler equations breaks down, and their integral form (equivalent to the conservation laws) needs to be used. At a shock front, this yields the Rankine-Hugoniot jump conditions that relate the upstream and downstream states of the fluid. These relations show that the specific entropy of the gas always increases at a shock front, implying that in the shock layer itself the gas dynamics can no longer be described as inviscid. In turn, this also implies that the discretized SPH equations we derived above can not correctly describe a shock, simply because they keep the entropy strictly constant.

One thus must allow for a modification of the dynamics at shocks and somehow introduce the necessary dissipation. This is usually accomplished in SPH by an artificial viscosity. Its purpose is to dissipate kinetic energy into heat and to produce entropy in the process. The usual approach is to parameterize the artificial viscosity in terms of a friction force that damps the relative motion of particles. Through the viscosity, the shock is broadened into a resolvable layer, something that makes a description of the dynamics everywhere in terms of the differential form possible. It may seem a daunting task though to somehow tune the strength of the artificial viscosity such that just the right amount of entropy is generated in a shock. Fortunately, this is however relatively unproblematic. Provided the viscosity is introduced into the dynamics in a conservative fashion, the conservation laws themselves ensure that the right amount of dissipation occurs at a shock front.

What is more problematic is to devise the viscosity such that it is only active when there is really a shock present. If it also operates outside of shocks, even if only at a weak level, the dynamics may begin to deviate from that of an ideal gas.

The viscous force is most often added to the equation of motion as

$$\left.\frac{\mathrm{d}\mathbf{v}_i}{\mathrm{d}t}\right|_{\mathrm{visc}} = -\sum_{j=1}^{N} m_j \Pi_{ij} \nabla_i \overline{W}_{ij} , \tag{304}$$

where

$$\overline{W}_{ij} = \frac{1}{2}\left[W_{ij}(h_i) + W_{ij}(h_j)\right] \tag{305}$$

denotes a symmetrized kernel, which some researchers prefer to define as $\overline{W}_{ij} = W_{ij}([h_i + h_j]/2)$. Provided the viscosity factor Π_{ij} is symmetric in i and j, the viscous force between any pair of interacting particles will be antisymmetric and along the line joining the particles. Hence linear momentum and angular momentum are still preserved. In order to conserve total energy, we need to compensate the work done against the viscous force in the thermal reservoir, described either in terms of entropy,

$$\left.\frac{\mathrm{d}A_i}{\mathrm{d}t}\right|_{\mathrm{visc}} = \frac{1}{2}\frac{\gamma - 1}{\rho_i^{\gamma-1}} \sum_{j=1}^{N} m_j \Pi_{ij} \mathbf{v}_{ij} \cdot \nabla_i \overline{W}_{ij} , \tag{306}$$

or in terms of thermal energy per unit mass,

$$\left.\frac{\mathrm{d}u_i}{\mathrm{d}t}\right|_{\mathrm{visc}} = \frac{1}{2}\sum_{j=1}^{N} m_j \Pi_{ij} \mathbf{v}_{ij} \cdot \nabla_i \overline{W}_{ij} , \tag{307}$$

where $\mathbf{v}_{ij} = \mathbf{v}_i - \mathbf{v}_j$. There is substantial freedom in the detailed parametrization of the viscosity Π_{ij}. The most commonly used formulation of the viscosity is

$$\Pi_{ij} = \begin{cases} \left[-\alpha c_{ij}\mu_{ij} + \beta\mu_{ij}^2\right]/\rho_{ij} & \text{if } \mathbf{v}_{ij}\cdot\mathbf{r}_{ij} < 0 \\ 0 & \text{otherwise,} \end{cases} \tag{308}$$

with

$$\mu_{ij} = \frac{h_{ij}\,\mathbf{v}_{ij}\cdot\mathbf{r}_{ij}}{\left|\mathbf{r}_{ij}\right|^2 + \epsilon h_{ij}^2}. \tag{309}$$

Here h_{ij} and ρ_{ij} denote arithmetic means of the corresponding quantities for the two particles i and j, with c_{ij} giving the mean sound speed, and $\mathbf{r}_{ij} \equiv \mathbf{r}_i - \mathbf{r}_j$. The strength of the viscosity is regulated by the parameters α and β, with typical values in the range $\alpha \simeq 0.5-1.0$ and the frequent choice of $\beta = 2\,\alpha$. The parameter $\epsilon \simeq 0.01$ is introduced to protect against singularities if two particles happen to get very close.

In this form, the artificial viscosity is basically a combination of a bulk and a von Neumann-Richtmyer viscosity. Historically, the quadratic term in μ_{ij} has been added to the original Monaghan-Gingold form to prevent particle penetration in high Mach number shocks. Note that the viscosity only acts for particles that rapidly approach each other, hence the entropy production is always positive definite.

7.4 New Trends in SPH

Smoothed particle hydrodynamics is a remarkably versatile and simple approach for numerical fluid dynamics. The ease with which it can provide a large dynamic range in spatial resolution and density, as well as an automatically adaptive resolution, are unmatched in Eulerian methods. At the same time, SPH has excellent conservation properties, not only for energy and linear momentum, but also for angular momentum. The latter is not automatically guaranteed in Eulerian codes, even though it is usually fulfilled at an acceptable level for well-resolved flows. When coupled to self-gravity, SPH conserves the total energy exactly, which is again not manifestly true in most mesh-based approaches to hydrodynamics. Finally, SPH is Galilean-invariant and free of any errors from advection alone, which is another advantage compared to Eulerian mesh-based approaches.

Thanks to its completely mesh-free nature, SPH can easily deal with complicated geometric settings and large regions of space that are completely devoid of particles. Implementations of SPH in a numerical code tend to be comparatively simple and transparent. At the same time, the scheme is characterized by remarkable robustness. For example, negative densities or negative temperatures, sometimes a problem in mesh-based codes, can not occur in SPH by construction. Although shock waves are broadened in SPH, the properties of the post-shock flow are correct.

The main disadvantage of SPH is its limited accuracy in multi-dimensional flows (e.g. Agertz et al. 2007; Bauer and Springel 2012). One source of noise originates in the approximation of local kernel interpolants through discrete sums over a small set of nearest neighbors. While in 1D the consequences of this noise tend to be reasonably benign, particle motion in multiple dimensions has a much higher degree of freedom. Here the mutually repulsive forces of pressurized neighboring particle pairs do not easily cancel in all dimensions simultaneously, especially not given the errors of the discretized kernel interpolants. As a result, some 'jitter' in the particle motions readily develops, giving rise to velocity noise up to a few percent of the local sound speed. This noise seriously messes up the accuracy that can be reached with the technique, especially for subsonic flow, and also leads to a slow convergence rate.

Another problem is the relatively high numerical viscosity of SPH. To reduce the numerical viscosity of SPH in regions away from shocks, several studies have recently advanced the idea of keeping the functional form of the artificial viscosity, but making the viscosity strength parameter α variable in time (Morris 1997; Dolag et al. 2005; Rosswog 2005). Adopting $\beta = 2\alpha$, one may for example evolve the parameter α individually for each particle with an equation such as

$$\frac{\mathrm{d}\alpha_i}{\mathrm{d}t} = -\frac{\alpha_i - \alpha_{\max}}{\tau_i} + S_i, \tag{310}$$

where S_i is some source function meant to ramp up the viscosity rapidly if a shock is detected, while the first term lets the viscosity exponentially decay again to a prescribed minimum value $\alpha_{\min}$ on a timescale τ_i. So far, mostly simple source functions like $S_i = \max[-(\nabla \cdot \mathbf{v})_i, 0]$ and timescales $\tau_i \simeq h_i/c_i$ have been explored and the viscosity α_i has often also been prevented from becoming higher than some prescribed maximum value $\alpha_{\max}$. It is clear that the success of such a variable α scheme depends critically on an appropriate source function. The form above can still not distinguish purely adiabatic compression from that in a shock, so is not completely free of creating unwanted viscosity. More advanced parameterizations that try to address this problem have therefore also been developed (Cullen and Dehnen 2010).

Particularly problematic in SPH are also fluid instabilities across contact discontinuities, such as Kelvin-Helmholtz instabilities. These are usually found to be suppressed in their growth. Recent new formulations of SPH try to improve on this either through different kernel shapes combined with a much larger number of smoothing neighbors (e.g. Read and Hayfield 2012), through artificial thermal conduction at contact discontinuities to reduce pressure force errors and spurious surface tension there (e.g. Price 2008), or by alluding to a pressure-based formulation where the density is estimated only indirectly from a kernel-interpolated pressure field (Hopkins 2013). These developments appear certainly promising. At the moment many new ideas for improved SPH formulations are still advanced, and new implementations are published regularly. While many problems of SPH have been addresses by these new schemes, so far they have not yet been able to cure the relatively large gradient errors in SPH, suggesting that the convergence rate of them is still lower than that of comparable mesh-based approaches (e.g. Hu et al. 2014).

8 Moving-Mesh Techniques

8.1 Differences Between Eulerian and Lagrangian Techniques

It has become clear over recent years that both Lagrangian SPH and Eulerian AMR techniques suffer from fundamental limitations that make them inaccurate in certain regimes. Indeed, these methods sometimes yield conflicting results even for basic calculations that only consider non-radiative hydrodynamics (e.g. Frenk et al. 1999; Agertz et al. 2007; Tasker et al. 2008; Mitchell et al. 2009). SPH codes have comparatively poor shock resolution, offer only low-order accuracy for the treatment of contact discontinuities, and suffer from subsonic velocity noise. Worse, they appear to suppress fluid instabilities under certain conditions, as a result of a spurious surface tension and inaccurate gradient estimates across density jumps. On the other hand, Eulerian codes are not free of fundamental problems either. They do not produce Galilean-invariant results, which can make their accuracy sensitive to the presence of bulk velocities (e.g. Wadsley et al. 2008; Tasker et al. 2008). Another concern lies in the preference of certain spatial directions in Eulerian hydrodynamics, which can make poorly resolved disk galaxies align with the coordinate planes (Dubois et al. 2014).

There is hence substantial motivation to search for new hydrodynamical methods that improve on these weaknesses of the SPH and AMR techniques. In particular, we would like to retain the accuracy of mesh-based hydrodynamical methods (for which decades of experience have been accumulated in computational fluid dynamics), while at the same time we would like to outfit them with the Galilean-invariance and geometric flexibility that is characteristic of SPH. The principal idea for achieving such a synthesis is to allow the mesh to move with the flow itself. This is in principle an obvious and old idea(Braun and Sambridge 1995 ; Gnedin 1995; Whitehurst 1995; Mavriplis 1997; Xu 1997; Hassan et al. 1998; Pen 1998; Trac and Pen 2004), but one fraught with many practical difficulties. In particular, mesh tangling (manifested in 'bow-tie' cells and hourglass like mesh motions) is the traditional problem of such attempts to simulate multi-dimensional hydrodynamics in a Lagrangian fashion.

8.2 Voronoi Tessellations

We here briefly describe a new formulation of continuum hydrodynamics based on an unstructured mesh that overcomes many of these problems (Springel 2010a). The mesh is defined as the Voronoi tessellation of a set of discrete mesh-generating points, which are in principle allowed to move freely. For the given set of points, the Voronoi tessellation of space consists of non-overlapping cells around each of the sites such that each cell contains the region of space closer to it than to any of the other sites. Closely related to the Voronoi tessellation is the Delaunay tessellation, the topological dual of the Voronoi diagram. Both constructions can also be used for natural neighbor interpolation and geometric analysis of cosmic structures (e.g. van de Weygaert 1994; Pelupessy et al. 2003; van de Weygaert and Schaap 2009). In 2D, the Delaunay tessellation for a given set of points is a triangulation of the plane, where the points serve as vertices of the triangles. The defining property of the Delaunay triangulation is that each circumcircle around one of the triangles of the tessellation is not allowed to contain any of the other mesh-generating points in its interior. This empty circumcircle property distinguishes the Delaunay triangulation from the many other triangulations of the plane that are possible for the point set, and in fact uniquely determines the triangulation for points in general position. Similarly, in three dimensions, the Delaunay tessellation is formed by tetrahedra that are not allowed to contain any of the points inside their circumspheres.

As an example, Fig. 30 shows the Delaunay and Voronoi tessellations for a small set of points in 2D, enclosed in a box with imposed periodic boundary conditions. The midpoints of the circumcircles around each Delaunay triangle form the vertices of the Voronoi cells, and for each line in the Delaunay diagram, there is an orthogonal face in the Voronoi tessellation.

The Voronoi cells can be used as control volumes for a finite-volume formulation of hydrodynamics, using the same principal ideas for reconstruction, evolution and averaging (REA) steps that we have discussed earlier in the context of Eulerian techniques. However, as we will discuss below it is possible to consistently include the mesh motion in the formulation of the numerical steps, allowing the REA-scheme to become Galilean-invariant. Even more importantly, due to the mathematical properties of the Voronoi tessellation, the mesh continuously deforms and changes its topology as a result of the point motion, without ever leading to the dreaded mesh-tangling effects that are the curse of traditional moving-mesh methods.

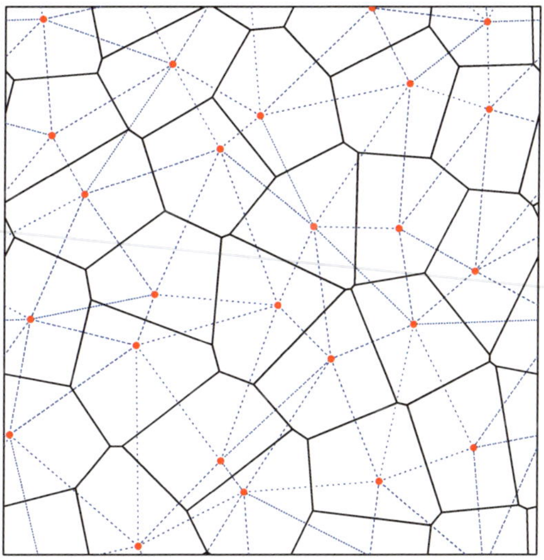

Fig. 30 Example of a Voronoi and Delaunay tessellation in 2D, with periodic boundary conditions. The *red circles* show the generating points of the Voronoi tessellation, which is drawn with *solid lines*. Its topological dual, the Delaunay triangulation, is overlaid with thin *dashed lines*

8.3 Finite Volume Hydrodynamics on a Moving-mesh

As already discussed earlier in Sect. 6.4, the Euler equations are conservation laws for mass, momentum and energy that take the form of a system of hyperbolic partial differential equation. They can be written in the compact form

$$\frac{\partial \mathbf{U}}{\partial t} + \nabla \cdot \mathbf{F} = 0, \tag{311}$$

which emphasizes their character as conservation laws.

Following the *finite-volume* strategy, we describe the fluid's state by the cell-averages of the conserved quantities for these cells. In particular, integrating the fluid over the volume V_i of cell i, we can define the total mass m_i, momentum p_i and energy E_i contained in the cell as follows,

$$\mathbf{Q}_i = \begin{pmatrix} m_i \\ \mathbf{p}_i \\ E_i \end{pmatrix} = \int_{V_i} \mathbf{U}\,\mathrm{d}V. \tag{312}$$

With the help of the Euler equations, we can calculate the rate of change of $\mathbf{Q}_i$ in time. Converting the volume integral over the flux divergence into a surface integral over the cell results in

$$\frac{\mathrm{d}\mathbf{Q}_i}{\mathrm{d}t} = -\int_{\partial V_i} \left[\mathbf{F}(\mathbf{U}) - \mathbf{U}\mathbf{w}^T\right] \mathrm{d}\mathbf{n}. \tag{313}$$

Here $\mathbf{n}$ is an outward normal vector of the cell surface, and $\mathbf{w}$ is the velocity with which each point of the boundary of the cell moves. In Eulerian codes, the mesh is taken to be static, so that $\mathbf{w} = 0$, while in a fully Lagrangian approach, the surface would move at every point with the local flow velocity, i.e., $\mathbf{w} = \mathbf{v}$. In this case, the right hand side of Eq. (313) formally simplifies, because then the first component of $\mathbf{Q}_i$, the mass, stays fixed for each cell. Unfortunately, it is normally not possible to follow the distortions of the shapes of fluid volumes exactly in multi-dimensional flows for a reasonably long time, or in other words, one cannot guarantee the condition $\mathbf{w} = \mathbf{v}$ over the entire surface. In this case, one needs to use the general formula of Eq. (313). As an aside, we note that one conceptual problem of SPH is that these surface fluxes due to the $\mathbf{w}$-term are always ignored.

The cells of our finite volume discretization are polyhedra with flat polygonal faces (or lines in 2D). Let $\mathbf{A}_{ij}$ describe the oriented area of the face between cells i and j (pointing from i to j). Then we can define the averaged flux across the face i-j as

$$\mathbf{F}_{ij} = \frac{1}{A_{ij}} \int_{A_{ij}} \left[\mathbf{F}(\mathbf{U}) - \mathbf{U}\mathbf{w}^T\right] \mathrm{d}\mathbf{A}_{ij}, \tag{314}$$

and the Euler equations in finite-volume form become

$$\frac{\mathrm{d}\mathbf{Q}_i}{\mathrm{d}t} = -\sum_j A_{ij}\mathbf{F}_{ij}. \tag{315}$$

We obtain a manifestly conservative time discretization of this equation by writing it as

$$\mathbf{Q}_i^{(n+1)} = \mathbf{Q}_i^{(n)} - \Delta t \sum_j A_{ij}\hat{\mathbf{F}}_{ij}^{(n+1/2)}, \tag{316}$$

where the $\hat{\mathbf{F}}_{ij}$ are now an appropriately time-averaged approximation to the true flux $\mathbf{F}_{ij}$ across the cell face. The notation $\mathbf{Q}_i^{(n)}$ is meant to describe the state of the system at step n. Note that $\hat{\mathbf{F}}_{ij} = -\hat{\mathbf{F}}_{ji}$, i.e., the discretization is manifestly conservative.

Evidently, a crucial step lies in obtaining a numerical estimate of the fluxes $\hat{\mathbf{F}}_{ij}$. We employ the MUSCL-Hancock scheme (van Leer 1984, 2006; Toro 1997) already discussed in Sect. 6.7, which is a well-known and relatively simple approach for obtaining second-order accuracy in space and time. This scheme is used in several state-of-the art Eulerian codes (e.g. Fromang et al. 2006; Mignone et al. 2007; Cunningham et al. 2009). In its basic form, the MUSCL-Hancock scheme involves a slope-limited piece-wise linear reconstruction step within each cell, a first order prediction step for the evolution over half a timestep, and finally a Riemann solver to estimate the time-averaged inter-cell fluxes for the timestep. After the fluxes have been applied to each cell, a new averaged state of the cells is constructed. This sequence of steps in a timestep hence follows the general REA approach.

Figure 31 gives a sketch of the geometry involved in estimating the flux across the face between two Voronoi cells. Truly multidimensional Riemann solvers have been developed recently (Wendroff 1999; Brio et al. 2001; Balsara 2010), but it is unclear whether they can be readily adapted to our complicated face geometry. We therefore follow the common approach and calculate the flux for each face separately, treating it as an effectively one-dimensional problem. Since we do not work with Cartesian meshes, we cannot use operating splitting to deal with the individual spatial dimensions, hence we apply an *unsplit* approach. For defining the Riemann problem normal to a cell face, we rotate the fluid state into a suitable coordinate system with the x'-axis normal to the cell face (see sketch). This defines the left and right states across the face, which we pass to an (exact) Riemann solver, following Toro (1997). Once the flux has been calculated with the Riemann solver, we transform it back to the lab frame. In order to obtain Galilean-invariance and stable upwind behavior, the Riemann problem needs to be solved *in the frame of the moving face*.

In the moving-mesh hydrodynamical scheme implemented in the AREPO[2] code (Springel 2010a), each timestep hence involves the following basic steps:

---

[2]Named after the enigmatic word AREPO in the Latin palindromic sentence *sator arepo tenet opera rotas*, the 'Sator Square'.

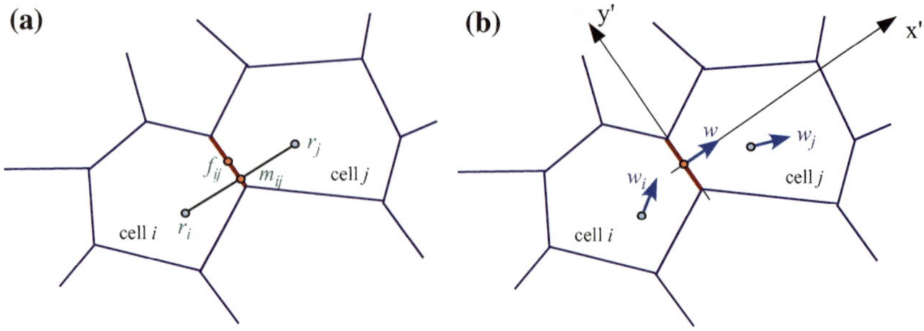

Fig. 31 Sketch of a Voronoi mesh and the relevant geometric quantities that enter the flux calculation across a face. In **a**, we show the mesh-generating points $\mathbf{r}_i$ and $\mathbf{r}_j$ of two cells i and j. The face between these two cells has a center-of-mass vector $\mathbf{f}_{ij}$, which in general will be offset from the mid-point m_{ij} of the two points. In **b**, we illustrate the two velocity vectors $\mathbf{w}_i$ and $\mathbf{w}_j$ associated with the mesh-generating points. These are normally chosen equal to the gas velocity in the cells, but other choices are allowed too. The motion of the mesh-generating points uniquely determines the motion of the face between the cells. Only the normal velocity $\mathbf{w}$ is however needed for the flux computation in the rotated frame x', y'

1. Calculate a new Voronoi tessellation based on the current coordinates $\mathbf{r}_i$ of the mesh generating points. This also gives the centers-of-mass $\mathbf{s}_i$ of each cell, their volumes V_i, as well as the areas A_{ij} and centers $\mathbf{f}_{ij}$ of all faces between cells.
2. Based on the vector of conserved fluid variables $\mathbf{Q}_i$ associated with each cell, calculate the 'primitive' fluid variables $\mathbf{W}_i = (\rho_i, \mathbf{v}_i, P_i)$ for each cell.
3. Estimate the gradients of the density, of each of the velocity components, and of the pressure in each cell, and apply a slope-limiting procedure to avoid overshoots and the introduction of new extrema.
4. Assign velocities $\mathbf{w}_i$ to the mesh generating points.
5. Evaluate the Courant criterion and determine a suitable timestep size Δt.
6. For each Voronoi face, compute the flux $\hat{\mathbf{F}}_{ij}$ across it by first determining the left and right states at the midpoint of the face by linear extrapolation from the cell midpoints, and by predicting these states forward in time by half a timestep. Solve the Riemann problem in a rotated frame that is moving with the speed of the face, and transform the result back into the lab-frame.
7. For each cell, update its conserved quantities with the total flux over its surface multiplied by the timestep, using Eq. (316). This yields the new state vectors $\mathbf{Q}_i^{(n+1)}$ of the conserved variables at the end of the timestep.
8. Move the mesh-generating points with their assigned velocities for this timestep.

Full details for each of these different steps as well as test problems can be found in Springel (2010a). Recently, a number of science applications involving fairly large calculations with AREPO have been carried out that demonstrate the practical advantages of this technique for applications in galaxy formation and evolution (e.g. Greif et al. 2011a, b; Vogelsberger et al. 2012, 2013, 2014; Marinacci et al. 2014; Pakmor et al. 2014).

9 Parallelization Techniques and Current Computing Trends

Modern computer architectures offer ever more computational power that we ideally would like to use to their full extent for scientific purposes, in particular for simulations in astrophysics. However, unlike in the past, the speed of individual compute cores, which may be viewed as serial computers, has recently hardly grown any more (in stark contrast to the evolution a few years back). Instead, the number of cores on large supercomputers has started to increase exponentially. Even on laptops and cell-phones, multi-core computers have become the norm rather than the exception.

However, most algorithms and computer languages are constructed around the concepts of a serial computer, in which a stream of operations is executed sequentially. This is also how we typically think when we write computer code. In order to exploit the power available in modern computers, one needs to change this approach and adopt parallel computing techniques. Due to the large variety of computer hardware, and the many different technical concepts for devising parallel programs, we can only scratch the surface here and make a few basic remarks about parallelization, and some basic techniques that are currently in wide use for it. The interested student is encouraged to read more about this in textbooks and/or in online resources.

9.1 Hardware Overview

Let's start first with a schematic overview over some of the main characteristics and types of current computer architectures.

9.1.1 Serial Computer

The traditional model of a computer consists of a central processing unit (CPU), capable of executing a sequential stream of load, store, and compute operations, attached to a random access memory (RAM) used for data storage, as sketched in Fig. 32. Branches and jumps in this stream are possible too, but at any given time, only one operation is carried out. The operating system may still provide the illusion

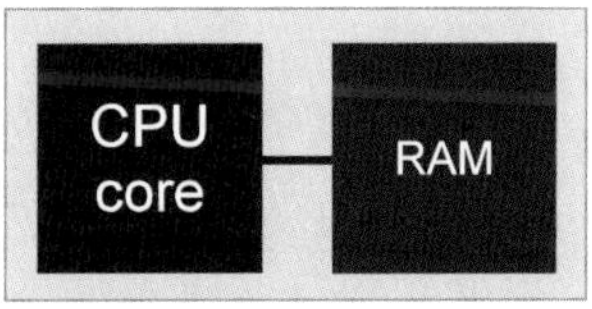

Fig. 32 Simple schematic sketch of a serial computer—most traditional computer languages are formulated for this type of machine

that several programs are executed concurrently, but in this case this is reached by time slicing the single compute resource.

Most computer languages are built around this model; they can be viewed as a means to create the stream of serial operations in a convenient way. One can in principle also by-pass the computer language and write down the machine instructions directly (assembler programming), but fortunately, modern compilers make this unnecessary in scientific applications (except perhaps in very special circumstances where extreme performance tuning is desired).

9.1.2 Multi-core Nodes

It is possible to attach multiple CPUs to the same RAM (see Fig. 33), and, especially in recent times, computer vendors have started to add multiple cores to individual CPUs. On each CPU and each core of a CPU, different programs can be executed concurrently, allowing real parallel computations.

In machines with uniform memory access, the individual cores can access the memory with the same speed, at least in principle. In this case the distinction between a CPU and a core can become confusing (and is in fact superfluous at some level), because it is ambiguous whether "CPU" refers to a single core, or all the cores on the same die of silicon (it's hence best to simple speak about cores to avoid any confusion).

9.1.3 Multi-socket Compute Nodes

Most powerful compute servers feature these days a so-called NUMA (non-uniform memory access) architecture. Here the full main memory is accessible by all cores, but not all parts of it with the same speed. The compute nodes usually feature individual multi-core CPUs, each with a dedicated memory bank attached (see Fig. 34). Read and write operations to this part of physical memory are fastest, while accessing the other memory banks is typically noticeably slower and often involves going through special, high-bandwidth interprocessor bus systems.

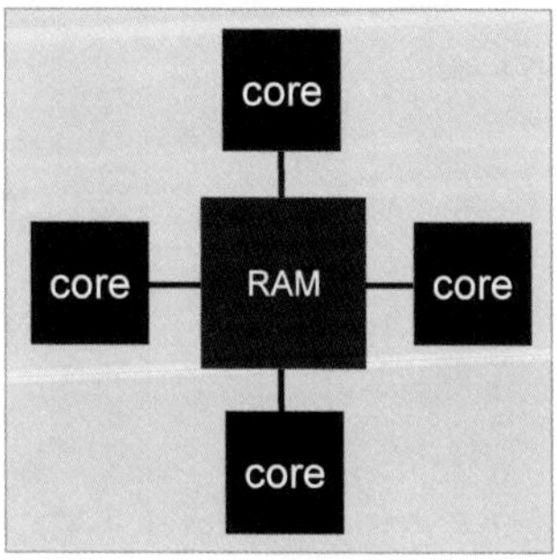

Fig. 33 Multi-core computer with shared memory

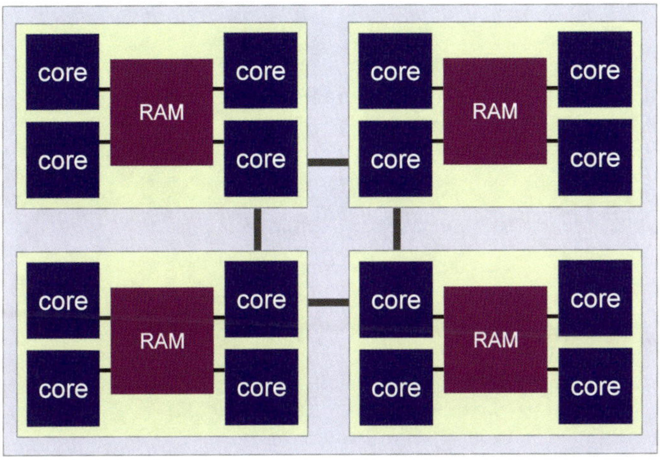

Fig. 34 Non-uniform memory access computer. Here multiple sockets contain several processor dies, each with multiple cores. The total memory is split up into banks, which can be accessed with maximum speed by the processor associated with it, and with a reduced speed from different processors

In such machines, maximum compute performance is only reached when the data that is worked on by a core resides on the "right" memory bank. Fortunately, the operating system will normally try to help with this by satisfying memory requests out of the closest part of physical memory, if possible. It is then also advantageous to tell the operating system to "pin" execution of a process or thread to a certain physical core, so that it is not rescheduled to run on another core from which the already allocated data may be accessible only with slower speed.

9.1.4 Compute Clusters

Very powerful supercomputers used in the field of high-performance computing (HPC) can be formed by connecting many compute nodes through a fast communication network, as sketched in Fig. 35. This can be standard gigabit ethernet in some cases, but usually much faster (and more expensive) communication networks such as infiniband are employed. The leading supercomputers in the world are of this type, and currently typically reach several 10^5 cores in total, with the first machines surpassing even 10^6 cores. Towards the end of the decade, when exaflop machines (capable of carrying out 10^{18} floating point operations per second) are expected, this may even grow to 10^8 or beyond. How to use these machines efficiently for *interesting* science problems, which tend to be tightly coupled and not amenable to unlimited levels of computational concurrency, is however still an unsolved problem.

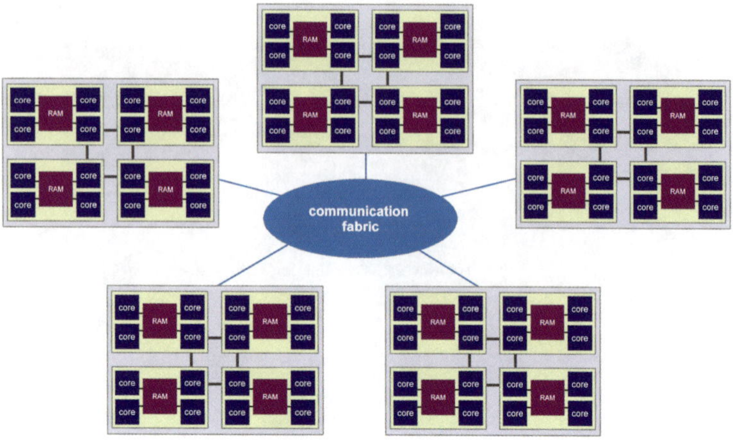

Fig. 35 In large high-performance supercomputers, one typically connects a large number of powerful compute nodes (often of NUMA type) through a very fast dedicated communication network

9.1.5 Device Computing

A comparatively new trend in scientific computing is to augment classical compute nodes with special accelerator cards that are particularly tuned to floating point calculations. These cards have much simpler, less flexible compute cores, but the transistors saved on implementing chip complexity can be spent on building more powerful compute engines that can execute many floating point operations in parallel. Graphics processing units (GPUs) have been originally developed with such a design just for the vector operations necessary to render graphics, but now their streaming processors can also be used for general purpose calculations. For certain applications, GPUs can be much faster than ordinary CPUs, but programming them is harder.

In so-called hybrid compute nodes (Fig. 36), one has one or several ordinary CPUs coupled to one or several GPUs, or other accelerator cards such as the new Intel Xeon Phi. Of course, these hybrid nodes can be clustered again with a fabric to form powerful supercomputers. In fact, the fastest machines in the world are presently of this type.

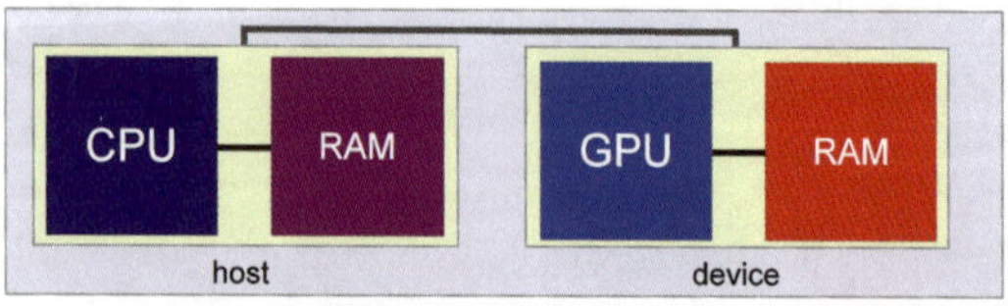

Fig. 36 Hybrid compute node. An accelerator device (a GPU, or an Intel Xeon Phi card) is connected to an ordinary compute node through a fast bus system. Usually, host CPU and computing device each have their own RAM

9.1.6 Vector Cores

Another hardware aspect that should not be overlooked is that single compute cores are actually increasingly capable to carry out so-called vector instructions. Here a single instruction (such as addition, multiplication, etc.) is applied to multiple data elements (a vector). This is also a form of parallelization, allowing the calculation throughput to be raised significantly. Below is an example that calculates $x = a \cdot b$ element by element for 4-vectors a and b using Intel's Advanced Vector Instructions (AVX). These can be programmed explicitly through *intrinsics* in C, which are basically individual machine instructions hidden as macros or function calls within C. (Usually, one does not do this manually though, but rather hopes the compiler emits such instructions somehow automatically).

```
#include <xmmintrin.h>

void do_stuff(void)
{
  double a[4], b[4], res[4];

  __m256d x = _mm256_load_pd(a);
  __m256d y = _mm256_load_pd(b);

  x = _mm256_mul_pd(x, y);

  _mm256_store_pd(res, x);
}
```

The current generation of the x86 processors from Intel/AMD features SSE/AVX instructions that operate on vectors of up to 256 bits. This means that 4 double-precision or 8 single precision operations can be executed with such an instruction, roughly in the same speed that an ordinary double or single-precision operation takes. So if these instructions can be used in an optimum way, one achieves a speed-up by a factor of up to 4 or 8, respectively. On the Intel Xeon Phi chips, the vector length has already doubled again and is now 512 bits, hence allowing another factor of 2 in the performance. Likely, we will see even larger vector lengths in the near future.

9.1.7 Hyperthreading

A general problem in exploiting modern computer hardware to its full capacity is that accessing main memory is very much slower than doing a single floating point operation in a compute core (note that moving data also costs more energy than doing a floating point calculation, which is becoming an important consideration too). As a result, a compute core typically spends a large fraction of its cycles waiting for data to arrive from memory.

The idea of hyperthreading as implemented in CPUs by Intel and in IBM's Power architecture is to use this wait time by letting the core do some useful work in the meantime. This is achieved by "overloading" the compute core with several execution streams. But instead of letting the operating system toggle between their execution, the hardware itself can switch very rapidly between these different "hyperthreads". Even though there is still some overhead in changing the execution context from one thread to another, this strategy can still lead to a substantial net increase in the calculational throughput on the core. Effectively, to the operating system and user it appears as if there are more cores (so called virtual cores) than there are real physical cores. For example, the IBM CPU on the Bluegene/Q machine has 16 physical cores with 4-fold hyperthreading, yielding 64 virtual cores. One may then start 64 threads in the user application. Compared with just starting 16 threads, one will then not get four times the performance, but still perhaps 2.1 times the performance or so, depending on the particular application.

9.2 Amdahl's Law

Before we discuss some elementary parallelization techniques, it is worthwhile to point out a fundamental limit to the parallel speed-up that may be reached for a given program. We define the speed up here as the ratio of the total execution time without parallelization (i.e., when the calculation is done in serial) to the total execution time obtained when the parallelization is enabled.

Suppose we have a program that we have successfully parallelized. In practice, this parallelization is never going to be fully perfect. Normally there are parts of the calculation that remain serial, either for algorithmic reasons, due to technical limitations, or we considered them unimportant enough that we have not bothered to parallelize those too. Let us call the residual serial fraction f_s, i.e., this is the fraction of the execution time spent in the corresponding code parts when the program is executed in ordinary serial mode.

Then Amdahl's law (Amdahl 1967) gives the maximum parallel speed up as

$$\text{maximum parallel speed up} = \frac{1}{f_s}. \tag{317}$$

This is simply because in the most optimistic case we can at most assume that our parallelization effort has been perfect, so that the time spent in the parallel parts approaches zero for a sufficiently large number of cores. The serial time remains unaffected by this, however, and does not shrink at all. The lesson is a bit sobering: Achieving large parallel speed-ups, say beyond a factor of 100 or so, also requires extremely tiny residual serial fractions. This is sometimes very hard to reach in practice, and for certain problems, it may even be impossible.

9.3 Shared Memory Parallelization

Shared memory parallelization can be used to distribute a computational work-load on the multiple available compute cores that have access to the same memory, which is where the data resides. UNIX processes are *isolated* from each other—they usually have their own protected memory, preventing simple joint work on the same memory space (data exchange requires the use of files, sockets, or special devices such as /dev/shm). But, a process may be split up into multiple execution paths, called *threads*. Such threads share all the resources of the parent process (memory, files, etc.), and they are the ideal vehicle for efficient shared memory parallelization.

Threads can be created and destroyed manually, e.g., with the pthreads-library of the POSIX standard. Here is a simple example how this could be achieved:

```
#include <pthread.h>

void do_stuff(void)
{
  int i, threadid = 1;
  pthread_attr_r attr;
  pthread_t mythread;
  pthread_attr_init(&attr);
  pthread_create(mythread, &attr, evaluate, &threadid);

  for(i = 0; i < 100; i++)
    some_expensive_calculation(i);
}

void *evaluate(void *p)
{
  int i;

  for(i = 100; i < 200; i++)
    some_expensive_calculation(i);
}
```

Here the two loops in lines 11/12 and 19/20 will be carried out simultaneously (i.e., in parallel) by two different threads of the same parent process. While certainly doable in principle this style of parallel programming is a bit cumbersome if one has to do it regularly—fortunately, there is a convenient alternative (see below) where much of the thread logistics is carried out by the compiler.

9.3.1 OpenMP and Its Fork-Join Model

A simpler approach for shared memory programming is to use the OpenMP standard, which is a language/compiler extension for C/C++ and Fortran. It allows the programmer to give simple hints to the compiler, instructing it which parts can be executed in parallel sections. OpenMP then automatically deals with the thread creation and destruction, and completely hides this nuisance from the programmer. As a result, it becomes possible to parallelize a code with minimal modifications, and the modified program can still be compiled and executed without OpenMP as a serial code. Here is how the example from above would like like in OpenMP:

```
#include <omp.h>

void do_stuff(void)
{
  int i;

#pragma omp parallel for
  for(i = 0; i < 200; i++)
    some_expensive_calculation(i);
}
```

This is obviously a lot simpler. We see here an example of so-called loop-level parallelism with OpenMP. In practice, one simply puts a special directive for the compiler in front of the loop. That's basically all. The OpenMP compiler will then automatically wake up all available threads at the beginning of the loop (the "fork"), it will then distribute the loop iterations evenly onto the different threads, and they are then executed concurrently. Finally, once all loop iterations have completed, the threads are put to sleep again, and only the master thread continues in serial fashion. Note that this will only work correctly if there are *no dependencies* between the different loop iterations, or in other words, the order in which they are carried out needs to be unimportant. If everything goes well, the loop is then executed faster by a factor close to the number of threads.

This illustrates the central idea of OpenMP, which is to let the programmer identify sections in a code that can be executed in parallel and annotate these to the compiler. Whenever such a section is encountered, the program execution is split into a number of threads that work in a team in parallel on the work of the section. Often, this work is a simple loop whose iterations are distributed evenly on the team, but also more general parallel sections are possible. At the end of the parallel section, the threads join again onto the master thread, the team is dissolved, and serial execution resumes until the next parallel section starts. Normally, the number of threads used in each parallel section is constant, but this can also be changed through calls of OpenMP runtime library functions. In order for this to work in practice, one has to do a few additional things:

- The code has to be compiled with an OpenMP capable compiler. This feature often needs to be enabled with a special switch, e.g., with gcc,

```
gcc -fopenmp ...
```

  needs to be used.
- For some more advanced OpenMP features accessible through calls of OpenMP-library functions, one should include the OpenMP header file

```
#include <omp.h>
```

- In order to set the number of threads that are used, one should set the `OMP_NUM_THREADS` environment variable before the program is started. Depending on the shell that is used (here bash is assumed), this can be done for example through

```
export OMP_NUM_THREADS=8
```

  in which case 8 threads would be allocated by OpenMP. Normally one should then also have at least eight (virtual) cores available. The `omp_get_num_threads()` function call can be used inside a program to check how many threads are available.

9.3.2 Race Conditions

When OpenMP is used, one can easily create hideous bugs if different threads are allowed to modify the same variable at the same time. Which thread wins the "race" and gets to modify a variable first is essentially undetermined in OpenMP (note that the exact timings on a compute core can vary stochastically due to "timing noise" originating in interruptions from the operating system), so that subsequent executions may each time yield a different result and seemingly produce non-deterministic behavior. A simple example for incorrect code with this problem is the following double-loop:

```
  int i, j;
#pragma omp parallel for
  for(i = 0; i < N; i++)
    {
      for(j = 0; j < N; j++)
        {
          do_stuff(i, j);
        }
    }
```

Here the simple OpenMP directive in the outer loop will instruct the `i`-loop to be split up between different threads. However, there is only one variable for `j`, *shared* by all the threads. They are hence not able to carry out the inner loop independent from each other! What is needed here is that each thread gets its own copy of `j`, so that the inner loop can be executed independently. This can be achieved by either adding a `private(j)` clause to the OpenMP directive of the outer loop, like this:

```
  int i;
#pragma omp parallel for private(j)
  for(i = 0; i < N; i++)
    {
      for(j = 0; j < N; j++)
        {
          do_stuff(i, j);
        }
    }
```

or by exploiting the scoping rules of C for the variable `j`, declaring it in the loop body:

```
  int i;
#pragma omp parallel for
  for(i = 0; i < N; i++)
    {
      int j;
      for(j = 0; j < N; j++)
        {
          do_stuff(i, j);
        }
    }
```

9.3.3 Reductions

Another common construct in code are reductions that build, e.g., large sums or products, such as attempted incorrectly in this example:

```
  int count = 0;
#pragma omp parallel for
  for(i = 0; i < N; i++)
    {
      if(complicated_calculation(i) > 0)
        count++;
    }
```

Again, even though here the loop is nicely parallelized by OpenMP, we may nevertheless get an incorrect result for `count`. This is because the increment of this variable is not necessarily carried out as a single instruction. It basically involves a read from memory, an addition of 1, and a write back. If now two threads happen to arrive at this statement at essentially the same time, they will both read `count`, increment it, and then write it back. But in this case the variable will end up being incremented only by one unit and not by two, because one of the threads is ignorant of the change of `count` by the other and overwrites it. We have here another example for a race conditions.

There are different solutions to this problem. One is to serialize the increment of `count` by putting a so-called lock around it. Since we here have a simple increment of a variable, a particularly fast look—a so-called atomic instruction—is possible. This can be done through:

```
   int count = 0;
#pragma omp parallel for
   for(i = 0; i < N; i++)
     {
       if(complicated_calculation(i) > 0)
         {
#pragma omp atomic
           count++;
         }
     }
```

But this can still cost substantial performance: If one or several threads arrive at the statement protected by the atomic lock at the same time, they have to wait and do the operation one after the other.

A better solution would be to have private variables for `count` for each thread, and only at the end of the parallel section add up the different copies to get the global sum. OpenMP can generate the required code automatically, all that is needed is to add the clause `reduction(+:count)` to the directive for parallelizing the loop:

```
   int count = 0;
#pragma omp parallel for reduction(+:count)
   for(i = 0; i < N; i++)
     {
       if(complicated_calculation(i) > 0)
          count++;
     }
```

This shall suffice for giving a flavor of the style and the concepts of OpenMP. A more detailed description of the OpenMP standard can for example be found in various textbooks, and good online tutorials.[3]

9.4 Distributed Memory Parallelization with MPI

To use multiple nodes in compute clusters, OpenMP is not sufficient. Here one either has to use special programming languages that directly support distributed memory models (for example UPC, Co-Array Fortran, or Chapel), or one turns to the "Message Passing Interface" (MPI). MPI has become the de-facto standard for programming large-scale simulation code.

MPI offers library functions for exchanging messages between different processes running on the same or different compute nodes. The compute nodes do not necessarily have to be physically close, in principle they can also be loosely connected over the internet (although for tightly coupled problems the message latency makes this unattractive). Most of the time, the same program is executed on all compute cores (SPMD, "single program multiple data"), but they operate on different data such that the computational task is put onto many shoulders and a parallel speed up is achieved. Since the MPI processes are isolated from each other, all data exchanges necessary for the computations have to be explicitly programmed—this makes this approach much harder than, e.g., OpenMP. Often substantial program modifications and algorithmic changes are needed for MPI.

Once a program has been parallelized with MPI, it may also be augmented with OpenMP. Such hybrid parallel code may then be executed in different ways on a cluster. For example, if one has two compute nodes with 8 cores each, one could run the program with 16 MPI tasks, or with 2 MPI tasks that each using 8 OpenMP threads, or with 4 MPI tasks and 4 OpenMP threads each. It would not make sense to use 1 MPI task and 16 OpenMP threads, however—then only one of the two compute nodes could be used.

9.4.1 General Structure of an MPI Program

A basic template of a simple MPI program in C looks as follows:

[3] https://computing.llnl.gov/tutorials/openMP.

```
#include <mpi.h>

int main(int argc, char **argv)
{
   MPI_Init(&argc, &argv);
    .
    .
   /* now we can send/receive message to other MPI ranks */
    .
    .
   MPI_Finalize();
}
```

- To compile this program, one would normally use a compiler wrapper, for example `mpicc` instead of `cc`, which sets pathnames correctly such that the MPI header files and MPI library files are found by the compiler.
- For executing the MPI program, one will normally use a start-up program such as `mpirun` or `mpiexec`. For example, the command

  ```
  mpirun -np 8 ./mycalc
  ```

  could be used to launch 8 instances of the program `mycalc`.

If a normal serial program is augmented by `MPI_Init` in the above fashion, and if it is started multiple times with `mpirun -np X`, it will simply do multiple times exactly the same thing as the corresponding serial program (unless they somehow synchronize their work through modifying common files). To change this behavior and achieve non-trivial parallelism, the execution paths taken in each copy of the program need to become different. This is normally achieved by making it explicitly depend on the *rank* of the MPI task. All the N processes of an MPI program form a so-called communicator, and they are labelled with a unique rank-id, with the values $0, 1, 2, \ldots, N-1$. MPI processes can then send and receive message from different ranks using these IDs to identify each other.

The first thing an MPI program normally does is therefore to find out how many MPI processes there are in the "world", and what the rank of the current instance of the program is. This is done with the function calls

```
int NTask, ThisTask;

MPI_Comm_size(MPI_COMM_WORLD, &NTask);
MPI_Comm_rank(MPI_COMM_WORLD, &ThisTask);
```

The returned integers `NTask` and `ThisTask` then contain the number of MPI tasks and the rank of the current one, respectively.

9.4.2 A Simple Point to Point Message

With this information in hand, we can then exchange simple messages between two different MPI ranks. For example, a send of a message consisting of 50 integers from rank 5 to rank 7 could be programmed like this[4]:

```
  int data[50], result[50]

  if(ThisTask == 5)
    MPI_Send(data, 50, MPI_INT,
// buffer, size, type
             7, 12345,
// destination, message tag
             MPI_COMM_WORLD);      // communicator id

  if(ThisTask == 7)
    MPI_Recv(result, 50, MPI_INT, // buffer, size, type
             5, 12345,
// destination, message tag
             MPI_COMM_WORLD, MPI_STATUS_IGNORE); //id, status
```

Here one sees the general structure of most send/recv calls, which always decompose a message into an "envelope" and the "data". The envelope describes the rank-id of sender/receiver, the size and type of the message, and a user-specified tag (this is the '12345' here), which can be used to distinguish messages of the same length.

Through the if-statements that depend on the local MPI rank, different execution paths for sender and receiver are achieved in this example. Note that if something goes wrong here, for example an MPI rank posts a receive but the matching send does not occur, the program will deadlock, where one or several of the MPI tasks gets stuck in waiting in vain for messages that are not sent. This is one of the many new types of bugs one has to cope with in distributed parallel programs.

It is also possible to make MPI communications non-blocking and achieve asynchronous communication in this way. The MPI-2 standard even contains some calls for one-sided communication operations that do not always require direct involvement of both the sending and receiving sides.

9.4.3 Collective Communications

The MPI standard knowns a large number of functions that can be used to conveniently make use of commonly encountered communication patterns. For example, there are calls for *broadcasts* which send the same data to all other MPI tasks in the same communicator. There are also *gather* and *scatter* operations that collect data elements from all tasks, or distribute them as disjoint sets to the other tasks. Finally,

---

[4]We note that normally one would of course not hard-code the rank numbers like this, but rather design the communication such that the program can run with different numbers of MPI tasks.

there are *reduction* function that allow one to conveniently calculate sums, minima, maxima, etc., over variables held by all MPI tasks in a communicator.

A detailed description of all these possibilities is way passed the scope of these brief lecture notes. Please check out a textbook (e.g. Pacheco 1997) or some of the online resources[5] on MPI if you want to get detailed information about MPI and start to program in it.

References

Agertz, O., Moore, B., Stadel, J., Potter, D., Miniati, F., Read, J., Mayer, L., Gawryszczak, A., Kravtsov, A., et al. Nordlund, Å., Pearce, F., Quilis, V., Rudd, D., Springel, V., Stone, J., Tasker, E., Teyssier, R., Wadsley, J., & Walder, R. 2007, MNRAS, 380, 963

Amdahl, G. M. 1967, in Proceedings of the April 18–20, 1967, Spring Joint Computer Conference, AFIPS '67 (Spring) (New York, NY, USA: ACM), 483–485

Atkinson, K. 1978, An introduction to numerical analysis (Wiley)

Bagla, J. S. 2002, Journal of Astrophysics and Astronomy, 23, 185

Balsara, D. S. 2010, Journal of Computational Physics, 229, 1970

Balsara, D. S., Rumpf, T., Dumbser, M., & Munz, C.-D. 2009, Journal of Computational Physics, 228, 2480

Barnes, J. & Hut, P. 1986, Nature, 324, 446

Bauer, A. & Springel, V. 2012, MNRAS, 423, 2558

Bertone, G., Hooper, D., & Silk, J. 2005, Phys. Rep., 405, 279

Binney, J. & Tremaine, S. 1987, Galactic dynamics (Princeton University Press)

Binney, J. & Tremaine, S. 2008, Galactic Dynamics: Second Edition (Princeton University Press)

Brandt, A. 1977, Mathematics of Computation, 31, 333

Braun, J. & Sambridge, M. 1995, Nature, 376, 655

Briggs, W. L., Henson, V. E., & McCormick, S. F. 2000, A Multigrid Tutorial, EngineeringPro collection (Society for Industrial and Applied Mathematics (SIAM, Philadelphia))

Brio, M., Zakharian, A. R., & Webb, G. M. 2001, Journal of Computational Physics, 167, 177

Campbell, J. E. 1897, Proc Lond Math Soc, 28, 381

Chandrasekhar, S. 1943, ApJ, 97, 255

Cooley, J. W. & Tukey, J. W. 1965, Math. Comp., 19, 297

Courant, R., Friedrichs, K., & Lewy, H. 1928, Mathematische Annalen, 100, 32

Cullen, L. & Dehnen, W. 2010, MNRAS, 408, 669

Cunningham, A. J., Frank, A., Varnière, P., Mitran, S., & Jones, T. W. 2009, ApJS, 182, 519

Dehnen, W. 2000, ApJ, 536, L39

Dehnen, W. 2002, Journal of Computational Physics, 179, 27

Dehnen, W. & Aly, H. 2012, MNRAS, 425, 1068

Diniz, P., da Silva, E., & Netto, S. 2002, Digital Signal Processing: System Analysis and Design (Cambridge University Press)

Dolag, K., Vazza, F., Brunetti, G., & Tormen, G. 2005, MNRAS, 364, 753

Dubois, Y., Pichon, C., Welker, C., Le Borgne, D., Devriendt, J., Laigle, C., Codis, S., Pogosyan, D., Arnouts, S., Benabed, K., Bertin, E., Blaizot, J., Bouchet, F., Cardoso, J.-F., Colombi, S., de Lapparent, V., Desjacques, V., Gavazzi, R., Kassin, S., Kimm, T., McCracken, H., Milliard, B., Peirani, S., Prunet, S., Rouberol, S., Silk, J., Slyz, A., Sousbie, T., Teyssier, R., Tresse, L., Treyer, M., Vibert, D., & Volonteri, M. 2014, ArXiv e-prints: 1402.1165

Eckart, C. 1960, Physics of Fluids, 3, 421

[5]For example: https://computing.llnl.gov/tutorials/mpi.

Field, G. B. 1965, ApJ, 142, 531
Frenk, C. S., White, S. D. M., Bode, P., Bond, J. R., Bryan, G. L., Cen, R., Couchman, H. M. P., Evrard, A. E., Gnedin, N., Jenkins, A., Khokhlov, A. M., Klypin, A., Navarro, J. F., Norman, M. L., Ostriker, J. P., Owen, J. M., Pearce, F. R., Pen, U.-L., Steinmetz, M., Thomas, P. A., Villumsen, J. V., Wadsley, J. W., Warren, M. S., Xu, G., & Yepes, G. 1999, ApJ, 525, 554
Fromang, S., Hennebelle, P., & Teyssier, R. 2006, A&A, 457, 371
Gauss, C. F. 1866, Nachlass: Theoria interpolationis methodo nova tractata (Earl Friedrich Gauss, Werke, Band 3, Gottingen: Koniglichen Gesellschaft der Wissenschaften), pp. 265–303
Gingold, R. A. & Monaghan, J. J. 1977, MNRAS, 181, 375
Gnedin, N. Y. 1995, ApJS, 97, 231
Goldstein, H. 1950, Classical mechanics (Addison-Wesley)
Greif, T. H., Springel, V., White, S. D. M., Glover, S. C. O., Clark, P. C., Smith, R. J., Klessen, R. S., & Bromm, V. 2011a, ApJ, 737, 75
Greif, T. H., White, S. D. M., Klessen, R. S., & Springel, V. 2011b, ApJ, 736, 147
Hairer, E., Lubich, C., & Wanner, G. 2002, Geometric numerical integration, Springer Series in Computational Mathematics (Springer, Berlin)
Harten, A., Lax, P. D., & Van Leer, B. 1983, SIAM Review, 25, 35
Hassan, O., Probert, E. J., & Morgan, K. 1998, International Journal for Numerical Methods in Fluids, 27, 41
Hernquist, L. 1987, ApJS, 64, 715
Hernquist, L., Bouchet, F. R., & Suto, Y. 1991, ApJS, 75, 231
Hernquist, L., Hut, P., & Makino, J. 1993, ApJ, 402, L85
Hernquist, L. & Katz, N. 1989, ApJS, 70, 419
Hockney, R. W. & Eastwood, J. W. 1988, Computer simulation using particles (Bristol: Hilger)
Hopkins, P. F. 2013, MNRAS, 428, 2840
Hu, C.-Y., Naab, T., Walch, S., Moster, B. P., & Oser, L. 2014, ArXiv e-prints: 1402.1788
James, R. A. 1977, Journal of Computational Physics, 25, 71
Kirkwood, J. G. 1946, J. Chem. Phys., 14, 180
Klypin, A. A. & Shandarin, S. F. 1983, MNRAS, 204, 891
Knebe, A., Green, A., & Binney, J. 2001, MNRAS, 325, 845
Kolmogorov, A. N. 1941, Proceedings of the USSR Academy of Sciences, 32, 16
Landau, L. D. & Lifshitz, E. M. 1959, Fluid mechanics (Course of theoretical physics, Oxford: Pergamon Press)
LeVeque, R. J. 2002, Finite volume methods for hyperbolic systems (Cambridge University Press)
Lucy, L. B. 1977, AJ, 82, 1013
Makino, J., Fukushige, T., Koga, M., & Namura, K. 2003, PASJ, 55, 1163
Marinacci, F., Pakmor, R., & Springel, V. 2014, MNRAS, 437, 1750
Mavriplis, D. J. 1997, Annual Review of Fluid Mechanics, 29, 473
Mignone, A., Bodo, G., Massaglia, S., Matsakos, T., Tesileanu, O., Zanni, C., & Ferrari, A. 2007, ApJS, 170, 228
Mitchell, N. L., McCarthy, I. G., Bower, R. G., Theuns, T., & Crain, R. A. 2009, MNRAS, 395, 180
Miyoshi, T. & Kusano, K. 2005, Journal of Computational Physics, 208, 315
Mo, H., van den Bosch, F. C., & White, S. 2010, Galaxy Formation and Evolution (Cambridge University Press)
Monaghan, J. J. 1992, ARA&A, 30, 543
Morris, J. 1997, Journal of Computational Physics, 136, 41
Ollivier-Gooch, C. F. 1997, Journal of Computational Physics, 133, 6
Pacheco, P. S. 1997, Parallel Programming with MPI (Morgan Kaufmann Publishers, San Francisco)
Pakmor, R., Marinacci, F., & Springel, V. 2014, ApJ, 783, L20
Peacock, J. A. 1999, Cosmological Physics (Cambridge University Press)
Pelupessy, F. I., Schaap, W. E., & van de Weygaert, R. 2003, A&A, 403, 389
Pen, U.-L. 1998, ApJS, 115, 19
Pope, S. B. 2000, Turbulent Flows (Cambridge University Press)

Power, C., Navarro, J. F., Jenkins, A., Frenk, C. S., White, S. D. M., Springel, V., Stadel, J., & Quinn, T. 2003, MNRAS, 338, 14
Press, W. H., Teukolsky, S. A., Vetterling, W. T., & Flannery, B. P. 1992, Numerical recipes in C. The art of scientific computing (Cambridge: University Press, 1992, 2nd ed.)
Price, D. J. 2008, Journal of Computational Physics, 227, 10040
Price, D. J. 2012, in Astronomical Society of the Pacific Conference Series, Vol. 453, Advances in Computational Astrophysics: Methods, Tools, and Outcome, ed. R. Capuzzo-Dolcetta, M. Limongi, & A. Tornambè, 249
Pringle, J. E. & King, A. 2007, Astrophysical Flows (Cambridge University Press)
Rankine, W. J. M. 1870, Philosophical Transactions of the Royal Society of London, 160, 277–288
Read, J. I. & Hayfield, T. 2012, MNRAS, 422, 3037
Read, J. I., Hayfield, T., & Agertz, O. 2010, MNRAS, 405, 1513
Renardy, M. & Rogers, R. 2004, An Introduction to Partial Differential Equations, Texts in Applied Mathematics (Springer)
Rosswog, S. 2005, ApJ, 634, 1202
Runge, C. 1895, Math. Ann., 46, 167
Rusanov, V. V. 1961, J. Comput. Math. Phys. USSR, 1, 267
Saad, Y. 2003, Iterative Methods for Sparse Linear Systems: Second Edition (Society for Industrial and Applied Mathematics)
Saha, P. & Tremaine, S. 1992, AJ, 104, 1633
Salmon, J. K. & Warren, M. S. 1994, J. Comp. Phys., 111, 136
Shu, F. H. 1992, The physics of astrophysics. Volume II: Gas dynamics. (University Science Books, Mill Valley, CA)
Springel, V. 2005, MNRAS, 364, 1105
Springel, V. 2010a, MNRAS, 401, 791
Springel, V. 2010b, ARA&A, 48, 391
Springel, V. & Hernquist, L. 2002, MNRAS, 333, 649
Springel, V., Yoshida, N., & White, S. D. M. 2001, New Astronomy, 6, 79
Stadel, J. G. 2001, PhD thesis, University of Washington
Stoer, J. & Bulirsch, R. 2002, Introduction to Numerical Analysis, Texts in Applied Mathematics (Springer)
Stone, J. M., Gardiner, T. A., Teuben, P., Hawley, J. F., & Simon, J. B. 2008, ApJS, 178, 137
Strang, G. 1968, SIAM J. Numer. Anal., 5, 506
Tasker, E. J., Brunino, R., Mitchell, N. L., Michielsen, D., Hopton, S., Pearce, F. R., Bryan, G. L., & Theuns, T. 2008, MNRAS, 390, 1267
Teyssier, R. 2002, A&A, 385, 337
Toro, E. 1997, Riemann solvers and numerical methods for fluid dynamics (Springer)
Trac, H. & Pen, U.-L. 2004, New Astronomy, 9, 443
van de Weygaert, R. 1994, A&A, 283, 361
van de Weygaert, R. & Schaap, W. 2009, in Lecture Notes in Physics, Berlin Springer Verlag, Vol. 665, Data Analysis in Cosmology, ed. V. J. Martínez, E. Saar, E. Martínez-González, & M.-J. Pons-Bordería, 291–413
van Leer, B. 1984, SIAM J. Sci. Stat. Comput., 5, 1
van Leer, B. 2006, Communications in Computational Physics, 1, 192
Vogelsberger, M., Genel, S., Sijacki, D., Torrey, P., Springel, V., & Hernquist, L. 2013, MNRAS, 436, 3031
Vogelsberger, M., Genel, S., Springel, V., Torrey, P., Sijacki, D., Xu, D., Snyder, G., Bird, S., Nelson, D., & Hernquist, L. 2014, Nature, 509, 177
Vogelsberger, M., Sijacki, D., Kereš, D., Springel, V., & Hernquist, L. 2012, MNRAS, 425, 3024
Wadsley, J. W., Stadel, J., & Quinn, T. 2004, New Astronomy, 9, 137

Wadsley, J. W., Veeravalli, G., & Couchman, H. M. P. 2008, MNRAS, 387, 427
Wendroff, B. 1999, Computers & Mathematics with Applications, 38, 175
White, S. D. M., Frenk, C. S., & Davis, M. 1983, ApJ, 274, L1
Whitehurst, R. 1995, MNRAS, 277, 655
Xu, G. 1995, ApJS, 98, 355
Xu, G. 1997, MNRAS, 288, 903

Index

Y. Revaz et al. (eds.), *Star Formation in Galaxy Evolution: Connecting Numerical Models to Reality*, Saas-Fee Advanced Course 43,
DOI 10.1007/978-3-662-47890-5

Printed by Printforce, the Netherlands